The Biological Chemistry of Iron

NATO ADVANCED STUDY INSTITUTES SERIES

Proceedings of the Advanced Study Institute Programme, which aims at the dissemination of advanced knowledge and the formation of contacts among scientists from different countries

The series is published by an international board of publishers in conjunction with NATO Scientific Affairs Division

A	Life Sciences	Plenum Publishing Corporation
B	Physics	London and New York
C	Mathematical and Physical Sciences	D. Reidel Publishing Company Dordrecht, Boston and London
D	Behavioural and Social Sciences	Martinus Nijhoff Publishers
E	Engineering and Materials Sciences	The Hague, London and Boston
F	Computer and Systems Sciences	Springer Verlag
G	Ecological Sciences	Heidelberg

Series C – Mathematical and Physical Sciences

Volume 89 – The Biological Chemistry of Iron

The Biological Chemistry of Iron

A Look at the Metabolism of Iron and Its Subsequent Uses in Living Organisms

Proceedings of the NATO Advanced Study Institute held at Edmonton, Alberta, Canada, August 23 - September 4, 1981

edited by

H. BRIAN DUNFORD
Dept. of Chemistry, University of Alberta, Edmonton, Canada

DAVID DOLPHIN
Dept. of Chemistry, University of British Columbia, Vancouver, Canada

KENNETH N. RAYMOND
Dept. of Chemistry, University of California, Berkeley, U.S.A.

and

LARRY SIEKER
Dept. of Molecular Structure, University of Washington, Seattle, U.S.A.

D. Reidel Publishing Company

Dordrecht : Holland / Boston : U.S.A. / London : England

Published in cooperation with NATO Scientific Affairs Division

Library of Congress Cataloging in Publication Data

CIP

NATO Advanced Study Institute (1981: Edmonton, Alta.)
The biological chemistry of iron.

(NATO advanced study institute series. Series C, mathematical and physical sciences, v. 89)
Includes index.
1. Iron proteins–Congresses. 2. Iron–Metabolism–Congresses. I. Dunford, H. Brian. II. North Atlantic Treaty Organization. Division of Scientific Affairs. III. Title. IV. Series.
QP552.I67N37 1981 574.19'214 82-9843
ISBN-13: 978-94-009-7884-3 e-ISBN-13: 978-94-009-7882-9
DOI: 10.1007/978-94-009-7882-9

Published by D. Reidel Publishing Company
P.O. Box 17, 3300 AA Dordrecht, Holland

Sold and distributed in the U.S.A. and Canada
by Kluwer Boston Inc.,
190 Old Derby Street, Hingham, MA 02043, U.S.A.

In all other countries, sold and distributed
by Kluwer Academic Publishers Group,
P.O. Box 322, 3300 AH Dordrecht, Holland

D. Reidel Publishing Company is a member of the Kluwer Group

CONTENTS

PREFACE

The results of a NATO Advanced Study Institute (ASI) entitled "Coordination Chemistry Environments in Iron-Containing Proteins and Enzymes - Including Smaller Molecules and Model Systems" are summarized in this book. The ASI was held in the Province of Alberta, Canada, from August 23 to September 4, 1981. The first half of the conference was held on the campus of the University of Alberta, Edmonton, and the second half at the Overlander Lodge, Hinton.

Two other conferences had the greatest impact upon the planning for this ASI. One was a NATO ASI held in Tomar, Portugal in September of 1979, entitled "Metal Ions in Biology". Among the organizers for that conference were Allen Hill and Antonio Xavier; we are happy to acknowledge their beneficial influence on our subsequent conference. The other most influential conference was one organized by Ralph Wilkins and Dennis Darnell entitled "Methods for Determining Metal Ion Environments in Proteins" which was held in Las Cruces, New Mexico, U.S.A., January 10-12, 1979. The Las Cruces conference invited lectures were published as Volume 2 of "Advances in Inorganic Biochemistry", G. Eichhorn and L. Marzilli, editors. Most of the physical techniques used to probe metal ion environments are eloquently described in the latter volume. The undersigned organizers for the Alberta ASI made two decisions which shaped its format: first, to narrow the scope of "Metal Ions in Biology" to iron-containing systems. Second, to emphasize a description of the results obtained by investigation of the biological systems, and not the physical techniques used to probe the systems. One exception is Mössbauer spectroscopy, a technique not described in the proceedings of the Las Cruces conference, which is uniquely suited to iron-containing systems. This volume contains an excellent description of both the technique and the type of results which can be obtained from it.

The title of this book "The Biological Chemistry of Iron" is simplicity itself. It also is misleading, because justice to the title would require a monumental multivolume series. Nevertheless, we have used it for the sake of brevity. To the best of our knowledge this volume is the first to attempt to describe in some detail both sides of the story of the biology of iron. One side is the gathering of iron, its storage and transport, in other words the metabolism of iron. The other is the utilization of the iron by living systems.

Most animal systems use most of their iron to transport or store oxygen. Myoglobin and hemoglobin immediately come to mind. However, these molecules have been intensively studied and well

H. B. Dunford et al. (eds.), The Biological Chemistry of Iron, ix–xii.

described elsewhere (see for example "Hemoglobin and Myoglobin in their Reactions with Ligands" by E. Antonini and M. Brunori, North-Holland, Amsterdam - a book which is still of great value although it was published in 1971. We do not mean to imply that research on these molecules is anywhere near complete.) There are many references to hemoglobin in this volume, but we describe in detail a less well-known oxygen transport molecule, hemerythrin. Another well known heme protein, cytochrome *c* is discussed briefly, but we have concentrated more on the remarkable four-heme molecule cytochrome c_3. With these types of decisions the conference was held to a viable two week format.

Most of our interest centered on iron-containing enzymes and model systems, as essential for aerobic life as oxygen transport. It would appear that we may be approaching the point where we have enough structural and mechanistic information to elucidate completely the mechanism of an enzyme reaction. One may judge for oneself how well the criteria of complete mechanistic information as defined by biologists, chemists and/or physicists are being met for the enzymes discussed in this volume. Many important iron enzymes are not discussed; it was our concern to obtain some depth for which we paid a price in breadth of coverage.

The participants at the conference ranged from graduate students to senior investigators, from theoretical chemists to medical doctors, from microbiologists to physicists. The mix of participants was truly interdisciplinary providing an accurate match to the types of research required to push back the research frontier on the biological chemistry of iron.

This book is organized in sections which we describe briefly. Speakers (as distinct from coauthors) are also listed. Section A, the introduction, contains the keynote lecture by Allen Hill. It also contains two lectures which were organized as an "inorganic chemistry teach-in" to review the basic principles of inorganic and physical chemistry which must apply equally to non-biological and biological iron. The first, by Ralph Wilkins, describes ligand reaction dynamics. The second, by Chris Reed, discusses the influence of structure, spin state, ligand equilibria, coordination number, redox potential and oxidation state. The remainder of the book might be summarized as follows. It is concerned with the elucidation of ways in which the biological environment and the iron (or in a few cases, copper or molybdenum) influence each other according to the basic principles of physics and chemistry.

Iron metabolism is summarized in Section B. Robert Crichton describes the iron storage protein, ferritin. The iron transport protein, transferrin, is the topic of Philip Aisen. Ken Raymond and Günther Winkelmann discuss those unique iron-grabbers for microorganisms, the siderophores, as well as the chemistry of analogous molecules. And last, but not least, Rowe Byers describes iron uptake and distribution in cultured beating heart muscle cells.

Section C is for cytochromes, particularly as viewed using NMR spectroscopy. These are the topic of António Xavier. In Section D, the topic is the oxygen transport molecule, hemerythrin; the dynamics and mechanisms of its reactions (Ralph Wilkins) and its three-dimensional structure (Larry Sieker and Ron Stenkamp).

The topic of Section E is iron-sulfur centers, clusters and enzymes. We are indebted to José Moura for coordinating this section. José also gave the introductory talk. Hydrogenases and nitrogenases are described in detail: hydrogenases by Hans Grande and Jean LeGall and nitrogenases by Cees Veeger and Vincent Huynh. In the latter talk, Mössbauer spectroscopy is described.

Section F is for ferrous and ferric hemes. Heme model studies are documented. Magnetic complexities in hemes (Robert Scheidt), resonance energies (Philip George), models for peroxidases and cytochrome P-450 (David Dolphin), theoretical calculations on model cytochrome P-450 (Gilda Loew) and hemes of hydroporphyrins (Chris Chang) provide an excellent background for a look at the heme-containing enzymes themselves.

In the final part, Section G, heme enzymes are the topic. An introduction to the peroxidases is provided by Brian Dunford, Rick Rutter, Ron Wever and Nils Ellfolk. The nature of the ligand in the fifth coordination position of the heme iron of horseradish peroxidase is scrutinized intensively by Gerd LaMar and Teizo Kitagawa, as is Compound I of horseradish and yeast cytochrome *c* peroxidases by Brian Hoffman. Cytochrome P-450 is the topic for Gerald Wagner and Volker Ullrich. With increasing enzyme complexity, catalase is introduced by Peter Jones. The three-dimensional structure of beef liver catalase is outlined by Nobuo Tanaka and Thomas Reid III. Finally the *pièce de resistance* of heme enzymes, cytochrome *c* oxidase is reached. Its subunits (Howard Mason) and EXAFS structural results on its copper (Robert Scott) are the aspects treated here.

Philip George was honorary chairman for the conference. He was the hardest working honorary chairman any of us have seen. Not only did he act as chairman for events, both scientific and social, but he is a contributor to this volume. Our hats are off to you, Philip. The only criticism we can make of your pioneering research is that it was ahead of its time.

Acknowledgements are in order to the following organizations and persons: to the University of Alberta and the Overlander Lodge for providing first-rate conference facilities. In particular we thank Therese Roberts and her helpers at Lister Hall; Stockwell Day, his family and his staff at the Overlander Lodge. The grant from NATO Scientific Affairs Division made the conference possible. In addition, the financial help of the Alberta Heritage Foundation for Medical Research is gratefully acknowledged as well as that from the Natural Sciences and Engineering Research Council of Canada, an anonymous donor and Smith-Klein Pharmaceutical. Social events sponsored by the City of Edmonton Business Development Office, (thank you Al Bleiken and Paul Ouimet) and a hospitality grant from the Province of Alberta contributed in a large way to the success of the conference. A large fraction of this volume was typed in camera-ready form by Lavine Shupenia and Jacki Jorgensen; they now will have an audience of appropriate size to view their handiwork. Special thanks go to Jacki Jorgensen and Dee Dunford who looked after many of the local arrangements.

The Editors: Brian Dunford, Edmonton, Alberta
David Dolphin, Vancouver, British Columbia
Kenneth Raymond, Berkeley, California
Larry Sieker, Seattle, Washington

The bombardier beetle (genus Brachinus*) in action. The beetle uses catalase and hydrogen peroxide to generate its own rocket fuel. It also uses peroxidase and hydroquinones to generate quinones which are ejected by the rocket fuel at high temperatures. The quinones are the active ingredient used to repel its enemies. The ejection occurs with an "audible detonation." Reproduced with permission of* Science *and Dr. Thomas Eisner.*

(Aneshanley, D.J. and Eisner, T.: 1969, *Science* 165, pp. 61-63.)

SECTION A

INTRODUCTION

IRON: AN ELEMENT WELL-FITTED FOR ITS TASK?

H. Allen O. Hill

Inorganic Chemistry Laboratory, South Parks Road,
Oxford, OX1 3QR, England.

The chemistry of iron relevant to biology is described. The role of the protein in modifying this chemistry to accomplish biological tasks is stressed. The special relationship between the chemistry of iron and that of oxygen is emphasised.

The element with which we are concerned bestrides (Figure 1) the periodic table like a collossus. Why iron? The second simplest

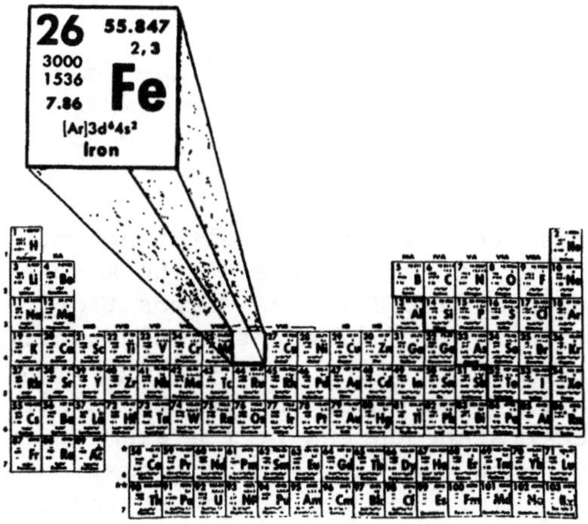

Figure 1. An idiosyncratic view of the Periodic Table.

H. B. Dunford et al. (eds.), The Biological Chemistry of Iron, 3–12.

answer to this question is: because it is there. The cosmic abundance of the elements has been interpreted (1) in terms of the burning of hydrogen and helium nuclei and this provides an adequate explanation for the observed decrease (Figure 2) in abundance with increasing atomic mass. However the first row

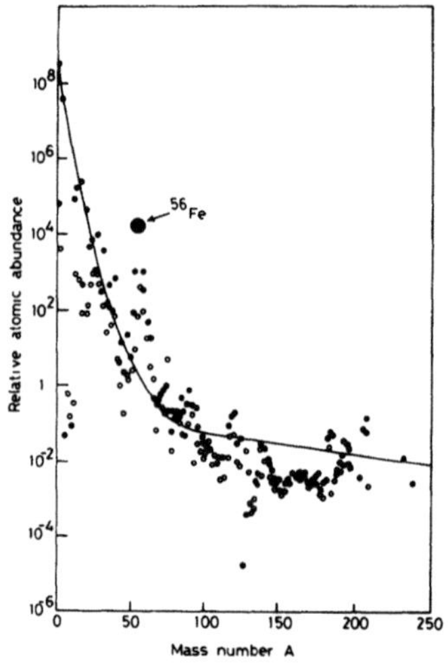

Figure 2. Relative cosmic abundance of nuclides normalized to 10,000 atoms of silicon. The solid curve gives calculated abundances for successive (n,γ) capture process. From reference 2.

transition elements are present in greater amounts than can be accounted for on this basis. Of these, the most abundant is iron-56 and this is reflected in the composition of the earth's crust. Of course, abundance is not the only factor which influences the utility of an element to biological systems; accessibility or availability are important additional criteria. The existence of concentrated sources of the element may also have encouraged use. Though geochemical fractionation of the elements is presumably responsible for the concentration of elements, biological processes have made a contribution to the concentration of iron ores through the agency of the so-called 'iron bacteria'.

Given the presence of the elements, how well-fitted is it to the task? First we must ask: what is required of it? Since Life can be defined as a series of redox reactions, one task to which we know the transition elements are well-adapted is that of taking part in electron transfer reactions. There are two further constraints: given water as the solvent, the redox potentials are bounded by the formal potentials of dihydrogen, ($H_2/2H^+$) and dioxygen, ($O_2/2H_2O$) and require any redox-active system to operate approximately within these limits. In addition, since the solvent and many of the potential ligands are Lewis acids, there may have been advantages in having a *cationic* redox active centre. We are interested, therefore, in the factors that influence redox potentials. The inexorable rise in the *effective* nuclear charge is partially responsible for the increase (Figure 3) in the potentials of the $M^{n+}/M^{(n-1)+}$ couples as the first period is transversed. The ionisation potentials (Figure 4) of the atoms in the configuration $3d^{n-2}\ 4s^1 \rightarrow 3d^{n-2}$ adequately

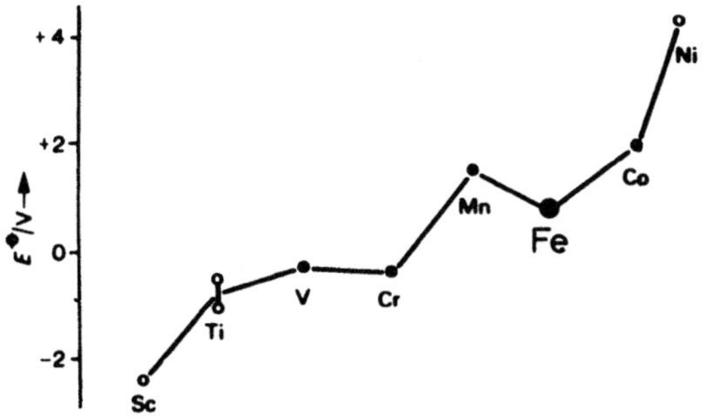

Figure 3. The standard potentials of the first-row transition elements M^{3+}/M^{2+}.

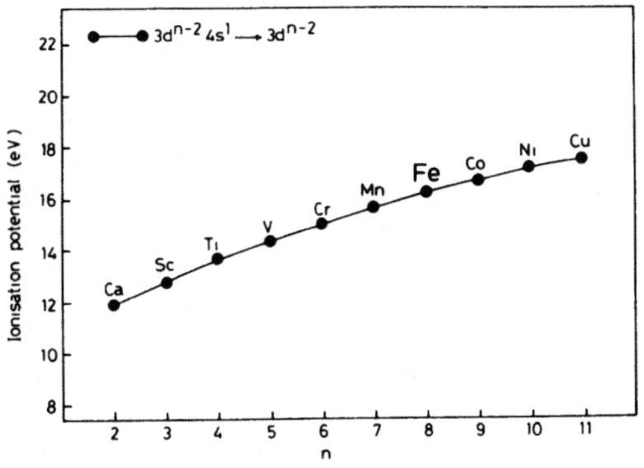

Figure 4. The ionisation potentials of the first row transition elements, $3d^{n-2}\ 4s^1 \rightarrow 3d^{n-2}$.

reflects this trend. A second important factor is the unexpectedly low sublimation energy of the iron (Figure 5) which makes a

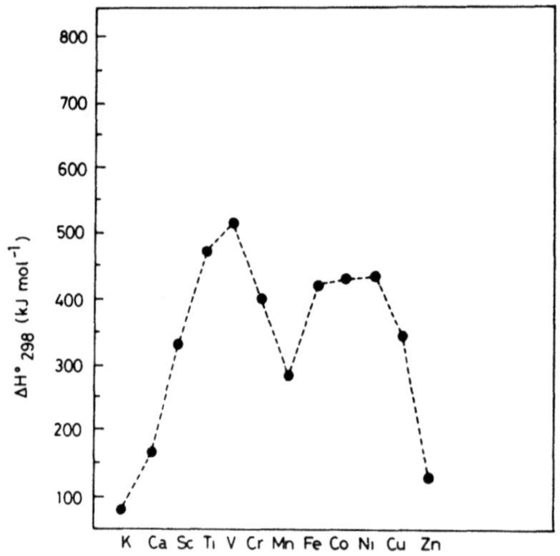

Figure 5. The enthalpy of atomisation of the first row transition elements.

significant contribution towards the relatively easy accessibility of iron(II) in aqueous solution. Whereas the simplest interpretation of the strength of bonding in metals might have led one to expect a maximum at a d^5s^1 atomic configuration, i.e., chromium (this expectation is indeed fulfilled in the third transition series at tungsten), the loss of exchange energy between the atomic and metallic states, leads (3) to a decreased binding energy in the middle of the first transition row.

These immutable properties of the element are beyond the reach of biological systems. Fortunately the redox properties of an element and its ions are markedly sensitive to the immediate environment of the latter. We are principally concerned with the relative stabilities of the iron(II) and iron(III) ions. Negatively-charged ligands are a major influence; in aqueous solution, and *in the absence of any other factors*, these stabilise Fe(III) relative to Fe(II). This effect is usually reflected in the entropy term associated with the increase in the number of microstates available to the solvent released from the constraint of the trivalent ion. A wide range of negatively-charged ligands is open to biological systems and a certain degree of 'fine-tuning' is possible through ligands of different basicity. This may be expressed directly or, within the confines of an active site, by differential hydrogen bonding to the ligand atom. The selectivity for iron(III), the normal state in an oxygenated environment, is best observed (4) in the siderophores.

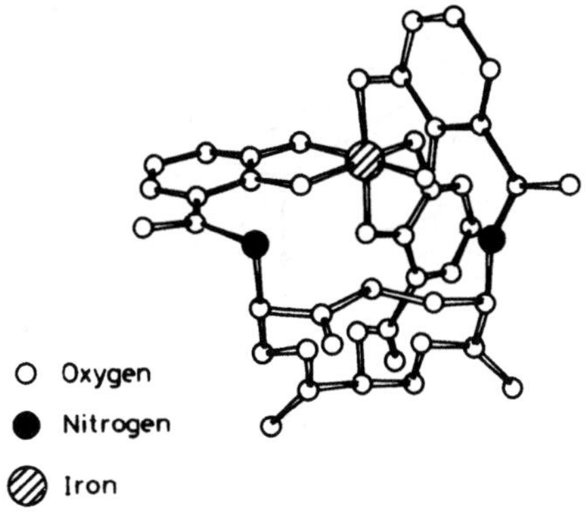

Figure 6. The structure of iron(III) enterobactin based on that of the analogous (5) gallium(III) complex.

For example, enterobactin (Figure 6) is elegantly designed to select iron(III). It provides a maximum of six negatively-charged ligands, though three would suffice, arranged octahedrally, with the chelate effect fully-expressed. With such basic ligands, most Fe(III)/Fe(II) potentials will be pH-dependent, like the $Fe(III)_{aq}/Fe(II)_{aq}$ itself. (Though the standard potential of Fe(III)/Fe(II) is 0.77V, at pH 7 hydrolysis leads to a marked decrease in the potential. The complex situation is well-described by a Poubaix diagram, Figure 7). The advantage of having a tri-positively charged cation, with its high Lewis acidity, is complicated by the perennial pest of polymerisation of iron(III) in solution, the hydroxide ion acting as a bridging ligand. Such aggregation is exploited by biological systems with the iron-sulphur proteins providing (7) the most ubiquitous examples. In these, sulphide, S^{2-}, is used as the bridging ligand. Other examples, or potential examples, are OH^-, OR^-, SH^-, SR^- and NHR^- e.g., imidazolate.

A more difficult problem is to *raise* the potential. It is here that the coincidence of the electronic configuration and the nuclear stability of nucleide 56 is at its most striking. In the presence of certain ligands, the so-called strong field ligands, a number of options are open. The symmetry properties of the d-orbitals, which provide (8) a useful elementary description of

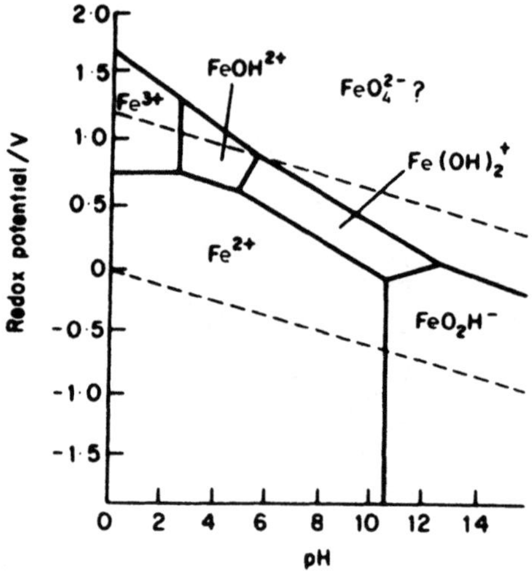

Figure 7. A Poubaix diagram for iron, from reference 6.

the eigenfunctions of the iron, are such that in a ligand environment that is octahedrally-arranged, they belong to two symmetry types, t_{2g} and e_g, of which the former lies lowest in energy, (Figure 8).

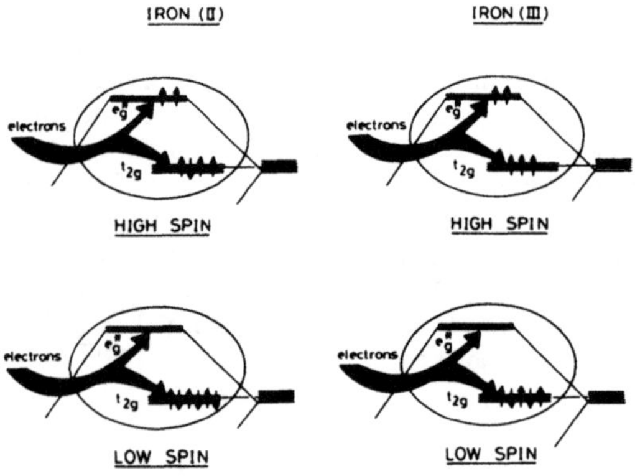

Figure 8. A schematic representation of the one-electron energy levels in high- and low-spin iron(II) and iron(III) complexes.

It interacts less with the ligands, being either non-bonding or π-bonding in the terminology of molecular orbital theory. The e_g set is the result of strong interaction with the ligands, being σ-antibonding in type. The energy separation $|t_{2g}-e_g|$ is described by a one-electron parameter, which is some measure of the interaction of the ligands with the metal ion. This simple description neglects or avoids consideration of electron repulsion effects.

In an octahedral environment, *two* states exist for *each* oxidation state, low- and high-spin Fe(III) and Fe(II), in which the electronic configurations are approximated by, (Figure 8), t_{2g}^5, $t_{2g}^3e_g^2$ and t_{2g}^6, and $t_{2g}^4e_g^2$. There are good reasons to believe that the reorganisation of the electron distribution associated with the change from high- to low-spin *iron(II)* is particularly favourable. This makes a significant contribution to its stabilisation relative to iron(III) and consequently increases the formal potential. One striking feature of the porphyrin ligand is its ability to elicit *both* spin states for *both* oxidation states. The alteration of the properties of the porphyrin ligand itself and especially the additional 'fine-tuning' possible through the axial ligand(s) makes it possible to cover essentially the whole potential range between those of dihydrogen and dioxygen. There are even further subtleties to be exploited. In symmetries lower than octahedral, intermediate spin states are possible for, once again, *both* oxidation states. Indeed, the porphyrin ligand is such that it brings (9) both oxidation states close to a spin-triple point. Iron complexes in these different spin states will have different structural, thermodynamic and kinetic properties. Even the detailed nature of the electronic configuration associated with the different spin states may be important. For example, electron transfer between low-spin Fe(III) t_{2g}^5 and low-spin Fe(II), t_{2g}^6 or between high-spin Fe(III), $t_{2g}^3e_g^2$ and high-spin Fe(II), $t_{2g}^4e_g^2$, is likely to be fast since, to a first approximation, the population of the orbitals that interact strongly with the ligands remains constant. This is not the case, if, for example the oxidation of high-spin Fe(II), $t_{2g}^4e_g^2$ leads to the formation of low-spin Fe(III), t_{2g}^5. In this instance a concomitant structural change must ensue. This could be, and perhaps is, exploited in those processes in which effect of reaction at the iron must be relayed to more distant parts of the protein or membrane.

As a negatively-charged ligand, porphyrins can be expected to stabilise higher oxidation states. In simple inorganic chemistry, oxide is perhaps the best ligand at doing so as in the well-characterised ferrates. There is now good evidence (10) that these two ligands in conjunction allow the formation of the Fe(IV) and perhaps Fe(V) ferryl complexes in peroxidases and oxidases. However the porphyrin ligand itself has redox chemistry and the identification (10) of Compound I as Fe(IV) porphyrin radical cation is quite convincing. The full expression of the redox chemistry of iron might therefore require the direct assistance of the ligand and even residues within the protein.

The peripheral perturbation supplied by the protein is capable of almost infinite variation. The recent identification (11) of hydrogen bonding between dioxygen and the distal histidine in myoglobin is one elegant example. It might be required to stabilise the delocalisation of charge on to the dioxygen. Another is provided by the postulated (12) role of the neighbouring arginine and histidine residues in cytochrome c peroxidase, which perhaps assist (Figure 9) the proton transfer and O-O bond cleavage. Just as no iron is an island, so

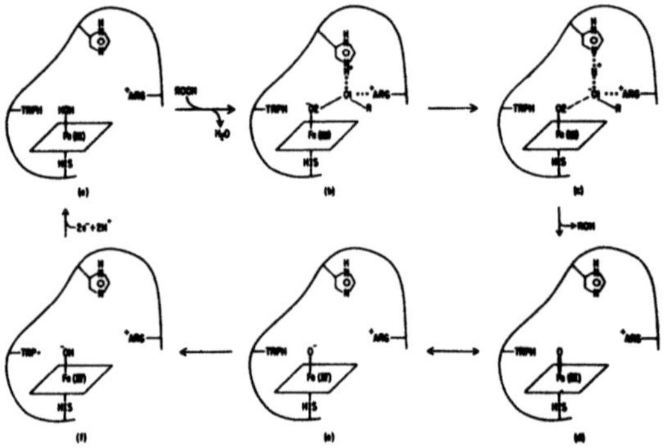

Figure 9. A postulated mechanism (12) for the reaction of cytochrome c peroxidase with peroxides

most proteins act in concert with other. Identifying the role, if any, of the iron in protein-protein interaction is not a trivial problem. It is difficult to imagine a more elegant example (13) than that which leads to the cooperativity in hemoglobin where the difference between the spatial requirements of five- and six-coordinate iron, the electronic configurations associated with high and low spin iron, or simply the altered charge on the iron or some combination of all three factors leads to allosterism at its most alluring.

Many of the redox reactions central to life lead inexorably to the reduction of dioxygen. Here also iron has a crucial role to play. The four electron reduction of dioxygen to water *without the release of any intermediate reduction products* requires a site that can bind dioxygen *and the intermediate reduction products* until they are discharged as water. A schematic view of the mitochondrial and membrane-bound cytochrome oxidase is shown in Figure 10. It appears likely that the iron of heme a_3

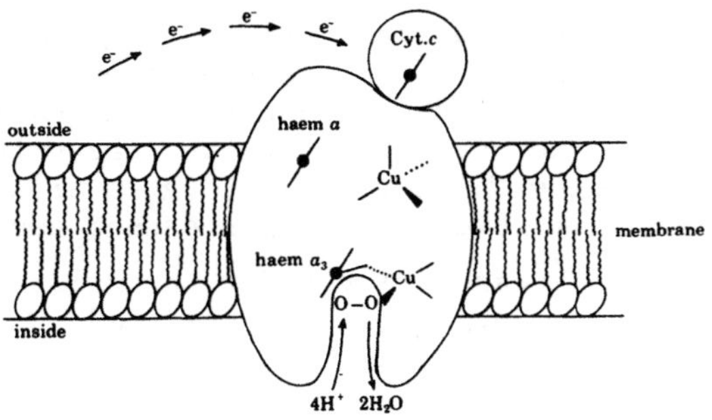

Figure 10. A schematic version (14) of the function of mitochondrial cytochrome c oxidase.

requires the assistance of a neighbouring redox site and an attendant copper ion is the most likely candidate. It is not yet known whether both metal centres are required for the containment of the reduction products of dioxygen.

Obviously iron is an essential element for all aerobic organisms (and most anaerobic ones). The consequences of iron deficiency are well-known (15) though the relationships between these and the biochemical processes requiring iron are less well-understood. Regretably it is possible to have too much iron and the deleterious effects, even toxicity, of uncontrolled, misplaced or excessive iron have been well described. How could this be so? It is possible that iron could replace other metal ions at active sites of enzymes; it could change the expression of the genetic information. It is (14) most likely that toxicity results from just those reactions for which iron, *when its chemistry is properly controlled*, is so well-suited, i.e., its redox chemistry and interactions with dioxygen. When control is absent or impaired, then it is conceivable that the reduction products of oxygen will be produced or high oxidation states of iron formed. These powerful oxidants will lead to cell damage, perhaps especially of the cell membranes. Fortunately defences exist and here again, iron-containing proteins, the superoxide dismutases and the peroxidases have an important role to play. If these defences are absent or overcome then the consequences of iron toxicity ensue.

Many of the features discussed above in relation to the chemistry of iron are also relevant to the chemistry of copper. Only the electronic features that are reflected in different spin states are absent. The redox couple of most relevance is that of copper(II)/copper(I). Many of the same roles are played by copper proteins though, as far as we know, non-proteinaceous ligands are not required to allow copper to play most of the parts proper to iron. It appears not to be as widely used as iron; indeed in instances of copper-deficiency some organisms synthesise iron-containing substitute proteins. Whatever the reason for the dominance of iron in biological systems; its greater abundance, its lower toxicity or some role that copper cannot adopt:

"Gold is for the mistress - silver for the maid -
Copper for the craftsman, cunning at his trade."
"Good!" said the Baron, sitting in his hall,
"But Iron - Cold Iron - is master of them all."

Rudyard Kipling, *Cold Iron*

REFERENCES

(1) Evans, R.D., 1955, "The Atomic Nucleus",McGraw-Hill, New York.
(2) Phillips, C.S.G. and Williams, R.J.P., 1966, "Inorganic Chemistry", Oxford University Press, Oxford.
(3) Griffiths, J.S., 1956, *J. Inorg. Nucl. Chem.* 3, 15-23.
(4) Raymond,K., This volume, pp. 85-106.
(5) Llinás, M., Wilson, D.M., and Neilands, J.B. 1973, *Biochemistry*, 12, 3836-3841.
(6) Bowen, H.J.M., 1979, "Environmental Chemistry of the Elements" Academic Press, p.26.
(7) Bezkorovainy, A., 1980, 'The Biochemistry of Nonheme Iron", Plenum Press, New York.
(8) Cotton, F.A., 1971, "Chemical Applications of Group Theory", Second Edition., J. Wiley & Sons, New York.
(9) Reed, C.A., This volume, pp. 25-42.
(10) Dolphin, D., This volume, pp. 283-294.
(11) Phillips, S.E.V., and Schoenborn, B.P., 1981, *Nature*, 204 81-82.
(12) Poulos, T.L., and Kraut, J., 1980, *J. Biol. Chem.*, 255, 8199-8205.
(13) Perutz, M., 1978, *Sci. Amer.* 238, 92-104.
(14) Bothwell, T.H., Charlton, R.W., Cook, J.D., and Finch, C.A., 1979, "Iron Metabolism in Man", Blackwells Scientific Publications, Oxford.

SUBSTITUTION AND ELECTRON TRANSFER IN METAL COMPLEXES--PARTICULARLY THOSE OF IRON

Ralph G. Wilkins

Chemistry Department, New Mexico State University, Las Cruces, NM 88003

The mechanisms of substitution and electron transfer reactions of metal complexes are described, with reference to the chemistry of iron(II) and iron(III). Tables of rate constants for substitution reactions of low spin and high spin iron(II) and iron(III) complexes; of redox reactions of iron complexes; and of low spin-high spin iron(II) and iron(III) equilibria are included.

I. INTRODUCTION

In this contribution, the mechanisms of complex ion reactions will be outlined, emphasizing those relevant to the subject of this Conference. Illustrative examples will be chosen mainly from the chemistry of iron. The advent of rapid reaction techniques (1,2) has been important in the development of this research area since the redox and substitution reactions of the metal ions span rate constant differences of almost 20 orders of magnitude.

II. SUBSTITUTION MECHANISMS

A. General Principles

The replacement of one ligand coordinated to a metal ion by a free ligand in solution is a fundamental process in complex ion chemistry, In the substitution [1] (where charges are omitted for simplicity)

$$M - X + Y \rightleftharpoons M - Y + X \qquad [1]$$

either ligand X or Y may be a water molecule, in which case a formation or dissociation (hydrolysis) reaction results. Both re-

H. B. Dunford et al. (eds.), The Biological Chemistry of Iron, 13–24.

placed (X) and replacing (Y) ligand may be a water molecule, leading to coordinated water-solvent exchange which is, in principle at least, one of the simplest substitution processes. Finally, neither X nor Y need be water, although it should be recognized that such direct ligand interchange may proceed through the intervention of solvent water.

There are basically three mechanisms for substitution (3,4):

$$\left.\begin{array}{l} M - X \rightleftharpoons M + X \\ M + Y \rightleftharpoons M - Y \end{array}\right. \quad (D) \qquad [2]$$

$$Y + M - X \rightleftharpoons Y - M - X \rightleftharpoons Y - M + X \quad (A) \qquad [3]$$

$$Y + M - X \rightleftharpoons Y...M - X \rightleftharpoons Y - M...X \rightleftharpoons Y - M + X \quad (I) \qquad [4]$$

In the Dissociative, *D* mechanism, the important step is the breakage of the M-X bond to give an intermediate of reduced coordination number. The entering ligand, Y, plays a minor role. In the Associative, *A* mechanism, the entering ligand enters in an important mechanistic step, producing an intermediate of higher coordination number. Finally, the Interchange, *I* mechanism, involves a concerted path in which X is moving away from the inner coordination sphere while Y is moving into it. This may take place within an outer-sphere or collision complex, but the important point is that no modification of the coordination number of M occurs during the process. If movement by X precedes (to a degree) movement by Y, an interchange (dissociative), I_d category is appropriate, whereas an interchange (associative) mechanism, I_a, means that there is a substantial bonding by both X and Y to M during the interchange. It is relatively easy to distinguish an *A* from a *D* mechanism by the pronounced effect in the former of different entering ligands on the rate characteristics, and concomitant generally negative $\Delta S^{\neq}$ and $\Delta V^{\neq}$ values (when neutral ligands are involved (5)). It is however very difficult to distinguish between an *A* and an I_a and between a *D* and I_d mechanism and the differentiation of even I_a and I_d mechanisms presents problems. Dissociative mechanisms tend to dominate substitution in octahedral complexes, which characterize the majority of iron complexes. Associative mechanisms on the other hand are more apt to occur in reactions of complexes of lower coordination number, e.g., the well-documented *A* mechanisms for substitution in square planar platinum(II) complexes (6).

Substitution processes are usually studied with one reactant in excess (say Y in reaction [1]). First-order formation (k_{obs}) of M-Y is invariably observed, and at low concentrations of Y, k_{obs} is directly proportional to [Y]. The quotient $k_{obs}/[Y]$ is the second-order substitution rate constant. This behaviour would be consistent with any of the three mechanisms [2]-[4]. Occasionally, as [Y] is increased, $k_{obs}/[Y]$ decreases until a constant value of k_{obs} is reached, unchanged by further increases in [Y]. This is sometimes referred to as *saturation kinetics* and conforms to the rate law [5].

$$\frac{-d[M\text{-}X]}{dt} = k_{obs}[M\text{-}X] = \frac{a[M\text{-}X][Y]}{1 + b[Y]} \qquad [5]$$

This rate law is very unlikely to arise from an *A* mechanism (6). Criteria for a *D* mechanism, and its possible differentiation from an I_d mechanism have been detailed (7), but the distinction remains very difficult since the question of *D* vs. I_d or *A* vs. I_a is very much one of degree (3,4).

B. Iron(II) and Iron(III)-Rate and Electronic Configuration

A sample of substitution reactions of Fe(II) and Fe(III) complexes, with comments, is shown in Tables 1 and 2 (7 - 18). High spin octahedral Fe(II) complexes, including the $Fe(H_2O)_6^{2+}$ ion, have the electronic configuration $t_{2g}^4e_g^2$. They undergo no

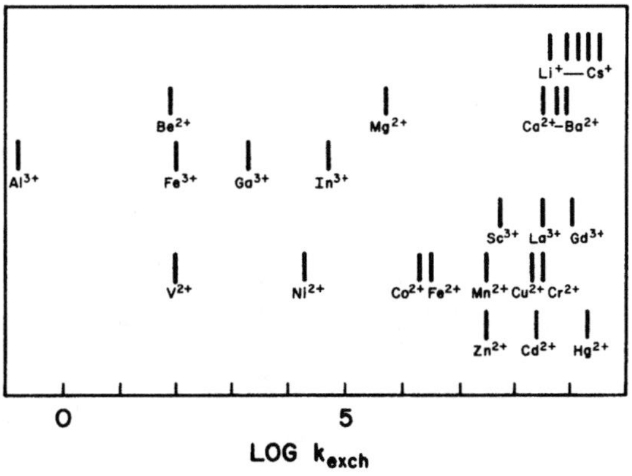

Figure 1. Approximate Labilities of Coordinated Waters

loss of crystal field stabilization energy (CFSE) when converted into the 5-coordinated species which might represent the activated complex for a dissociative mechanism. Low spin octahedral Fe(II)

Table 1. *Rate constants for a number of substitution reactions of iron(II)*

Reactants		$k, M^{-1}s^{-1}$	Comments	Ref.
High Spin Fe(II)				
$Fe(H_2O)_6^{2+}$	H_2O	4.4×10^{6} [a]	Evidence for partial I_a character from $\Delta V^{\neq}$ ($3.8 cm^3 mol^{-1}$) compared with that for $Ni(H_2O)_6^{2+}$ ($7.2 cm^3 mol^{-1}$) which is definitely I_d	8
	F^-	1.4×10^{6}		6
	NO	6.2×10^{5}		6
	bipy	1.6×10^{5}		9
	terpy	0.8×10^{5}		9
$Fe(terpy)(H_2O)_3^{2+}$	terpy	$>10^{7}$	Complex and ligand may 'stack' in forming adduct, before substitution	9,10
Low Spin Fe(II)				
$Fe(CN)_5OH_2^{3-}$	$(NH_3)_5CoN$ 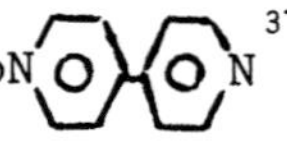 N^{3+}	5.5×10^{3} 2.6×10^{-3} [a]	Intramolecular electron transfer follows formation of (3-)(3+) bridged adduct	11
	wide variety of neutral ligands	2×10^{2} — 4×10^{2}	I_d or D mechanism	12
$Fe(CN)_5(3,5\text{-}Me_2py)^{3-}$	CN^-, imidazole, pyrazine	1.3×10^{3} [b]	D preferred to I_d on basis of $\Delta V^{\neq} = 21 \pm 1\ cm^3 mol^{-1}$ at high ligand concentrations. Saturation kinetics	7

[a] s^{-1} [b] limiting k_{obs} at high [ligand], value independent of ligand

Table 2. *Rate constants for a number of substitution reactions of iron(III)*

Reactants		$k, M^{-1}s^{-1}$	Comments	Ref.
High Spin Fe(III)				
$Fe(H_2O)_6^{3+}$	H_2O	1.6×10^{2} [a]	I_a mechanism favored since apparent k dependency on nature of ligand	13,14
	SO_4^{2-}	2.3×10^3		
	Cl^-	4.8		
	NCS^-	90		
	$Cl_3CCO_2^-$	63		
$Fe(H_2O)_5OH^{2+}$	H_2O	1.4×10^{5} [a]	I_d mechanism favored since simple correlation of k with ligand charge. Marked labilizing effect of coordinated OH^-.	13,15
	SO_4^{2-}	1.1×10^5		
	Cl^-	5.5×10^3		
	NCS^-	5×10^3		
	$Cl_3CCO_2^-$	7.8×10^3		
$Fe(TPPS)(H_2O)_n$ [b]	H_2O	1.4×10^{7} [a]	Large enhancement of water exchange rate over $Fe(H_2O)_6^{3+}$	16
$Fe(HEDTRA)(H_2O)^-$ [c]	SCN^-	$>10^7$		17
Low Spin Fe(III)				
$Fe(CN)_5OH_2^{2-}$	$Co(CN)_6^{3-}$	$<10^{-7}$	Both $Fe(CN)_5OH_2^{2-}$ and $Fe(CN)_5NH_3^{2-}$ are very inert but anation catalyzed by corresponding Fe(II) salts	18

[a] s^{-1} [b] TPPS = mesotetrakis(p-sulfonatophenyl)porphine

[c] HEDTRA = N-(hydroxyethyl)ethylenediamine-N,N',N'-triacetate

complexes (t_{2g}^6) on the other hand will lose considerable CFSE when converted into the 5-coordinated complex and dissociative reactions of low spin Fe(II) complexes are therefore likely to be much slower than their high spin counterparts. Examination of Table 1 shows that this difference is generally observed. Similar considerations apply to substitution in Fe(III) complexes where, *in general*, high spin Fe(III) is more labile than low spin Fe(III), Table 2. This approach has been fully detailed in Basolo and Pearson's book (19). The position of the aquated Fe(II) and Fe(III) ions relative to the lability of a number of other metal ions is shown in Figure 1 (20-22). Studies of iron complexes of coordination number other than six are sparse (6) and this is unfortunate since a geometry other than regular octahedral for iron in proteins is likely to be generally observed.

C. Substitution by Multidentate Ligands

Comparison of the rate constants for the replacement of coordinated water in bivalent transition metal ions by unidentate and multidentate ligands indicates that to a reasonable approximation, the replacement of the first H_2O by the chelating ligand is rate-determining. The evidence is overwhelming for many reactions of Ni(II) and Co(II) and strongly suggestive for the other bivalent metal ions, including Fe(II). The corollary of this approximation is that total dissociation of a multidentate ligand from any metal may be *very slow*. The dissociation rate constant may be orders of magnitude less than that for first bond rupture, and this is basically the reason for the marked enhanced stability of chelates and metal proteins over complexes containing only unidentate ligands. The dissociation of a chelate may be speeded up by the addition of another multidentate ligand. The latter forms a ternary species which can dissociate relatively easily (6,22). A complicated example of this effect is in the important process of apoprotein formation by treatment of a metal protein with a strong (usually chelating) ligand (23). The ligand-assisted removal of iron from transferrin by using tricatechols is deduced to occur via a ternary complex from the form of the rate law (24).

III. OXIDATION-REDUCTION MECHANISMS

A. General Principles

It is easy to visualize the several steps in a single electron transfer reaction between an oxidant O^+ and a reductant R to give O and R^+. An adduct is first formed (*precursor complex*). Within this complex electron transfer occurs to yield a *successor complex*, $O.R^+$, which then breaks apart to give the products:

$$O^+ + R \rightleftarrows O^+.R \qquad [7]$$

$$O^{+}.R \rightleftharpoons O.R^{+} \qquad [8]$$

$$O.R^{+} \rightleftharpoons O + R^{+} \qquad [9]$$

If the overall electron transfer is more rapid than the ligand loss from O^+ or R, then there will be no interpenetration of the inner coordination sphere of one reactant by another in forming $O^+.R$ or $O.R^+$. Such a process is referred to as an *outer sphere* redox reaction (25). An example of this would be the electron transfer between the oxidized and reduced iron clusters shown in reaction [10]:

$$*\,Fe_4S_4(SR)_4^{\,2-} + Fe_4S_4(SR)_4^{\,3-} \rightleftharpoons *\,Fe_4S_4(SR)_4^{\,3-} + Fe_4S_4(SR)_4^{\,2-} \qquad [10]$$

This, being extremely rapid (Table 3) and involving no net chemical reaction (i.e. O = R), must be measured by NMR line broadening (26).

A second type of mechanism is encountered when metal-ligand bonds break as a result of the formation of $O^+.R$, the so-called *inner sphere* redox reaction. Thus the reduction by $Fe(H_2O)_6^{2+}$ of $Co(NH_3)_5SCN^{2+}$ generates a transient Fe(III) species $FeNCS^{2+}$ which can only result from the formation of a bridged adduct and thus from an inner sphere redox process (28):

$$(NH_3)_5\overset{III}{Co}SCN^{2+} + \overset{II}{Fe}(H_2O)_6^{2+} \rightleftharpoons (NH_3)_5\overset{III}{Co}SCN\overset{II}{Fe}(H_2O)_5^{4+} + H_2O \qquad [11]$$

$$(NH_3)_5\overset{III}{Co}SCN\overset{II}{Fe}(H_2O)_5^{4+} \rightleftharpoons (NH_3)_5\overset{II}{Co}SCN\overset{III}{Fe}(H_2O)_5^{4+} \qquad [12]$$

$$(NH_3)_5\overset{II}{Co}SCN\overset{III}{Fe}(H_2O)_5 \rightleftharpoons (NH_3)_5\overset{II}{Co}^{2+} + (H_2O)_5\overset{III}{Fe}NCS^{2+} \qquad [13]$$

With the outer-sphere redox reaction, the formation and breakdown of the adducts $O^+.R$ and $O.R^+$ are most likely to be extremely rapid, in which case the electron transfer, [8], is the rate-determining step. Similarly, for the inner sphere reaction [11]-[13], the formation of the precursor complex [11] and the breakdown of the successor complex [13] are also rapid processes, since the Fe(II)-OH_2 and Co(II)-NCS bonds are very labile (Figure 1). Once again the intramolecular electron transfer [12] is rate-determining and the observed second-order rate constant k_{obs} will be composite (K_7k_8 or $K_{11}k_{12}$). With judicious choice of reactants it is possible to observe the build up of precursor complex. This shows as curvature in the plot of k_{obs} v. [R], where k_{obs} is the pseudo first-order rate constant, using excess R. If K is the equilibrium

Table 3. *Rate constants for a number of electron transfer reactions involving iron*

Reactants	$k, M^{-1}s^{-1}$	Comments	Ref.
Reaction [10], R = $p\text{-}C_6H_4Me$	2.8×10^6 ($\Delta H^{\neq}=3.6$, $\Delta S^{\neq}=-17$)	Outer sphere, faster than between oxidized and reduced form of protein containing one 4Fe site	26
$Fe(CN)_5PPh_3^{2-} + Fe(CN)_6^{4-}$	8×10^4 ($\Delta H^{\neq}=3.3$, $\Delta S^{\neq}=-25$)	Outer sphere, conforms well to Marcus relationship	27
$Fe(H_2O)_6^{2+} + Co(NH_3)_5SCN^{2+}$	0.12	Inner sphere, $FeNCS^{2+}$ immediate product	28
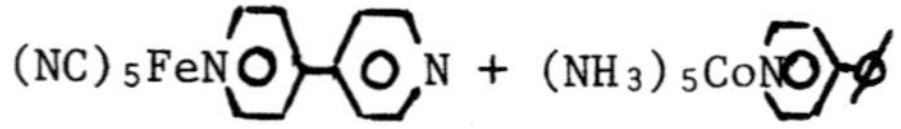	2.6×10^{-3}	Internal electron transfer rate constant (s^{-1}) following adduct formation	29
$Co(EDTA)^{2-} + Fe(CN)_6^{3-}$	4.5	Bridged adduct formed ($K=8 \times 10^2 M^{-1}$) but electron transfer occurs directly by outer sphere mechanism	30
$Fe(CN)_6^{3-} + Cophen_3^{2+}$	6×10^6	k(calc.) = $2.1 \times 10^3 M^{-1}s^{-1}$. Corrections for electrostatic effects necessary	36
$Fe(6\text{-}Mepy)(py)_2tren^{2+}$	$4 \times 10^5 s^{-1}$	For low spin → high spin; $8 \times 10^6 s^{-1}$ in reverse direction	39
$Fe(acac_2trien)^{+}$	$1.6 \times 10^8 s^{-1}$	For low spin → high spin; $3.2 \times 10^8 s^{-1}$ in reverse direction	41

constant for the formation of the precursor complex, and k the precursor → successor rate constant,

$$k_{obs} = \frac{kK[R]}{1 + K[R]} \qquad [14]$$

Such behavior has been observed for an outer sphere reaction (29) and an inner sphere reaction (11), see entries in Tables 3 and 1 respectively. Observation of an adduct however does not mean necessarily its participation in the electron transfer process. It may represent a "dead-end" adduct, see entry in Table 3 (30).

B. Outer-Sphere and Inner-Sphere Reactions--Theoretical Considerations

The outer sphere redox reaction involves no bond breaking or making and has been much examined by theoretical chemists (31). Since internuclear distances cannot change during the electron transfer (Franck-Condon principle), the bond distances in O^+ and R in [8] must each adjust to some "common state" prior to the electron transfer. It is the free energy that is required to change the atomic coordinates from their equilibrium values to those in the activated complex which must be calculated in any theory. A good deal of success has been realized in calculating the rate constant of an outer-sphere redox reaction from *nonkinetic* parameters of the reactants, with the Marcus approach (32) the most popular. For another approach refer to ref. 33. In some respects more useful than the direct calculations have been the estimation from Marcus theory of rate constants for net "cross-reactions", using experimental data for relevant self electron transfer reactions and oxidation potentials (34). For example, consider reaction [15] which because of the nature of reactants is undoubtedly an outer sphere reaction

$$\text{Fe(II)cyt-c} + \text{Cophen}_3^{3+} \rightleftharpoons \text{Fe(III)cyt-c} + \text{Cophen}_3^{2+} \quad k_{12}, k_{21}, K_{12} \qquad [15]$$

The forward rate constant k_{12} can be estimated from the relationship [16]

$$k_{12} = (k_{11}k_{22}K_{12}f_{12})^{\frac{1}{2}} \qquad [16]$$

$$\text{where } \log f = \frac{(\log K_{12})^2}{4\log(k_{11}k_{22}/10^{22})} \qquad \text{(often } f \sim 1) \qquad [17]$$

and k_{11} and k_{22} are the rate constants for self-exchange, reactions [18] and [19] (data from ref. 34),

$$*Fe(II)cyt\text{-}c + Fe(III)cyt\text{-}c \rightleftharpoons *Fe(III)cyt\text{-}c + Fe(II)cyt\text{-}c \quad [18]$$

$(k_{11} = 1.2 \times 10^3 M^{-1}s^{-1}$, $E° = 260mV)$

$$*Cophen_3^{2+} + Cophen_3^{3+} \rightleftharpoons *Cophen_3^{3+} + Cophen_3^{2+} \quad [19]$$

$(k_{22} = 40M^{-1}s^{-1}$, $E° = 370$ mV)

Since log K_{12} (=16.9(0.37-0.26)) = 1.86, from [16], k_{12} (calc.) = $2.0 \times 10^3\ M^{-1}s^{-1}$, which is in excellent agreement with the experimental value of $1.5 \times 10^3\ M^{-1}s^{-1}$. The relationship [16] has been derived also by simple thermodynamic considerations (35). A very large number of systems involving inorganic, bioinorganic and organic molecules, as well as excited inorganic states, have been subjected to this type of treatment. Agreement between calculated and observed rate constants is often within an order or two of magnitude, particularly if the two self exchanges are similar in charge type and nature of ligands (34). However see entry, ref. 36, in Table 3. The relationship has been used (31) to obtain rate constants difficult to measure, to estimate unknown E° values and infer details of electron pathways with proteins (37).

The emphasis in the study of inner-sphere redox reactions has been on the function of the bridge (38). Literally hundreds of different bridges have been investigated for their effect on electron mobility and on the site of electron attack. This is accomplished by changing the geometry, length, degree of conjugation, chelate character, nature and number of bridging groups (16,31,38).

Certain complexes of Fe(II) and Fe(III) exist in low and high spin forms in equilibrium. The very large intersystem crossing rate constants (Table 3) have been measured by Raman laser temperature-jump and ultrasonic absorption techniques (39-42). Some calculations yield rate constants within an order of magnitude of experimental values (43). A number of Fe(III) proteins exist in spin equilibria. Kinetic measurements on metmyoglobin hydroxide (42) and methemoglobin azide (44) indicate similar dynamics to simple Fe(III) complexes.

REFERENCES

1. Hammes, G.G. (Ed.): 1974, "Investigation of Rates and Mechanisms of Reactions, Part II: Investigation of Elementary Reaction Steps in Solution and Very Fast Reactions," Wiley, New York, N.Y.
2. Hammes, G.G.: 1980, in "Methods for Determining Metal Ion Environment in Proteins," Darnall, D.W. and Wilkins, R.G. (Eds.), pp. 237-263, Elsevier/North Holland, New York, N.Y.
3. Langford, C.H., and Gray, H.B.: 1965, "Ligand Substitution Processes," Benjamin, New York, N.Y.
4. Swaddle, T.W.: 1974, *Coordn. Chem. Revs.*, *14*, pp. 217-268.

5. Palmer, D.A., and Kelm, H.: 1981, *Coordn. Chem. Revs., 36*, pp. 89-153.
6. Wilkins, R.G.: 1974, "The Study of Kinetics and Mechanism of Reactions of Transition Metal Complexes," Allyn and Bacon, Inc., Boston.
7. Sullivan, T.R., Stranks, D.R., Burgess, J., and Haines, R.I.: 1977, *J. Chem. Soc. (Dalton)*, pp. 1460-1466.
8. Ducommun, Y., Newman, K.E., and Merbach, A.E.: 1980, *Inorg. Chem. 19*, pp. 3696-3703.
9. Holyer, R.H., Hubbard, C.D., Kettle, S.F.A., and Wilkins, R.G.: 1965, *Inorg. Chem. 4*, pp. 929-935; 1966, *Inorg. Chem. 5*, pp. 622-625.
10. Cayley, G.R., and Margerum, D.W.: 1974, *Chem. Commun.*, pp. 1002-1004.
11. Gaswick, D., and Haim, A.: 1974, *J. Amer. Chem. Soc. 96*, pp. 7845-7846. Jwo, J.J., Gaus, P.L., and Haim, A.: 1979, *J. Amer. Chem. Soc. 101*, pp. 6189-6197.
12. Macartney, D.H., and McAuley, A.: 1981, *Inorg. Chem. 20*, pp. 748-751.
13. Grant, M., and Jordan, R.B.: 1981, *Inorg. Chem. 20*, pp. 55-60.
14. Dodgen, H.W., Liu, G., and Hunt, J.P.: 1981, *Inorg. Chem. 20*, pp. 1002-1005.
15. Mentasti, E.: 1979, *Inorg. Chem. 18*, pp. 1512-1515.
16. Ostrich, I.J., Liu, G., Dodgen, H.W., and Hunt, J.P.: 1980, *Inorg. Chem. 19*, pp. 619-621.
17. Ogino, H., Shimura, M., and Tanaka, N.: 1979, *Inorg. Chem. 18*, pp. 2497-2501.
18. James, A.D., Murray, R.S., and Higginson, W.C.E.: 1974, *J. Chem. Soc., (Dalton)*, pp. 1273-1278.
19. Basolo, F., and Pearson, R.G.: 1967, "Mechanisms of Inorganic Reactions," Wiley, New York, N.Y.
20. Diebler, H., Eigen, M., Ilgenfritz, G., Maass, G., and Winkler, R.: 1969, *Pure Appl. Chem. 20*, pp. 93-115.
21. Wilkins, R.G.: 1975, in "The Nature of Seawater," Goldberg, E.D. (Ed.), pp. 397-415, Dahlem Konferenzen.
22. Margerum, D.W., Cayley, G.R., Weatherburn, D.C., and Pagenkopf, G.K.: 1978, in "Coordination Chemistry," Martell, A.E. (Ed.), American Chemical Society, Washington, D.C., pp. 1-220.
23. Harrington, P.C., and Wilkins, R.G.: 1980, *J. Inorg. Biochem. 12*, pp. 107-118.
24. Carrano, C.J., and Raymond, K.N.: 1979, *J. Amer. Chem. Soc. 101*, pp. 5401-5404.
25. Taube, H., Myers, H., and Rich, R.L.: 1953, *J. Amer. Chem. Soc. 75*, pp. 4118-4119; Taube, H., and Myers, H.: 1954, *J. Amer. Chem. Soc. 76*, pp. 2103-2111.
26. Reynolds, J.G., Coyle, C.L., and Holm, R.H.: 1980, *J. Amer. Chem. Soc. 102*, pp. 4350-4355.
27. Stasiw, R., and Wilkins, R.G.: 1969, *Inorg. Chem. 8*, pp. 156-157.
28. Fay, D.P., and Sutin, N.: 1970, *Inorg. Chem. 9*, pp. 1291-1293.

29. Gaus, P.L., and Villaneuva, J.L.: 1980, *J. Amer. Chem. Soc.* *102*, pp. 1934-1938.
30. Rosenheim, L, Speiser, D., and Haim, A.: 1974: *Inorg. Chem.* *13*, pp. 1571-1575.
31. Cannon, R.D.: 1980, "Electron Transfer Reactions," Butterworths, London.
32. Marcus, R.A.: 1960, *Discuss. Faraday Soc. 29*, pp. 21-30; 1963, *J. Phys. Chem. 67*, pp. 853-857; Marcus, R.A., and Sutin, N.: 1975, *Inorg. Chem. 14*, pp. 213-216.
33. Chien, J.C.W., Gibson, H.L., and Dickinson, L.C.: 1978, *Biochemistry 17*, pp. 2579-2584.
34. Chou, M., Creutz, C., and Sutin, N.: 1977, *J. Amer. Chem Soc.* *99*, pp. 5615-5623, where data for a number of outer sphere reactions are reviewed, and failings fully discussed.
35. Newton, T.W.: 1968, *J. Chem. Educ. 45*, pp. 571-575; Ratner, M.A., and Levine, R.D.: 1980, *J. Amer. Chem. Soc. 102*, pp. 4898-4900.
36. Haim, A., and Sutin, N.: 1976, *Inorg. Chem. 15*, pp. 476-478.
37. Wherland, S., and Pecht, I.: 1978, *Biochemistry 17*, pp. 2585-2591. Wherland, S., and Gray, H.B.: 1977, in "Biological Aspects of Inorganic Chemistry," Cullen, W.R., Dolphin, D., and James, B.R. (Eds.), pp. 289-368, Wiley, New York, N.Y.
38. Haim, A.: 1975, *Accs. Chem. Res. 8*, pp. 264-272.
39. Hoselton, M.A., Drago, R.S., Wilson, L.J. and Sutin, N.: 1976, *J. Amer. Chem. Soc. 98*, pp. 6967-6969.
40. Reeder, K.A., Dose, E.V., and Wilson, L.J.: 1978, *Inorg. Chem.* *17*, pp. 1071-1075.
41. Binstead, R.A., Beattie, J.K., Dose, E.V., Tweedle, M.F., and Wilson, L.J.: 1978, *J. Amer. Chem. Soc. 100*, pp. 5609-5614; Binstead, R.A., Beattie, J.K., Dewey, T.G., and Turner, D.H.: 1980, *J. Amer. Chem. Soc. 102*, pp. 6442-6451.
42. Dose, E.V., Tweedle, M.F., Wilson, L.J., and Sutin, N.: 1977, *J. Amer. Chem. Soc. 99*, pp. 3886-3888.
43. Buhks, E., Navon, G., Bixon, M., and Jortner, J.: 1980, *J. Amer. Chem. Soc. 102*, pp. 2918-2923.
44. Perutz, M.F., Sanders, J.K.M., Chenery, D.H., Noble, R.W., Pennelly, R.R., Fung, L.W.M., Ho, C., Giannini, I., Pörschke, D., and Winkler, H.: 1978, *Biochemistry, 17*, pp. 3640-3652.

OXIDATION STATES, REDOX POTENTIALS AND SPIN STATES

Christopher A. Reed

University of Southern California, Los Angeles
California, U.S.A. 90007

The status of iron as the most important and versatile metal in biological redox chemistry can be traced to a unique underlying inorganic chemistry. Essential to understanding this chemistry are the basic concepts of oxidation state, coordination number, spin state and ligand field effects. Their comprehension leads to an appreciation of active site structure and function, and invites a discussion of the factors which contribute to the wide range of iron redox potentials.

1. INTRODUCTION

Although the evolutionary selection of iron in biological systems probably arises from its high natural abundance in the earth's crust, one would be hard pressed to choose a single metal ion more uniquely suited for the diverse role it takes in biological redox chemistry. Its only existing rival is copper and it has been argued (1) that the dominance of iron over copper is a result of earlier availability. It is theorized that the more accessible negative potentials of redox-active iron sulfides in the strongly reducing "primordial soup" made it more readily solublized. However, it might equally well be argued that the intrinsic chemical properties of iron made it the singularly attractive choice at the initiation of life and the fittest to survive as the persistent redox metal throughout evolution. Its versatility of oxidation state, redox potential, coordination number, spin state, ligand type, ligand equilibria, ligand dynamics and structure gives iron its special characteristics.

H. B. Dunford et al. (eds.), The Biological Chemistry of Iron, 25–42.

This paper views these properties as the framework of fundamental inorganic chemistry upon which iron biochemistry is based. In this way one can appreciate the diversity, and subtlety of iron biochemistry, while at the same time, recognize that this prodigious and seemingly complex chemistry can ultimately be understood in terms of simple coordination chemistry principles. Of course, the simplicity is rarely perceived without the benefit of years of hindsight. But enough is now known about hemoglobin, cytochrome c, ferredoxins, certain siderophores and their model compounds to justify this view. One should be encouraged that research into the still vast area of poorly understood iron biochemistry, particularly with the seemingly complex multielectron enzymes, will eventually lead to a straightforward description of the iron chemistry which is involved. At the present time we are witnessing the early stages of this process with enzymes such as the cytochromes P-450 and the peroxidases. On the other hand, a quite primitive knowledge of structure and mechanism exists with most of the oxidases and reductases. There is every reason to believe that a good background of iron coordination chemistry will be helpful in elucidating and understanding the many remaining problems in iron biochemistry, particularly if one recognizes that only rarely is fundamentally novel coordination chemistry uncovered in biological systems.

2. OXIDATION STATES

By virtue of its position in the middle of the first transition series of the periodic table iron has a partially filled *3d* valence shell and can exhibit many oxidation states. Non-biological iron shows oxidation states from iron(-II) (e.g. in $Fe(CO)_4^{2-}$) up to iron(VI) (e.g. in the ferrate ion FeO_4^{2-}) while biological iron is presently known in three states: iron(II) (ferrous), iron(III) (ferric) and iron(IV). The d^n configurations of these states are d^6, d^5, and d^4 respectively. The diversity of oxidation state arises because to a first approximation (i.e. σ-bonding only) transition elements can participate in bonding to other atoms without using the occupied *3d* orbitals, thereby leaving them free to enjoy variable occupancy. But any implication that the *d* orbitals are unimportant is totally misleading. They are, in fact, responsible for most of the subtlety and complexity of iron chemistry. Their involvement in the chemistry of iron is the key to understanding structure, spin states and magnetic properties. These are discussed in the latter part of this chapter.

An oxidation state, it should be remembered, is not a measurable physical parameter. It is assigned by an arbitrary, though usually logical, set of rules for organizational and heuristic purposes. Frequently it is useful for rationalizing and predicting structure, bond polarity and integral electron properties such as magnetic states. It is defined as the formal charge left on a metal when the ligands have been removed in their normal or closed shell electron configurations. For example, carbonmonoxyhemoglobin, HbCO, is an iron(II) complex since a formal charge of +2 is left on the metal once CO and imidazole are removed as neutral ligands and the heme group is removed as a ligand of -2 charge. It is preferable to use Roman numerals to designate oxidation states since they represent an assigned *formal* charge, not an actual charge. With the typically "innocent" nitrogen and oxygen donor ligands of biochemistry there is rarely any ambiguity about oxidation states. Problems do, however, arise when assigned charges are taken literally. It is important to appreciate the limitations of the oxidation state formalism as an electron bookkeeping device and a guide to qualitative chemical properties.

The celebrated case of oxyhemoglobin, HbO_2, has been much argued, frequently unnecessarily so. An iron(III) superoxo formulation is more useful than an iron(II) dioxygen formulation for rationalizing the structure and bond polarity of the FeO_2 moiety (2). The superoxide ion, $\cdot O_2^-$, is not a closed shell configuration for a dioxygen ligand but can be considered a "normal" configuration for monodentate dioxygen complexes. The closed shell configuration of dioxygen is of course the peroxide ion, O_2^{2-}, and is the formalism used in oxyhemocyanin. The iron(III) assignment for HbO_2 which results from the superoxide formalism is useful for rationalizing and classifying such properties as resonance Raman, infrared and Mössbauer spectral parameters. Antiferromagnetic coupling or π-bonding between the low-spin d^5 iron and the superoxo ligand is invoked to rationalize the diamagnetic ground state. The Fe(III)(O_2^-) formulation implies an Fe-O_2 bond polarity which transfers electron density from iron(II) to dioxygen. The recently observed H-bonding from the distal histidine to the terminal oxygen atom in oxymyoglobin is consistent with this (3). However, it is a mistake to think that the oxidation state implies a charge of -1 on the O_2 ligand. Yet, it is surprizing how often this mistake is made. No one would argue in HCl, where the oxidation states are obviously usefully assigned H(I) and Cl(-I) that there is actually a charge of +1 on H and a charge of -1 on Cl. Nor would one argue that the actual charge density on iron in HbO_2 amounts to +3. It may, in fact, be closer to +2. Actual charge separations between ligands and iron will always be fractional and it is rare when oxidation states are useful in discussing calculated or measured electron populations. Thus, it is a non-issue to challenge the utility of oxidation states when they fail to reflect true electron populations. That is not what they were designed for. Despite its limitations, the oxidation state formalism remains as the basis of the organizational framework of inorganic chemistry, especially redox chemistry. Moreover, as inorganic chemistry's most simple yet most useful "theories", it provides a qualitative basis for comparing many physical properties. It is unparalleled in rationalizing the structure of transition metal complexes.

A further area of possible utility (or ambiguity) with iron oxidation states is in thiolate bonding. Iron(III) thiolate ligation from cysteinate is common in iron proteins (e.g. ferredoxins, cytochrome P-450) and circumstances may arise where some iron(II) thiolate radical character (RS·) might be usefully invoked. For example, resonance structures could be written:

$$Fe^{III}(RS^-) \longleftrightarrow Fe^{II}(RS\cdot) \quad (1)$$

Some contribution of the Fe(II)(RS·) formulation is useful to rationalize the strongly electron donating capability of the thiolate ligand and its tendency to act as a reducing agent, presumably giving thiolate radicals (which form a disulfide, RSSR):

$$Fe^{III}(RS^-) \rightleftharpoons Fe^{II} + RS\cdot \quad (2)$$

It is important to remember that the resonance structures, related by a *double-headed* arrow in eq. (1), are quite distinct from the chemical equilibrium in eq. (2). Resonance structures are simply different oxidation state formalisms (or valence bond formalisms) of a single, unique, chemical entity and they allow chemists to describe bonding as weighted averages of different integral electron formulations. The only provisos are that each contributor must have the same *overall* spin state and that any redistribution of electrons between metal and ligand must use orbitals of like symmetry. Application of this principle to the so-called iron(I) porphyrins (e.g. $[Fe(TPP)]^-$, where TPP is the dianion of tetraphenylporphyrin) has recently led (4) to their reformulation as resonance hybrids with significant spin-coupled porphyrin radical anion character (written $TPP\cdot^-$):

$$[Fe^{I}(TPP)]^- \longleftrightarrow [Fe^{II}(TPP\cdot)]^-$$

Interesting arguments arise concerning the ground state of this molecule but these will not be discussed here because the iron(I) d^7 state is presently unknown in biochemical systems.

A related but more important case is the oxidation state of the horseradish peroxidase intermediate, HRP Compound I, formed by treatment of the iron(III) heme resting state with a peroxide (P = porphyrin, L = axial ligand, probably imidazole from histidine):

$$[Fe^{III}(P)(L)]^+ + ROOH \longrightarrow [Fe^{IV}(O)(P\cdot)(L)]^+ + ROH$$

resting state HRP I

Compound I is two electrons oxidized above the iron(III) resting state and the early literature used the designation Fe(V) as a label of convenience. The most reasonable formulation is, however, that of an iron(IV) oxo complex (a so-called ferryl species) with the porphyrin as a radical cation, written $P\cdot^+$. The iron (IV) assignment rationalizes the very low isomer shift (δ = 0.08 mm/s) in the Mössbauer spectrum (5). The need to assign the oxidation state of the porphyrin as that of a radical cation comes from the UV-VIS spectrum which shows broad $\alpha\beta$ bands and a low intensity Soret band (6). More recently, ENDOR measurements have directly confirmed the presence of a porphyrin radical (7). This is discussed by Hoffman in this volume.

The one-electron reduction of HRP Compound I gives Compound II which, by virtue of its normal iron porphyrin UV-VIS spectrum and the similarity of its Mössbauer spectrum to Compound I, is given a straight iron(IV) porphyrin formulation:

$$[Fe^{IV}(O)(P\cdot)(L)]^+ + AH_2 \longrightarrow [Fe^{IV}(O)(P)(L)]^+ + H^+ + AH\cdot$$

Compound I Compound II

This shows that in the case of HRP it is easier to oxidize the iron from III to IV than it is to oxidize the porphyrin to a porphyrin radical cation. It is interesting to contrast this situation with other iron porphyrins having different axial ligation. For example, $[FeCl(TPP)]^+$, long considered an iron(IV) species (8), has recently been reformulated as an iron(III) radical cation (9). So unlike HRP II, it is easier to oxidize TPP to $TPP\cdot^+$ than iron(III) to iron(IV) in FeCl(TPP). In other words the relative redox potentials of the iron and the porphyrin are reversed in HRP and FeCl(TPP). The crucial difference lies in the axial ligation, oxo/imidazole in one case and chloride in the other. This illustrates the profound effect that axial ligation can have on iron redox potentials. The iron(IV) assignments in Compounds I and II support the assumed oxo (or ferryl) ligation since it is well known that oxo ligation is uniquely suited to stabilizing high oxidation states of the transition metals. The possibility that the axial ligand is a hydroxide ion cannot, however, be rigourously ruled out at present. The choice between Fe=O and $[FeOH]^+$ ligation will simply depend on the pK_a of the iron center which in turn will be critically dependent upon how well the immediate protein and solvent environment stabilizes (or destabilizes) an overall positive charge on the iron heme complex.

Finally, there are the fractional or averaged oxidation states of the delocalized four-iron ferredoxins. Mössbauer measurements, and other spectroscopic techniques, indicate that within the time scale of such experiments the four irons in an $[Fe_4S_4(RS)_4]^{n-}$ cluster are all very nearly equivalent (10). Thus, in the n = -3 cluster of a reduced bacterial ferredoxin, $[Fe_4S_4(RS)_4]^{3-}$, (also written $[Fe_4S_4^*]^+$) the charge left on iron after removing the ligands as RS^- and S^{2-} is $[Fe_4]^{+9}$. The state is written as $3Fe^{II} \cdot Fe^{III}$ or more accurately $Fe^{2.25}$ as a *formal* charge. By contrast, the two-iron sulfur protein centers have so-called trapped iron(II) and iron(III) valencies in $[Fe_2S_2(SR)_4]^-$ which are *not* in resonance (11). The sites are structurally different and any valency interchange between the two iron sites, if it in fact occurs, would be a chemical reaction and involve a structural change. The delocalization in the four-iron clusters means that individual oxidation states of iron are not very useful. Rather, the state of the Fe_4S_4 cluster is the important redox unit.

3. REDOX POTENTIALS

A measure of the stability of a particular oxidation state can be derived from redox potentials. It is important to remember that a redox potential simply expresses relative thermodynamic stability. It does not necessarily give a reliable indication of aerobic stability, thermal stability or electron transfer reaction rates since such properties depend upon *kinetic* as well as *thermodynamic* factors. By convention, redox potentials are quoted on the E_0' scale (i.e., versus the hydrogen electrode at pH 7) rather than as standard potentials on the E_0 scale (i.e. versus the Standard Hydrogen Electron at pH 1). Also by convention, they are reduction potentials. The more *positive* the potential, the more stable the *reduced* species. Thus, the cytochromes c with their markedly positive potentials of some +260 mV have the ferrous state quite stabilized relative to the ferric. By contrast, the ferredoxins typically have quite negative potentials (-200 to -550 mV). They are easily oxidized and thus make good reducing agents.

The tuning of iron redox potentials is vital for iron redox components to take their appropriate thermodynamic place in biological electron transfer processes. Usually, it is the iron(II/III) valency change which is involved by the iron(III/IV) redox change is also known in the hydroperoxidases and is postulated in intermediates of other multielectron enzymes such as cytochrome P-450 and cytochrome c oxidase. Iron(II/III) potentials can vary by several hundred millivolts even within a particular class of iron proteins such as the cytochromes or iron-sulfur proteins. Understanding the factors which lead to this wide variation is perhaps the most interesting problem in the study of redox potentials.

There are, however, a few caveats to consider when comparing literature values of redox potentials. Most are measured by indirect methods of variable reliability and may not be true equilibrium potentials. Many vary dramatically with pH, ionic strength and temperature. Ligand equilibria alter redox potentials according to the Nernst equation. For example, the aqueous iron(II/III) potential is +0.77 V at pH 1 but shifts to +0.11 V at pH 7 because of aquo ligand hydrolysis to hydroxide. It is surprising how often the pH 1 value is used as some apparently meaningful bench mark by which to compare metalloprotein potentials at pH 7. Sample purity, aggregation problems or detergent phenomena can also be sources of problems in the actual measurement. Furthermore, an *in vivo* enzyme may have a different potential than the same enzyme *in vitro* because of environment or solvation effects. Finally, comparison of redox potentials can only be semiquantitative at best when different media, electrolytes and reference potentials are used. The common practice of adjusting potentials derived from non-aqueous media versus a calomel reference to aqueous potentials versus the standard hydrogen electrode must never be considered more than an approximation. It is much safer to compare relative differences of potential within one particular system with relative differences in another. Any serious attempt at quantitative consideration of redox potentials must take these problems into account.

The single most obvious factor affecting iron redox potentials is ligand type. Properties such as charge, ligand field effects, stereochemistry, inductive effects and soft/hard characteristics all influence the relative stability of the iron(II) versus the iron (III) state, and hence the iron(II/III) redox potential. It is often difficult to separate out each of these influences in iron proteins or even in model compounds. However, some of them can be illustrated with inorganic complexes of iron. For example, negatively charged ligands such as F^- shift the E_0 value of the hexaquo ion from +0.77 to 0.40 V, thereby stabilizing the ferric state. This is expected from charge considerations. Conversely, neutral ligands tend to move the potential more positive and stabilize the ferrous state, particularly if they are strong field ligands such as o-phenanthroline (E_0 = 1.14 V) (12). Strong field ligands which form octahedral (or pseudooctahedral) low-spin complexes tend to stabilize the ferrous state because ligand field stabilization energy (LFSE) is maximized in the low-spin d^6 configuration. On the other hand, high-spin tetrahedral complexes have much weaker ligand field effects and the high-spin d^5 configuration of the ferric state gains stability relative to the ferrous by maximizing the spin exchange energy. The combined effects of charge, ligand field effects and stereochemistry provide a qualitative rationale for the quite negative potentials of the high-spin pseudotetrahedral rubredoxins (ferric stabilized) while the most positive potentials are usually seen in the low-spin pseudooctahedral cytochromes having neutral axial ligands (e.g., cytochromes c). Exceptions to these generalities provide an insight into the other factors which can affect iron redox potentials.

Inductive effects are also important. Ligands with electron withdrawing substituents will stabilize a lower oxidation state and there is frequently a correlation between ligand pK_a and redox potential. For example, in the hemes of *a*, *b*, and *c* type cytochromes the respective formyl, vinyl and alkyl thioether peripheral side-chains give positive shifts in the iron(II/III) redox potential formyl > vinyl > alkyl (13). The magnitude of this effect is, however, quite small — probably no more than ~100 mV. A related but more substantial effect, is the lowering of the tetrapyrrole ring oxidation potential in siroheme compared to heme. This is discussed by Chang in this volume.

One of the few reliable quantitative estimations of an axial ligand effect is the comparison of iron(II/III) redox potentials of a bis-imidazole ligated heme with an imidazole-thioether ligated heme. Studies show that if all other factors are kept essentially constant the substitution of one imidazole for a thioether ligand in a low-spin bis-imidazole iron porphyrin complex gives rise to about +160 mV increase in the iron(II/III) potential (14,15). The soft thioether ligand stabilizes the ferrous state relative to the ferric state (which has a much lower affinity for thioethers).

A related situation exists in the thoroughly studied cytochrome P-450 from *P. putida*. Upon substrate binding there is a low to high spin state change which is accompanied by a +170 mV increase in redox potential (16). This is consistent with the loss of a hard ligand which stabilizes the low-spin ferric state of the resting enzyme. This appears to be part of an enzyme activation mechanism since it shifts the potential to a value which is accessible to the native reductant, putidaredoxin.

But there is another very important factor affecting iron redox potentials. This is the nature of the immediate environment surrounding the redox center. There are two dramatic illustrations of this. Firstly, referring to the thioether axial ligand effect discussed above one might naively think that the histidine/methionine ligated cytochrome c ($E_o' \simeq +260$ mV) might have a redox potential about 160 mV more positive than the bis-histidine ligated cytochrome c_3 ($E_o' \simeq -200$ mV). The actual difference is ~460 mV, not 160, despite the heme groups being otherwise chemically identical. The additional 300 mV must arise from the ability of cytochrome c to discourage formation of the positive charge generated in the ferric state. This is widely ascribed to hydrophobic protein encapsulation of the deeply buried heme (17). This is akin to generating a positively charged species in a low dielectric medium. Moreover, the propionic acid side chains of the heme are atypically buried in the protein and are not free to ionize. By contrast, the heme groups of cytochrome c_3 are much more exposed to the aqueous medium (18), the propionic acid side chains are more free to ionize, and thus, the immediate heme environment can more readily accommodate the overall positive charge of the ferric state. This is akin to the stabilization of ions by solvation in a high dielectric medium. Kassner (17) has devised an interesting experiment which shows that a heme complex in benzene can have about a 300 mV more positive potential than in water. It is clear that the ability of the protein and solvation environment (19) around the redox center to stabilize (or destabilize) charge is a critical determinant of redox potential.

The second notable case of protein manipulation of redox states and redox potentials is the celebrated case of the high potential iron sulfur protein HiPIP. Bacterial four-iron ferredoxins operate between the $[Fe_4S_4]^{+1/+2}$ oxidation levels at quite negative potentials, typically -200 to -550 mV. On the other hand, according to the "three state" hypothesis (20), HiPIP operates between the $[Fe_4S_4]^{+2/+3}$ states at a positive potential (ca. +350 mV). Since the cores of the Fe_4S_4 clusters are chemically identical it is apparent that the immediate protein environment exercises a major influence over the redox state, and hence the redox potential. The factors which might contribute to this include NH···S hydrogen bonding, protein constraint of geometry, electrostatic effects and dielectric effects (21). These factors are presumed to make the $[Fe_4S_4]^{+1}$ state of HiPIP, and the $[Fe_4S_4]^{+3}$ state of bacterial ferredoxins, inaccessible. Definitive characterization of a super-reduced HiPIP or a super-reduced ferredoxin is lacking. There is one report of a synthetic cluster existing in all three states (22) and the potential difference between the +1/+2 and +2/+3 states is very large (1.3 V). This indicates how inaccessible the three states might be in any particular protein.

Not only do protein environment factors dictate which pair of redox states is accessible but, as with cytochromes, the redox potentials of Fe_4S_4 proteins vary considerably within each class. A range of 650 mV has been catalogued for the ferredoxins and a range of 750 mV has been reported for the HiPIP type proteins (21). Since these ranges comprise values taken under a variety of conditions (pH, buffer type, etc.) and since some may be artifactual due to environmental or chemical interference, they should be taken *con granulo salis*. However, it is clear that wide ranges do exist and again, these are ascribed to hydrogen bonding, geometrical constraints, electrostatic and/or dielectric effects. Model complexes typically have lower redox potentials than proteins, an observation which is usually discussed in terms of H-bonding and dielectric effects of the surrounding solvent (21-23).

In summary, it is clear that the immediate environment of a redox center plays a crucial role in determining its redox potential, let alone its rate and mechanism of electron transfer. To identify the precise origin of potential differences between chemically similar species remains a difficult problem. Specific protein dipole orientations, the location of ionizable residues, H-bonding, and the degree of solvent exposure are all factors which can be important in stabilizing or destabilizing charge. By coupling these environmental influences with structural constraints and the ligand manipulation discussed earlier, it can be appreciated that biological systems have an impressive versatility for tuning redox potentials.

Finally, a topical issue is whether cytochrome oxidase has an iron(IV) resting state at the heme a_3 active site (24). Its redox potential has been measured at about +300 mV and must, of course, lie between +260 mV of cytochrome c and +810 mV of oxygen. Normally, a hemoprotein iron(III/IV) potential lies in the region of +1000 mV (e.g. in peroxidases). Protein environment effects, such as those discussed above, could lower this into the operating region but any factor which lowers the III/IV potential would almost certainly lower the II/III potential well below 260 mV. Since a two electron redox change at heme a_3 is required in the iron(IV) oxidase model it is difficult to see how this could occur within a 260-810 mV range, let alone both at about 300 mV as is measured. An unprecedented environmental change during redox, such as a complete charge separation, would be necessary to both lower the iron(III/IV) potential but raise the iron(II/III) potential. Two-electron redox changes at a single metal site (or even a Fe_4S_4 cluster) typically occur in discrete steps separated by several hundred millivolts.

4. SPIN STATES

In principle, there are at least four ways that biological systems might exploit the spin states of iron. Firstly, because of spin-spin correlation, a paramagnetic iron center has a kinetic advantage when interacting with another paramagnetic species. This may be important for the rapid reaction of dioxygen with hemoglobin or cytochrome oxidase. Secondly, a *d* orbital containing an unpaired electron has a defined locus and its orientation may be critical for determining the rate and mechanism of electron transfer. Thirdly, the interrelationship of structure and spin state means that a spin state change might be exploited to effect a structural change. The controversial tension theory of hemoglobin cooperativity invokes this principle. Conversely, the absence of a spin state change may be important in maintaining structural integrity (e.g., in low-spin cytochromes). Fourthly, metal-ligand bond strengths and labilities are affected by spin state. As a consequence, spin states must be important in determining reaction rates which involve iron-ligand bond fission. The overriding difficulty with spin states is proving that their properties are purposefully exploited by biological systems. In a number of cases it appears that spin states may be merely the inconsequential result of ligand structure about the iron atom. Regardless of their function, however, spin states are an extremely useful indicator of structure and oxidation state.

To understand spin states it is important to recognize that they are dictated solely by structure. More specifically, it is the number, type and stereochemistry of the ligands about iron which determine the relative energies of the *d* orbitals, and hence the spin state. The splitting of the *d* orbital energies for various stereochemistries can be derived from simple electrostatic considerations using crystal field theory. In essence, orbitals whose lobes point directly at the ligands are raised in energy relative to those whose lobes point between the ligands. The familiar splittings of the five *d* orbitals in tetrahedral and octahedral ligand fields are seen in Figure 1.

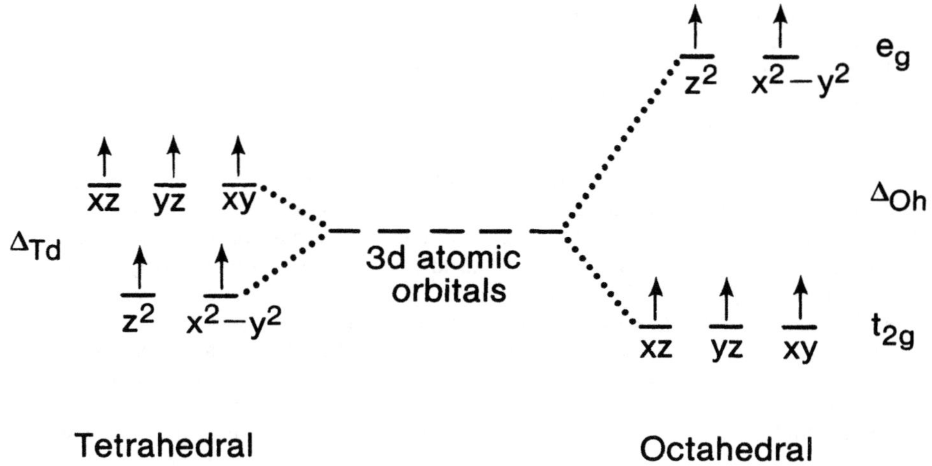

Figure 1. Splitting of the 3d orbitals in tetrahedral (left) and octahedral (right) ligand fields. High-spin d^5 ferric configurations (↑) are illustrated in both cases.

This is equivalent to a σ-bonding only molecular orbital picture. The orbitals which are raised in energy are antibonding with respect to metal-ligand σ-bonding. Those lowered in energy are σ-nonbonding, but may participate in π-bonding.

In a tetrahedral ligand field there are only four ligands and none of the orbitals have all of their lobes pointing directly at the ligands. As a consequence, the splitting Δ_{Td} is rather small and is never large enough in biological iron to cause preferential spin pairing in the lower lying orbitals. Thus, as illustrated for d^5 iron(III) in Figure 1, iron is always high-spin in a tetrahedral environment. Iron(III) will have five unpaired electrons and a magnetic moment of 5.9 BM. Similarly, tetrahedral iron(II) is d^6 and will have four unpaired electrons. A typical example is the $[Fe(SR)_4]^{-1/-2}$ site of rubredoxin.

For an octahedral steroechmistry the six ligands give rise to a much larger splitting Δ_{Oh} ($\Delta_{Oh} \simeq 9/5\ \Delta_{Td}$). Even so, as exemplified by the siderophores, the ligand field splitting is frequently not large enough to cause spin-pairing in the t_{2g} orbitals prior to filling the e_g orbitals (see right side of Figure 1). In fact, all naturally occurring biological iron is high spin except certain six-coordinate hemoproteins. Strictly speaking, such heme stereochemistries are not octahedral but tetragonal distortions thereof. They deserve special attention because only with hemoproteins and iron prophyrin complexes are the subtleties and complexities of iron spin states revealed.

The spin states of iron porphyrins and hemoproteins will be discussed mainly for the d^5 configuration of iron(III). Similar principles apply to the d^6 configuration of iron(II), in fact with less complexity, simply by adding an additional electron to the lowest energy d orbital.

To a very good approximation the spin state of heme iron is determined by the ligand field strength of the axial ligand(s) (25). Ligand field strength, it should be remembered, is simply a measure of the ability of a ligand to cause a splitting of the d orbital energies. There is frequently no correlation with ligand binding energies. In the case of axial ligands (defined to lie on the z axis) in a heme complex (the x, y, axes pass through the porphinato nitrogen atoms) the strong field ligands are those that are good σ-donors and/or π-acceptors. Strong σ-donors raise the antibonding d_{z^2} orbital and π-acceptors lower the d_{xy}, d_{yz} orbitals. The chelating porphyrin is a relatively strong field ligand in the xy plane and splits the antibonding $d_{x^2-y^2}$ well apart from the d_{xy}. These splittings are illustrated in Figure 2.

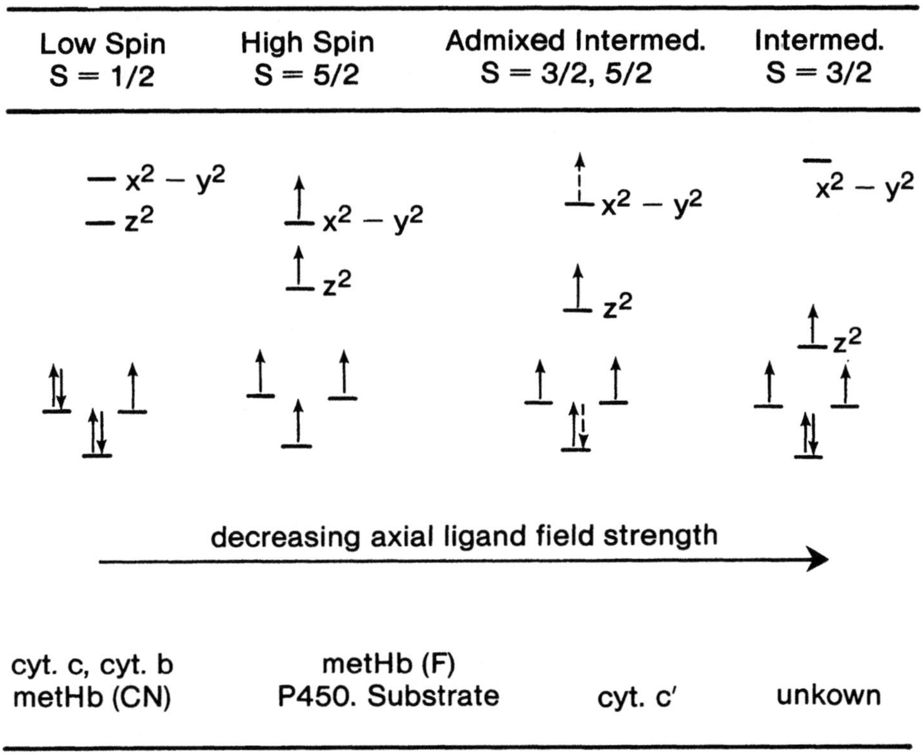

Figure 2. A qualitative representation of the presumed d orbital splittings for the spin states of ferric hemes as a function of axial ligand (25). Below each configuration are hemoprotein examples.

From left to right the axial ligand field is decreasing by virtue of ligand type and coordination number. The main effect is that the d_{z^2} orbital drops in energy from left to right. Accompanying this there are some compensatory changes in the relative positions of the other orbitals (and the precise ordering of the lowest orbitals is frequently not known with certainty). The Figure illustrates how the low, high and intermediate states arise sequentially from decreasing axial ligand field strength. Typical examples of low-spin $S = \frac{1}{2}$ states are ferricytochrome *c*, oxidized *b*-type cytochromes, and metHb(CN). For the iron(II) oxidation state the analogous d^6 $S = 0$ states are found. Examples of high-spin $S = 5/2$ states are metHb(F) and oxidized, substrate-bound cyt. P450. For iron(II), deoxyhemoglobin is the best known example of a high-spin $S = 2$ state.

For iron(III), when the axial ligand field strength is at the borderline between low and high spin, a spin equilibrium situation can arise. This is a thermal equilibrium between chemically distinct structures and there are many known cases. For example, at room temperature metHb(H_2O) is approximately 22% low-spin and 78% high-spin (26). As the temperature is decreased the fraction of low-spin species increases. Model complexes for such equilibria are discussed by Scheidt in this volume.

Referring again to Figure 2, further weakening of the axial ligand field from the high-spin state brings one into the admixed intermediate S = 3/2, 5/2 regime. This is a difficult state to conceptualize because the familiar integral electron model of *d* orbital population breaks down. The state can, however, be pictorially represented by non-integral occupation of the $d_{x^2-y^2}$ and d_{xy} orbitals as illustrated. The state arises with weak axial ligand field strengths which are borderline between high and intermediate spin states. Unlike the *thermal* equilibrium discussed above, selection rules allow a *quantum mechanical* admixing of the S = 5/2 and S = 3/2 states when their energy separation is of similar magnitude to the spin-orbit coupling constant (27). In theory, this gives rise to a continuum of discrete admixed states and any particular example is a single discrete structural entity, not a mixture of two distinct species. The S = 3/2, 5/2 admixed intermediate spin state is real in weak field ligated iron(III) porphyrins such as Fe($OClO_3$)(TPP) (28) and there is good EPR evidence for its existence in certain cytochromes *c'* (27). It remains a poorly understood state of controversial existence and unknown relevance to biological electron transfer.

Perhaps the most important chemical aspect of spin state is the interplay of structure and spin state (25). They are always related and thus, the magnetic state of an iron-containing biomolecule always has information about its structure. Some of the more detailed aspects of structure change with spin state change are examined by Scheidt in this volume. Summarized here is perhaps the most celebrated case of spin state/structure interrelationship, that of hemoglobin cooperativity of oxygen binding (29).

Deoxyhemoglobin (Hb) is an S = 2 high-spin iron(II) complex with axial coordination of an imidazole from the so-called proximal histidine of this square pyramidal stereochemistry and is modelled by Fe(TPP)(2-MeIm). The structural manifestation of the high-spin state is that the $Fe-N_{porph}$ bonds are long (~2.08 Å) and the iron atom is displaced by ~0.5 Å from the porphyrin plane (25). These features can be rationalized by the occupation of the $d_{x^2-y^2}$ orbital. The $Fe-N_{Im}$ bond is also long (~2.10 Å) and this can be rationalized by d_{z^2} occupation. Upon oxygen binding to the sixth coordination site the iron becomes low spin and there is a very significant contraction of these coordination parameters. From the X-ray structures of 'picket fence' porphyrin model complexes (30) it is known that the Fe-N_{porph} distance in HbO_2 must decrease to ~1.99 Å and the out-of-plane iron displacement to $\lesssim$ 0.2 Å. The axial Fe-N_{Im} distance probably also decreases to ~2.05 Å. These changes can be rationalized by the depopulation of the $d_{x^2-y^2}$ and d_{z^2} orbitals in the low-spin state. (This rationale is independent of the oxidation state assignment in HbO_2). Other likely stereochemical changes at the heme occurring between Hb and HbO_2 are less doming and more ruffling of the porphinato core, and less asymmetric tilting of the proximal imidazole. It is the thesis of the Perutz-Hoard 'trigger' mechanism of cooperativity, now modified to a constraint mechanism (29), that these stereochemical changes at one heme can be propagated via the protein to influence the properties of neighboring hemes. In particular, the spin-state induced change of location of the proximal histidine relative to the heme is believed to be the most important stereochemical change (31). However, this issue is contentious and the relative importance of particular stereochemical changes remains unresolved. Moreover, the relative importance of stereochemical versus electrostatic factors has yet to be resolved. What can be stated categorically is that a spin state change will always cause a structural change. The magnitudes of such structural changes are fairly predictable by considering whether the $d_{x^2-y^2}$ and d_{z^2} orbitals are occupied (25). But the extent to which spin-state/structure relationships are actually exploited by biological systems remains arguable.

Finally, mention must be made of antiferromagnetic coupling, a commonly observed phenomenon in metalloproteins having paramagnetic metal ions in close proximity. Low magnetic moments, nonexistent EPR signals and other magnetically dependent spectroscopic probes give clues to the existence of antiferromagnetic coupling. Such magnetic states always have implications for structure. Strong antiferromagnetic coupling usually implies metal ions separated only by single-atom bridges. For example, $[Fe_4S_4(SR)_4]^{3-}$ clusters have magnetic moments at 4.2 K which approach that expected of an overall S = ½ system per four-iron cluster (32). If magnetically isolated, the four high-spin iron atoms would show magnetic moments equivalent to three S = 2 systems and one S = 5/2 system of a $3Fe^{II}\cdot Fe^{III}$ oxidation level, a total of 17 unpaired electrons! Similarly, oxyhemerythrin is diamagnetic even though the Mössbauer spectrum identifies the state of iron as high-spin ferric. Structurally, bridging carboxylate and possibly bridging dioxygen are implicated at the diiron active site (33).

Coupling is not always strong. For example, in Compound I of HRP there is only very weak coupling between the S = 1 iron(IV) center and the S = ½ porphyrin radical. Numerically, the extent of coupling is given by J (negative for antiferromagnetic coupling) and this is a measure of the energy separation between the spin-paired and spin-free states. For HRP I $J \simeq -2\ cm^{-1}$ and this means that the system behaves rather like an overall S = 3/2 system, except at very low temperatures (34). This contrasts very sharply with $[Fe^{III}Cl(TPP\cdot)]^{+}$ which has very strong antiferromagnetic coupling ($J > -200\ cm^{-1}$) between the S = 5/2 iron(III) and the S = ½ porphyrin radical. The magnetic moment at room temperature is 5.1 BM indicative of an overall S = 2 system (9).

Moderate antiferromagnetic coupling ($J \simeq -150\ cm^{-1}$) is observed in $[Fe_2S_2(SR)_4]^{2-}$ systems (35). Electronic delocalization between the iron centers is considerably less in the two-iron centers than in the four-iron clusters. These contrasting results serve to illustrate how little is understood about the detailed factors which control the degree of antiferromagnetic interaction. Orbital overlap mediated by bridging ligands (so-called superexchange) is clearly very important. However, the amount of energy involved is frequently rather small (a few hundred cm^{-1}) and calculations can rarely be applied to such situations with any reliability.

For the most part, antiferromagnetic coupling may be simply the result of structural proximity of magnetic centers. There is little evidence for intrinsic biological importance.

5. CONCLUSION

The basic principles of oxidation state and spin state provide a fundamental framework for discussing and understanding the structure of iron-containing biomolecules. More than providing a consistent language, it is becoming possible to describe apparently complex biochemistry in terms of straightforward coordination chemistry principles.

REFERENCES

1. Harrison, P.M. and Hoare, R.J.: 1980, "Metals in Biochemistry" Chapman and Hall: London, pp. 10-12.
2. Reed, C.A. and Cheung, S.K.: 1977, *Proc. Nat. Acad. Sci. USA.*, 74, pp. 1780-1784.
3. Phillips, S.E.V. and Schoenborn, B.P.: 1981, *Nature*, 292, pp. 81-82.
4. Reed, C.A.: 1982, in "Electrochemical and Spectrochemical Studies of Biological Redox Components", ACS Advances in Chemistry Series, (Kadish, K.M. ed.,) in press.

5. Schulz, C.E., Devaney, P.W., Winkler, H., Debrunner, P.G., Doan, N., Chiang, R., Rutter, R. and Hager, L.P.: 1979 *FEBS Lett.*, 103, pp. 102-105.
6. Dolphin, D., Forman, A., Borg, D.C., Fajer, J. and Felton, R.H.: 1971, *Proc. Nat. Acad. Sci. USA.*, 68, pp. 614-618.
7. Hoffman, B.M., Roberts, J.E., Brown, T.G., Kang, C.H. and Margoliash, E.: 1979, *Proc. Nat. Acad. Sci. USA.*, 76, pp. 6132-6136. See also this volume.
8. Felton, R.H., Owen, G.S., Dolphin, D., Forman, A., Borg, D.C. and Fajer, J.: 1973, *Ann. N.Y. Acad. Sci.*, 206, pp. 504-514.
9. Gans, P., Marchon, J-C., Reed, C.A. and Regnard, J-R.: 1981, *Nouv. J. Chemie*, 5, pp. 203-204.
10. Holm, R.H., Averill, B.A., Herskovitz, T., Frankel, R.B., Gray, H.B., Siiman, O. and Grunthaner, F.J.: 1974, *J. Am. Chem. Soc.*, 96, pp. 2644-2646.
11. Mascharak, P.K., Papaefthymiou, G.C., Frankel, R.B. and Holm, R.H.: 1981, *J. Am. Chem. Soc.*, 103, pp. 6110-6116.
12. Ochiai, E., 1977, "Bioinorganic Chemistry. An Introduction", Allyn and Bacon: Boston, pp. 60-61.
13. Falk, J.E.: 1964, "Porphyrins and Metalloporphyrins", Elsevier: New York, pp. 67-71.
14. Mashiko, T., Marchon, J-C., Musser, D.T., Reed, C.A., Kastner, M.E. and Scheidt, W.R.: 1979, *J. Am. Chem. Soc.*, 101, pp. 3653-3655.
15. Marchon, J-C., Mashiko, T., Reed, C.A.: 1982, "Interactions Between Iron and Proteins in Oxygen and Electron Transport", Ho, C., Easton, W.A., Collman, J.P., Gibson, Q.H., Leigh, J.S., Margoliash, E., Moffat, J.K. and Scheidt, W.R. Eds., Elsevier: New York, in press.
16. Sligar, S.G. and Gunsalus, I.C.: 1976, *Proc. Nat. Acad. Sci. USA.*, 73, pp. 1078-1082.
17. Kassner, R.J.: 1972, *Proc. Nat. Acad. Sci. USA.*, 69, pp. 2263-2267.
18. Haser, R., Pierrot, M., Frey, M., Payan, F., Astier, J-P., Bruschi and LeGall, J.: 1979, *Nature*, 282, pp. 806-810.
19. Stellwagen, E.: 1978, *Nature*, 275, pp. 73-74.
20. Carter, C.W.: 1977, "Iron-sulfur Proteins", Academic Press: New York, Vol III, pp. 157-204.
21. Sweeney, W.V. and Rabinowitz, J.C.: 1980, *Ann. Rev. Biochem.*, 49, pp. 139-161.
22. DePamphilis, B.V., Averill, B.A., Herskovitz, T., Que, L. Jr. and Holm, R.H.: 1974, *J. Am. Chem. Soc.*, 96, pp. 4159-4167.
23. Maskiewicz, R. and Bruice, T.C.: 1978, *J.C.S. Chem. Comm.*, pp. 703-704.
24. Seiter, C.H.A. and Angelos, S.G.: 1980, *Proc. Nat. Acad. Sci. USA.*, 77, pp. 1806-1808.
25. Scheidt, W.R. and Reed, C.A.: 1981, *Chem. Rev.*, 81, in press.

26. Perutz, M.F.: 1979, *Ann. Rev. Biochem.*, 48, pp. 327-386.
27. Maltempo, M.M. and Moss, T.H.: 1976, *Quart. Rev. Biophys.*, 9, pp. 181-215.
28. Reed, C.A., Mashiko, T., Bentley, S.P., Kastner, M.E., Scheidt, W.R., Spartalian, K. and Lang, G.: 1979, *J. Am. Chem. Soc.*, 101, pp. 2948-2958.
29. Perutz, M.F.: 1980, *Proc. Roy. Soc. London*, B208, pp. 135-162.
30. Jameson, G.B., Molinaro, F.S., Ibers, J.A., Collman, J.P., Brauman, J.I., Rose, E. and Suslick, K.S.: 1980, *J. Am. Chem. Soc.*, 102, pp. 3224-3237.
31. Baldwin, J. and Chothia, C.: 1979, *J. Mol. Biol.*, 129, pp. 175-220.
32. Laskowski, E.J., Reynolds, J.G., Frankel, R.B., Foner, S., Papaefthymiou, G.C. and Holm, R.H.: 1979, *J. Am. Chem. Soc.*, 101, pp. 6562-6570.
33. Loehr, J.S. and Loehr, T.M.: 1979, *Adv. Inorg. Biochem.*, 1, pp. 235-252.
34. Schulz, C.E., Devaney, P.W., Winkler, H., Debrunner, P.G., Doan, N., Chiang, R., Rutter, R. and Hager, L.P.: 1979, *FEBS Lett.*, 103, pp. 102-105.
35. Gillum, W.O., Frankel, R.B., Foner, S. and Holm, R.H.: 1976, *Inorg. Chem.*, 15, pp. 1095-1100.

SECTION B

IRON METABOLISM

FERRITIN - THE STRUCTURE AND FUNCTION OF AN IRON STORAGE PROTEIN

Robert R. Crichton

Unité de Biochimie, Université Catholique de Louvain,
B-1348 Louvain-la-Neuve, Belgium.

The requirement for an iron storage protein is discussed and the distribution and role of ferritin in iron metabolism are reviewed. The structure of the protein shell and of the iron contained within it is described and the molecular mechanisms which may be involved in ferritin iron deposition and mobilisation are passed in review. A cellular view of ferritin function in which interactions between ferritin and the extracellular transport protein, transferrin, the supply of iron for mitochondrial haem synthesis and exchange between ferritin and a low molecular weight transit pool are considered concludes this review.

INTRODUCTION

The importance of iron in biological systems has already been emphasized sufficiently in this volume for it to be superfluous to insist on it here. However, it is evident that if iron is needed as an essential component of numerous proteins, its metabolism, as a logical consequence of its solution chemistry under physiological conditions, requires both transport and storage facilities. The former will be dealt with by my colleagues in the articles which follow. The latter is the subject of the present review in which we discuss the way in which ferritin functions in the regulation of intracellular iron metabolism both at the molecular and the cellular level.

The requirement for an iron storage protein is evident in view of the pronounced tendency of iron in aqueous solution at neutral pH to oxidise and hydrolyse with formation of essentially insoluble iron hydroxides (indeed as we shall see the iron in

H. B. Dunford et al. (eds.), The Biological Chemistry of Iron, 45–61.

ferritin is mostly present as a ferric oxyhydroxide). However, as I shall endeavour to underline in this presentation, ferritin is much more than just a cul-de-sac, a one way street leading to nowhere in the dynamic pathways of iron within cells but a major protagonist in the regulation of iron metabolism. We can summarise the principal attributes of an iron storage protein succinctly in one sentence. It should enable the storage of iron within the cell in a soluble, non-toxic form from which iron can be mobilised as and when (and indeed we might even add where) it is required within the cell and by exchange of its iron with the extracellular transport form (transferrin) permit the redistribution of iron between cells as a function of their needs.

DISTRIBUTION OF FERRITIN AND ITS ROLE IN IRON METABOLISM

The first successful purification of ferritin was from horse spleen (1) and it is the ferritin from this source which has remained, during the forty-six years since its discovery, the benchmark for all subsequent research. Although the work of Granick and Kuhn and their collaborators rapidly established the idea that ferritin was <u>the</u> major iron storage protein of mammals (2, 3) located essentially in a limited number of organs and tissues, subsequent studies have confirmed its presence in virtually all types of mammalian cells (4) and in serum (5, 6), in other non-mammalian vertebrates, in molluscs and arthropods, in plants, in fungi and most recently in bacteria (7-15). This extensive distribution, covering essentially the complete spectrum of living organisms suggests that iron is an element of sufficient importance for life that its conservation in an appropriate storage form has developed rather early in evolution. However, we must regretfully add that if ferritins were one of the first storage proteins to evolve, we know at present little or nothing about their role in iron metabolism in organisms other than mammals.

The function of ferritin in the overall regulation of iron metabolism has been studied almost exclusively in mammals. Figure 1 presents a summary of the daily exchange of iron between the different compartments in man (16). The most striking feature of this scheme is that iron metabolism, like many other phenomena in daily life, is essentially conservative. From our diet we accumulate about 1 mg of iron/day, which represents almost exactly our capacity to eliminate iron from the system. The consequences of this equilibrium for man are clear, and often lethal. Too little iron (either by lack of supply or excess iron loss) and we become anaemic ; too much (either by excess dietary absorption or by administration of iron, for example by blood transfusion) and we become iron overloaded. Dietary iron is absorbed in the intestine where it may be deposited in ferritin in the mucosal cells, or

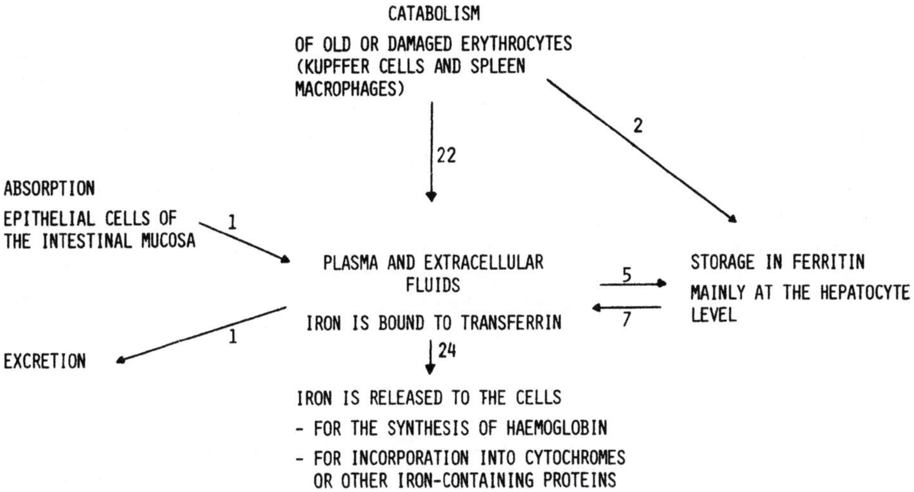

Figure 1 Daily exchange of iron between the different compartments in man (adapted from (16))

traverse the mucosal cells to the plasma where it is transported by transferrin to the rest of the organism. As can be seen from the figure the greatest part of the flux of iron through the transferrin pool results from the breakdown and recovery of haemoglobin iron in the reticuloendothelial system and its reutilisation for the synthesis of haemoglobin and other iron-containing proteins. A small amount of haem iron passes to the liver where it is stored in ferritin and there is a daily exchange of 5 mg into and 7 mg out of the hepatic ferritin, which is predominantly located in the parenchymal cells (hepatocytes) of the liver. The reticuloendothelial cells of liver (the Kupffer cells), and of other tissues, contain the bulk of the other iron storage substance, haemosiderin. Haemosiderin is an insoluble form of iron which is visible as yellow intracellular granules (mostly if not exclusively intralysosomal) by light microscopy and which gives an intense Prussian blue reaction. It is generally considered to be a product of intracellular degradation and aggregation of ferritin which by electron microscopy, X-ray diffraction, Mössbauer spectroscopy and measurements of its magnetic susceptibility resembles the iron core of ferritin. We shall not further consider haemosiderin here but refer the reader to reference (16) pp. 320-321 where the nature and distribution of haemosiderin are succinctly reviewed. As described in (17) it is likely that there is ferritin present in all cell types of mammals with the largest amounts being found in hepatocytes and in cells of the reticuloendothelial system. We shall return to the role of ferritin in iron metabolism in a later section. In normal adult man ferritin and haemosiderin together represent 25 % of the total body iron.

FERRITIN STRUCTURE

The ferritin molecule consists of a hollow, roughly spherical protein shell (apoferritin) of external diameter 12 nm and internal diameter 7 nm within which one or more crystallites of micellar inorganic iron of approximate composition $(FeOOH)_8 \cdot (FeO.OPO_3H_2)$ are deposited (18-20). The iron content of ferritin is variable and can attain a maximum value of 4,300-4,500 atoms of iron such that the iron core represents approximately one half of the molecular mass of full ferritin (21). We now consider in more detail the primary structure and three-dimensional organisation of the apoferritin protein shell, the structure of the micellar iron core and the heterogeneity of different tissue ferritins.

The Structure of Apoferritin

The best characterised apoferritin is that of horse spleen which can be readily obtained in crystalline form in excellent yields (typically > 1 g protein/kg wet weight of tissue). The amino acid sequence of this protein has recently been determined in our laboratory (22) and is presented in figure 2.

```
                    5                   10                  15
NAC-SER-SER-GLN-ILE-ARG-GLN-ASN-TYR-SER-THR-GLU-VAL-GLU-ALA-ALA-
                    20                  25                  30
    VAL-ASN-ARG-LEU-VAL-ASN-LEU-TYR-LEU-ARG-ALA-SER-TYR-THR-TYR-
                    35                  40                  45
    LEU-SER-LEU-GLY-PHE-TYR-PHE-ASP-ARG-ASP-ASP-VAL-ALA-LEU-GLU-
                    50                  55                  60
    GLY-VAL-CYS-HIS-PHE-PHE-ARG-GLU-LEU-ALA-GLU-GLU-LYS-ARG-GLU-
                    65                  70                  75
    GLY-ALA-GLU-ARG-LEU-LEU-LYS-MET-GLN-ASN-GLN-ARG-GLY-GLY-ARG-
                    80                  85                  90
    ALA-LEU-PHE-GLN-ASP-LEU-GLN-LYS-PRO-SER-GLN-ASP-GLU-TRP-GLY-
                    95                  100                 105
    THR-THR-LEU-ASP-ALA-MET-LYS-ALA-ALA-ILE-VAL-LEU-GLU-LYS-SER-
                    110                 115                 120
    LEU-ASN-GLN-ALA-LEU-LEU-ASP-LEU-HIS-ALA-LEU-GLY-SER-ALA-GLN-
                    125                 130                 135
    ALA-ASP-PRO-HIS-LEU-CYS-ASP-PHE-LEU-GLU-SER-HIS-PHE-LEU-ASP-
                    140                 145                 150
    GLU-GLU-VAL-LYS-LEU-ILE-LYS-LYS-MET-GLY-ASP-HIS-LEU-THR-ASN-
                    155                 160                 165
    ILE-GLN-ARG-LEU-VAL-GLY-SER-GLN-ALA-GLY-LEU-GLY-GLU-TYR-LEU-
                    170
    PHE-GLU-ARG-LEU-THR-LEU-LYS-HIS-ASP-
```

Figure 2 The amino acid sequence of the apoferritin from horse spleen (Reproduced from ref. 22 with permission from Elsevier-North Holland Biomedical Press)

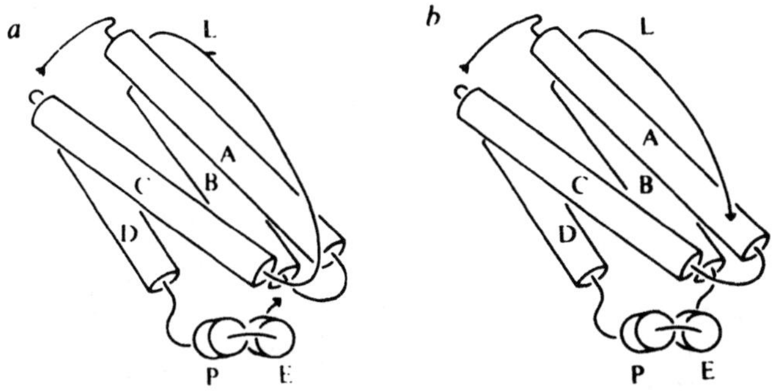

Figure 3 Two alternative conformations of the apoferritin subunit A-B-L-C-D-P-E (a) and A-C-D-P-E-B-L (b) (Reproduced from ref. 28 with permission from Macmillan Journal Ltd.)

The 174 amino acid residues of the subunit correspond to a molecular weight of 19.824, which is somewhat greater than that determined by physico-chemical methods (23-25). However, the molecular weight of the apoferritin protein shell, composed of 24 polypeptide chains, is in good agreement with physico-chemical estimations (summarised in reference (20) p. 88). The accord with the X-ray diffraction data (26) is particularly satisfying (480,000 compared with 476,000).

Horse spleen apoferritin crystallises from $CdSO_4$ in the space group F 432 with a = 18.4 nm ; there is one subunit per asymmetric unit. The X-ray diffraction studies initiated in the late 50's by Harrison (26) have resulted in an electron density map at a resolution of 0.28 nm (27, 28). Each subunit consists of a bundle of four long helices (A-D) lying parallel or anti-parallel to one another, with two shorter helices (P and E) and regions of irregular conformation (figure 3). The connections between the helices are assured by non-helical sequences which correspond to regions of the map where the electron density is weakest and where it is difficult to trace the polypeptide chain inambiguously. The helices A, B, C, D, P and E contain respectively 28, 25, 28, 20, 7 and 10 residues, giving a total of 117 residues (68 % of the sequence). There is an irregularity near the middle of helix B and the N-terminal sections of helices C and D would seem to be best fitted by one or two turns of 3_{10} helix. Good density is also observed for the sharp turn between C and D, the long loop L, which together with the two long helices A and C is located at the exterior of the apoferritin protein shell (figure 4), and for the connections between D-P to E. The four long helices of one subunit are related by pseudo 222 symmetry. In the quaternary structure pairs of subunits related by molecular diads lie with their long axes anti-parallel. The short helices P and E form the channels around the fourfold axes of symmetry through which iron, and other low molecular weight substances, can gain access to the

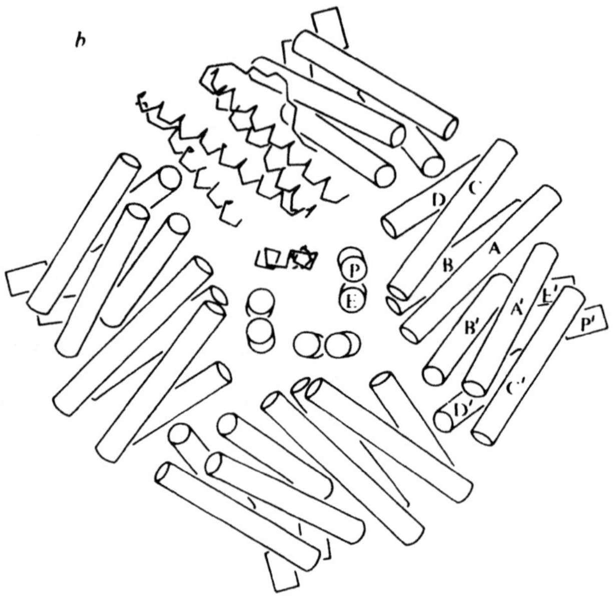

Figure 4 The molecular packing around a moleculer four-fold axis of symmetry showing four dimers (Reproduced from ref. 22 with permission from the American Chemical Society)

interior of the apoferritin subunit. From model building a short stretch of anti-parallel β-pleated sheet has been found joining the loops L from neighbouring subunits in a region around a two-fold axis. Four hydrogen bonds are formed involving a total of 5 residues in each subunit ; the pleated sheet structure lies on the outside of the apoferritin protein shell at a region of intermolecular contact.

Two alternative connectivities between the helices have been suggested (28), namely A-B-L-C-D-P-E and A-C-D-P-E-B-L (figure 3). Both from prediction of the secondary structure of the subunit on the basis of its amino acid sequence (22) and from fitting of the sequence to the electron density map (P.M. Harrison, R.R. Crichton et al., in preparation) the order in figure 3a would seem to be most likely. An extended N-terminal sequence of about 10 residues (mostly polar) precedes helix A which contains 28 amino acids and is particularly rich in hydrophobic and aromatic amino acids. A short non-helical section connects helix A to helix B. This helix of 25 residues is interesting from several points of view since it contains a contiguous sequence (56-60) of charged amino acid residues, which would be expected to destabilise the helix and probably accounts for the irregularity observed near the middle of the helix in the X-ray structure. Helix B is also interesting for the large number of Glu residues, two groups of which respectively Glu 53 and Glu 57, and Glu 57, 60 and 63 are on the same side of the helix and might form metal binding sites, and for the presence at the N-terminal end of the helix of one of the two cysteine residues of the subunit (Cys 48). From the end of helix B the loop L, which contains a predominance of charged and polar residues, as well as one of the two prolines present in the sequence, passes on the outer surface of the protein to connect with helix C. This helix, 28 residues long, and like A located at the exterior of the protein is similar to A in having a preponderance of hydrophobic residues. The sharp turn between helix C and D contains the second proline residue present in the sequence. The 20 residue long helix D is interesting in that, like helix B, it has a very high content of charged residues (8 compared with 10 in B) and once again a potential metal binding site may be formed by Asp 127 and Glu 130 which are disposed on the same side of the helix. It also contains the second cysteine residue of the subunit which is again located at the N-terminal end of the helix at position 126. Helices B and D are anti-parallel and thus the two cysteine residues are a considerable distance apart. A non-helical turn leads into helix P (7 residues), another connects P to helix E and the sequence ends with a short C-terminal segment presumably buried in the interior of the apoferritin shell. It is interesting to point out that in human spleen apoferritin (Wustefeld, C. and Crichton, R.R., unpublished observations) there are two additional proline residues at positions 157 (in the PE turn) and 93 (just at the beginning of helix C).

The packing of four long α-helices found in the apoferritin subunit is similar to that observed in several other proteins such as myohaemerythrin, haemerythrin, cytochrome c', cytochrome b_{562} and tobacco mosaic virus (28).

The Structure of the Iron Core

The iron core of ferritin has been the object of a recent investigation using EXAFS (29) and we discuss here the model proposed on the basis of these studies. It has been well established from magnetic susceptibility and Mössbauer spectroscopy that the supramagnetic behaviour of Fe^{3+} in ferritin is attributable to exchange coupling, which, in conjunction with the stoichiometry, implies that the Fe^{3+} ions in ferritin are connected by O^{2-} and OH^- bridges. Low angle X-ray scattering data from iron polymers, in particular the polymer of Spiro and Saltman (30) which has a composition of $Fe_4O_3(OH)_4(NO_3)_3 \cdot 1.5\ H_2O$, a molecular weight of about 150,000 and a diameter of about 70 Å, very similar to the iron core of ferritin, led to the proposition of a tetrahedral arrangement of O^{2-} and OH^- ligands around Fe^{3+} ions (31) which has been challenged (32, 33). EXAFS studies on ferritin and on an

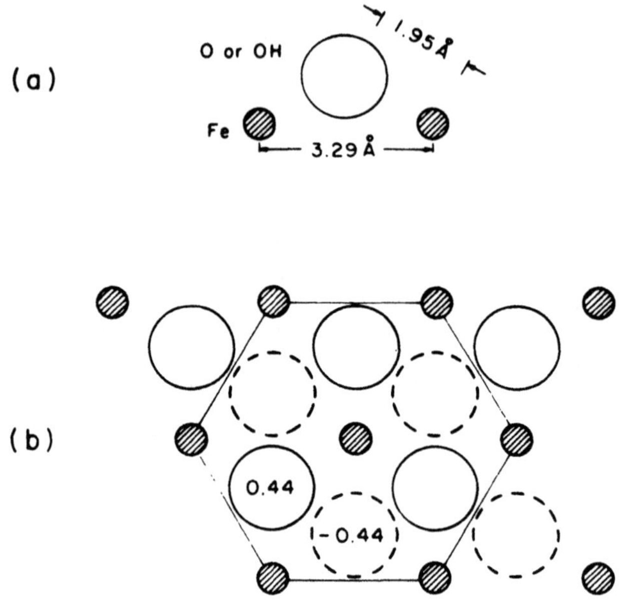

Figure 5 (a) Average structural unit determined by the EXAFS measurements
(b) Idealised model structure constructed using the unit in (a)
(Reproduced from ref. 29 with permission from the American Chemical Society)

Fe-glycine complex lead to the conclusion that the iron atoms of the ferritin core are surrounded by 64 ± 0.6 oxygens at a distance of 1.95 ± 0.02 Å, most probably in a distorted octahedral arrangement. Each iron atom has 7 ± 1 iron neighbours at a distance of 3.29 ± 0.05 Å. A model for the ferritin core is proposed on the basis of these results (figure 5) as a layered arrangement where the iron atoms are localised at the interstices between two closely packed layers of oxygen ; the compact Fe-O-Fe layers are only weakly bound to adjacent layers. In essence all three-dimensional structures are ruled out by combining the EXAFS results with the known stoichiometry and the density of the micelle. Only the two-dimensional layered structure of figure 5b accounts for all of the experimental results. The authors suggest that phosphorous atoms terminate the sheets into strips (figure 6a) whose width naturally accounts for the size of the micelle of about 70 Å (figure 6b). The length of the strip depends on the amount of iron in the micelle and the phosphate presumably interacts with the protein shell. This may account for the rather large number of basic amino acids which we find in the interior of the

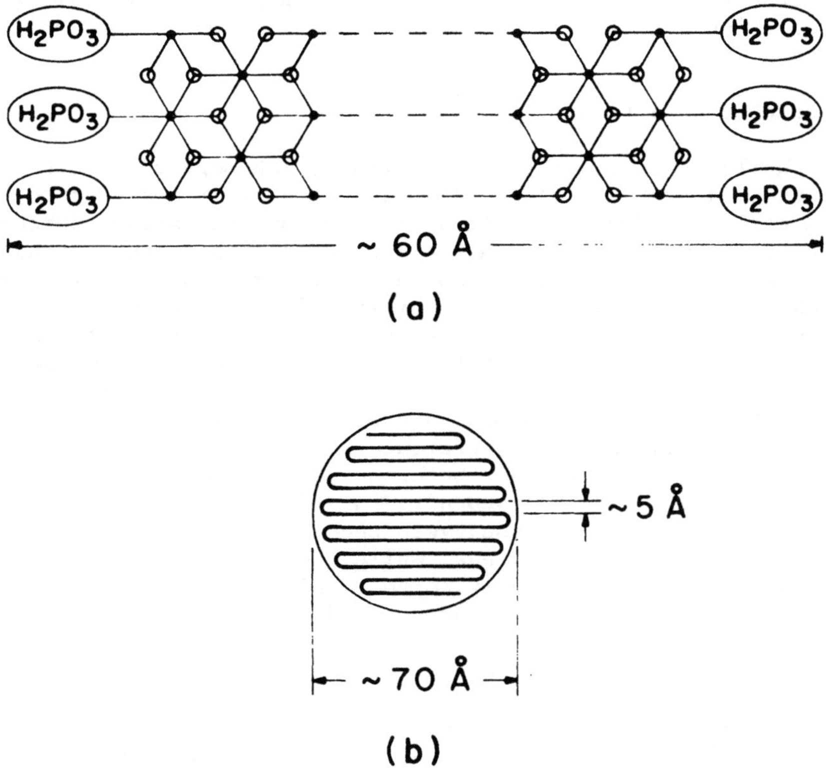

Figure 6 (a) One way of using the phosphorous atoms to terminate the two-dimensional sheets of Figure 5b into a strip. The solid circles are iron and the larger open circles are O or OH. From the stoichiometry there are 9 Fe per P, giving a width of 60 Å. The length of the strip depends on the amount of iron in the micelle.
(b) Schematic drawing of the folding of the strip into a 70 Å diameter micelle
(Reproduced from ref. 29 with permission from the American Chemical Society)

apoferritin protein shell. High resolution electron microscopy studies on the disrupted protein shell remain associated with their core crystallites (34) , this, in parallel with previous results (35), suggests that there is a strong attachment of the iron cores to the protein shell, and in view of the volume of evidence which indicates that the phosphate of the iron core is situated at the surface of the core, this presumably implies an interaction between the phosphate and the inner surface of the protein shell.

Microheterogeneity of Ferritins

While it is clear that ferritin is inhomogeneous with regard to its iron content, other heterogeneities in its structure have been observed. Thus, ferritins from various mammalian sources have been shown to yield multiple bands on isoelectric focussing (reviewed in 36). Moreover the range of components of different pI observed are characteristic of the tissue of origin for a given species and there appears to be a well established correlation between the content of two subunits of molecular weights 19,000 (called L on account of its importance in liver) and 21,000 (H in view of the large amounts present in heart) and the range of pI values observed in different tissues (37, 38). It is proposed that apoferritin is a heteropolymer composed of different proportions of these two subunits in different tissues (36, 37) ; the analogy with other isoenzymes such as lactate dehydrogenase is immediately apparent. Thus, while the apoferritin of horse spleen is almost exclusively composed of homopolymers of the L subunit (and presumably the amino acid sequence established for this apoferritin represents the L subunit) in other tissues such as liver, kidney or heart we may expect to find a much larger proportion of heteropolymers of the type $L_{24}H_0, L_{23}H_1$, ... to L_1H_{23}, L_0H_{24}. To date we do not have any data available on the sequence of the H subunit, although peptide mapping and amino acid analyses suggest that it is probably quite similar to the L subunit (37, 38). The functional implications of this heterogeneity are not yet clear although some correlations exist for human and horse ferritins between the pI of ferritin fractions and their rate of iron uptake (39, 40).

FERRITIN FUNCTION

Iron Deposition in Ferritin

As indicated earlier the function of an iron storage protein is both to take up iron and to release its iron as the requirements of the cells and tissues dictate. This necessity explains why there is a heterogeneity in ferritin iron content, since under normal physiological conditions, a latent iron storage capacity must always be available. In this context it is interesting also to note that apoferritin synthesis responds to increasing iron concentrations in the milieu, an augmentation in apoferritin synthesis being observed when excess iron is administered (reviewed in 41).

When we consider how iron might be deposited in the apoferritin molecule three mechanisms may be considered. First we might envisage that ferric iron complexed with an appropriate biological chelator might penetrate the channels between the subunits and be hydrolysed and incorporated into the core micelle. While some evidence does exist for such a process (42-44) the rate at which it occurs suggests that it contributes only to a small extent in ferritin iron uptake. The two other mechanisms (45, 46) both part from the premise that iron enters the apoferritin shell as Fe^{2+} and that in the interior of the protein shell the iron is oxidised, followed by hydrolysis and incorporation into the micellar core. Both models concur in attributing to the protein a catalytic role in the oxidation of iron in the initial phase of iron deposition. Thereafter they differ in that the model of (45) implies that once the iron core has begun to be deposited, subsequent oxidation of Fe^{2+} occurs on the surface of the ferric oxyhydroxide micelle whereas in the alternative version the protein is assumed to continue in its role as catalyst of iron oxidation (46).

The mechanism which has been proposed (47, 48) for the initial steps in iron oxidation (and which may be involved in subsequent cycles of iron oxidation) is summarised in figure 7.

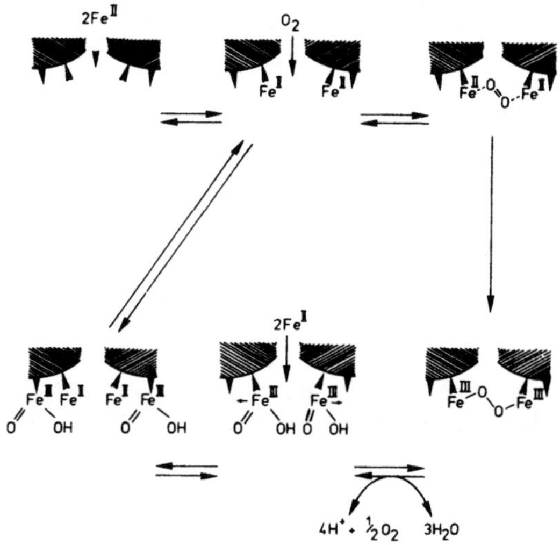

Figure 7 A mechanism for the oxidation and hydrolysis of iron catalysed by apoferritin (from 48).
(Reproduced with permission from Elsevier Sequoia S.A.)

Fixation of two Fe^{2+} ions to iron binding sites located close to an axis of two-fold symmetry on adjacent subunits is followed by binding of a dioxygen molecule and oxidation of the iron with formation of a bridged peroxo-complex. Thereafter we suggest that subsequent attack by Fe^{2+} ions might transform the peroxo-complex to a μ-oxo complex which could then hydrolyse to yield the FeO.OH which is the final product. The nature of the catalytic site has been studied by chemical modification and it seems clear that carboxyl groups are involved (49). Recently a tryptic peptide containing residues 53-59 (figure 2) has been identified as part of the catalytic site (50). This peptide contains Glu 53, Glu 56 and Glu 57.

The subsequent steps of iron deposition remain controversial. The kinetics of the incorporation of iron are complex (45, 46, 50-52) and have been interpreted in terms of the two models described above, the biphasic nature of the kinetics being attributed respectively to an autocatalytic effect of the iron micelle on the rate of iron oxidation or to a heteronucleation effect of the micelle on the release of iron from the catalytic sites. The rate of iron incorporation and indeed of iron release is a function of the amount of iron already present in the molecule and maximum rates are observed for ferritins with 1,000-1,500 atoms/molecule (53, 54). The inhibition of iron deposition in ferritin by a number of transition metal ions such as Zn^{2+}, Cu^{2+}, Co^{2+}, VO^{2+}, Ni^{2+}, Tb^{3+} and Cr^{3+} (50, 55, 56) may explain the effects of some of these metals on iron metabolism.

It is likely that most of the phosphate present in the ferritin core is loosely bound at the surface of the micelle and studies of its incorporation and release suggest that it is not present at the intracellular site of iron incorporation into ferritin but is added subsequently (57).

Iron Mobilisation from Ferritin

If our understanding of the way in which iron is deposited in ferritin are somewhat confused, our understanding of the pathways of iron mobilisation from ferritin are less clear. We know that *in vitro* iron can be released from ferritin by reduction in the presence of appropriate chelators and that in the absence of reducing agents direct chelation of ferritin iron is slow with the exception of several chelators that we do not find in mammalian cells (59). None of the potential biological reductants such as ascorbate, glutathione or cysteine is capable of mobilising ferritin iron at physiologically significant rates (58). However, under anaerobic conditions, reduced flavins, either prepared by chemical reduction (60) by the use of a coupled NADH-FMN system (61) or by an NADH-flavin oxidoreductase (62), mobilise ferritin iron in the presence of bipyridyl very rapidly. The most detailed study (61) was carried out under anaerobisis with the NAD(P)H-flavin oxidoreductase from Beneckea harveyi to generate a constant steady state concentration of reduced flavin. The authors concluded that (i) the system released iron rapidly and quatitatively in the presence of bipyridyl (ii) Sepharose-linked dihydroflavins did not mediate iron release (suggesting that the reduced flavin must penetrate the protein shell to reduce iron) (iii) no specific binding sites for flavins were detected, although dihydroflavins gave saturation kinetics with very high apparent K_m's (iv) the rate of iron release was maximal for ferritins of an iron content of 1,200 atoms/molecule. For dihydroriboflavin electron transport appears to be rate determining whereas for dihydroFMN and dihydroFAD diffusion of the dihydroflavin through the protein channel is slow. The overall stoichiometry as expected is 2 Fe^{3+} reduced per equivalent of dihydroflavin. In studies on iron mobilisation from ferritin by chelators (59) it was found that in the presence of small amounts of FMN iron release was markedly enhanced for Fe II chelators, whereas with Fe III chelators such as desferrioxamine B and rhodotorulic acid an almost total inhibition was observed. It was suggested that the flavin (probably photoreduced) blocked access of the Fe III chelators to the ferritin iron.

While this mechanism is certainly effective in vitro the evidence that it intervenes in vivo is poor (63-65). The concentration of free flavin in cells is low - there is to date no mammalian counterpart to the oxidoreductase of B. harveyi, which is unusual in that it generates dihydroflavins free in solution, and it is uncertain that dihydroflavins would be free to diffuse and have sufficiently long lifetimes in an anaerobic mammalian cell to be effective. Two solutions to this problem could be proposed. Firstly it might be possible that ferritin contains an associated flavin - there is some recent evidence for this (66), such that dihydroflavins could be generated in situ on the protein itself. Alternatively, as we will see in the next and final section, reduced flavins might be generated at the places in the cell where release of ferritin iron is required.

FUNCTION OF FERRITIN - A CELLULAR VIEW

It seems appropriate to end this review where we began, namely with the role of ferritin in the overall regulation of iron metabolism. To do this we have grouped together here some reflections on the exchange of iron (in both directions) with the extracellular carrier transferrin as well as the exchange of iron between the storage form and the sites of intracellular need.

Transferrin-ferritin iron exchange is unlikely to occur by a direct interaction between the two proteins and it seems most likely that ferritin takes up iron from transferrin after the transferrin iron has been released, and presumably reduced. Recent studies suggest that transferrin uptake by cells occurs by receptor-mediated endocytosis and by fluid phase endocytosis (67, 68) and that iron release is achieved in the secondary lysosomes formed by fusion of the endocytic vacuole with lysosomes ; the acidic pH of the lysosome is particularly favourable for transferrin iron release. Thereafter the iron can pass from the lysosome into the cytoplasm and be incorporated into ferritin. The release of ferritin iron within the cell for incorporation into the transport form might either occur at the plasma membrane, the iron passing from the storage to the transport form associated with a biological chelator or else be released elsewhere in the cell and pass into a low molecular weight transit pool (69) which is assumed to be in equilibrium both with ferritin and transferrin iron.

However, ferritin iron may also be used for the synthesis of iron-containing enzymes within the cell and while some authors have claimed that ferritin is an obligatory intermediate on the pathway of iron from transferrin to haem synthesis in reticulocytes (70, 71) recent studies on rat liver demonstrate that ferritin iron can be mobilised by mitochondria using a suitable substrate such as succinate to reduce exogenous FMN ; the reducing equivalents are drained from that respiratory chain at the level of ubiquinone (72, 73). Ferritin iron can also be used by rat liver mitochondria for the ferrochelatase reaction, the final step in haem synthesis (74). Recent results suggest that the intersection between ferritin and mitochondria is quite specific (75). Thus we may be on the verge of finding out that the secret of how to get iron out of ferritin within the cell lies in its interaction with membranes.

REFERENCES

(1) Laufberger, V.: 1937, Bull. Soc. Chim. Biol.19, pp. 1575-1582.
(2) Kuhn, R., Sörenson, N.A. and Birkhofer, L.: 1940, Ber. Chem. Ges.73B, pp. 823-837.
(3) Granick, S.: 1943, J. Biol. Chem.149, pp. 157-167.
(4) Arora, R.S., Lynch, E.C., Whitley, C.E., Alfrey, C.P.: 1970, Texas Rep. Biol. Med.28, pp. 189-196.
(5) Mazur, A. and Shorr, E.: 1950, J. Biol. Chem.182, pp. 607-627.
(6) Jacobs, A., Miller, E., Worwood, M., Beamish, M.R. and Wardrop, C.A.: 1972, Brit. Med. J. iv, pp. 206-208.

(7) Kato, T., Shinjo, S. and Shimada, S.: 1968, J. Biochem.63, pp. 170-175.
(8) Brown, D.R. and Caston, J.D.: 1962, Develop. Biol.5, pp. 445-450.
(9) Towe, K.M., Lowenstorm, H.A. and Nesson, M.H.: 1963, Science 142, pp. 63-64.
(10) Roche, J., Bessis, M., Breton-Gorius, J. and Stralin, H.: 1961, Compt. Rend. Acad. Scien.252, pp. 3886-3887.
(11) Crichton, R.R., Ponce-Ortiz, Y., Koch, M.H.J., Parfait, R. and Stuhrmann, H.B.: 1978, Biochem. J.171, pp. 349-356.
(12) Hyde, B.B., Hodge, A.J., Kahn, A. and Bimstiel, M.L.: 1963, J. Ultrastruc. Res.9, pp. 248-258.
(13) Seckbach, J.: 1968, J. Ultrastruc. Res.22, pp. 413-423.
(14) David, C.N. and Easterbrook, K.: 1971, J. Cell. Biol.48, pp. 15-28.
(15) Stiefel, E.I. and Watt, C.D.: 1979, Nature 279, pp. 81-83.
(16) Bothwell, T.H., Charlton, R.W., Cook, J.D. and Finch, C.A.: 1979, "Iron Metabolism in Man", Blackwell, London, pp. 327-349.
(17) Linder, M.C. and Munro, H.N.: 1973, Amer. J. Path.72, pp. 263-282.
(18) Clegg, G.A., Fitton, J.E., Harrison, P.M. and Treffry, A.: 1980, Prog. Biophys. Molec. Biol.36, pp. 56-86.
(19) Aisen, P. and Listowsky, I.: 1980, Ann. Rev. Biochem.49, pp. 357-387.
(20) Crichton, R.R.: 1973, Struct. and Bond.17, pp. 67-134.
(21) Fischbach, F.A. and Anderegg, J.W.: 1965, J. Mol. Biol.14, pp. 458-473.
(22) Heusterspreute, M. and Crichton, R.R.: 1981, FEBS Lett.129, pp. 322-329.
(23) Crichton, R.R. and Bryce, C.F.A.: 1980, FEBS Lett.6, pp. 121-124.
(24) Bryce, C.F.A. and Crichton, R.R.: 1971, J. Biol. Chem.246, pp. 4198-4205.
(25) Björk, I. and Fish, W.W.: 1971, Biochemistry 10, pp. 2844-2848.
(26) Harrison, P.M.: 1959, J. Mol. Biol.1, pp. 69-80.
(27) Banyard, S.H., Stammers, D.K. and Harrison, P.M.: 1978, Nature 271, pp. 282-284.
(28) Clegg, G.A., Stansfield, R.F.D., Bourne, P.E. and Harrison, P.M.: 1980, Nature 288, pp. 298-300.
(29) Heald, S.M., Stern, E.A., Bunker, B., Hott, E.M. a,d Hott, S.L.: 1979, J. Am. Chem. Soc.101, pp. 67-73.
(30) Spiro, T.G., Allterton, S.E., Bils, R., Renner, J., Saltman, P. and Terzis, A.: 1966, J. Am. Chem. Soc.88, pp. 2721-2726.
(31) Brady, G.W., Kurkjian, C.R., Lyden, E.F.X., Rolin, M.B., Saltman, P., Spiro, T.G. and Terzis, A.: 1968, Biochemistry 7, pp. 2185-2192.
(32) Gray, H.B.: 1971, Adv. Chem. Ser.100, pp. 365-368.

(33) Holt, E.M., Holt, S.L., Tucher, W.F., Asplund, R.O. and Watson, K.J.: 1974, J. Am. Chem. Soc.96, pp. 2621-2623.
(34) Massover, W.M.: 1978, J. Mol. Biol.123, pp. 721-726.
(35) Stuhrmann, H.B., Haas, J., Ibel, K., Koch, M.H.J. and Crichton, R.R.: 1976, J. Mol. Biol.100, pp. 399-413.
(36) Drysdale, J.W.: 1977, Ciba Symp.51, pp. 41-67.
(37) Arosio, P., Adelman, T.G. and Drysdale, J.W.: 1978, J. Biol. Chem.253, pp. 4451-4458.
(38) Ostuka, S., Maruyama, H. and Listowsky, I.: 1981, Biochemistry, in press.
(39) Wagstaff, M., Worwood, M. and Jacobs, A.: 1978, Biochem. J. 173, pp. 969-977.
(40) Russell, S.M. and Harrison, P.M.: 1978, Biochem. J.175, pp. 91-104.
(41) Zahringer, J., Baliga, B.S. and Munro, H.N.: 1976, Proc. Natl. Acad. Sci. U.S.A.73, pp. 857-861.
(42) Hoy, T.G. and Harrison, P.M.: 1976, Brit. J. Haematol.33, pp. 497-504.
(43) Treffry, A. and Harrison, P.M.: 1979, Biochem.J.181, pp. 709-716.
(44) Konopka, K., Mareschal, J.C. and Crichton, R.R.: 1980, Biochem. Biophys. Res. Commun.96, pp. 1408-1413.
(45) Macara, I.G., Hoy, T.G. and Harrison, P.M.: 1972, Biochem. J.126, pp. 151-162.
(46) Bryce, C.F.A. and Crichton, R.R.: 1973, Biochem.J.133, pp. 301-309.
(47) Crichton, R.R. and Roman, F.: 1978, J. Mol. Catal.4, pp. 75-82.
(48) Crichton, R.R., Roman, F., Roland, F., Pâques, E., Pâques, A. and Vandamme, E.: 1980, J. Mol. Catal.7, pp. 267-276.
(49) Wetz, K. and Crichton, R.R.: 1976, Eur. J. Biochem.61, pp. 545-550.
(50) Vandamme, E.: 1980, Doctoral thesis, Université Catholique de Louvain.
(51) Pâques, E.P., Pâques, A. and Crichton, R.R.: 1979, J. Mol. Catal.5, pp. 363-375.
(52) Pâques, E.P., Pâques, A. and Crichton, R.R.: 1980, Eur. J. Biochem.107, pp. 447-453.
(53) Harrison, P.M., Hoy, T.G., Macara, I.G. and Hoare, R.J.: 1974, Biochem. J.143, pp. 445-451.
(54) Russell, S.M. and Harrison, P.M.: 1978, Biochem. J.175, pp. 91-104.
(55) Macara, I.G., Hoy, T.G. and Harrison, P.M.: 1973, Biochem. J.135, pp. 343-348.
(56) Wauters, M., Michelson, A.M. and Crichton, R.R.: 1978, FEBS Lett.91, pp. 276-280.
(57) Treffry, A. and Harrison, P.M.: 1978, Biochem. J.171, pp. 313-320.
(58) Dognin, J. and Crichton, R.R.: 1975, FEBS Lett.54, pp. 234-236.

(59) Crichton, R.R., Roman, F. and Roland, F.: 1980, J. Inorg. Biochem.13, pp. 305-316.
(60) Sirivech, S., Frieden, E. and Osaki, S.: 1974, Biochem. J. 143, pp. 311-315.
(61) Crichton, R.R., Wauters, M. and Roman, F.: 1975 in "Proteins of Iron Storage and Transport in Biochemistry and Medicine" (Crichton, R.R. ed.) pp. 287-294, North Holland, Amsterdam.
(62) Jones, T., Spencer, R. and Walsh, C.: 1978, Biochemistry 17, pp. 4011-4017.
(63) Osaki, S. and Sirivech, S.: 1971, Fed. Proc. Am. Soc. Exp. Biol.30, Abstract 1292.
(64) Sirivech, S., Driskell, J. and Frieden, E.: 1977, J. Nutr. 107, pp. 739-745.
(65) Zaman, Z. and Verwilghen, R.: 1977, Biochem. Soc. Trans.7, pp. 306-308.
(66) Mareschal, J.C., Mingeot, M.P. and Crichton, R.R.: 1980, 13th FEBS Meeting, Jerusalem, abstract.
(67) Octave, J.-N., Schneider, Y.-J., Crichton, R.R. and Trouet, A.: 1981, Eur. J. Biochem.115, pp. 611-618.
(68) Karin, M. and Mintz, B.: 1981, J. Biol. Chem.256, pp. 3245-3252.
(69) Jacobs, A.: 1977, Ciba Symp.51, 91-106.
(70) Speyer, B. and Fielding, J.: 1979, Brit. J. Haematol.42, pp. 255-267.
(71) Nunez, M.T., Glass, J. and Robinson, S.H.: 1978, Biochim. Biophys. Acta 509, pp. 170-180.
(72) Ulvik, R. and Romslo, I.: 1979, Biochim. Biophys. Acta 588, pp. 256-271.
(73) Ulvik, R. and Romslo, I.: 1980, Biochim. Biophys. Acta 635, pp. 457-469.
(74) Ulvik, R. and Romslo, I.: 1978, Biochim. Biophys. Acta 541, pp. 251-262.
(75) Ulvik, R., Romslo, I., Roland, F. and Crichton, R.R.: 1981, Biochim. Biophys. Acta, in press.

CHEMISTRY AND PHYSIOLOGY OF THE TRANSFERRINS

Philip Aisen

Albert Einstein College of Medicine of
Yeshiva University

Iron in aqueous solution, near neutral pH and atmospheric oxygen tension, seeks the ferric state as its thermodynamically stable form. Because of the insolubility of ferric hydroxide, however, organisms at all levels of biological development have been obliged to elaborate specific iron-binding molecules to maintain this essential metal in soluble, bioavailable form. In the vertebrate world, the function of iron transport is given to the transferrins, a class of two-sited iron-binding proteins distinguished by a specific interdependence of anion and metal-ion binding. The mechanisms of iron-binding and release by the transferrins, and whether the two-sited nature of the protein serves a biologic function, are still unsettled questions in the biochemistry of iron metabolism.

A. INTRODUCTION: THE AQUEOUS CHEMISTRY OF IRON

The properties of iron in aqueous solution are dominated by two reactions: one-electron oxidation-reduction cycle of the Fe(II)/Fe(III) couple, and hydrolysis of iron to form complex and insoluble hydroxides. In virtually all organisms the metabolism of iron is largely determined by the opportunities and hazards these reactions present.

1. Redox properties of iron

At the pH of most physiological fluids, and in the presence of oxygen, the Fe(II)/Fe(III) couple is strongly poised toward the ferric state [1]. Under reducing conditions, however, particularly when Fe(II)-stabilizing ligands are available, the

H. B. Dunford et al. (eds.), The Biological Chemistry of Iron, 63–83.

ferrous state may prevail. Most oxygen-bearing "hard" (non-polarizable) ligands, including biologically ubiquitous complexing groups such as water, OH^-, CO_3^{2-}, COO^- and PO_4^{3-}, promote the higher oxidation level, while many "softer" ligands offering nitrogen or sulfur atoms to iron are more favorable to the ferrous state [1,2]. Thus, a number of iron enzymes are known in which the metal rests in its lower oxidation state [3], and iron in the reducing environment of the inner membrane of mitochondria is maintained in the ferrous form [4]. One function of ascorbic acid may be to maintain the ferrous state of iron in prolyl and lysyl hydroxylases [3], enzymes intimately involved in the laying down of collagen during wound healing.

A variety of oxidizing species may act *in vitro* to convert ferrous iron to ferric iron, but it is generally believed (although direct experimental evidence is scant) that the physiological electron acceptor is ordinary molecular dioxygen. The rate at which oxidation of ferrous iron occurs is dependent on a number of variables, the most important of which are oxygen tension, the availability of Fe(III) chelators to sequester and thus stabilize ferric iron, and the presence of oxidative catalysts [1]. The simple autoxidation of Fe(II) in aqueous solution probably entails the following reactions:

$$\mathrm{Fe(II)} + \mathrm{OH^-} \rightleftarrows (\mathrm{FeOH})^+ \quad (1)$$

$$\mathrm{O_2} + \mathrm{OH^-} \rightleftarrows (\mathrm{O_2OH})^- \quad (2)$$

$$\mathrm{Fe(OH)^+} + (\mathrm{O_2OH})^- \rightleftarrows \mathrm{Fe(OH)_2} + \mathrm{O_2^-} \quad (3)$$

$$\text{Overall:}\quad \mathrm{Fe(II)} + \mathrm{O_2} + 2\mathrm{OH^-} \rightleftarrows \mathrm{Fe(III)(OH^-)_2} + \mathrm{O_2^-} \quad (4)$$

Because alkaline pH promotes an increased oxidation rate, and stabilizes the ferric state in insoluble ferric hydroxides, the autoxidation of ferrous iron in a physiological milieu is apt to be particularly rapid. Experimental confirmation of this is often inadvertently obtained when rusty precipitates inevitably form in standing solutions of ferrous iron.

The ease which spontaneous autoxidation occurs may be further enhanced by agents which complex Fe(III). Many common laboratory sequestering agents, such as EDTA, and the biologically ubiquitous citrate and phosphate ions, are quite effective in facilitating autoxidation of Fe(II) to Fe(III) [5]. This "chelate effect" may contribute to the so-called "ferroxidase" activity of iron-binding proteins such as transferrin and ferritin, which bind Fe(II) weakly if at all, but which have great avidity for Fe(III). Conversely, chemical species with soft or intermediate ligands, such as imidazole, may tend to favor Fe(II). The biological importance of such ligand effects has perhaps received less attention than it deserves.

The cupric ion acts as a true catalyst in promoting the autoxidation of Fe(II) by reactions presumed to involve the following steps [6]:

$$Fe(II) + Cu(II) \rightarrow Fe(III) + Cu(I) \qquad (5)$$

$$Cu(I) + O_2 \rightarrow Cu(II) + O_2^{\overline{\cdot}} \qquad (6)$$

The catalytic effectiveness of cupric ion is enhanced when it is incorporated into the plasma copper protein ceruloplasmin [7]. Whether ferroxidases as such are required for the regulation of iron metabolism *in vivo*, however, remains a contentious question [8].

Because iron shuttles between ferrous and ferric states during its storage and transport by almost all organisms, redox reactions have a critical role in the metabolism of the metal [9]. Many agents will reduce ferric iron to the more tractable ferrous form. Ascorbate, a variety of thiols, and hydroxylamine are commonly employed in the laboratory as reductants of ferric iron. Mobilization of iron from the hematite-like core of ferritin may be effected by physiological reductants such as reduced pyridine nucleotides and flavins, with flavins most effective *in vitro* [10]. Reductants, although relatively ineffective by themselves, also function synergistically with iron-chelating agents to facilitate release of iron from transferrin, at least *in vitro* [11].

A striking example of the uses of autoxidation of Fe(II) in biological systems is provided by the acidophilic bacterium *thiobacillus ferrioxidans*. This interesting organism grows at pH 2, and derives most of its metabolic energy from the oxidation of ferrous iron:

$$2Fe(II) + \tfrac{1}{2}O_2 + 2H^+ \rightarrow 2Fe(III) + H_2O \qquad (7)$$

Since the interior of the bacterial cell is at pH 5, the protons consumed during the reaction are replenished by an energy-yielding flow of replacement protons along the proton concentration gradient from the acidic environment to the interior of the bacterial cell [12]. Such coupling of proton transport to ATP production may be one of the simplest and most direct validations of the chemiosmotic model for which Mitchell received the Nobel prize.

The redox chemistry of iron may also pose a hazard to the organism. Superoxide, a presumed product of the autoxidation of ferrous iron (Reaction 4) may act to generate the noxious hydroxyl radical via the Haber-Weiss reaction (which may be catabolized by iron or its complexes):

$$O_2^{\cdot -} + H_2O_2 \rightarrow O_2^{\cdot} + OH^- + OH\cdot \qquad (8)$$

Ferrous iron itself may produce hydroxyl radical by the Fenton pathway:

$$Fe(II) + H_2O_2 \rightarrow Fe(III) + OH^- + OH\cdot \qquad (9)$$

The hydroxyl radical, or perhaps a derivative of it such as the ferryl ion, $(FeO)^{2+}$, is thought to be a principal culprit in iron toxicity [13], and may also be a participant in the mechanism by which bleomycin exerts its antitumor activity [14].

2. Hydrolysis

Both Fe(II) and Fe(III) are subject to hydrolysis with the formation of insoluble iron hydroxides [15]. The solubility of ferrous hydroxide, however, is sufficiently large so that at pH 7.4 an equilibrium concentration of 10^{-2} M Fe(II) can be maintained. In contrast, the solubility of ferric hydroxide is so low that the equilibrium concentration of Fe(III) at physiological pH is no greater than about 10^{-18} M, corresponding to only a few hundred free ferric ions per ml. Because the thermodynamically favored state of iron under aerobic conditions is Fe(III), most organisms have been compelled to evolve Fe(III)-sequestering agents to solubilize iron and maintain it in bioavailable form. Indeed, to Neiland's famous dictum that life and iron are inseparable may be added the corollary that life, iron and iron-binding molecules are also inextricably linked. In the vertebrate world, this linkage is maintained by the transferrins and ferritin.

The hydrolysis of ferrous ion is essentially a single-step process, with only negligible concentrations of species intermediate between aquated Fe(II) and insoluble $Fe(OH)_2$. In contrast, the hydrolysis of ferric ion involves a complex sequence of reactions still only poorly understood. Each step is a process involving deprotonation, condensation, or nucleation [15,16]. Among the intermediates identified in these reactions are $Fe(OH)_2^+$, $Fe_2(OH)_2^{4+}$, $FeOH_2^{4+}$, $FeOH^{2+}$ and $Fe_3(OH)_4^{5+}$. Some of these species, particularly $FeOH^{2+}$, may be more reactive in complex formation than the aquated ferric ion itself. Even at pH 7.4, however, the reactions of hydrolysis occur very slowly so that years may be taken to achieve equilibrium [15]. This probably accounts for the reactivity of ferric ion, including its binding to transferrin, in physiological fluids, in seeming defiance of thermodynamic considerations.

The stable form of precipitated ferric hydroxide is

ferric oxohydroxide, FeO(OH). Hydrolysis of Fe(III) *in vitro* begins with formation of monomeric and dimeric species, followed by growth to polymers which function as substrates for continued nucleation of FeO(OH) until a stable size is achieved [16]. The aqueous physical chemistry of hydroxide formation *in vitro* may be mimicked, at least in part, by events in the interior of the ferritin molecule.

3. Biological ligand exchange reactions

Because of the tightness with which biological chelators must bind Fe(II) to maintain it to soluble and bioavailable form, a simple dissociative mechanism is precluded in the exchange of ferric iron among physiological binding agents [17]. More likely, the transfer of iron from one chemical species to another occurs by progressive exchange of ligands.

Exchange reactions have been described in the transfer of iron from transferrin to simpler molecules which can bind ferric iron more tightly than the protein [18,19,20]. Such reactions, although strongly favored thermodynamically, may proceed with inordinate slowness. In the case of iron release from transferrin to desferrioxamine, for example, days may be required to observe appreciable exchange of iron. This kinetic inertness is familiar to the clinician administering desferrioxamine for the treatment of iron overload. To maintain adequate exposure of the chelating agent to the physiological reservoir of iron, the chelating drug is best given by slow infusion which minimizes excretion before chelation [21]. The kinetic barrier to iron exchange may be overcome, however, at least *in vitro*. Kinetically active iron-binding species, such as citrate, nitrilotriacetate and ketomalonate, although themselves much weaker iron-binders than desferrioxamine, act as true catalysts or mediators to reduce the time required for iron transfer from transferrin to desferrioxamine [18,19]. Catechol-based chelators, in contrast to the hydroxamate-based chelators exemplified by desferrioxamine, may be kinetically as well as thermodynamically competent in removing iron from transferrin [20]. Such agents, perhaps in combination with reducing agents as proposed by Kojima and Bates [8], may offer new promise in the treatment of iron overload.

B. EVOLUTIONARY AND STRUCTURAL BIOCHEMISTRY OF THE TRANSFERRINS

1. Definitions and description

In the vertebrate world, the task of transporting iron in bioavailable and non-hazardous form is given to the transferrins. The transferrins constitute a class of two-sited iron-binding proteins of vertebrates, predominantly but not exclusively extracellular in distribution. The specific sites will accept a variety of multivalent transition metal and lanthanide ions [22]. Of these, however, Fe(III) appears to be most tightly bound by the protein, and is the only species for which transferrin is definitively known to be a physiological carrier. The binding of iron is tight, but reversible, and the protein undergoes many cycles of iron transport before succumbing to its catabolic fate. Probably the most distinctive feature of iron-binding by the transferrins is its dependence on concomitant binding of a suitable anion. One anion is obligated for each metal ion bound. The cooperativity between metal- and anion-binding is dramatic: strong binding of neither is observed in the absence of the other [23-25]. Ordinarily the anion preferred by the transferrins is carbonate or bicarbonate (some uncertainty persists about which of these derivatives of carbonic acid is actually bound), but in the absence of carbon dioxide and its derivatives a variety of other anions may satisfy the protein. The interdependence of anion and metal ion binding at the specific sites of the transferrins is sufficiently unique so that it may be taken as a defining characteristic: the transferrins are distinguished from other metalloproteins by the linkage of metal- and anion-binding functions.

Three major kinds of transferrins have been identified: serum transferrin or serotransferrin, the iron-transporting protein of the blood circulation; ovotransferrin, isolable from egg white and formerly known as conalbumin; and lactoferrin or lactotransferrin, originally isolated from milk but since identified in a variety of extracellular fluids and secretions as well as in granulocytes [22]. Each of these transferrins is composed of a single polypeptide chain of molecular weight 75,000-80,000 on which are disposed two similar but not identical binding sites. Transferrins are glycoproteins, with carbohydrates arranged in branching chains; the number of chains, their constituent sugars, and the number of branch points in each chain, varies with the kind of transferrin and the species from which it is obtained [26-30]. Apart from the carbohydrate chains, ovotransferrin and serum transferrin from the domestic fowl are identical in structure [31], but the lactoferrins and serum transferrins differ appreciably in primary sequence [32].

In addition to these well-studied transferrins, specialized mucosal transferrins, believed to function in the transmucosal transport of absorbed iron, have been reported. They differ from serum transferrin in immunological properties [33], amino acid composition [34] and, in the case of the guinea pig protein, in being composed of two subunits [33].

2. Hydrodynamic properties

Perhaps because of the relative ease with which they can be obtained in highly purified form, their great solubility in water, their stability, and their amenability to rigorous physiochemical investigation, the transferrins have been subjects of numerous studies of their solution properties. Although there are several fine points of discordance, most investigations have yielded generally consistent models. When treated as a prolate ellipsoid of revolution the axial ratio of human serum transferrin is 2.1 - 2.5; that of ovotransferrin is 2.5 [35]. Minor changes in hydrodynamic properties are seen on passing from the iron-free to the iron-loaded form. More interesting, and somewhat more difficult to explain in detail, are the metal-induced changes in susceptibility to denaturation by heat or exposure to urea [36, 37], vulnerability to proteolytic attack [36], surface charge distribution [38], accessibility of reactive side chains [35], and resistance to hydrogen-tritium exchange [39]. Compared to the apoprotein, the iron-bearing form appears to resist unfolding, preserving its native structure in the face of strong perturbants.

3. Amino acid composition and sequence

Apart from the absence of sulfhydryl groups, there are no peculiarities in the amino acids of human serum transferrin or other transferrins [35]. The sequence of human transferrin is largely elucidated [40]; there are 676 amino acid residues, with valine as the N-terminal residue and proline as the C-terminal residue. Two doubly-branched (biantennary) heteropolysaccharide (glycan) chains are found in the C-terminal half of the protein [26,40]. A small proportion of human transferrin molecules may carry a triantennary polysaccharide chain replacing one of the usual biantennary chains [30]. Substantial homologies in amino acid sequence are found among transferrins from different species and, despite the absence of immunological cross-reactivity, between human serum transferrin and human lactoferrin [30].

C. THE SPECIFIC SITES OF TRANSFERRIN

Even with the amino acid sequence of human transferrin largely in hand, the determination of three-dimensional structure by x-ray crystallography is not sufficiently detailed for the active sites of the protein to be distinguished. Accordingly, what is known of their structure is derived from hydrogen ion titration, chemical modification and spectroscopic studies.

1. The role of tyrosyl residues

In now classical studies, Warner and Weber [41] clearly

demonstrated that three protons were released from ovotransferrin for each Fe(III) bound, and two protons for each Cu(II). Since these protons were released even at a pH as high as 10, tyrosyl residues were implicated as metal-binding ligands at each site. Although protons released from ionizable groups of the protein are not readily distinguishable from protons released by hydrolysis of Fe(III), the correctness of Warner and Weber's inference has been amply confirmed by spectrophotometric titration [42,43], a variety of chemical modification studies, resonance-enhanced Raman spectroscopy, ultraviolet difference spectroscopy, NMR spectroscopy, CD spectroscopy, and most recently, combined chemical modification and NMR spectroscopy [30,44]. Thus, the transferrins have emerged as prototypes of the iron-tyrosyl class of proteins.

Some uncertainty persists, however, about the number of tyrosyl residues participating in the binding of particular metal ions. Differing spectroscopic approaches, and studies with a variety of tyrosyl-modifying reagents, have failed to yield agreement on whether two or three tyrosyl-derived oxygen ligands are coordinated to specifically bound metal ions. Quite possibly the number depends on the particular metal ion bound, or, less likely, on the particular site to which binding occurs.

2. Nitrogen ligands

Comparative hydrogen ion titration studies detected two more protons ionizing near pH 6 for each empty site of apotransferrin than for each occupied site of diferric transferrin [45]. On this basis the suggestion was advanced that two histidyl residues participate in metal-binding at each site. Subsequent chemical modification studies, in which histidyl residues were either alklyated or photooxidized [30], demonstrated the need for intact histidyl residues in the metal-combining functions of the transferrins, although once again there was no agreement on whether one or two residues were present at each active site.

Electron paramagnetic resonance (EPR) spectroscopy of the transferrins, using isotopically pure Cu(II) to probe the active sites, have also been revealing. Superhyperfine splittings, due to interaction of the electron spin of the bound copper with the nuclear spin of a single nitrogen ligand, are evident in the spectrum of cupric-transferrins under physiological conditions [46]. Pulsed EPR studies have identified this ligand as coming from the imidazole ring of a histidyl residue [47]. Hypotheses are available to explain why only a single histidyl ligand is observable by EPR spectroscopy, while most chemical modification studies incriminate two histidyl residues at each site [48], but a definitive reconciliation of disparate conclusions probably must await detailed structural analysis by x-ray crystallography.

3. Other ligands

Two types of studies have established that water is coordinated to specifically bound metal ions of transferrin. First, nuclear relaxation is much faster in solvent water protons of solutions of ferric and cupric transferrins, indicating that the inner coordination sphere of metal ions at the active sites is accessible to exchangeable water [49]. Then, EPR linewidths in the cupric complexes of transferrin are appreciably narrower in D_2O solution than in H_2O solution [46]. This effect indicates that solvent molecules visit metal ions at the active site, since water protons are more effective EPR linebroadeners than are deuterons.

No involvements of tryptophanyl residues, or amino or carboxyl groups, could be demonstrated by a variety of chemical modification studies [30].

Although the importance of anion-binding by transferrin has long been appreciated, only recently has the obligatory anion been demonstrated to be directly ligated to specifically bound metal ions. High-resolution NMR studies with ^{13}C-labeled carbonate occupying the anion-binding site have shown that the ^{13}C resonance is readily observable when the carbonate is associated with a diamagnetic metal ion, but is no longer detectable when associated with paramagnetic Fe(III) [50,51]. Distance calculations indicate that the anion is directly coordinated to the metal ion. Similarly, pulsed EPR studies of Cu(II) complexes of transferrin bearing ^{13}C-labeled oxalate as the anion demonstrate a modulation of the copper signal attributable to interactions of the electron spin with a ^{13}C nucleus, again indicating the ligation of the anion to the metal ion [47].

4. The anion-binding site and the role of the anion

Carbonate (or bicarbonate) specifically bound to transferrin is much more labile than its metal ion partner, exchanging over a period of days with ambient bicarbonate [52]. Exchange may be catalyzed by other anions, most of which are also acceptable to the anion-binding site [52-54]. Anion exchange is a simple first-order process in monoferric transferrin, but in doubly-occupied transferrins the sites exchange at different rates [52]. Carbonate is the physiological occupant of the anion-binding site, and is the anion apparently preferred by the protein, but in its absence a variety of other anions may also meet the needs of transferrin [23,55,56]. Binding anions include oxalate, malonate, thioglycolate, glycinate, salicylate, nitrilotriacetate, EDTA and others. These binding or "synergistic" anions all display two functional groups, at least one of which must be a carboxylate. The second electron-withdrawing group is within 6.3 Å of the carboxylate [56]. From an examination of structure binding relationships in a large series of anions, Schlabach and Bates estimated the anion-binding site to be 3 Å deep, 6 Å wide, and 4-6 Å (or more) in length [56]. The occupancy of the anion-binding site profoundly affects the symmetry and strength of the ligand field about the metal ion [57].

Although exchangeable, the anion is bound sufficiently strongly so that interlocking linkage to protein and metal may be inferred [56]. A likely ligand for the anion is the positively-charged guanidinium group of arginine [58], or, perhaps, the ε-amino group of lysine [51]. Chemical modification of arginine residues in human serum transferrin and chicken ovotransferrin substantially depresses metal-binding activity, with detailed kinetic and statistical analyses suggesting that one critical arginine is involved at each metal-binding site [58]. It is likely, therefore, that the anion stabilizes metal-protein binding by providing an additional, indirect linkage of metal to protein, thus impeding access of water to the metal ion. In the absence of a suitable anion, hydrolysis proceeds at the expense of metal-protein binding.

The sequence of events in formation of ternary metal-transferrin-anion complexes is not well understood. Because weak but specific binding of anions to transferrin appears to occur in the absence of metal ions, it has been suggested that anion-binding precedes metal-binding [8,59]. In keeping with this mechanism is the observation that bicarbonate can displace Fe-EDTA from the ternary Fe-transferrin-EDTA complex [60]. Evidently, the physiological anion disrupts the EDTA-protein bond, and the displaced EDTA is released still complexed with iron, which it then releases to the binary transferrin-carbonate complex to yield Fe-transferrin-carbonate.

D. PHYSIOLOGY OF THE TRANSFERRINS

1. Serum transferrin

The function of serum transferrin as an iron carrier, mandated by the hydrolytic propensities of Fe(III) and the insolubility of iron hydroxides at physiological pH, has long been known. Transferrin is the vehicle which delivers iron to immature erythroid cells where hemoglobin is synthesized [61]. It is an indispensible growth factor for most cell lines grown in culture [62-64], presumably because of its iron-donating abilities. Transferrin has also been shown to be a specific iron donor for hepatocytes [65], and so it is generally believed (but not yet proven) that it is required for virtually all actively growing and metabolizing cells.

Transferrin has little affinity for ferrous iron [8,66]. By stabilizing and solubilizing iron in the ferric state, transferrin may also function to minimize cycles of iron autoxidation in which noxious derivatives of oxygen, such as superoxide, singlet oxygen and hydroxyl radical, may be generated. When unsaturated with iron, therefore, transferrin may function as a circulating "iron buffer." Tissue damage in disorders of iron overload may be particularly severe when transferrin is saturated with the metal and non-transferrin iron complexes appear in the circulation [67].

The interaction of transferrin with cells. Three distinct events occur when transferrin serves as an iron donor for cells: (1) Transferrin binds to a specific receptor on the plasma membrane of the cell. Attempts to isolate and characterize this receptor have failed to produce concordant findings [68], and it is not yet known whether the same receptor is expressed by all cells dependent on transferrin for iron. Treatment of cells with proteases abolishes their ability to interact with transferrin, presumably because of degradation of the receptor [65,69,70]. (2) The protein relinquishes its iron to the cell. It has been heatedly argued, and just as heatedly denied, that this process requires internalization of transferrin to the interior of the cell [68]. Whatever the mechanism and site of iron release, the process is dependent on metabolic energy [71] and cell integrity [72]. (3) The iron-depleted protein is released from the cell for another round of iron transport. Unlike many other plasma transport proteins, transferrin is conserved during its interaction with cells. The entire protein-to-cell cycle can be completed within 15-30 seconds [66,68].

The mechanism of iron release. At physiological pH and pCO_2, the apparent stability constants for the binding of iron by transferrin are near 10^{22} M^{-1} [73]. Even if iron-binding is taken to be a diffusion-limited process, the rate at which iron release occurs spontaneously cannot exceed 10^{-12} sec^{-1}, or about once in many hundreds of years [74]. Evidently, an iron-dependent cell must provide a specific mechanism for releasing iron from transferrin, without damaging the protein during the short time transferrin dwells with the cell. Among the possibilities for such a mechanism are reduction, chelation, reductive chelation, and protonic or enzymic attack on the anion. Since substitution of non-physiological anions for carbonate in transferrin may greatly impede the iron-transferrin functions of the protein [75-77], it seems likely that the anion has an intimate role in the physiological processes of iron release.

The mechanism of iron-binding. Despite extensive studies *in vitro*, almost nothing is known of the mechanisms underlying the loading of iron by transferrin *in vivo*. Since iron is believed to be released from the storage protein ferritin in the ferrous state [10], and may also be absorbed from the gut in the ferrous state, it has been argued that "ferroxidases" are involved in the binding of iron by transferrin [7,9]. Ceruloplasmin has been identified as one such ferroxidase [7]; a less well characterized plasma copper protein [78] and intestinal xanthine oxidase [79] have also been claimed to have ferroxidase activity. However, autoxidation of ferrous iron, particularly in the presence of ferric iron stabilizing agents, may occur spontaneously and rapidly in the absence of ferroxidases. Also, as previously pointed out, the aquated cupric ion itself may be a potent catalyst for the autoxidation of ferrous iron [6]. Finally, simple agents such as thioglycolate may be much more active than ceruloplasmin in catalyzing the "oxidative fixation" of ferrous iron by transferrin [8]. For these reasons it may be best to consider the role of ferroxidases in the binding of iron by transferrin as still unsettled.

In vitro, iron is readily bound by transferrin when presented as a ferric chelate, as Fe(II) in the presence of oxygen and bicarbonate, or, but less reliably, as aquated Fe(III) [80]. In the last instance, reactions of hydrolysis and nonspecific polynuclear complex formation compete with reactions of specific binding, so that most investigators have turned away from simple ferric salts for loading transferrin with iron.

When, as practiced by many workers, iron is presented to transferrin (Tf) as ferric nitrilotriacetate (FeNTA) in the presence of bicarbonate a likely reaction sequence for binding of iron to each site is:

$$Fe^{3+}\text{-NTA} + Tf \xrightarrow[\text{(rapid; milliseconds)}]{} Fe^{3+}\text{-Tf-NTA} \qquad (10)$$

$$Fe^{3+}\text{-Tf-NTA} + HCO_3^- \xrightarrow[\text{(slow; seconds)}]{} Fe^{3+}\text{-Tf-}CO_3^- + H^+ + NTA \qquad (11)$$

If iron is offered in the ferrous state, again in the presence of bicarbonate, a probable sequence is [8]:

$$Tf + HCO_3 \xrightarrow[\text{(slow)}]{} Tf\text{-}CO_3 + H^+ \qquad (12)$$

$$Fe^{2+} + Tf\text{-}CO_3 \xrightarrow[\text{(slow)}]{} Fe^{2+}\text{-Tf-}CO_3 \qquad (13)$$

$$Fe^{2+}\text{-Tf-}CO_3 + O_2 \xrightarrow[\text{(rapid)}]{} Fe^{3+}\text{-Tf-}CO_3 + O_2^{\overline{\cdot}} + H^+ \qquad (14)$$

Since transferrin may encounter both ferrous and ferric iron during its wanderings in the organism, it is evident that it is prepared to cope with either state of the metal.

2. Lactoferrin and ovotransferrin

Relatively little is known about the biological roles of lactoferrin and ovotransferrin. Because of their iron-sequestering abilities, and the bacteriostatic consequences thereof, it is generally believed that both of these proteins function in protecting their parent organisms from microbial invasion [81,82]. The occurrence of lactoferrin in white blood cells (neutrophiles) may be related to its ability to catalyze the Haber-Weiss reaction, generating microbicidal hydroxyl radical [83]. Lactoferrin may also participate in the regulation of intestinal transport of iron but the sign of its effect is not clear [84,85]. The abundance of ovotransferrin in egg white may be related to its bacteriostatic activity, but its ability to serve as an iron source for hemoglobin synthesis by chick embryo erythrocytes suggests that it may have a role in iron transport as well [86].

E. THE TWO-SITED NATURE OF THE TRANSFERRINS

Modern, two-sited transferrins almost certainly arose, during the course of molecular evolution, by a mechanism involving duplication and fusion of a gene for an ancestral single-sited protein [40]. Yet, many attempts at finding a species in which the single-sited precursor persists have been fruitful. The inference that the two-sited protein offers some biological advantage to account for its survival at the expense of its single-sited ancestor is almost inescapable. Attempts to explicate this advantage, by experiment and reason, have as yet yielded no convincing insight. These attempts have considered both the chemistry and biology of the protein.

F. THE TWO-DOMAIN HYPOTHESIS OF TRANSFERRIN STRUCTURE

The two-domain hypothesis of transferrin structure suggests that each binding site of the protein resides in its own domain [22,87]. The two domains are independent, or nearly so, and correspond to the C-terminal and N-terminal halves of the intact molecule. Just as apotransferrin is far more susceptible to denaturation than iron-loaded protein, so may either domain be selectively denatured by heat or by urea when it is empty, while its occupied partner may be little affected by these perturbants. Since the two domains are neither structurally nor topologically identical, a different electrophoretic mobility is displayed by the four possible molecular species of transferrin (apotransferrin, Fe_N-transferrin, transferrin-Fe_C and Fe_2-transferrin) as the domains are selectively denatured in the urea-polyacrylamide gel system introduced by Makey and Seal [88]. Thus, it is possible to determine the distribution of iron in the binding sites of transferrin in incompletely saturated preparations [73], or even in blood [89,90]. Preliminary x-ray crystallographic studies reveal a bilobar structure of the transferrin molecule (Fig. 1) with the lobes connected by a thin region of

Fig. 1: Balsa wood model of the crystal structure of transferrin at 6 Å resolution. The courtesy of Dr. Beatrice Gorinsky and colleagues of the Department of Crystallography, Birbeck College, University of London, in providing this photograph is acknowledged with pleasure and gratitude.

join [91]. Very likely, then, these lobes correspond to the two domains, with the site of proteolytic disruption in the joining segment.

The most compelling evidence for the two domain-model comes from studies in which the domains are proteolytically separated from each other, with preservation of iron-binding activity in each fragment [92,93]. Spectroscopic studies of cleaved transferrins, clearly and sometimes dramatically display differences between the sites not appreciated in holoprotein (Fig. 2).

1. Distinguishability of the sites

After two decades of controversy [68] it is now clear that the sites of most, if not all, of the transferrins differ in their iron-binding properties. In ovotransferrin these differences are displayed in the ultraviolet, Mossbauer and EPR

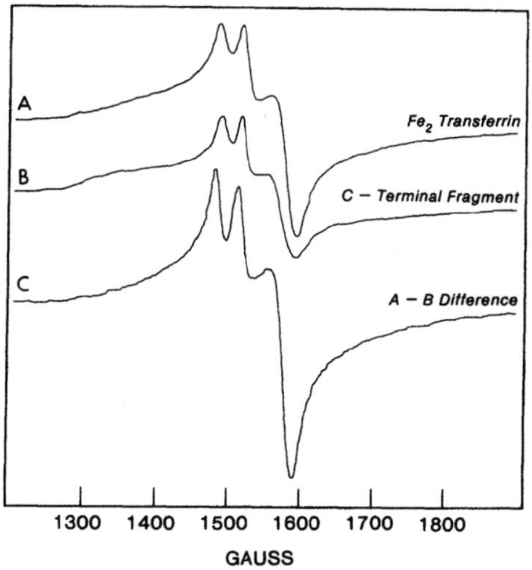

Fig. 2: EPR spectra of the individual sites of transferrin.
A. Diferric transferrin.
B. C-terminal fragment of transferrin.
C. Difference spectrum, A-B, representing the site in the N-terminal domain.
All spectra are normalized to the same concentration of iron.

features of each site [93,94], while in serum transferrin differences are seen in the anion-binding requirements [46], susceptibilities to conformational change [95,96], EPR properties [73], resistance to acid [73,97,98] and strengths of iron-binding of the two sites [73]. In sum, these findings offer a view of transferrins as proteins with two chemically inequivalent but independent sites. The provocative question is whether the differences between the sites have physiological consequences.

The simplest, most appealing conjecture, first clearly stated by Fletcher and Huehns [99], is that the two sites of transferrin have distinctive roles in iron metabolism. One site is presumed to direct its iron preferentially to immature, hemoglobin-synthesizing red blood cells and the iron-requiring cells of the placenta, while the other site is oriented toward liver and intestinal mucosal cells. Attempts to validate or disprove the Fletcher-Huehns hypothesis have generated an abundant and contradictory literature.

Most of the reported studies have been conducted *in vitro*, and depend on preincubating uniformly iron-labeled transferrin with reticulocytes until a desired amount of initially protein-bound iron has been delivered to the cells, or by selectively removing iron from one site, the so-called acid-labile site, of the protein. The partially iron-depleted transferrin is then recovered, and its ability to donate iron to specific target tissues is assessed by comparing it to fresh, randomly-labeled protein at the same overall saturation with iron. Alternatively, the preincubated protein may be reloaded with another isotope of iron, and the uptakes of each isotope by the cell or tissue of interest may be compared. Studies of these kinds have generally failed to establish a convincing and coherent view of whether the sites of transferrin are functionally equivalent [68].

Since *in vitro* studies may be criticized for failing to reproduce all conditions which prevail *in vivo*, as well as for methodological shortcomings, several laboratories have turned to *in vivo* systems in attempts to discern whether the two sites of transferrin are functionally distinguishable.

Studies from two different groups, using rat and rabbit models, respectively, have yielded essentially concordant results [100,101]. Iron is cleared from diferric transferrin in essentially pairwise fashion, implying that the two sites of iron-loaded transferrin are at least kinetically equivalent (Fig. 3). Since the bone marrow is the recipient of some 85-90% of transferrin-bound iron, the likelihood is that the two sites are functionally equivalent as well. Furthermore, the two possible monoferric transferrins (four in the rat, which has two circulating isotransferrins) were also kinetically equivalent. Thus, these recent investigations seem to indicate that the two sites of transferrin function indistinguishably as iron donors *in vivo*.

Nevertheless, careful analyses based on the Makey-Seal urea-gel electrophoresis method have shown that iron in normal human plasma is not distributed equally among the two sites of transferrin in the circulation [89,90]. Rather, the weaker binding B-site, in the N-terminal domain of the protein, is preferentially occupied. Thus, neither thermodynamic nor statistical (random) factors determine the distribution of iron. To complicate matters further, we have observed, in a few patients with aberrations of iron metabolism that the A-site is apt to be preferentially occupied when iron turnover is increased.

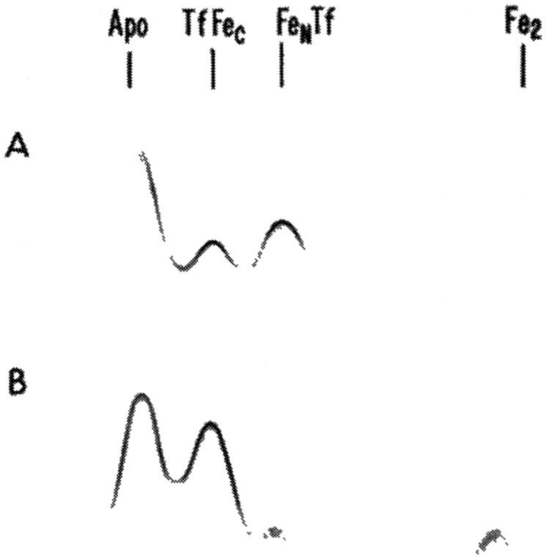

Fig. 3: Crossed immunoelectrophoresis patterns of transferrin in human serum.
A. Normal subject.
B. Patient with hemolytic anemia.

No clear and convincing explanation, then, is available for the biological advantage offered by a circulating two-sited carrier of iron. Perhaps a difference in function is displayed at the level of iron-loading, in a "supply side" control of iron economy. Preliminary studies hint that this may be the case. Zapolski and Princiotto injected iron-labeled heat-damaged erythrocytes into the rat, and studied the appearance of label in circulating transferrin [102]. About 2/3 of the recycled iron had been directed to the acid-resistant A-site (C-terminal domain) of transferrin, thus suggesting that site differences in iron-loading exist *in vivo*. If the two-sitedness of transferrin has regulatory significance, then, it may be at the level of iron-loading, rather than iron release. For the moment, however, the hypothesis that the sites of transferrin are functionally distinguishable is best considered "not proven."

G. PROSPECTS AND PERSPECTIVES

Even after three decades of intensive study by physicists, structural chemists, biochemists and physiologists the role of transferrins in the regulation of iron metabolism remains a fascinating problem. Is the serum protein simply a passive vehicle for iron transport, in the circulation, across cell membranes and through cells, or does it exert a more subtle effect in the control of iron absorption, storage and transport? How is iron loaded by transferrin *in vivo*, and how and where in the cell is iron unloaded? What are the functions of ovotransferrin and lactoferrin? Do their antibacterial properties reflect more than their iron-sequestering abilities? How do transferrins interact with the receptors, and how are transferrin receptors regulated in number and function? Why is iron toxic, and what is the role of transferrin in suppressing this toxicity? Whatever the answers to these questions prove to be, the biochemistry of iron metabolism remains an investigative arena where problems proliferate at an even greater rate than solutions.

REFERENCES

1. Aisen, P. (1977) In *Iron Metabolism,* Ciba Found. Symp. 51, pp. 1-14.
2. May, P.M. and Williams, D.R. (1980) In *Iron in Biochemistry and Medicine II* (A. Jacobs & M. Worwood, eds.) Academic Press, London, pp. 1-28.
3. Wrigglesworth, J.M. and Baum, H. (1980) see Ref. 2, pp. 29-86.
4. Romslo, I. and Husby, P. (1980) *Int. J. Biochem.* 12, pp. 709-712.
5. Harris, D.C. and Aisen, P. (1973) *Biochim. Biophys. Acta* 329, pp. 156-158.
6. Kurimura, Y. and Murakami, K. (1969) *Bull. Chem. Soc. Japan* 42, pp. 2715-2717.
7. Osaki, S., Johnson, D.A. and Frieden, E. (1966) *J. Biol. Chem.* 241, pp. 2746-2751.
8. Kojima, N. and Bates, G.W. (in press) *J. Biol. Chem.*
9. Frieden, E. (1973) *Nutr. Rev.* 31, pp. 41-44.
10. Jones, T., Spencer, R. and Walsh, C. (1978) *Biochemistry* 17, pp. 4011-4017.
11. Kojima, N. and Bates, G.W. (1979) *J. Biol. Chem.* 254, pp. 8847-8854.
12. Ingledew, W.J., Cox, J.C. and Halling, P.J. (1977) *FEMS Microbiol. Lett.* 2, pp. 193-197.
13. Tien, M., Svingen, B.A. and Aust, S.D. (1981) *Fed. Proc.* 40, pp. 179-182.
14. Burger, R.M. *et al* (1980) *J. Biol. Chem.* 255, pp. 11832-11838.
15. Baes, C.F., Jr. and Mesmer, R.E. (1976) *The Hydrolysis of Cations,* John Wiley, New York, pp. 226-237.
16. Dousma, J. and DeBruyn, P.L. (1979) *J. Coll. Interfac. Sci.* 72, pp. 314-320.
17. Aisen, P. and Leibman, A. (1968) *Biochem. Biophys. Res. Commun.* 30, pp. 407-413.
18. Pollack, S. *et al* (1976) *Br. J. Haemat.* 34, pp. 235-239.
19. Pollack, S., Vanderhoff, G. and Lasky, F. (1977) *Biochim.*

Biophys. Acta 497, pp. 481-487.
20. Carrano, C.J. and Raymond, K.N. (1979) *J. Am. Chem. Soc.* 101, pp. 5401-5404.
21. Propper, R.D. (1980) *Ann. N.Y. Acad. Sci.* 344, pp. 375-383.
22. Aisen, P. and Listowsky, I. (1980) *Ann. Rev. Biochem.* 49, pp. 357-393.
23. Aisen, P. *et al* (1967) *J. Biol. Chem.* 242, pp. 2484-2490.
24. Price, E.M. and Gibson, J.F. (1972) *Biochem. Biophys. Res. Commun.* 46, pp. 646-651.
25. Bates, G.W. and Schlabach, M.R. (1975) *J. Biol. Chem.* 250, pp. 2177-2181.
26. Dorland, L. *et al* (1977) *FEBS Lett.* 77, pp. 15-20.
27. Prieels, J.-P. *et al* (1978) *Proc. Nat. Acad. Sci., USA* 75, pp. 2215-2219.
28. Leger, D. *et al* (1978) *FEBS Lett.* 93, pp. 255-260.
29. Dorland, L. *et al* (1979) *Eur. J. Biochem.* 100, pp. 569-574.
30. Hatton, M.W.C. *et al* (1979) *Biochem J.* 181, pp. 633-638.
31. Williams, J. (1968) *Biochem. J.* 108, pp. 57-67.
32. Metz-Boutigue, M.H. *et al* (1980) *Biochim. Biophys. Acta* 622, pp. 308-314.
33. Pollack, S. and Lasky, F. (1976) *J. Lab. Clin. Med.* 87, pp. 670-679.
34. Huebers, H. *et al* (1976) *Eur. J. Biochem.* 66, pp. 447-455.
35. Bezkorovainy, A. (1980) *Biochemistry of Nonheme Iron.* Plenum Press, New York, pp. 127-206.
36. Azari, P.R. and Feeney, R.E. (1958) *J. Biol. Chem.* 232, pp. 293-299.
37. Azari, P.R. and Feeney, R.E. (1961) *Arch. Biochem. Biophys.* 92, pp. 44-52.
38. Lane, R.S. (1971) *Biochim. Biophys. Acta* 243, pp. 193-202.
39. Ulmer, D.D. (1969) *Biochim. Biophys. Acta* 181, pp. 305-310.
40. MacGillivray, R.T.A., Mendez, E. and Brew, K. (1977) In *Proteins of Iron Metabolism* (E.B. Brown, P. Aisen, J. Fielding and R.R. Crichton, eds.) Grune and Stratton, New York, pp. 133-151.
41. Warner, R.C. and Weber, I. (1953) *J. Am. Chem. Soc.* 75, pp. 5094-5101.
42. Wishnia, A., Weber, I. and Warner, R.C. (1961) *J. Am. Chem. Soc.* 83, 2071-2080.
43. Aasa, R. and Aisen, P. (1968) *J. Biol. Chem.* 243, pp. 2399-2404.
44. Geoghegan, K.F., Borresen, T. and Feeney, R.E. (1980) *J. Biol. Chem.* 255, pp. 11429-11434.
45. Hazen, E.E., Jr. (1962) Ph.D. Thesis, Harvard University.
46. Zweier, J.L. and Aisen, P. (1977) *J. Biol. Chem.* 252, pp. 6090-6096.
47. Zweier, J.L. *et al* (1979) *J. Biol. Chem.* 254, pp. 3512-3515.
48. Froncisz, W. and Aisen, P. (submitted for publication).
49. Koenig, S.H. and Schillinger, W.E. (1969) *J. Biol. Chem.* 244, pp. 6520-6526.

50. Harris, D.C., Gray, G.A. and Aisen, P. (1974) *J. Biol. Chem.* 249, pp. 5261-5264.
51. Zweier, J.L., Wooten, J.B. and Cohen, J.S. (1981) *Biochemistry* 20, pp. 3505-3510.
52. Aisen, P. *et al* (1973) *Biochemistry* 12, pp. 3679-3684.
53. Egyed, A. (1975) *Biochim. Biophys. Acta* 411, pp. 349-356.
54. Martinez-Medellin, J. and Schulman, H.M. (1973) *Biochem. Biophys. Res. Commun.* 53, pp. 32-38.
55. Campbell, R.F. and Chasteen, N.D. (1977) *J. Biol. Chem.* 252, pp. 5996-6001.
56. Schlabach, M.R. and Bates, G.W. (1975) *J. Biol. Chem.* 250, pp. 2182-2188.
57. Aisen, P., Pinkowitz, R.A. and Leibman, A. (1973) *Ann. N.Y. Acad. Sci.* 222, pp. 337-346.
58. Rogers, T.B., Borreson, T. and Feeney, R.E. (1978) *Biochemistry* 17, pp. 1105-1109.
59. Woodworth, R.C. *et al* (1975) In *Proteins of Iron Storage and Transport in Biochemistry and Medicine* (R.R. Crichton, ed.) North-Holland Publishing Co., Amsterdam, pp. 39-50.
60. Rogers, T.B., Feeney, R.E. and Meares, C.F. (1977) *J. Biol. Chem.* 252, pp. 8108-8112.
61. Hemmaplardh, D. and Morgan, E.H. (1974) *Biochim. Biophys. Acta* 373, pp. 84-99.
62. Guilbert, L.J. and Iscove, N.N. (1976) *Nature* 263, pp. 594-595.
63. Rudland, P.S. *et al* (1977) *Biochem. Biophys. Res. Commun.* 75, pp. 556-562.
64. Hutchings, S.E. and Sato, G.H. (1978) *Proc. Nat. Acad. Sci. USA* 75, pp. 901-904.
65. Young, S.P. and Aisen, P. (1980) *Biochim. Biophys. Acta* 633, pp. 145-153.
66. Gaber, B.P. and Aisen, P. (1970) *Biochim. Biophys. Acta* 221, pp. 228-233.
67. Graham, G. *et al* (1979) *Am. J. Hemat.* 6, pp. 207-217.
68. Aisen, P. see Ref. 2, pp. 87-129.
69. Jandl, J.H. *et al* (1959) *J. Clin. Invest.* 38, pp. 161-185.
70. Hemmaplardh, D. and Morgan, E.H. (1976) *Biochim. Biophys. Acta* 426, pp. 385-398.
71. Kailis, S.G. and Morgan, E.H. (1977) *Biochim. Biophys. Acta* 464, pp. 389-398.
72. Morgan, E.H. (1976) *Biochem. J.* 158, pp. 489-491.
73. Aisen, P., Leibman, A. and Zweier, J.L. (1978) *J. Biol. Chem.* 253, pp. 1930-1937.
74. Aisen, P. and Leibman, A. (1968) *Biochem. Biophys. Res. Commun.* 32, pp. 220-226.
75. Aisen, P. and Leibman, A. (1973) *Biochim. Biophys. Acta* 304, pp. 797-804.
76. Egyed, A. (1973) *Biochim. Biophys. Acta* 304, pp. 805-813.
77. Schulman, H.M., Martinez-Medellin, J. and Sidloi, R. *Biochim. Biophys. Acta* 343, pp. 529-534.
78. Topham, R.W. and Frieden, E. (1970) *J. Biol. Chem.* 245, pp.

6698-6705.
79. Topham, R.W., Woodruff, J.H. and Walker, M.C. (1981) *Biochemistry* 20, pp. 319-324.
80. Workman, E.F., Jr., Graham, G. and Bates, G.W. (1975) *Biochim. Biophys. Acta* 399, pp. 254-264.
81. Weinberg, E.D. (1978) *Microbiol. Rev.* 42, pp. 45-66.
82. Griffiths, E., Rogers, H.J. and Bullen, J.J. (1980) *Nature* 284, pp. 508-510.
83. Ambruso, D.R. and Johnston, R.B., Jr. (1981) *J. Clin. Invest.* 67, pp. 352-360.
84. deVet, B.J.C.M. and van Gool, J. (1974) *Acta Med. Scand.* 196, pp. 393-402.
85. Cox, T.M. *et al* (1979) *Biochim. Biophys. Acta* 588, pp. 120-128.
86. Williams, S.C. and Woodworth, R.C. (1973) *J. Biol. Chem.* 248, pp. 5848-5853.
87. Donovan, J.W. see Ref. 40, pp. 179-186.
88. Makey, D.G. and Seal, U.S. (1976) *Biochim. Biophys. Acta* 453, pp. 250-256.
89. Leibman, A. and Aisen, P. (1979) *Blood* 53, pp. 1058-1065.
90. Williams, J. and Moreton, K. (1980) *Biochem. J.* 185, pp. 483-488.
91. Gorinsky, B. *et al* (1979) *Nature* 281, pp. 157-158.
92. Brock, J.H. and Arzabe, F. (1976) *FEBS Lett.* 60, pp. 63-66.
93. Butterworth, R.M., Gibson, J.F. and Williams. J. (1975) *Biochem. J.* 149, pp. 559-563.
94. Aisen, P., Lang, G.A. and Woodworth, R.C. (1973) *J. Biol. Chem.* 248, pp. 649-653.
95. Zweier, J.L. (1978) *J. Biol. Chem.* 253, pp. 7616-7621.
96. Chasteen, N.D., White, L.K. and Campbell, K.L. (1976) *Biochemistry* 16, pp. 363-374.
97. Princiotto, J. and Zapolski, E.J. (1975) *Nature* 255, pp. 87-88.
98. Lestas, A.N. (1976) *Brit. J. Haemat.* 32, pp. 341-350.
99. Fletcher, J. and Huehns, E.R. (1968) *Nature* 215, pp. 584-586.
100. Huebers, H. *et al* (1981) *Blood* 57, pp. 218-228.
101. Groen, R. *et al* (in press) *Br. J. Haemat.*
102. Zapolski, E.J. and Princiotto, J. (1980) *Life Sciences* 27, pp. 739-745.

ACKNOWLEDGEMENT

The preparation of this manuscript, and the experimental work from the author's laboratory, were supported in part by Grant AM 15056 from the National Institutes of Health.

COORDINATION CHEMISTRY OF THE SIDEROPHORES AND RECENT STUDIES OF SYNTHETIC ANALOGUES

Kenneth N. Raymond and Thomas P. Tufano

Department of Chemistry, University of California,
Berkeley, California 94720

INTRODUCTION

For the past ten years we have studied the coordination chemistry of microbial iron transport compounds, called siderophores, and structurally related synthetic analogues. Our research has included X-ray structure determinations (1-3), electrochemical studies (4), isomer assignment by spectrophotometric techniques (5), potentiometric titrimetry (6), and, most recently, kinetic measurements of the rates of iron exchange between siderophores (7). This work has helped to elucidate the chemical details of microbial iron transport, as well as to provide information for the logical design of chelating agents for the treatment of iron overload in humans. This paper will focus on a review of results obtained in the latter three areas of study.

As a consequence of the extremely large number of enzymes and proteins in which it plays a central role, iron is an essential nutrient for all living things (8-10). Since the evolution of an oxidizing atmosphere on this planet, the availability of iron to aerobic organisms in an aqueous environment has been limited by the extreme insolubility of ferric hydroxide ($K_{sp} \approx 10^{-39}$) (11). With the equilibrium concentration of ferric ion at pH 7 of about 10^{-18} M, even diffusion limited transport would be many orders of magnitude too slow to provide sufficient amounts of iron to a microbial cell. This problem has been circumvented by microbes through the production of the siderophore chelating agents, which solubilize ferric ion and facilitate its transport into bacteria, molds, yeast and fungi. Since the discovery of the first siderophore (ferrichrome) by Neilands (12),

H. B. Dunford et al. (eds.), The Biological Chemistry of Iron, 85–105.

a number of such compounds have been found — first isolated as microbial growth factors (13).

In addition to the interesting mechanistic questions of how these organisms acquire iron via a siderophore-mediated system, there are several areas of practical clinical importance in which siderophores play a part. The importance of iron in the pathogenicity of certain bacterial infections has been emphasized by various workers (14,15). As the probable origin of this relationship, it has been suggested that siderophores are an important means for pathogenic bacteria to acquire iron from their hosts (16). Furthermore, iron toxicity in humans represents a serious worldwide problem (17, 18), either in its acute form (usually infant poisoning), or in chronic iron overload (as occurs in patients requiring massive, periodic blood transfusions for the treatment of blood disorders such as Cooley's anemia). Several siderophores or their analogues have been tested as potential antidotes through iron chelation therapy and Desferal®, a preparation of ferrioxamine B, is routinely used worldwide.

Siderophores can be broadly classified into three major structural types — the hydroxamates, thiohydroxamates, and the catecholates. The mode of coordination to ferric ion exhibited by these functional moieties is illustrated in Figure 1. In each

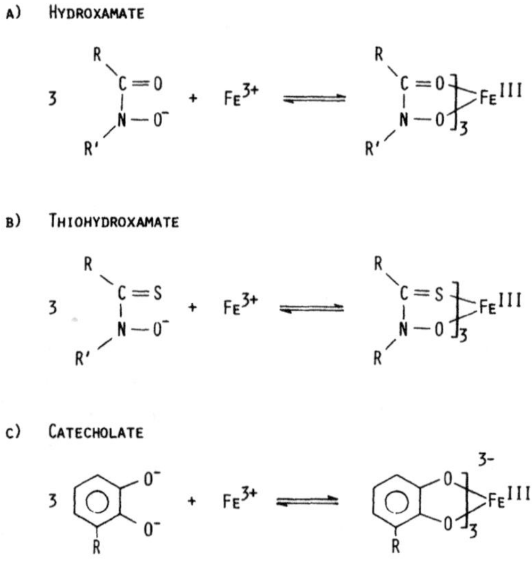

Figure 1. Reaction of simple hydroxamates, thiohydroxamates, and catecholates with ferric ion.

case deprotonation leads to the formation of an anionic bidentate ligand which can form very stable and selective complexes with ferric ion. Some examples of hydroxamate and catecholate siderophores derived from microbial cultures are shown in Figure 2.

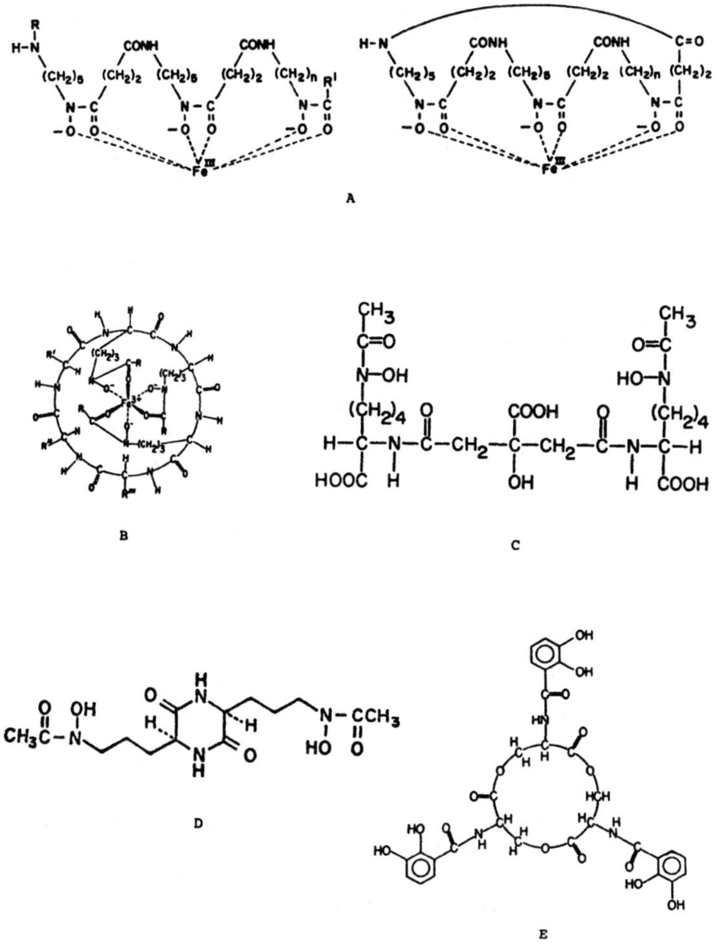

Figure 2. Structural formulas of representative siderophores. (A) Linear and cyclic ferrioxamines. When linear and $R = H$, $R' = CH_3$, $n = 5$, the compound is ferrioxamine B; when cyclic the compound is ferrioxamine E. (B) The ferrichromes. For $R' = R'' = R''' = H$ and $R = CH_3$, the compound is ferrichrome itself. (C) Aerobactin. (D) Rhodotorulic acid. (E) Enterobactin.

Note that each molecule makes use of two to three linked hydroxamate or catecholate moieties in complex formation, thus typically forming extremely stable, highly specific octahedral coordination compounds with iron(III).

The iron in these complexes has been shown to be high-spin d^5 ferric ion by magnetic susceptibility measurements, EPR, and Mössbauer spectroscopy (20).

In our research we have used metal substituted siderophore complexes to elucidate the mechanisms of iron transport into microbial cells and have studied the physiochemical properties of the iron siderophore complexes to delimit their iron release and exchange mechanisms.

In the first case, the substitution of chromic ion for ferric ion (which have the same size and charge) avoids the problems of ferric ion lability and lack of easily interpreted d-d bands in visible absorption spectra since (a) the kinetically inert d^3 chromic ion complexes preclude any isomerization or ligand exchange during transport experiments; the chromic siderophore complexes can be used as "biological probes" in studies of cellular iron transport; (b) the major features of the vis-uv and CD spectra of the chromic complexes are due to the well characterized d-d transitions of the metal ion, thus allowing the use of the chromic complexes as "stereochemical probes" in assigning the geometry at the metal site.

In the second case, we have examined the formation constants and kinetics of iron siderophore complexes and model synthetic analogues. Equilibrium formation constant studies are important in two major respects: (a) for ferric siderophore complexes these constants define the limits for conditions in which iron will be extracted from other biological ferric complexes in a host organism; (b) for synthetic ferric ion sequestering agents, this information can be used as a primary criterion for the rational design of stable and metal ion selective ligands for use in iron chelation therapy. These thermodynamic studies are in turn complemented by kinetic measurements of the rate of iron exchange between siderophores and model compounds. It has been assumed that the iron in these complexes is readily exchanged owing to the known substituted lability of high spin ferric ion. Since iron acquisition by microbes involves the exchange of the metal to and between siderophores, it is of considerable interest to clearly define in a systematic fashion the mechanism and relative rates of exchange for hydroxamate and catecholate ligands.

KINETICALLY INERT SIDEROPHORE COMPLEXES — USE AS STEREOCHEMICAL PROBES

Many mechanisms of siderophore-mediated microbial iron transport can be envisioned; Leong and Neilands (21) have summarized three limiting types. Mechanism 1 involves release of iron at the cell membrane without penetration of the iron complex or free ligand into the cell. Presumably the iron is released to a membrane-bound chelating agent which completes transport into the cell. The salient feature of mechanism 2 is the uptake of the iron siderophore complex followed by intracellular iron release and re-excretion of the free ligand. Mechanism 3 involves transport of the intact iron complex; but, once inside the cell, iron is released through chemical degradation of the ligand. The transport of siderophores across cell membranes evidently can be quite conformation dependent, since in some cases the ferric complexes are rapidly passed into the cell while metal-free siderophores are not. The goal of understanding this one-way transport of ferric ion across the cell membrane has led to the results discussed below.

In particular, three questions we would like to answer are: 1) Is the intact metal-siderophore complex always transported into the cell [as in the ferrichromes (22)] or is ferric ion transferred to a cell-wall-bound transport system [as in rhodotorulic acid mediated iron transport in *Rhodotorula pilimanae* (23)]? 2) How sensitive is the transport process to the exact shape or geometry of the iron-siderophore complex? 3) What are the detailed molecular geometries of various kinetically inert metal-siderophore complexes?

Answers to these questions may be obtained by studying the structure and stereochemistry (by single crystal X-ray diffraction and CD measurements) and the transport of specific coordination isomers of substitution-inert metal-siderophore complexes. Related work on metal siderophores has been reviewed recently (24,25). In this paper we present some representative results on studies of the stereochemical characterization of coordination isomers of microbial iron transport compounds and model hydroxamate and catecholate compounds.

Since hydroxamic acids are unsymmetrical bidentate ligands, both geometrical and optical isomers are possible in tris(hydroxamato)metal complexes. For an octahedral complex formed with three equivalent optically inactive hydroxamate anions, there are two geometric isomers possible, *cis* and *trans*. Each geometrical isomer consists of Δ and Λ optical isomers, forming a total of four possible isomers; Δ-*cis*, Λ-*cis*, Δ-*trans*, and Λ-*trans* (Figure 3). Often the optical isomers of a given geometric isomer are diastereomers because of the ligand optical

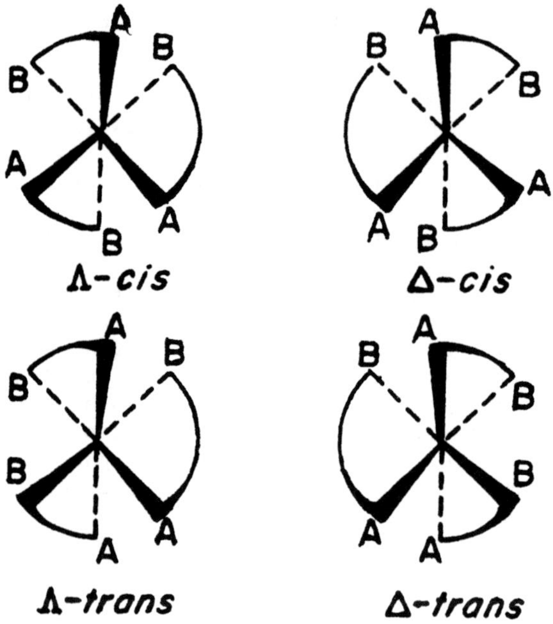

Figure 3. Geometrical and optical isomers of a simple tris-(chelate) complex with an unsymmetrical bidentate ligand.

activity. In such cases the four isomers theoretically can be separated, based on differences in their physical properties. In cases of trihydroxamate siderophores, more or fewer isomers are possible, depending on the structure and optical activity of the ligand.

The molecular structures of ferrichrome A and its aluminum-substituted complex (26a), ferrichrysin (20), ferrichrome (26b), ferrioxamine E (26c), ferrimycobactin P (26d) and ferric fusarinine (27) have been determined by X-ray crystallographic analysis. In these complexes the metal is bound by three hydroxamate groups in an octahedral configuration to give Λ-*cis* (ferrichrome, ferrichrome A, fusarinine) or Δ-*cis* (mycobactin) isomers. In the case of fusarinine, both Δ and Λ isomers are formed, depending upon crystallization conditions (27). A comparison of the CD spectrum of ferrichrome A with that of the resolved ferric tris(benzohydroxamato) complexes indicates that only the Λ isomer is present in solution (compare Figures 4 and 5) (28).

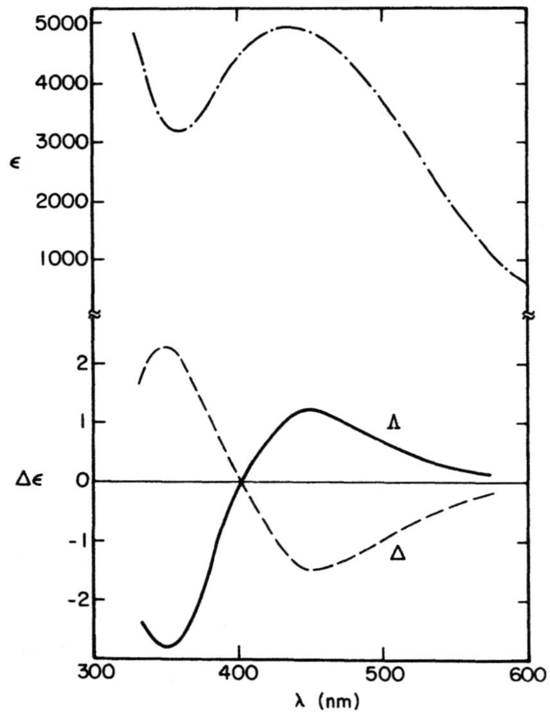

Figure 4. Absorption spectrum (—·—) and CD spectra of Δ (— — —) and Λ (——) tris(benzohydroxamato)-iron(III) in acetone solution.

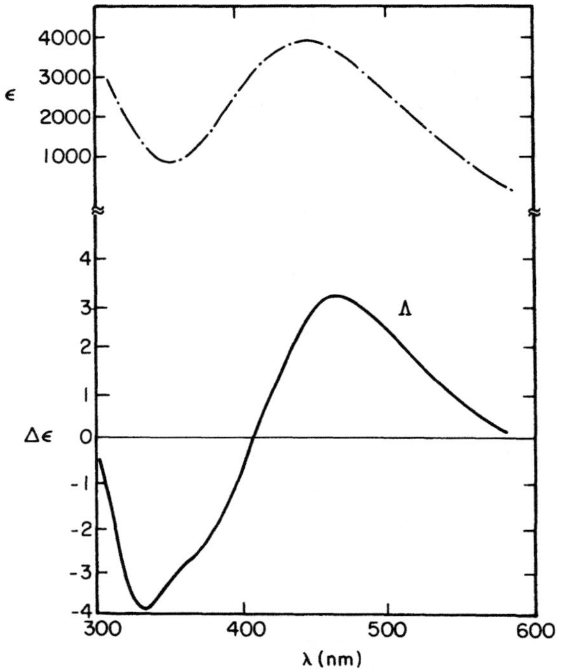

Figure 5. Absorption (—·—) and CD (——) spectra of ferrichrome A in aqueous solution.

Further support for the existence of one predominant isomer for ferrichrome comes from replacement of the ferric ion with chromic ion, where no evidence was found for separation of the additional coordination isomer (Δ) that might exist in an equilibrium mixture. The CD spectrum of chromic desferriferrichrome is shown in Figure 6; these data can be used to identify the coordination isomers of the siderophore complexes. The chromic complexes of the desferriferrichrome and desferriferrichrysin have identical CD spectra (29), which are the same as that of the Λ-*cis* isomer of tris(N-methyl-ℓ-menthoxyacethydroxamato)chromium(III), confirming the previous assignments (30).

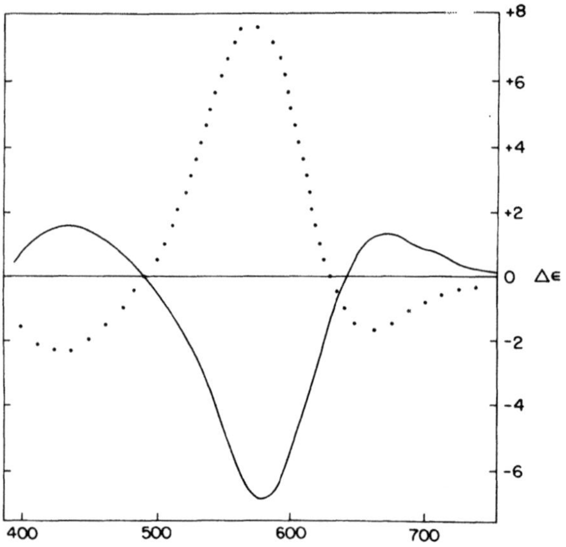

Figure 6. Circular dichroism spectra of Δ-$(NH_4)_3$ [Cr(enterobactin)] (——) and chromic desferriferrichrome (· · ·) (Δε in $L{\cdot}mol^{-1}$ cm^{-1}).

As with the ferrichromes, space-filling molecular models suggest that only two diastereomers, Δ-*cis* and Λ-*cis*, are possible for metal enterobactin complexes. The simple model complex, tris(catecholato)chromium(III), was prepared as a kinetically inert model complex for enterobactin. Although only partial resolution of solutions of $[Cr(cat)_3]^{3-}$ was achieved at neutral pH, complete resolution could be achieved at pH 13 (5). The CD of both $[Cr(ent)]^{3-}$ and Δ-$[Cr(cat)_3]^{3-}$ are essentially identical. The known crystal and molecular structure of the potassium salt of $[Cr(cat)_3]^{3-}$ (3), together with the established absolute configuration, allows the following assignment: the predominant isomer of the monomeric chromium enterobactin complex has a Δ-*cis* absolute configuration. The assignment is further supported by a comparison of the CD spectra of chromic desferriferrichrome and chromic enterobactin; they are clearly enantiomorphic (Figure 6). The similarity of the chromic and ferric complexes (3,31) allows this assignment to be extended to the ferric complexes as well. This is the isomer found also for the optically active hydroxamate siderophores mycobactin (26d), rhodotorulic acid (31) and one form of ferric fusarinine (27). Thus the Λ coordination isomer is not the only one recognized in microbial iron transport systems.

EQUILIBRIUM POTENTIOMETRIC STUDIES OF IRON BINDING

The possibility of using the metal-free siderophores or synthetic analogues as therapeutic agents for the treatment of iron overload associated syndromes or acute iron toxicity from poisoning (16) has prompted investigations into the formation constants of these ligands with ferric ion. The reaction of a trihydroxamate or tricatecholate chelate with iron can be represented by a formation constant, K_f where L is the fully deprotonated ligand (charges omitted).

$$Fe + L \underset{}{\overset{K_f}{\rightleftharpoons}} FeL \qquad (1)$$

The overall stability constants for several trihydroxamate and tricatecholate ligands are given in Table I. The values obtained are very large, making these compounds among the best iron chelating agents known. Also shown are pM values for ferric ion ($-\log[Fe^{3+}]$) which provide a better comparison of ligand binding under given conditions of pH, total ligand, and total metal concentrations.

Of all the naturally occurring siderophores, enterobactin exhibits the greatest thermodynamic stability with ferric ion (estimated formation constant of 10^{52} based upon spectrophotometric competition experiments with EDTA) (32). Enterobactin itself is not suitable for drug use in that its delicate cyclic triester backbone is susceptible to hydrolytic degradation to the monomer, 2,3-dihydroxybenzoylserine. The latter is a considerably poorer ligand for iron, as approximated by the pM for 2,3-dihydroxy-N,N-dimethylbenzamide (see Table I). In addition, since enterobactin is a growth factor for enteric bacteria, it is potentially hazardous if employed as a drug. However, enterobactin does serve as a structural archetype for the design of effective synthetic ferric ion sequestering agents. Several such enterobactin analogues are illustrated

Table I. Thermodynamic data for ferric complexes of biological ligands and related structural analogues. Equilibrium free metal ion concentrations expressed as pM = - log-$[Fe^{3+}]$.

Ligand	Log K_f	pM[a]	Ref.
Enterobactin	52	35.5	32
Desferrioxamine E	32.5	27.7	33
Desferrioxamine B	30.6	26.6	33
Desferrichrome	29.1	25.2	33
Aerobactin	22.9	23.3	6
Rhodotorulic acid	31.2[b]	21.9	4
MECAMS[c]	41	29.4	34
3,4-LICAMS[d]	41	28.5	34
3,3,4-CYCAMS[e]	38	24.9	34
2,3-Dihydroxy-N,N-dimethylbenzamide	-	∿15 [f]	35

[a]Calculated for 10 μM ligand, 1 μM Fe^{3+}, pH 7.4; [b]Stability constants per iron for the dimeric rhodotorulic acid complex; [c]N,N',N"-tris(2,3-dihydroxy-5-sulfobenzoyl)-1,3,5-triaminomethylbenzene; [d]N,N',N"-tris(2,3-dihydroxy-5-sulfobenzoyl)-1,5,10-triazadecane; [e]N,N',N"-tris(2,3-dihydroxy-5-sulfobenzoyl)-1,5,9-cyclotriazatridecane; [f]pM is below lower limit set by the K_{sp} of ferric hydroxide, indicating precipitation of $Fe(OH)_3$ under these conditions.

in Figure 7. Details of the synthesis of these ligands have appeared in the literature (36,37). Although the structural backbones differ greatly, each molecular type incorporates three catechol moieties to coordinate the iron with octahedral geometry, thus making use of the enhanced stability derived from the chelate effect (38). The high affinity of these ligands for ferric ion is demonstrated by the thermodynamic data presented in Table I. Sulfonation and carboxylation of the chelates has been routinely employed to increase aqueous solubility for physical measurements and biological test studies (36,39).

Aside from the demonstrated stability of these tricatechoylamide sequestering agents for iron, it has been assumed that the basic structural design of the ligands has rendered them highly selective for ferric ion as compared to other important biological trace metal ions. An important consideration for metal ion decorporation therapy is the relative affinity of the ligand for the target metal ion, as compared to its affinity for essential biological metal ions which are present

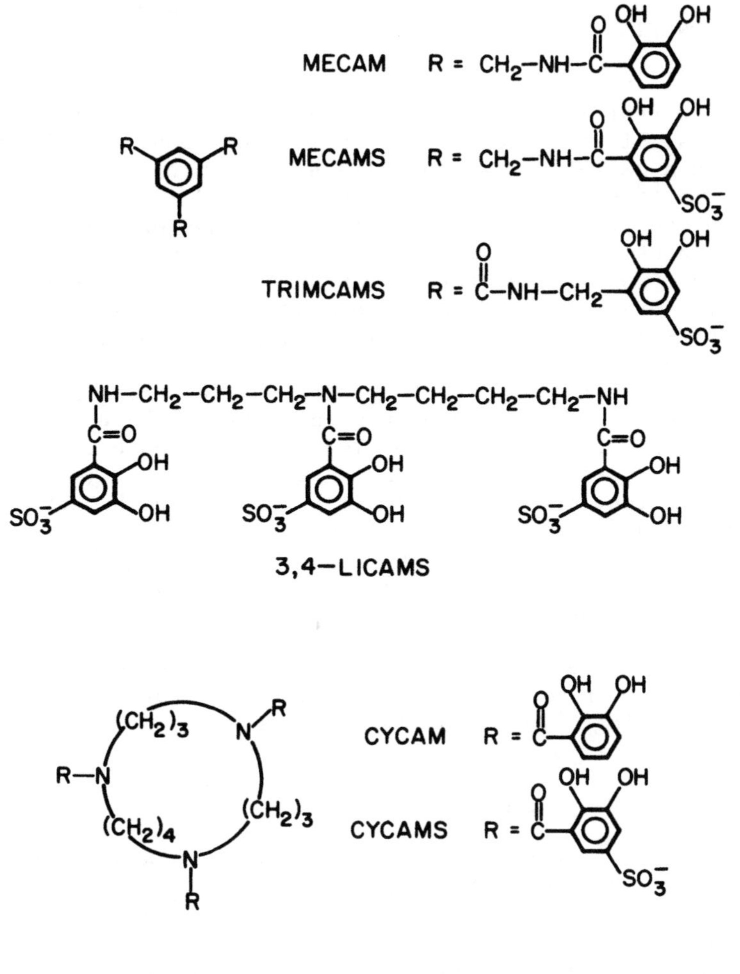

Figure 7. Structural formulas and acronyms of synthetic tri-catechoylamide sequestering agents.

in the body. A series of complexometric studies have examined the degree to which the newly synthesized compounds bind such biological ions (40). Potentiometric titration of MECAMS with various biological ions have revealed that neither Ca(II) nor Mg(II) are bound to any significant extent. The strongest complex is formed with Cu(II), but even this complex is approximately 10^{12} times less stable than the corresponding ferric chelate under physiological conditions.

KINETICS AND MECHANISM OF IRON EXCHANGE BETWEEN SIDEROPHORES AND SYNTHETIC ANALOGUES

It has been generally assumed in the literature of iron transport that iron is rapidly exchanged between siderophore complexes. In the one test of this assumption, Lovenberg *et al.* observed a half-time of approximately ten minutes for the exchange of $^{59}Fe^{3+}$ between ferric citrate and ferrichrome (41). This notion is consistent with the substitutional lability of high-spin d^5 iron(III) in coordination complexes (42). More recently, Emery and Hoffer have reported the rapid exchange of Ga(III) citrate with ferrichrome A and Fe(III) nitrilotriacetate with Ga(III) desferriferrichrome A (43). However, observations in our laboratory indicate that this assumption is not correct in general. For instance, the tris(hydroxamato) and tris(thiohydroxamato) complexes of high spin iron(III) have been resolved into their component optical isomers and have been found to be very stable in various non-aqueous solvents (28,44). Certainly siderophores of natural origin, with the enhanced structural stability of three linked hydroxamate or catecholate moieties, might be expected to exhibit unusually slow substitution kinetics for an iron(III) coordination complex, even though in an aqueous environment.

Kinetic studies of two types have been performed: 1) The exchange of iron between two ferric ion siderophores has been monitored using $^{55}Fe^{3+}$ labeling techniques. That is, the kinetics of the equilibration,

$$^{55}FeL + FeL' \rightleftharpoons FeL + {}^{55}FeL' \qquad (2)$$

have revealed interesting results for iron exchange between ferrioxamine B ($FeHDFO^{+}$) and ferrichrome A ($FeDFC^{3-}$) in comparison to that for ferrioxamine B and ferric(3,4-LICAMS). Such experiments provide mechanistic information about the exchange process, as well as a determination of the on-off rate for tightly bound siderophores at an iron center. 2) The kinetics and mechanism of iron removal from a siderophore complex to a free ligand has been examined using spectrophotometric techniques. Specifically, the removal of iron from ferrioxamine B to ethylenediaminetetraacetic acid (EDTA) and synthetic tricatechoylamide sequestering agents will be discussed. These studies have relevance to the proposed models for siderophore-mediated microbial iron transport. In particular, mechanisms 1 and 2 (discussed above) are characterized by steps involving iron release from a siderophore to either a membrane bound chelating agent or an intracellular ligand.

The kinetic profile for the reaction of ^{55}Fe-labeled ferrioxamine B with ferrichrome A is shown in Figure 8. The exchange

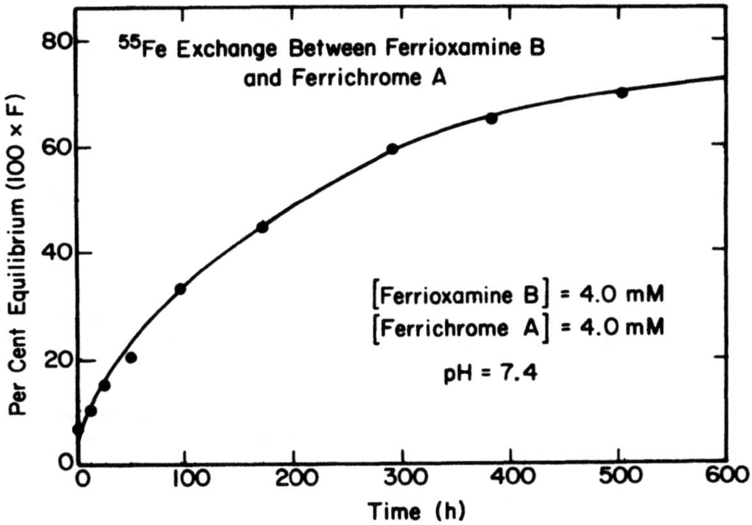

Figure 8. Reaction profile for exchange of iron between ^{55}Fe-labeled ferrioxamine B and ferrichrome A. The quantity F is the fractional degree to which equilibrium has been reached.

of iron between these two hexadentate, hydroxamate siderophores is an extremely slow process; the reaction is only 50% complete after 220 hours of incubation at 25°C and pH 7.4. The proposed mechanism (7) for the exchange involves a two-step chain reaction utilizing free ligand (present in slight excess in this study to ensure complexation of the ferric ion):

$$FeDFC^{3-} + H_4DFO^+ \rightleftharpoons FeHDFO^+ + H_3DFC^{3-} \quad (3)$$

$$^{55}FeHDFO^+ + H_3DFC^{3-} \rightleftharpoons {}^{55}FeDFC^{3-} + H_4DFO^+ \quad (4)$$

Implicit in the above mechanism is the fact that a mixed ligand ternary complex is formed during the exchange of a single iron atom from desferrichrome A to desferrioxamine B. Certainly, given the large formation constants of these ligands for ferric ion, an exchange process which involves the complete dissociation of either siderophore complex [forming $Fe(H_2O)_6^{3+}$], followed by rapid complexation by the competing ligand, can be eliminated. Such a process is also inconsistent with prior work on multidentate ligand exchange characterized by coordination chain reactions (45). Rather, the rate determining unraveling of one ligand from the inner coordination sphere, concomitant with complexation by the incoming ligand, is the predominant characteristic feature of the exchange process. In addition, predissociation is not required for initiation of the chain reaction mechanism in the presence of excess ligand.

In studies of the effect of pH on the rate of reaction (Figure 9), the rate of iron exchange, R, is greatly accelerated as a function of increasing hydrogen ion concentration. In

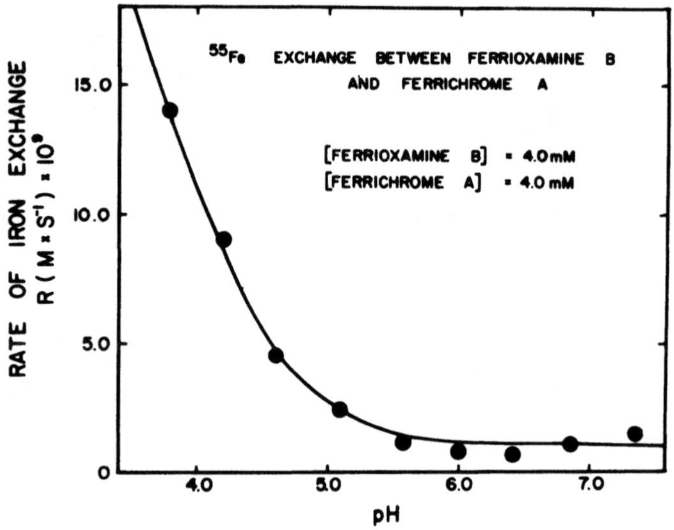

Figure 9. The dependence of the iron exchange rate, R, on pH.

view of the protonation behavior of ferrioxamine B and ferrichrome A,

$$FeHDFO^{+} + H^{+} + 2H_2O \underset{}{\overset{K_H}{\rightleftharpoons}} Fe(H_2DFO)(H_2O)_2{}^{2+} \quad (5)$$

$$FeDFC^{3-} + H^{+} + 2H_2O \underset{}{\overset{K_H'}{\rightleftharpoons}} Fe(HDFC)(H_2O)_2{}^{2-} \quad (6)$$

[assuming $K_H' \approx K_H = 10.0\ M^{-1}$ (46)] one would expect that exposing the inner iron coordination sphere to small, labile unidentate ligands (H_2O) through protonation would enhance the rate of subsequent miltidentate ligand exchange. Lowering the pH of reaction mixtures increases the equilibrium concentration of bis(aquo)bis(hydroxamato)iron(III) species, and, as a consequence, several proton dependent reaction pathways are operative.

In contrast to the slow exchange observed for tris(hydroxamato) complexes, iron exchange between ^{55}Fe-labeled ferric-(3,4-LICAMS) and ferrioxamine B (Figure 10) displays relatively rapid kinetics. Preliminary experiments reveal a half-time of approximately 10 hours for equimolar concentrations (1.0 mM) of the competing complexes and with no excess free ligand. Although, as before, the experiment involves exchange between two hexadentate ligands, the presence of the tricatechoylamide component in this experiment leads to more facile exchange. This is perhaps due to the formation of catalytic quantities of the much more labile Fe(II) complexes through the bis(catecholato)mono(semiquinone)iron(III) complex, which could be the vehicle for the observation of enhanced iron exchange rates.

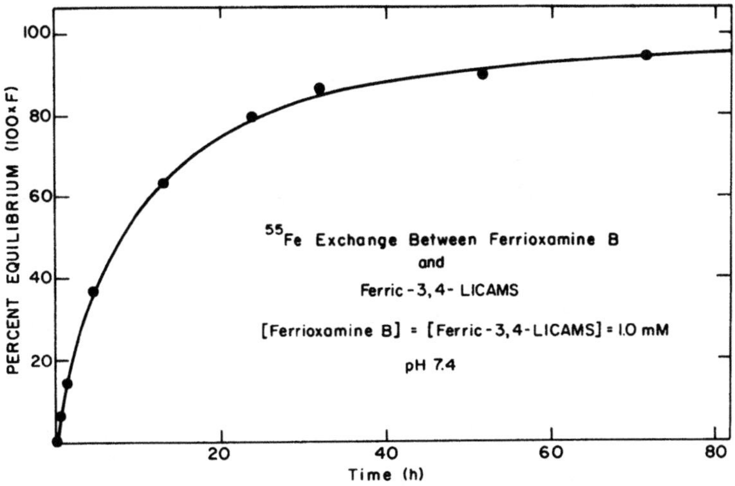

Figure 10. Kinetic profile for $^{55}Fe^{3+}$ scrambling between ferric(3,4-LICAMS) and ferrioxamine B.

The spectral changes accompanying iron removal from ferrioxamine B to EDTA are shown in Figure 11. The top trace (t = 1.0 min) corresponds essentially to the spectrum of unreacted ferrioxamine B. Subsequent spectra illustrate the decrease in absorbance at 428 nm due to $FeHDFO^+$, with a concomitant increase in the uv absorption (not shown) of the product, ethylenediaminetetraacetatoferrate(III) anion ($FeEDTA^-$). The isosbestic point at 352 nm indicates the presence of two absorbing species in solution as equilibrium is approached — the initial reactant, $FeHDFO^+$, and the final product, $FeEDTA^-$. The time elapsed between the first and last spectral traces is approximately 8 hours, at which time approximately 70% of the iron has been removed.

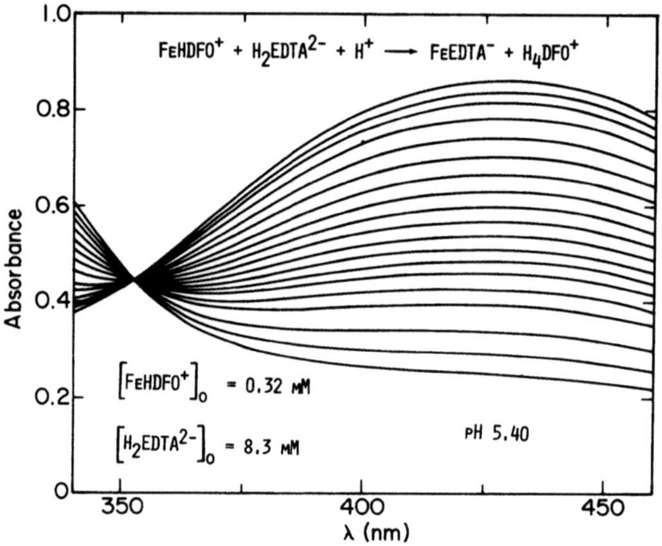

Figure 11. Spectral changes depicting iron removal from ferrioxamine B to EDTA.

The reaction is first order in complex and hydrogen ion concentrations (over the pH range 4.4 to 7.4) and gives evidence of saturation kinetics at large EDTA concentrations. A mechanism which is consistent with the kinetic observations is:

$$FeHDFO^+ + H^+ + 2H_2O \underset{k_{-1}}{\overset{k_1}{\rightleftharpoons}} Fe(H_2DFO)(H_2O)_2{}^{2+} \qquad (5)$$

$$Fe(H_2DFO)(H_2O)_2{}^{2+} + H_2EDTA^{2-} \xrightarrow{k_2} FeEDTA^- + H_4DFO^+ + 2H_2O \qquad (7)$$

From the known equilibrium for Eq. 5 and a Lineweaver-Burk analysis of the EDTA dependence, the following values for k_1, k_{-1} and k_2 are obtained:

$k_1 = 34(5)\ M^{-1}\cdot s^{-1}$

$k_{-1} = 3.4(9)\ s^{-1}$

$$k_2 = 200(10)\ M^{-1} \cdot s^{-1}$$

This mechanism illustrates the importance of opening the coordination sphere about the iron center for subsequent attack by the competing ligand. Protonation of one hydroxamate functionality (Eq. 5) leads to the momentary formation of a bis-(aquo)bis(hydroxamato)iron(III) species - a species known to exist in equilibrium at low pH (47). The primary coordination sphere is thereby relatively exposed, through the partial unwrapping of the hexadentate ligand and the presence of the labile unidentate aquo ligands. The EDTA then can gain a "coordination foothold" (48) on the iron center in forming a ternary complex of metal ion and entering and leaving ligands. In this respect, the detailed mechanism for iron releases from a siderophore complex compliments the results discussed in the ^{55}Fe exchange studies.

Tricatechoylamide sequestering agents (3,4-LICAMS and MECAMS) have also proved to be effective ligands in the removal of iron from ferrioxamine B (49). In particular, these ligands display faster iron removal kinetics compared to EDTA, as the values for the observed first-order rate constants in Table II illustrate. The 3,4-LICAMS and MICAMS can more effectively

Table II. Observed rate constants for the removal of iron from ferrioxamine B.

Ligand	$k_{obs}(s^{-1})$[a]
EDTA	3.5×10^{-7}[b]
3,4-LICAMS	1.9×10^{-3}
MECAMS	5.9×10^{-3}

[a]Calculated for ligand concentration of 10^{-3} M and pH 7.4: [b]Ligand concentration of 8.3×10^{-3} M.

compete (from a kinetic standpoint as well as a thermodynamic one) for ferrioxamine B bound iron at lower concentrations and at higher pH in comparison to EDTA. The kinetics reveal first order dependence on the concentration of both ferrioxamine B and tricatechoylamide. The pH dependence is more complicated as the data in Figure 12 for the MECAMS study indicates. Acid catalysis is evidenced at low pH - again reflecting the general role that hydrogen ion plays in these iron exchange and removal

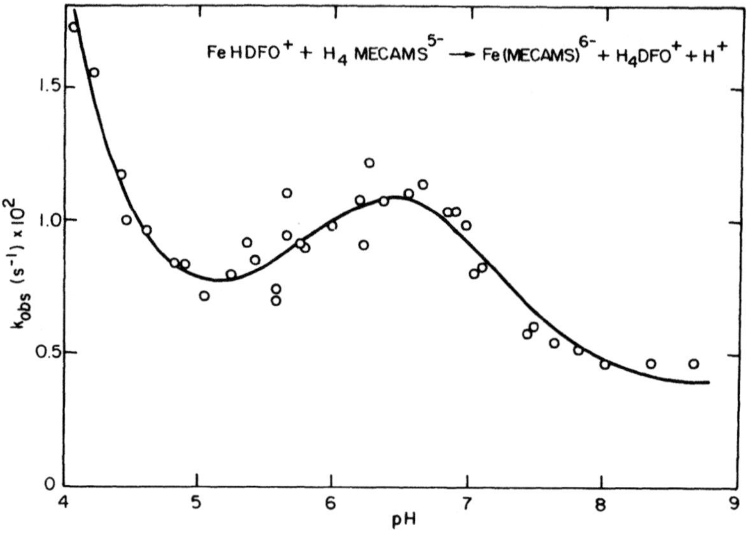

Figure 12. The dependence of the observed first-order rate constant for iron removal from ferrioxamine B (0.10 mM) to MECAMS (1.0 mM) in pH.

studies involving siderophores. The bell-shaped behavior observed between pH 5.5 and 7.0 is explained by considering the protonation behavior of the free ligand itself. The more protonated form of the ligand will be a poorer entering ligand and hence react more slowly. This effect is seen at low hydrogen ion concentration, but at low pH it is the accelerating effect of the protonation of the complex that is the controlling factor.

CONCLUDING REMARKS

The chromic substituted siderophore complexes can be prepared as kinetically inert complexes for use as stereochemical and biological probes in the study of microbial iron transport compounds. In this paper we have emphasized their use as chemical probes to assign the geometry of the metal coordination site by comparison of the vis-uv and CD spectra with model compounds. The biologically relevant isomers of ferrichrome and enterobactin among others have been so assigned.

The formation constants and observed pM values of the ferric siderophore complexes and related synthetic analogues represent a measure of their iron binding affinity in transport compounds. These values provide a useful thermodynamic criterion on which to base candidates for chelation therapy in the treatment of iron overload. Enterobactin has the highest affinity for ferric ion of any iron chelate tested, and this has engendered the synthesis of a series of enterobactin analogues, each utilizing the basic tricatecholate structural design. Furthermore, not only do the synthetic tricatechoylamide sequestering agents display high formation constants with ferric ion, but they also are selective for ferric ion over other biologically relevant metal ions.

The kinetics and mechanism of iron exchange between siderophores have revealed that iron bound in these complexes is not in general readily exchanged. However, catecholate complexes display more rapid reaction kinetics compared to the hydroxamate counterparts. The mechanisms outlined are in general consistent with prior observations in multidentate ligand exchange and imply catalysis in siderophore Fe exchange and uptake.

ACKNOWLEDGMENTS

We wish to acknowledge the contributions of our co-workers whose names appear in our referenced papers. The research summarized here is supported by the National Institutes of Health.

REFERENCES

1. Abu-Dari, K., Raymond, K.N.: 1980, *Inorg. Chem.* 19, pp. 2034-2040.
2. Raymond, K.N., Abu-Dari, K., Sofen, S.R.: 1980, *Adv. Chem. Ser.* No. 119, pp. 113-167.
3. Raymond, K.N., Isied, S.S., Brown, L.D., Fronczek, F.R., Nobert, J.H.: 1976, *J. Am. Chem. Soc.* 98, pp. 1767-1774.
4. Carrano, C.J., Cooper, S.R., Raymond, K.N.: 1979, *J. Am. Chem. Soc.* 101, pp. 599-604.
5. Isied, S.S., Kuo, G., Raymond, K.N.: 1976, *J. Am. Chem. Soc.* 98, pp. 1763-1767.
6. Harris, W.R., Carrano, C.J., Raymond, K.N.: 1979, *J. Am. Chem. Soc.* 101, pp. 2722-2727.
7. Tufano, T.P., Raymond, K.N.: *J. Am. Chem. Soc.,* in press.
8. Neilands, J.B.: "Microbial Iron Metabolism," Neilands, J. B., Ed., Academic Press, New York, 1974.
9. Neilands, J.B.: "Iron Metabolism, Ciba Foundation Symposium: 51," Elsevier, North Holland, Amsterdam, 1977.
10. With possible exception of lactic bacteria, where a closer look is probably in order.
11. Latimer, W.M.: "Oxidation Potentials," Prentice Hall, Englewood Cliffs, New Jersey, 1952.
12. Neilands, J.B.: 1952, *J. Am. Chem. Soc.* 74, p. 4846.
13. Lankford, C.E.: 1973, *CRC Crit. Rev. Microbiol.* 2, p. 273.

14. Kochan, I.: 1973, *Curr. Top. Microbiol. Immunol.* 60, p. 1.
15. Jones, R.L., Peterson, C.M., Grady, R.W., Kumbarai, T., Cerami, A., Graziano, J.H.: 1977, *Nature (London)* 267, p. 63.
16. See, for example, Glynn, A.A.: 1972, *Sym. Soc. Gen. Microbiol.* 22, p. 75, and Rogers, H.J.: 1973, *Infect. Immunol.* 7, p. 445.
17. Jacobs, A.: 1979, *Brit. J. Haematol.* 43, p. 1.
18. Hussain, M.A.M., Flynn, D.M., Gren, N., Hoffbrand, A.V.: 1977, *Lancet* 1, p. 977.
19. Andersen, W.F., Hiller, M.C.: "Development of Iron Chelators for Clinical Use," U.S. Department of Health Education, and Welfare, Bethesda, Maryland, 1975; DHEW Publication No. (NIH) 77-994.
20. Norrestam, R., Stensland, B., Brändén, C.I.: 1975, *J. Mol. Biol.* 99, p. 501.
21. Leong, J., Neilands, J.B.: 1978, *J. Bact.* 136, p. 69.
22. Emery, T.: 1971, *Biochemistry* 10, p. 1483.
23. Carrano, C.J., Raymond, K.N.: 1978, *J. Bact.* 136, pp. 69-74.
24. Raymond, K.N.: 1977, *Adv. Chem. Ser.* 162, pp. 33-54.
25. Raymond, K.N., Carrano, C.J.: 1979, *Acc. Chem. Res.* 12, pp. 183-190.
26. (a) Zalkin, A., Forrester, J.D., Templeton, D.H.: 1966, *J. Am. Chem. Soc.* 88, p. 1810; Loghry, R.A., van der Helm, D.: 1978, *Abs. Am. Cryst. Ass.*, Winter Meeting, Abs. PB2.
    (b) van der Helm, D., Baker, J.R., Eng-Wilmot, D.L., Hossain, M.B., Loghry, R.A.: 1980, *J. Am. Chem. Soc.* 102, p. 4224.
    (c) van der Holem, D., Poling, M.: 1976, *J. Am. Chem. Soc.* 98, p. 82.
    (d) Hough, E., Rogers, D.: 1974, *Biochem. Biophys. Res. Commun.* 57, p. 73.
27. Hossain, M.B., Eng-Wilmot, D.L., Loghry, R.A., van der Helm, D.: 1980, *J. Am. Chem. Soc.* 102, p. 5766.
28. Abu-Dari, K., Raymond, K.N.: 1977, *J. Am. Chem. Soc.* 99, pp. 2003-2005.
29. Leong, J., Raymond, K.N.: 1974, *J. Am. Chem. Soc.* 96, pp. 6628-6630.
30. Leong, J., Raymond, K.N.: 1974, *J. Am. Chem. Soc.* 96, pp. 1757-1762.
31. Carrano, C.J., Raymond, K.N.: 1978, *J. Am. Chem. Soc.* 100, pp. 5371-5374.
32. Harris, W.R., Carrano, C.J., Raymond, K.N.: 1979, *J. Am. Chem. Soc.* 101, pp. 2213-2214.
33. Martell, A.E., Smith, R.M.: "Critical Stability Constants," Vol. 3, Plenum Press, New York, 1977.
34. Harris, W.R., Raymond, K.N., Weitl, F.L.: 1981, *J. Am. Chem. Soc.* 103, pp. 2667-2675.

35. Harris, W.R., Raymond, K.N.: 1979, *J. Am. Chem. Soc.* 101, pp. 6534-6541.
36. Weitl, F.L., Harris, W.R., Raymond, K.N.: 1979, *J. Med. Chem.* 22, pp. 1281-1283.
37. Weitl, F.L., Raymond, K.N.: 1979, *J. Am. Chem. Soc.* 101, pp. 2728-2731.
38. Schwarzenbach, G.: 1952, *Helv. Chim. Acta* 35, p. 2344.
39. Weitl, F.L., Raymond, K.N., Durbin, P.W.: 1981, *J. Med. Chem.* 24, pp. 203-206.
40. Kappel, M.J., Raymond, K.N.: Manuscript in preparation.
41. Lovenberg, W., Buchanan, B.B., Rabinowitz, J.C.: 1963, *J. Biol. Chem.* 236, p. 3899.
42. Basolo, F., Pearson, R.G.: "Mechanisms of Inorganic Reactions," 2nd Ed., John Wiley and Sons, Inc., New York, New York, 1967, pp. 136, 152.
43. Emery, T., Hoffer, P.B.: 1980, *J. Nucl. Med.* 21, p. 935.
44. Abu-Dari, K., Raymond, K.N.: 1977, *Inorg. Chem.* 16, pp. 807-812.
45. Olson, D.C., Margerum, D.W.: 1963, *J. Am. Chem. Soc.* 85, p. 297.
46. Pecoraro, V.L., Raymond, K.N.: Unpublished results.
47. Schwarzenbach, G., Schwarzenbach, K.: 1963, *Helv. Chim. Acta* 46, p. 1390.
48. Rurabacher, D.B., Margerum, D.W.: 1964 , *Inorg. Chem.* 3, p. 382.
49. Tufano, T.P., Raymond, K.N.: Manuscript in preparation.

SPECIFICITY OF SIDEROPHORE IRON UPTAKE BY FUNGI

Günther Winkelmann
Institut für Biologie I, Mikrobiologie
Universität Tübingen

1. GENERAL BIOLOGICAL ASPECTS

Fungi, like all other organisms, need a continuous iron supply during mycelial growth. The amount of free ionic iron present in a neutral environment is limited by the low solubility product of iron hydroxide ($\sim 10^{-36}$). Layers of polymeric hydroxides were often observed on various surfaces and are even sometimes found associated with cell walls of fungi [1].

Polymeric iron hydroxides, however, are of no nutritional value unless solubilized by chelating agents excreted by microorganisms for iron uptake.

The iron chelating agents of fungi may represent simple organic acids, like citrate, or especially synthesized low molecular weight hydroxamate containing compounds called siderophores (formerly named siderochromes or sideramines). The diversity of fungal siderophores is impressive and underlines the importance of iron transport among microorganisms. Structures and properties of different microbial iron chelates are summarized by K. Raymond in a preceeding chapter. The reader is also referred to the earlier literature, reviewed by J.B. Neilands [2,3].

The present article will focus on the biological aspects of fungal siderophores and will primarily discuss the specificity of siderophores. Fungi of the *Ascomycetes* and *Basidiomycetes* are the main source of siderophores. Table 1. shows a compilation of the siderophores and their producing fungal strains according to the present knowledge [4,5,6,7,8,9,10,11,12].

H. B. Dunford et al. (eds.), The Biological Chemistry of Iron, 107–116.

Table 1.: Siderophores from *Ascomycetes* and *Basidiomycetes*

Siderophore	Strain		Ref.
Ferrichrome	*Aspergillus*	*niger*	4
		quadricinctus	7
	Penicillium	*adametzii*	12
		brefeldianum	12
		duclauxii	12
		parvum	21
		resticulosum	4
	Ustilago	*longissima*	10
		maydis	10
		scabiosae	10
		segetum	10
		sphaerogena	2
		trichophora	10
	Cintractia	*sorghi-vulgaris*	10
	Farysia	*olivacea*	10
	Moesziomyces	*penicillaria*	10
	Schizonella	*melanogramma*	10
	Sorosporium	*cenchri*	10
		saponaria	10
	Tilletia	*anomala*	10
	Protomyces	*macrosporus*	10
	Graphiola	*phoenicis*	10
Ferrichrome A	*Ustilago*	*maydis*	10
		scabiosae	10
		sphaerogena	2
	Sphacelotheca	*andropogenis*	10
	Schizonella	*melanogramma*	10
Ferrichrome C	*Cryptococcus*	*melibiosum*	9
	Aspergillus	*oryzae*	7
		duricaulis	7
	Neurospora	*crassa*	11
Ferricrocin	*Aspergillus*	*aureolus*	7
		auratus	7
		fischeri	7
		fumigatus	4,7
		stramenius	7
		unilateralis	7
		viridi-nutans	7
	Neurospora	*crassa*	11
Ferrichrysin	*Aspergillus*	*melleus*	4
		oryzae	
		terreus	4
Ferrirubin	*Penicillium*	*rugulosum*	12
		variabile	4

Table 1.: continued

Siderophore	Strain		Ref.
Ferrirubin	*Paecilomyces*	*varioti*	4
	Spicaria	*sp.*	4
Ferrirhodin	*Aspergillus*	*nidulans*	4
		versicolor	4
Fusigen	*Aspergillus*	*fumigatus*	5
(Fusarinin)	*Giberella*	*fujikuroi*	5
	Fusarium	*sp.*	5
		roseum	5
		cubense	5
	Paecilomyces	*varioti*	4
	Penicillium	*chrysogenum*	5
Triacetylfusigen	*Aspergillus*	*aureolus*	7
(Triacetylfusarinin)		*breviceps*	7
		fumigatus	7
	Penicillium	*javanicum*	12
		purpurogenum	12
Coprogen	*Neurospora*	*crassa*	8
	Penicillium	*camemberti*	4
		chysogenum	4
		citrinum	4
		notatum	4
		patulum	12
		urticae	4
Coprogen B	*Fusarium*	*sp.*	6
		dimerum	6
	Myrothecium	*roridum*	6
		striatisporum	6
		verrucaria	6
	Nectria	*cinnabarina*	6
	Neurospora	*crassa*	6
Rhodotorulic acid	*Leucosporidium*	*scottii*	9
	Rhodosporidium	*toruloides*	9
	Rhodotorula	*glutinis*	9
		gramminis	9
		rubra	9
		pilimanae	9
	Sporidiobolus	*johnsonii*	9
		ruinenii	9
	Sporobolomyces	*albo-rubescens*	9
		hispanicus	9
		pararoseus	9
		roseus	9
	Ustilago	*violacea*	10

It may well be that the diversity of siderophores and the number of producing organisms will increase in the near future, but what we need at present is a more detailed knowledge about the function of the siderophores. Why are so many different siderophores produced? Are all siderophores which are produced by a single strain involved in iron transport or do some of them have a regulatory function? Are non-transporting siderophores involved in iron storage or in an iron exchange mechanism on the membrane surface?

Most if not all fungi have developed sophisticated mechanisms to accumulate iron from the environment during vegetative growth. The most intensively studied mechanism is iron uptake mediated by siderophores. It is assumed that the biosynthesis of desferrisiderophores is repressed, when the iron content of the cells is high. After decreasing the iron content of the environment, the internal iron pool is decreased and this leads to a derepression of desferrisiderophore biosynthesis. By the use of this mechanism fungi can overcome iron deficiency and collect residual iron from the medium. Thus, siderophores solubilize iron hydroxides and extract iron from other organic ligands. One important property of siderophores is obviously their high stability constant for ferric iron. Another property, however, is their structure and conformation in order to be recognized and specifically taken up by the siderophore transport system, as illustrated in Fig. 1.

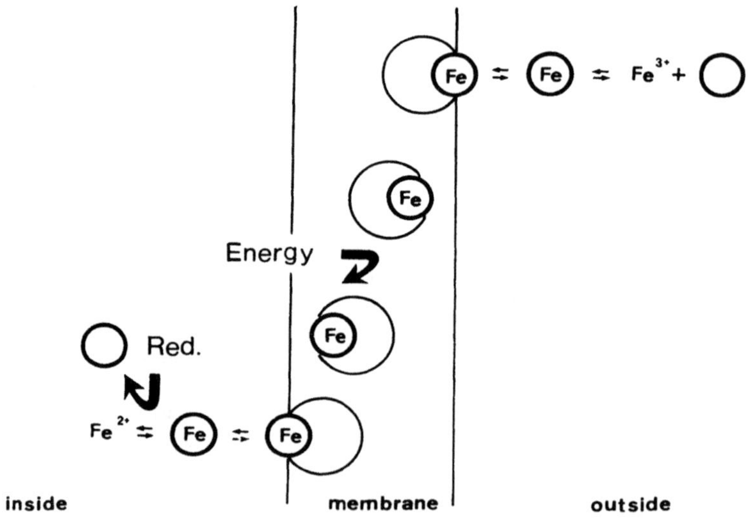

Fig. 1 Schematic illustration of the mechanism of siderophore mediated iron transport in fungi.

2. KINETIC STUDIES

Different strains of *Penicillium, Aspergillus, Neurospora* and *Ustilago* have been used for uptake of [^{55}Fe]-labelled siderophores. In general all strains tested so far have shown different siderophore uptake rates depending on the siderophore supplied. Summarizing these results, it is evident that a strain prefers those siderophores which are predominantly produced by this strain. Thus, the transport system seems to be closely connected with the biosynthetic capacity of the fungal strain. Kinetic studies with [^{55}Fe]-labelled siderophores on *Neurospora crassa* have revealed that the main siderophore, coprogen, was taken up most rapidly. Surprisingly, ferricrocin and ferrichrysin were taken up at comparable rates [13]. When these measurements were done, it was not known that ferricrocin was also a siderophore of *Neurospora crassa* [11]. Therefore, besides coprogen, ferricrocin and possibly ferrichrysin may also be regarded as strain specific siderophores.

Contrary to that, *Ustilago sphaerogena* excretes desferriferrichrome and desferriferrichrome A. Ferrichrome A, however, is taken up by *Ustilago sphaerogena* at only 8 % of the rate for ferrichrome [14], indicating that not all siderophores produced by a single strain function as optimal siderophores.

3. COMPETITION BETWEEN SIDEROPHORES

Siderophores are characterized by a high stability constant of approximately $K \sim 10^{30}$. The octahedrally coordinated iron, however, is kinetically labile and may exchange with other iron chelates present in the same aqueous environment. Iron exchange between different siderophores increases with time and is enhanced at low pH. Therefore iron exchange events render the interpretation of [^{55}Fe]-siderophore competition experiments difficult. Unequivocal results, however, are possible using [^{14}C]-labelled siderophores. In addition, iron exchange events are minimized by using a siderophore-non-producing strain. This was achieved by using a mutant of *Neurospora crassa arg-5 ota aga*, which is blocked in the biosynthesis of ornithine, a necessary constituent of most fungal siderophores. This mutant is unable to biosynthesize siderophores, if ornithine is omitted in the cultivation medium. As shown in Fig. 2, there is a competition between [^{55}Fe]-ferricrocin and [^{14}C]-coprogen (N-1-[^{14}C]-acetyl-coprogen B) during uptake by *N.crassa arg-5 ota aga*. [^{14}C]-coprogen uptake rates are decreased in the presence of increasing amounts of [^{55}Fe]-ferricrocin and vice versa. This experiment is a proof of siderophore competition during uptake by fungal cells, but it does not allow a determination of the number of transport systems involved during siderophore uptake. Competition for one uptake system does not exclude the existence of a second uptake system.

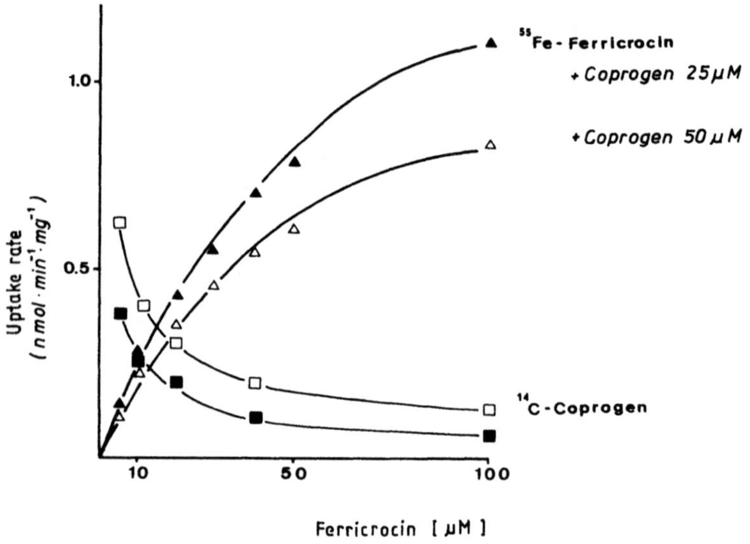

Fig. 2 Competition of siderophores in *N.crassa arg-5 ota aga.* [^{55}Fe]-ferricrocin uptake in the presence of 25 μM and 50 μM [^{14}C]-coprogen.

4. INHIBITION BY NON-UTILIZED SIDEROPHORES

Of all siderophores tested, the most significant inhibition of coprogen uptake was achieved in the presence of ferrirubin [13]. Ferrirubin is a ferrichrome-type siderophore produced by *Penicillium variabile*. Whilst ferrirubin functions as a normal siderophore in the producing strain, it revealed strong inhibition of coprogen uptake in *Neurospora crassa*. It is assumed that inhibition of coprogen uptake is due to a partial structural analogy of both molecules. Coprogen uptake by *N.crassa* was significantly reduced in the presence of ferrirubin ($K_i \sim 6 \times 10^{-8}$ M), although ferrirubin uptake was very low (~30 pmoles min^{-1}mg^{-1}). Furthermore, ferrirubin did not support the growth of the siderophore-free mutant of *N.crassa*. These observations suggest that inhibition of coprogen uptake by ferrirubin results from an occupation of coprogen transport sites within the cytoplasmic membrane by a non-utilized siderophore. Based on this assumption, the number of coprogen transport sites (~6000μ^{-2}) have been calculated from the amount of bound ferrirubin ($\sim 30 \cdot 10^{-12}$ moles mg^{-1}). Provided that the coprogen carrier is monomeric, the resulting turnover number for coprogen was 0.5 s^{-1}. In other words, one molecule of iron is taken up within 2 seconds, if coprogen is present in the medium.

5. SIDEROPHORES AS MEMBRANE REACTIVE IONOPHORES

The designation siderophore might suggest that there exist a functional conformity with the various ionophores found among microorganisms. Indeed the central iron is chelated by a chelator making the whole molecule more lipophilic, which may be examined by studying the partition coefficient. According to the original concept elaborated by Pressmann [15] ionophores are low molecular weight compounds forming selective ion complexes, which by their lipophilicity penetrate artificial or biological membranes. Besides a remarkable ion selectivity, ionophores often possess a high efficiency of translational velocity, e.g. cations translocated per valinomycin per second range from 30-3,000 in mitochondria and from 1-40 in human red cells [16]. However, the occurrence of specific ionophore receptors in biomembranes has not been reported until now. Contrary to that fungal siderophores represent ionophores which react with specific membrane receptors [17]. Competition between different siderophores as well as sensitivity towards various energy poisons may be regarded as a proof.

6. STEROSELECTIVITY OF SIDEROPHORE UPTAKE

A further confirmation of the view that siderophores react specifically with membrane receptors resulted from uptake kinetics with ferrichrome and enantio-ferrichrome [18]. The naturally occuring ferrichrome is composed of L-ornithine, the synthetic enantio-ferrichrome contains D-ornithine [19]. As the enantio-ferrichrome represents the mirror image of ferrichrome (Fig. 3) both molecules possess an opposite configuration about the iron atom, which could be confirmed by opposite cotton effects in the CD-spectra. Thus, the molecules represent Λ-cis-ferrichrome and Δ-cis-enantio-ferrichrome, according to the nomenclature of Leong and Raymond [20].

Transport studies with *Neurospora crassa* and *Aspergillus quadricinctus* [18] and with the ferrichrome producing *Penicillium parvum* [21] revealed that contrary to Λ-cis-ferrichrome the Δ-cis-enantio-ferrichrome is not recognized by the siderophore receptor in fungal membranes (Fig. 4). On the other hand Δ-cis-enantio-ferrichrome was taken up by *E.coli* at a rate of 50 %, relative to Λ-cis-ferrichrome [21], indicating that the ferrichrome receptor in bacteria is less stereoselective. However, stereospecificity of the bacterial enterobactin receptor has been reported [22], using enterobactin and enantio-enterobactin [23].

Since the molecular dimension of both enantiomeric siderophores and also their lipophilicity is absolutely identical, the differences in the transport behaviour may result from configurational differences of the cyclic hexapeptide and/or from differences in the iron coordination sphere. In order to solve these questions further studies with synthetic model siderophores are required.

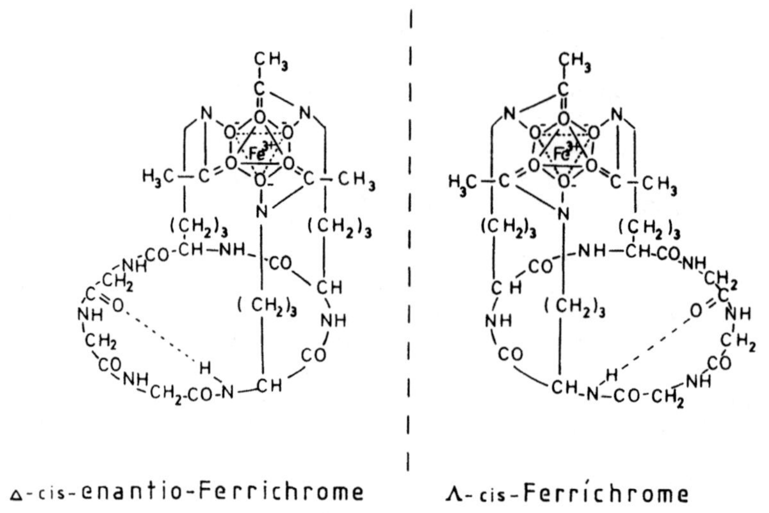

Fig. 3 Structures of enantiomeric ferrichromes showing an opposite chirality of the peptide backbone and an opposite configuration about the iron center (from Winkelmann and Braun, 1981).

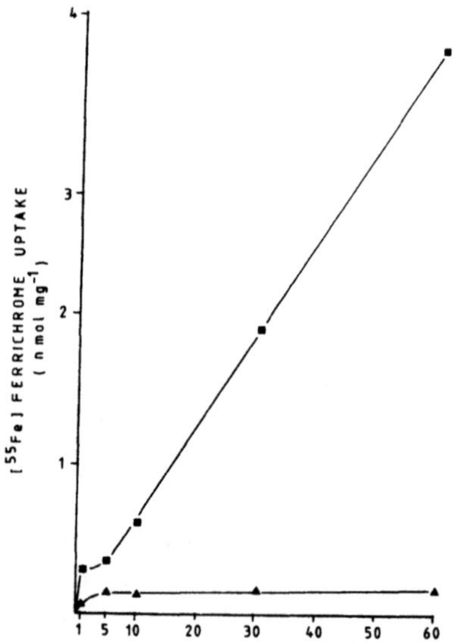

Fig. 4 Uptake of [^{55}Fe]-ferrichrome (■) and enantio-[^{55}Fe]-ferrichrome (▲) by *P.parvum* (from Winkelmann and Braun, 1981).

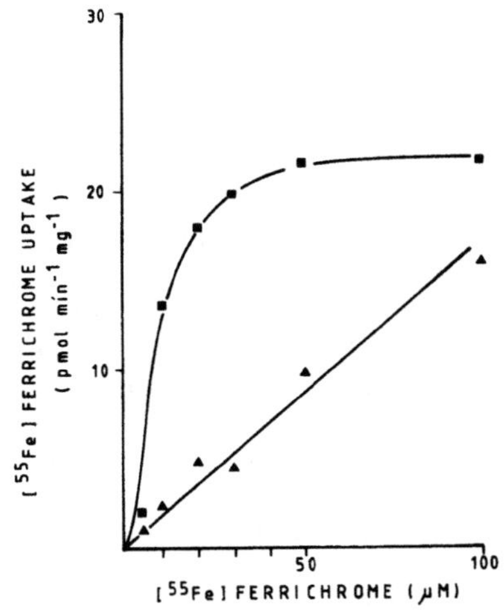

Fig. 5 Concentration dependent uptake of [^{55}Fe]-ferrichrome (■) and enantio-[^{55}Fe]-ferrichrome (▲) by *P.parvum* (from Winkelmann and Braun, 1981).

Earlier studies using the kinetically inert Cr-ferrichrome revealed that the Λ-cis-Cr-ferrichrome is at least one of the biologically active isomers during uptake by *Ustilago sphaerogena* [24]. On the other hand there is evidence that fusigen [25], rhodotorulic acid [26] and coprogen [27] exists predominantly as the Δ-cis isomer. The question therefore remains open if these siderophores represent an equilibrium between the Λ-cis and Δ-cis isomers in solution, from which only one isomer is selected for uptake by the fungal siderophore receptor.

In the case of ferrichrome and enantio-ferrichrome it is obvious that the peptide portion forces the iron center to adopt a predominantly Δ-cis or Λ-cis coordination. The biological results of uptake are evidence that a rapid isomerisation between theese coordination isomers was not involved. As shown in Fig. 5, the concentration dependent uptake of Δ-cis-ferrichrome is a diffusion-controlled process, while the Λ-cis-ferrichrome is assimilated by a receptor-mediated process, confirming the stereoselectivity during uptake of fungal siderophores.

ACKNOWLEDGEMENTS

I wish to thank R.B. Johnson for reading the manuscript. This work was supported by the Deutsche Forschungsgemeinschaft (SFB 76).

REFERENCES

[1] Winkelmann, G.: 1979, Arch. Microbiol. 121, 43-51.
[2] Neilands, J.B.: 1973, pp. 167-202. In: Eichhorn, G.L. (ed.) "Inorganic Biochemistry" Vol. 1, Elsevier, Amsterdam
[3] Neilands, J.B.: 1980, pp. 529-572. In: Iron in Biochemistry and Medicine, II. (A. Jacobs, and M. Worwood, eds.) Academic Press, London-New York.
[4] Zähner, H., Keller-Schierlein, W., Hütter, R., Hess-Leisinger, K., and Deer, A.: 1963, Arch. Mikrobiol. 45, 119-135.
[5] Diekmann, H.: 1967, Arch. Mikrobiol. 58, 1-5.
[6] Diekmann, H.: 1970, Arch. Mikrobiol. 73, 65-76.
[7] Diekmann, H. and Krezdorn, E.: 1975, Arch. Microbiol. 106, 191-194.
[8] Keller-Schierlein, W. and Diekmann, H.: 1970, Helv. Chim. Acta 53, 2035-2044.
[9] Atkin, C.L., Neilands, J.B., and Phaff, H.J.: 1970, J. Bacteriol. 103, 722-733.
[10] Demel, G. and Oberwinkler, F.: 1980, Paper presented at the Vth International Symposium on Yeasts in London, Canada, Pergamon Press Limited 1981.
[11] Horowitz, N.H., Charlang, G., Horn, G., and Williams, N.P.: 1976, J. Bacteriol. 127, 135-140.
[12] Winkelmann, G.: (unpublished results).
[13] Winkelmann, G.: 1974, Arch. Microbiol. 98, 39-50.
[14] Emery, T.: 1971, Biochemistry 10, 1483-1487.
[15] Pressman, B.C., Harris, E.J., Jagger, E.S., and Johnson, J.H.: 1967, Proc. Nat. Acad. Sci. 58, 1949-1956.
[16] Haynes, D.H., Wiens, T., and Pressman, B.C.: 1974, J. Membr. Biol. 18, 23-38.
[17] Müller, G. and Winkelmann, G.: 1981, FEMS Microbiol. Lett. 10, 327-331.
[18] Winkelmann, G.: 1979, FEBS Lett. 97, 43-46.
[19] Naegeli, H. and Keller-Schierlein, W.: 1978, Helv. Chim. Acta 61, 2088-2095.
[20] Leong, J. and Raymond, K.N.: 1975, J. Am. Chem. Soc. 97, 293-296.
[21] Winkelmann, G. and Braun, V.: 1981, FEMS Microbiol. Lett. (in press).
[22] Neilands, J.B., Erickson, T.J., and Rastetter, W.H.: 1981, J. Biol. Chem. 256, 3831-3832.
[23] Rastetter, W.H., Erickson, T.J., and Venuti, M.C.: 1980, J. Org. Chem. 45, 5011-5012.
[24] Leong, J., Neilands, J.B., and Raymond, K.N.: 1974, Biochem. Biophys. Res. Commun. 60, 1066-1071.
[25] Hossain, M.B., Eng-Wilmot, D.L., Loghry, R.A., and van der Helm, D.: 1980, J. Am. Chem. Soc. 102, 5766-5773.
[26] Carrano, C.J. and Raymond, K.N.: 1978, J. Am. Chem. Soc. 100, 5371-5374.
[27] Wong, G.B., Raymond, K.N., Matzanke, B., and Winkelmann, G.: In preparation

IRON UPTAKE AND INTRACELLULAR IRON DISTRIBUTION IN CULTURED RAT HEART CELLS: EFFECTS OF IRON CHELATORS.

B.R. Byers, C.V. Sciortino, P. Cox and P. Robinson

University of Mississippi Medical Center, Department of Microbiology, Jackson, MS 39216 USA.

ABSTRACT.

Primary cultures of beating rat heart muscle cells were shown to be a good model system for studies of iron uptake and intracellular iron metabolism. Iron uptake from several iron sources was rapid and ferritin was visualized (by electron microscopy) within 48-96 hours after addition of iron. Iron was distributed in various cellular fractions; an iron-containing cytosol component(s) (non-ferritin) that may represent a "pool" of iron was evident. The cultured rat myocytes were used to screen chelators that might become useful in therapy of human iron overload. One chelator, agrobactin, was significantly more effective than the drug deferoxamine in reduction of iron uptake from transferrin-iron by the cells; however, enterobactin increased iron uptake from this source. Deferoxamine inhibited incorporation of iron into ferritin and increased the amount of iron in the non-ferritin cytosol fraction. Therefore, some of these chelators may be useful probes in studies of intracellular iron metabolism.

INTRODUCTION.

The essentiality of iron in most (if not all) biological systems is undisputed and the complex interactions of mammalian iron metabolism have been the subject of many studies. In humans iron absorption occurs mainly in the small intestine; mucosal cells pass the iron into circulation where the metal binds with the iron transport protein, transferrin. The transferrin-iron complex delivers iron to cells (probably through specific transferrin receptors on the surfaces of these cells),

H. B. Dunford et al. (eds.), The Biological Chemistry of Iron, 117–124.

Figure 1. Relationships in intracellular iron metabolism. Modified from Jacobs (3).

although other forms of iron also may be assimilated. There is no specific physiological mechanism for excretion of excess iron and iron overload caused by abnormally increased absorption or by parenteral intake can result in tissue damage and sometimes death. For example, persons with Cooley's anemia (homozygous β-thalassemia) require lifelong blood transfusions with the inevitable complication of iron toxicity that eventually is fatal, often from congestive heart failure. These aspects of mammalian iron metabolism are discussed fully in several excellent publications (e.g., 1,2,3).

Although the knowledge of movement of iron within the whole animal is extensive, less is known about the intracellular pathways of iron. Hypotheses explaining the intracellular flow of iron have been published (4,5,6). Iron is seen to enter the cells from transferrin-iron (or other iron-containing molecules) where the metal becomes associated with an unidentified molecule(s) to form a "pool" of available iron (Figure 1). Iron is delivered from the pool to iron-requiring cellular systems and to the storage depot, ferritin. Iron may be mobilized from ferritin for metabolic use. Overload of the pool may lead to toxic deposition of iron and consequent cell damage.

CELLULAR IRON METABOLISM: MODEL SYSTEMS.

As discussed by Bailey-Wood, et. al. (7), much of the knowledge of intracellular iron metabolism is derived from studies of erythrocytes (which are atypical in several ways) and there is need for model systems of single cell types for study of iron uptake and intracellular iron distribution. They selected Chang cells, a stable laboratory cell line that can be cultivated in suspension. In a series of studies with Chang cells (see reference 8 and publications cited therein) iron uptake and its distribution to various cell fractions (including ferritin) were described.

TABLE 1

RADIOIRON UPTAKE BY CULTURED RAT MYOCYTES[1]

Source (1 μM Fe)	Iron Uptake[2] (pmol/10^6 cells)
[^{55}Fe]transferrins[3]	
Rat	43
Rabbit	35
Human	29
Equine	25
Bovine	5
$^{55}FeCl_3$	15
$^{55}FeNH_4$ citrate	11

[1] Data from (9,11).
[2] Determined at 24 hours.
[3] All transferrins were 100% iron saturated.

For our studies, we selected cultures of newborn rat myocardial cells. These cultures provided an excellent model system for study of iron metabolism because they 1) are primary cultures (experiments may be done with cells taken from the animals), 2) are 95-100% pure as to cell type, 3) are available in large numbers, and 4) can be monitored for normal physiology of heart muscle cells (i.e., cross-striations, rhythmic contractions, and specific enzymes). Furthermore, the use of myocardial cells is especially interesting because cardiac failure is one of the leading causes of death in patients with iron overload.

IRON UPTAKE BY MYOCARDIAL CELLS.

Administration of iron to experimental animals causes deposition of iron in several organs; however, one of the interesting aspects of this experimental approach is that the heart muscle cells in the animal load very little iron. This is not the case with heart muscle cells in culture. Primary cultures of rat cardiac muscle cells, prepared by described methods (9) from 2-5 day old Sprague-Dawley rats, incorporated iron from several radioiron sources (Table 1). Of the several species of transferrins tested, rat transferrin supported greatest uptake. Iron uptake from transferrin-iron was linear for about 3 hours after which net uptake ceased for about 3 hours, although uptake of iron resumed thereafter (11,12). Iron supplied as the various sources (Table 1) was assimilated into metabolism; radioiron was detected (by immunoprecipitation) in ferritin within 2 hours after addition of the radioiron source (10). Ferritin particles could be visualized (by electron microscopy) in the cultured muscle cells within 48-96 hours after addition of the iron source (10). Thus, the model system of cultured heart cells shows rapid iron uptake and ferritin synthesis.

INTRACELLULAR DISTRIBUTION OF IRON.

Rapid accumulation of radioiron to high levels by cultured heart muscle cells indicates that this system may be useful in definition of the cellular pathways of iron. Initial determinations of the intracellular distribution of iron were done by centrifugation procedures in which the radioiron-loaded cells first were disrupted by sonication and the resulting cell extract then was separated into a 34,000 x g pellet (membranes, mytochondria, etc.) and supernatant (cytosol). Radioiron in the cytosol ferritin was determined by immunoprecipitation. If the cultures were exposed to [^{55}Fe]transferrin for 6 hours, radioiron was located in the 34,000 x g pellet (35%), in the cytosol ferritin (24%) and in the non-ferritin cytosol fraction (41%) (Figure 2). If the radioiron source was removed at 6 hours and incubation continued, re-distribution of the label was noted during the following 12 hours. Between 6 and 10 hours the amount of radioiron in the 34,000 x g pellet increased while radioiron in the cytosol ferritin and non-ferritin fractions decreased (Figure 2). This suggests transfer of iron from the two cytosol fractions to the 34,000 x g pellet, although the radioiron loss from the cytosol components was greater than the increase in the 34,000 x g pellet. Therefore, there was a net loss (14%) of radioiron between 6 and 10 hours. Spontaneous loss of cell-associated radioiron reached 65% by 18 hours. Most of the radioiron had disappeared from the cytosol non-ferritin component(s) and significant decline was evident in the 34,000 x g pellet (Figure 2). Interestingly, the amount of radioiron in the cytosol ferritin remained relatively constant after an initial loss at 8 hours. One or more of the components in the cytosol non-ferritin fraction may represent the "pool" of iron, although some ferritin iron also was mobilized during the test period.

These fractionation procedures show the complexity of the system, but also suggest that studies in heart cells may contribute useful information on intracellular iron metabolism. Determination of the nature of the iron-binding component(s) in the cytosol non-ferritin fraction is under way; preliminary experiments suggest that this iron may be associated with a relatively high molecular weight substance(s).

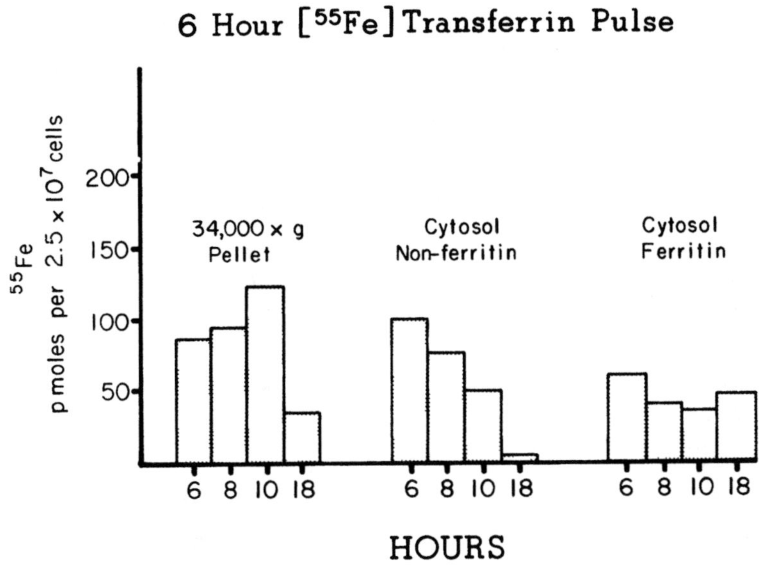

Figure 2. Intracellular distribution of radioiron in cultured rat myocytes. The cultures were labeled for 6 hours with [^{55}Fe]transferrin (1 μM Fe), the radioiron source was removed and radioiron distribution determined at 6,8,10 and 18 hours.

EFFECT OF CHELATORS ON IRON UPTAKE AND DISTRIBUTION.

Cultured rat myocytes have been used to develop an assay method for screening chelators that might become useful in chelation therapy for human iron overload (9) (Table 2). The present drug used for chelation therapy is deferoxamine (Desferal), a microbially produced secondary hydroxamate. At the concentration of chelators employed in these tests, deferoxamine reduced iron uptake from transferrin-iron to only 86% of controls without an added chelator. The microbial phenolate agrobactin was significantly more effective than deferoxamine; agrobactin lowered uptake to 43% of the controls. None of the other chelators tested here and in other unpublished studies was more effective than agrobactin in this assay system and agrobactin may have potential in treatment of iron overload disease. Tests of the effect of agrobactin on excretion of iron by iron-loaded animals are under way. Agrobactin is a threonyl peptide of spermidine, acylated with three residues of 2,3-dihydroxybenzoic acid. A synthetic tri-catechol derivative of

TABLE 2

EFFECT OF CHELATORS ON RADIOIRON UPTAKE FROM [^{55}Fe]TRANSFERRIN BY CULTURED RAT MYOCYTES[1]

Chelator[2]	% Uptake
None	100
Deferoxamine (Desferal)	86
Agrobactin	43
Aerobactin	86
2,3-dihydroxybenzoic acid	57
2,3-dihydroxybenzoylglycine	71
Fusarinine	100
Rhodotorulic acid	86
Schizokinen	91
Enterobactin	114
MECAM	83
LICAMS	100

[1]Data from (9) and Byers and Hillibert (unpublished).
[2]Chelators (20 μM) were added 6 hours after addition of [^{55}Fe]transferrin (1 μM Fe); uptake measured at 24 hours.

spermidine (LICAMS) (13) also was tested; interestingly, this chelator did not lower uptake of iron from transferrin-iron (Table 2).

The microbial phenolate enterobactin increased iron uptake from transferrin-iron, whereas a synthetic analog (MECAM) of enterobactin effected some reduction in uptake from this source (Table 2). MECAM lacks the ester bonds found in enterobactin but is a physiologically functional analog of enterobactin in certain microbial systems (14,15). The structural difference between the two chelators caused an alteration in the response of heart cells to these isosteric analogs.

The heart cell cultures are being used to determine the effects of certain chelators on the intracellular distribution of iron. For example, if cultures were labeled for 24 hours with [^{55}Fe]transferrin, in the presence and absence of deferoxamine, the chelator caused a distinct reversal in the amounts of radioiron present in the cytosol ferritin and non-ferritin fractions (Figure 3). Deferoxamine inhibited iron incorporation into ferritin but increased the relative amount of iron in the cytosol non-ferritin fraction. These data show that certain chelators may cause re-distribution of iron among the cell

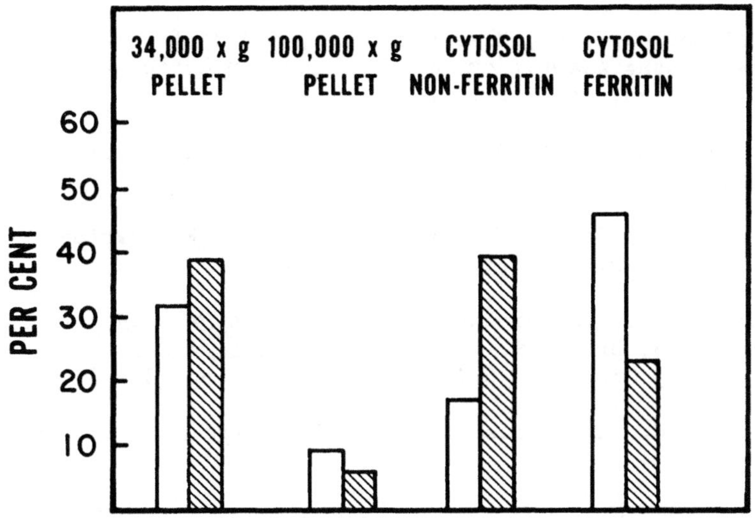

Figure 3. Intracellular distribution of radioiron in cultured rat myocytes labeled for 24 hours with [^{55}Fe]transferrin (1 μM Fe) with added deferoxamine (20 μM) (hatched bar) and without deferoxamine (open bar).

compartments and suggest that these compounds may be useful probes for studies of intracellular iron metabolism.

ACKNOWLEDGEMENTS.

This research was supported by a Grant from the Cooley's Anemia Foundation, New York, by Grant AM26163 from the National Institute of Arthritis, Metabolism and Digestive Diseases, National Institutes of Health, by a National Science Foundation Faculty Professional Development Award (to P.C.) and by a Research Fellowship from Research Corporation (to P.R.). J.B. Neilands, University of California, Berkeley, M. Venuti, Massachusetts Institute of Technology, and K.N. Raymond, University of California, Berkeley, are thanked for supplying chelators. Appreciation is given to D. Acosta, University of Texas, Austin, for assistance with the culture procedure.

REFERENCES.

1. Jacobs, A., and Worwood, M. (eds.): 1974, *Iron in Biochemistry and Medicine*, Academic Press, New York.

2. Muller-Eberhard, U., Miescher, P.A., and Jaffe, E.R. (eds.): 1977, *Iron Excess: Aberrations of Iron and Porphyrin Metabolism*, Grune and Stratton, New York.

3. Bezkorovainy, A.: 1980, *Biochemistry of Nonheme Iron*, Plenum Press, New York.

4. Jacobs, A.: 1977, in *Iron Metabolism*, CIBA Foundation Symposium, Elsevier, Amsterdam, pp. 91-106.

5. Trump, B. F., and Beresky, I.K.: 1977, in *Proteins of Iron Metabolism*, E. B. Brown, P. Aisen, J. Fielding, and R. R. Crichton (eds.), Grune and Stratton, New York, pp. 359-364.

6. Munro, H. N., and Linder, M. C.: 1978, *Physiol. Rev.* 58, pp. 317-396.

7. Bailey-Wood, R., White, G. P., and Jacobs, A.: 1975, *Br. J. Exp. Path.* 56, pp. 358-362.

8. Hoy, T. G. and Jacobs, A.: 1981, *Biochem. J.* 193, pp. 87-92.

9. Sciortino, C.V., Byers, B.R., and Cox, P.: 1980, *J. Lab. Clin. Med.* 96, pp. 1081-1085.

10. Cox, P. G., Harvey, N. E., Sciortino, C., and Byers, B. R.: 1981, *Am. J. Pathol.* 102, pp. 151-159.

11. Sciortino, C.V.: 1980, Ph.D. Dissertation, University of Mississippi Medical Center.

12. Sciortino, C.V., Cox, P.G., and Byers, B.R., manuscript submitted.

13. Harris, W.R., Raymond, K.N., and Weitl, F.L.: 1981, *J. Am. Chem. Soc.* 103, pp. 2667-2675.

14. Venuti, M.C., Rastetter, W.H. and Neilands, J.B.: 1979, *J. Med. Chem.* 22, pp. 123-124.

15. Lodge, J.S., Gaines, C.G., Arceneaux, J.E.L., and Byers, B.R.: 1980, *Biochem. Biophys. Res. Commun.* 97, pp. 1291-1295.

SECTION C

SOME PROPERTIES OF THE CYTOCHROMES

NMR STUDIES OF LOW-SPIN CYTOCHROMES

António V.Xavier, Isabel Moura, José J.G.Moura, M.Helena Santos and José Villalaín

Centro de Química Estrutural, Complexo Interdisciplinar, IST, Av.Rovisco Pais, 1000 Lisboa (Portugal) and Gray Freshwater Biological Institute, University of Minnesota P.O.Box 100, Navarre, Minn.55392 (U.S.A.)

1. INTRODUCTION

The understanding of the chemical and biochemical processes requires an extensive knowledge of molecular structures, thus the development of physical techniques which can provide structural information is of great importance. The X-ray diffraction method has been the most useful technique for obtaining the necessary parameters for the determination of the structure of molecules. However, the resolution obtained in large molecules (even up to 0.5 A of magnitude) may not be adequate to explain certain processes. Also, the X-ray method is restricted to the crystalline state and for the study of structures in solution other methods must be used.

Even though a structure may only exist for a very short time, it must be defined. On the one hand, X-ray crystallography can not hope to achieve this objective, for its time scale is hours rather than micro-seconds. On the other hand the fastest methods of inspection (10^{-10} sec) are not very structure specific, for they must use high energy radiation which may perturb the system during observation (e.g.,ultra violet spectroscopy). The examination of a structure which is in the micro-seconds range is possible through the use of low energy methods such as nuclear magnetic resonance (NMR) and electron spin resonance (ESR). Due to the comparatively small value of the energies involved in the resonance conditions, a subtle change in the system can be studied. Since the energy of nuclear magnetic resonance and the time of nuclear relaxation are

H. B. Dunford et al. (eds.), The Biological Chemistry of Iron, 127–141.

very much dependent on the environment of the nucleus NMR turns out to be a most useful technique for structural studies, particularly when applied in conjunction with other techniques.

Furthermore, NMR spectroscopy can provide information on the solution chemistry of specific residues of proteins as well as other thermodynamic parameters, which can be important for elucidating their function.

Cytochromes constitute the group of electron transfer proteins for which more structural data are available. The techniques used include just about all those available to molecular biophysicists. The X-ray structures have been obtained for different types of cytochromes and NMR has been of great importance to further our knowledge of their structure and solution chemistry.

In the cytochromes *c* the haem, a mesoporphyrin, is covalently bound to the polypeptide chain through thioether linkages provided by cysteinyl residues on an-CYS-...-CYS-HIS- sequence (it can also be bound by only one thioether linkage of a -CYS-HIS- sequence). This HIS residue is the fifth ligand to the haem iron. In the low--spin *c* type cytochromes the sixth ligand can be either a MET residue (e.g.,cytochrome *c*) or another HIS (cytochrome c_3). The range of redox potential values covered by *c* type cytochromes is very wide, going approx. from +400 to -100 mV for those with MET as the sixth ligand, and from approx. -150 to -350 mV for those with HIS as the sixth ligand.

In the *b* type cytochromes the haem, a protoporphyrin IX, is not covalently bound to the polypeptide chain. The fifth ligand to the iron is also a HIS residue and again the sixth ligand can be either a MET (e.g.,cytochrome b_5) or another HIS (cytochrome b_{562}). The redox potentials covered by the first type is approx. -200 to +350 mV and the only known member of the second, cytochrome b_{562}, has a redox potential of +100 mV.

This article will illustrate the application of NMR for obtaining detailed structural information of low-spin cytochromes in solution, to determine the mobility of individual amino-acid residues and to study the redox kinetics and mechanisms of electron transfer. Other examples will deal with the study of the effect of ionization and of the specific interactions of cytochromes with other electron transfer proteins. Rather than making a thorough revision of the work published it will concentrate in the examples most suitable to illustrate how to accomplish these studies.

2. TECHNICAL DETAILS

The NMR samples are typically 0.3 ml of 1-3 mM purified cytochrome. However, depending on the type of information required and instrumentation available, it might be possible to either work with smaller quantities or need more concentrated solutions. The protein is generally lyophilized from D_2O and the pH adjusted with DCl and NaOD. If one is interested in looking at the exchangeable NH protons the procedure is of course different and it is necessary to prepare the sample in variable concentrations of H_2O. The reduced samples are generally prepared by adding to deaerated solutions the required amount of either ascorbate or dithionite, depending on the redox potential of the cytochrome. Some cytochromes are partially reduced in solution and it is necessary to add minimal amounts of ferricyanide to obtain the fully oxidized spectrum. Spectra of intermediate stages of oxidation can be obtained by performing redox titrations.

The spectrometers used are of 200 MHz or greater and use FT. The techniques used include resonance enhancement methods (1), time-shared pulse spin-decoupling (2), multiplet selection and spin-echo double resonance, using Carr-Purcell pulse sequences (3,4), cross-saturation (5,6) and nuclear Overhauser enhancements (7,8).

3. ANALYSIS OF THE NMR SPECTRUM

The existence of two generally stable oxidation states for the low-spin cytochromes, one paramagnetic (oxidized) and the other diamagnetic (reduced), makes them particularly suitable to be studied by NMR techniques. Figure 1 shows the NMR spectra of oxidized and reduced cytochrome c_{551} isolated from *Pseudomonas perfectomarinus* (9).

The most striking difference between the two spectra is the range of chemical shifts covered by the resonances. In the spectrum of ferricytochrome *c* very large paramagnetic shifts are induced in the resonances of both the haem protons and those of the haem axial ligands. The range covered extends from approx. 35 to -45 ppm. The three proton intensity resonances at low field (32.9, 22.8, 17.2, and 12.3 ppm) belong to the haem methyl groups. The resonances at -10.2, 17.9, and -42.7 ppm are assigned to the bound MET.

In the spectrum of ferrocytochrome *c* the only resonances outside the main protein envelope are usually those of the haem meso protons and those of the bound MET shifted by the ring-current effect of the haem. The four haem meso proton resonances appear at low-field (8-10 ppm) and the MET proton resonances are observed at the high field end (four one-proton intensity resonances at -0.7, -1.1, -2.9, and -3.9 ppm, and one three-proton intensity resonance at -3.1 ppm).

The shifts induced by the intrinsic structural probe provided by

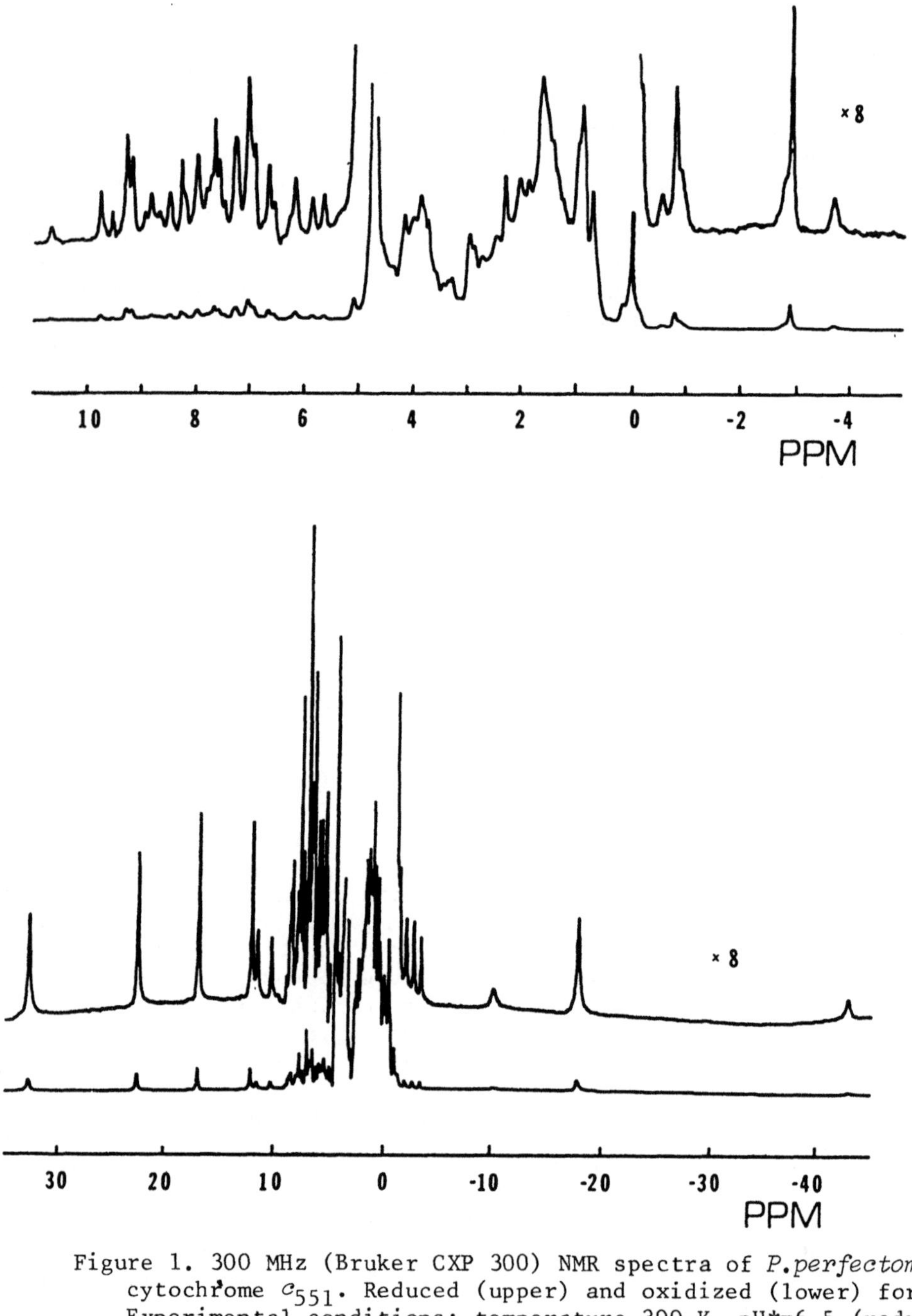

Figure 1. 300 MHz (Bruker CXP 300) NMR spectra of *P.perfectomar* cytochrome c_{551}. Reduced (upper) and oxidized (lower) forms Experimental conditions: temperature 300 K, pH*=6.5 (reduce and 6.9 (oxidized), 2048 scans. The concentration is 1 mM.

the haem in the oxidized (paramagnetic plus ring-current) and in the reduced state (ring-current only) are very useful both for assignments and for extracting geometrical information (10). This is particularly useful when the X-ray structure of the cytochrome is available. Here is another advantage for the NMR studies of these proteins, due to the large amount of data available (11).

Moore and Williams (12-17) have recently published a series of six articles on the NMR studies of several eukaryotic cytochromes *c*. This constitutes a very good example of the usefulness of the technique to study the structure of cytochrome *c* in solution. Horse cytochrome *c* is the cytochrome for which more NMR spectrum resonances have been assigned. The resonances of the haem group and the axial ligands (18-22),and of several exchangeable NH protons (22,24) have been assigned. The assignment of 33 resonances arising from 39 aromatic protons and 15 of 28 aliphatic protons of the ferrocytochrome, and of 18 resonances arising from 27 aromatic protons and 10 resonances of 24 aliphatic protons of the ferricytochrome have been done (12,13,21,22,23-29).

Using the accumulated information on the NMR spectra of *c* type cytochromes several properties of these proteins can be studied.

3.1. Electronic Structure of the Haem

The NMR lines of the haem protons were individually assigned for different *c* type cytochromes in the oxidized and reduced form (30). These assignments were obtained using an elegant combination of NOE and saturation transfer experiments (30,31). The differences between the paramagnetic shift induced in the individual haem methyl groups (Table 1), strongly dominated by the contact interaction with the unpaired electron,have been used to show that the spin density distribution is quite different in mamalian type cytochrome *c* and cytochrome c_{551} from *Pseudomonas aeruginosa* (32).

Table 1
Paramagnetic shifts of the haem methyl groups

Methyl group	Paramagnetic shift (32) cytochrome c_{551}	cytochrome *c*
Ring I	21.1	3.9
Ring II	9.6	27.3
Ring III	28.8	6.9
Ring IV	14.4	31.7

A combination of these NMR studies and X-ray crystallography data (33,34) has been taken to support a proposed mechanism which postulates that the electron transfer of cytochrome *c* is through the haem edge exposed to the solvent. Analysis of Table 1 shows that in the horse heart cytochrome *c* the high spin density is localized in the pyrrole rings II and IV while in cytochrome c_{551} it is in rings III and I. The X-ray structures show that in cytochrome c_{551}, ring III should be accessible to solvent, contrary to mammalian cytochromes *c*, where pyrrole ring II is the most exposed to solvent.

3.2. Characterization of the Haem Ligands

The assignement of the proton resonances of the haem axial ligands is very useful to identify the amino-acid residues which are bound to the iron. In the case of cytochrome c_3 it was from the comparison of the NMR spectra of proteins isolated from different *Desulfovibrio* species that it was shown that both the fifth and sixth ligands are HIS (35).

The utilization of MET as one of the axial ligands can easily be proved by the typical features present in the high-field region of the NMR spectrum of reduced cytochrome. The observation of a three-protons intensity singlet at approximately -3 ppm, together with four one-proton resonances, in this region of the spectrum is a unequivocal indication that methionine is coordinated to the haem. These typical features were used to prove that in cytochrome b_{562}, contrary to all other known *b* type cytochromes, the sixth ligand is a methionine residue (36).

The chirality of the sulfur ligand of the axial MET can be probed by measuring nuclear Overhauser effects between the protons of this residue and specific haem meso protons. Using this method it has been shown that the chirality of the sulfur ligand can either be R (e.g., tuna cytochrome *c*) or S (e.g., *Pseudomonas aeruginosa* cytochrome c_{551}) (37-39).

3.3. Structural Control of the Haem Redox Potentials

The control of the redox potentials of the haem by the polypeptide chain of cytochromes has been studied by several techniques. Here again, NMR has been useful in probing some of the features which might be important for this control.

Moore and Williams (40) have pointed out to a relationship between the redox potential of cytochromes *c* and the chemical shifts of the resonances of the methionine ligand in the spectrum of ferrocytochrome *c*. As stated before these resonances are shifted upfield to a region outside the main protein spectrum envelope by the ring-current effect of the haem. The induced shifts of these resonances, being dependent on their distance to the haem, increase with the decrease of the iron-sulfur bond lenghts. Thus, the correlation was interpreted as being due to an increase in the ability of the sulfur to donate electrons to the iron, leading to a stabilization of the oxidized state and hence a decrease in the redox potential.

A pH dependence of the mid-point oxidation reduction potential of the haem in the physiological pH range, has been reported for several prokaryotic cytochromes (41). This property was probed by an NMR study of cytochrome c_2 from *Rhodospirillum rubrum* (42) and cytochrome c_{551} from *Pseudomonas aeruginosa* (43) over a wide pH range. Two different ionization constants, $pK_a^{red} \simeq 6.2$ and $pK_a^{ox} \simeq 7.3$, were correlated to the pH dependence of the mid-point oxidation--reduction and associated with the ionization of a haem propionate (43). However, there is still some discussion for the possible origins of these ionizations (42). This fine tuning of the redox properties of the haem has the capability of functionning as a control mechanism for the electron exchange (43).

3.4. Mobility of the Side Chains and Stability of Cytochrome *c*

A recent important success of NMR is its application to the study of the mobility of the side chains of amino-acid residues and in particular to study the motion involving the rotation about the C_β-C_γ bonds of the aromatic side chains (44-46). A good example of this sort of study is described for a tyrosine residue in horse heart cytochrome *c*. Analysis of the NMR spectra of ferrocytochrome *c* at different temperatures,together with the values of exchange rate calculated by cross-saturation,and a comparison of the resonances assigned to this tyrosine residue with the computer simulated spectra generated for different values of k, are consis tent with a temperature dependent flip of this tyrosine residue about its C_β-C_γ bond (29,27).

This procedure was extended to measure flipping rates of given aromatic residues at various temperatures and calculate the enthalpy and entropy of activation for their rotations. Based on this study it has been suggested that the onset of the motion for PHE 46 and TYR 48 is cooperative (14). Similarly the results obtained have been used to probe the stability of cytochrome *c* and to investigate the dynamic characteristics of different regions of this protein (14,15).

4. STRUCTURAL STUDIES

In the NMR spectrum of globular protein with a given conformation each resonance may be shifted from its random coil position by a few parts per million. These shifts caused by the particular environment of the residue, secondary shifts, are very sensitive to conformational modifications. Thus, they are very useful for structural comparisons. Again, for cytochromes and in particular in the oxidized state, the presence of the haem makes this type of study even more sensitive. To complement this type of comparative study it is sometimes necessary to follow the temperature dependence of the paramagnetic shifted resonances. Other useful methods that can be used include the binding of extrinsic paramagnetic probes, ligands and other proteins (47).

For the case of horse heart cytochrome *c* another method used was the substitution of the iron atom by cobalt. A comparison between the NMR spectra of cobalticytochrome and ferrocytochrome *c* (both diamagnetic!) was the first evidence for an oxidation linked conformation change affecting the region about ILE 57 and extending to the PHE 10 region (48,49).

A comparative NMR study of tuna and horse heart cytochrome *c* shows that the two proteins have identical polypeptide folds (16). Using the NMR spectra parameters and the X-ray data reported for tuna cytochrome *c*, Moore and Williams have carried out a qualitative comparison between the solution and crystal structures of cytochromes *c*. In general terms there is a good agreement between the two structures (16).

The number of assignments already achieved for several eukaryotic cytochromes *c* with variable degrees of amino-acid sequence homologies allowed a comparative study to be carried out. Comparative studies of the NMR spectra of horse heart and tuna cytochromes *c* where the two sequences differ in 19 positions (17) have demonstrated that the two proteins have virtually identical main-chain folds. The structures in the immediate environment of the haem groups are virtually identical. However, local conformational differences have been depicted and shown to correlate with antigenic differences between the various proteins (17). The similarities in the solution structure are also apparent when the NMR spectra of cytochrome *c* from horse, donkey, cow, dog, rabbit, chicken and pigeon are compared

Altogether,the comparison between the solution structures of various eukaryotic cytochromes *c* confirms some generally accepted points regarding the structural effects of amino-acid substitutions (17):

1. There is no conformational change between horse and donkey cytochromes *c* which differ by only one amino-acid, (THR 47 for SER 47) which have side chain groups with similar size and chemistry. The NMR spectra of these two cytochromes is virtually identical as is evidenced by the difference spectrum given in

Figure 2, of reference 17. Only a three-proton intensity resonance, due to the methyl group of THR 47, is present, proving that all the secondary shifts are virtually the same.

2. The substitution of a polar for a non-polar residue produces a more wide-spread structural difference as shown when comparing the NMR spectra of donkey and cow cytochromes *c* (see Figure 2, of reference 17). These two cytochromes differ at two positions: residue 60, LYS (donkey)→GLY (cow) and residue 89, THR (donkey)→ →GLY (cow).

3. The differences observed in spectra of cytochromes *c* with several substitutions show that the structural modifications are cooperative.

The NMR spectral parameters, together with sequence data, have also allowed to probe the structural homology of many eukaryotic and diverse prokaryotic *c* type cytochromes (16). Mithocondrial cytochromes *c* are very similar to the cytochrome *c* from *Rhodospirillaceae*. For *P.aeruginosa* cytochrome c_{551} and *Euglena gracilis* cytochrome c_{552}, many features of the residues in the haem environment are similar to the above cytochromes, but the folding of the polypeptide chain is different. The haem environments of low-potential cytochromes, e.g., cytochrome c_{553} from *Desulfovibrio vulgaris*, show important differences when compared to the other cytochromes *c*.

All the *c* type monohaem cytochromes analysed so far have histidine and methionine residues as axial ligands. However, the location of this methionine residue may vary along the sequence.

5. ELECTRON EXCHANGE MECHANISMS

The difference between the spectra of oxidized and reduced cytochrome has been useful to determine the electron exchange rate. A limit for the rate can usually be obtained from the frequency difference of the resonance lines due to one and the same nucleus in the two oxidation states. Under suitable conditions the absolute values of the exchange rate can also be obtained.

There are three methods available for such studies, based respectively in measurements of line broadening, saturation transfer and spin-latice relaxation times (T_1) performed in solutions containing a mixture of the two oxidation states. The methods and relevant equations have been reviewed by Gupta and Mildvan (50).

The line broadening method can be used for suitable conditions of both fast and slow exchange conditions. It has been applied to determine the self exchange rate of cytochrome *c* ($k=2\times10^4$ $M^{-1}s^{-1}$) (51).

The saturation transfer method is applicable for the slow--exchange region provided that the electron exchange rate is of the order of $1/T_1$. It has the advantage of being applicable even when no appreciable broadening of resonance lines is observed and when one of the resonances is overlapping with others. The basis of the method is to saturate the resonance of one of the oxidation states by applying a selective pulse of radiofrequency radiation and measure the induced reduction of intensity in the resonance due to the same nucleus in the other oxidation state. Knowing the relevant spin-lattice relaxation time, the exchange rate can then be obtained (52,53). This method as well as that based on the measurement of T_1 only, has also been used to study the electron transfer rates between the two oxidation states of cytochrome *c* (51).

The application of these methods is fairly straightforward for cytochromes with only one haem and thus only two oxidation states. However, due to the resolution obtained by NMR it is sometimes possible to identify the intermediate oxidation states present in the redox equilibria of multihaem cytochromes (54,55). This is the case observed for the multihaem cytochrome c_3 (M.W.14,000 daltons) isolated from *Desulfovibrio gigas*. This cytochrome has four *c* type haems with different redox potentials. The haems are covalently bound to the polypeptide chain by two cysteinyl residues and has two HIS residues as axial ligands. The four haem irons are low-spin being paramagnetic in the oxidized form Fe(III) and diamagnetic in the reduced one Fe(II). The NMR analysis of a redox titration of this cytochrome will be used to illustrate the application of the method to study mechanisms of electron transfer.

In the redox equilibria between different oxidation states of multiredox center proteins such as cytochrome c_3, there are two one-electron transfer mechanisms to be considered: an intermolecular between haems of different molecules, and an intramolecular, between haems of the same molecule. The equilibria between the states of the same oxidation step is reached by intramolecular electron exchange. However, the equilibria between states of different steps involve an intermolecular electron exchange mechanism.

In the redox titration of the tetrahaem cytochrome c_3 (Figure 2) starting from the fully reduced state and going to the fully oxidized state, various populations of the possible 14 intermediate oxidation states are obtained. Four of these intermediate states have one of the haems oxidized, four have three of the haem oxidized, and six have two haems oxidized. The oxidation states can be grouped in 5 steps of different degrees of oxidation, according to the number of haems oxidized, as shown in Figure 2.

Since the redox potentials of the four haems are not widely different, (54), the oxidation of the fully reduced molecules does not proceed in the order of obtaining all the molecules with the

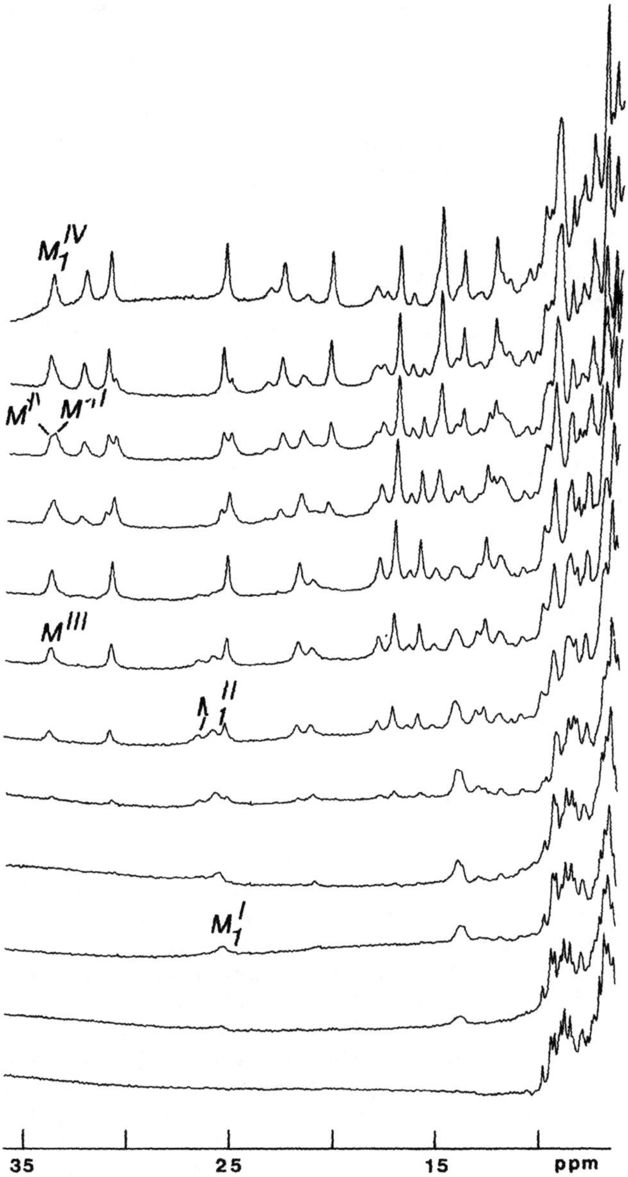

Figure 2. Oxidation reduction titration pattern of *D.gigas* cytochrome c_3 starting with the fully reduced spectrum (bottom), at 273 K, pH*=8.0 and protein concentration ≃3 mM. The Figure shows only the low-field region of the 300 MHz NMR spectra. The M_I^{III} and M_I^{IV} resonances are not well resolved but can be separated at different experimental conditions (see Figure 3 of reference 55).

haem with lowest redox potential (haem 1) oxidized, then all the molecules with haem 1 and 2 oxidized, and so on for haems 3 and 4. The populations of the sixteen states are governed by the 32 Nernst equations of the different redox equilibria. At a given stage in the redox titration, different amounts of intermediate steps of oxidation are present. The populations of the oxidation states can be calculated from the combination of all the Nernst equations.

Table 2
Oxidation states of cytochrome c_3

	Step 0	Step I	Step II	Step III	Step IV
Number of haems oxidized	0	1	2	3	4
Number of oxidation states in each step	1 (fully reduced)	4	6	4	1
Resonance due to haem methyl group 1 (see Figure 2)	M_1^0 *	M_1^I	M_1^{II}	M_1^{III}	M_1^{IV}

*At ≃ 3 ppm (not shown in Figure 2).

Similarly to the case of monohaem cytochromes, the resonances due to the same nucleus in different oxidation states (in slow exchange) can be assigned by saturation transfer methods. The assignments are exemplified for the case of one methyl group of haem 1 of *D.gigas* cytochrome c_3 (Figure 2 and Table 2). The different shifts observed for the same methyl group in the different steps of oxidation result from the cooperative nature of the total paramagnetic effect induced by the oxidation of the different haems. Each haem methyl group is shifted by two types of paramagnetic shifts. These are:those due to the intrinsic effect induced by the ferric ion of its own haem, Δ_{int},of contact and pseudo-contact origin, and those due to extrinsic effects induced by the neighbour haems, Δ_{ext}, of pseudo-contact origin only (10). It is also important to notice that the Δ_{int} being dominated by the contact contribution are usually quite large (10). The Δ_{ext} are much smaller and, depending on the specific structure,may be negligible. The total paramagnetic shift of any oxidation state depends on the oxidation state of each particular haem. Thus, a network of cross--saturation experiments, performed at different stages of oxidation is necessary in order to identify each set of resonances (55).

Once the assignments are completed,analysis of Figure 2 can be used to probe the kinetics of intermolecular electron exchange. This figure shows the low-field region of the NMR spectra of solutions at increasing values of redox potentials, starting with the spectra of the fully reduced solution (lower spectrum). The resonances were earlier assigned to be due to haem methyl groups and the identified resonances (M_1^i) were shown to belong to the same haem methyl group by saturation transfer experiments. Resonance M_1^I appears at the begining of the titration, then grows through a maximum of intensity and disappears. The large paramagnetic shift of this resonance (the resonance due to the fully reduced state, M_1^0 appears at approx. 3 ppm and is not shown),together with the fact that it appears very early in the reoxidation,indicates that M_1^I is due to a methyl group of haem 1. M_1^{II} appears after M_1^I and as is true for M_1^{III} follows a similar pattern to that of M_1^I. M_1^{IV} appears at the latest stage of the titration and grows to its final intensity in the spectrum of the fully oxidized solution (top spectrum). A similar behaviour can be observed for other haem methyl resonances shown in Figure 2.

The profile described above implies that the five M_1^i resonances are due to the five steps of increasing oxidation level present during the redox titration (M_1^0 is in the diamagnetic region at ≃3 ppm). For a 3 mM sample at 273 K, the individual resonances have a constant shift showing that the intermolecular electron exchange rate between the steps is slow in the NMR time scale. The upper limit for the intermolecular exchange is $k_{inter} < 5 \times 10^4\ M^{-1}\ s^{-1}$.

The analysis done above is however insuficient to probe the intramolecular electron exchange resulting from the equilibria between the oxidation states within each step. Even when this exchange is fast, the chemical shift of the average resonances remains constant. Indeed the ratios of these populations remain constant throughout the titration, as they are only dependent on the relative values of the midpoint redox potential of the haems. By changing the pH of the solution these population ratios can be changed, as the absolute redox potentials of the haems is pH dependent in a similar way as previously described for *P.aeruginosa* cytochrome c_{551}. The chemical shifts of the assigned resonances to the intermediate oxidation steps,show a much larger pH dependence than the resonance of the same methyl group in the fully oxidized state. The large shifts observed are a consequence of a large change in the relative contribution of states with intrinsic shifts and prove that the resonances are weighted average due to fast intramolecular electron exchange between at least two haems. A full analysis of the redox titration, performed at different pH values, indicates that the four haems can be grouped in two pairs and that for each pair there is fast intramolecular electron exchange ($k_{intra} > 5 \times 10^4\ s^{-1}$).

6. PROTEIN-PROTEIN INTERACTION

Specific interactions between proteins can be followed by NMR. An example of this type of studies is that of the interaction between cytochrome c_3 and other electron transfer proteins (54,56).

The titration of *D.gigas* cytochrome c_3 with ferredoxin shows that there is an alteration of the chemical shifts and the linewidths of only a few of the resonances of the spectrum of the cytochrome. The perturbations are a consequence of a specific complexation which, being distroyed by increasing the ionic strenght of the solution, should be of electrostatic origin.

Another process to probe this interaction is by following the NMR spectra during a redox titration of a solution with both proteins. A comparison of the NMR spectra of such a titration with those of an idential titration with cytochrome c_3 alone, shows that the electron exchange rate between the haems of different molecules (intermolecular electron exchange) is altered. Indeed, contrary to what is shown in Figure 2, the resonances assigned to the intermediate oxidation steps (e.g., M_1^{I}, M_1^{II} and M_1^{III}) are not observed in the spectra of the bound protein. Only one resonance is observed for each haem methyl group. This resonance starts appearing very broad. As the sample is more and more oxidized it sharpens up and moves to its final position, indicating that the intermolecular exchange is slower than for cytochrome c_3 alone.

Acknowledgements: We thank Drs.J.LeGall, G.Moore, and R.J.P.Williams for discussions and NIH (GM 25879)for support.JV thanks MEC (Spain). AVX thanks INIC,JNICT and the Gulbenkian Foundation (Portugal).

REFERENCES

1. Campbell,I.D., *et al.*: 1973, J.Mag.Res. 11, p.172.
2. Jenson,J.P., *et al.*: 1973, J.Am.Chem.Soc. 95, p.618.
3. Campbell,I.D., *et al.*: 1975, FEBS Lett. 57, p.96.
4. Campbell,I.D., and Dobson,C.M.: 1975, J.Chem.Comm., p.750.
5. Gupta,R.K. and Redfield,A.G.:1970, Science 169, p.1204.
6. Hoffman,R.A. and Forsen,S.: 1966, Prog.NMR Spect. 1, p.15.
7. Noggle,J.H. and Schirmer,R.E.:1971,"The NOE", Acad.Press, N.Y.
8. Richarz,R. and Wuthrich,K.: 1978, J.Mag.Res. 30, p.147.
9. LeGall,J. *et al.*, unpublished results.
10. Wuthrich,K.: 1976, "NMR in Biological Research:Peptides and Proteins", North-Holland Publ.Co., Amsterdam.
11. Timkovich,R.: 1979,"The Porphyrins"(ed.Dolphin,D.) 7, Academic Press, N.Y., p.241.
12. Moore,G.R. and Williams,R.J.P.: 1980,Eur.J.Biochem. 103, p.493.
13. Moore,G.R. and Williams,R.J.P.: 1980,Eur.J.Biochem. 103, p.503.
14. Moore,G.R. and Williams,R.J.P.: 1980,Eur.J.Biochem. 103, p.513.
15. Moore,G.R. and Williams,R.J.P.: 1980,Eur.J.Biochem. 103, p.523.

16.Moore,G.R. and Williams,R.J.P.:1980,Eur.J.Biochem. 103, p.533.
17.Moore,G.R. and Williams,R.J.P.:1980,Eur.J.Biochem. 103, p.543
18.Wutrich,K.:1969, Proc.Natl.Acad.Sci.USA 63, p.1071.
19.McDonald,C.C. *et al.*: 1969,Biochem.Biophys.Res.Commun. 36, p.442.
20.Redfield,A.G. and Gupta,R.K.:1971, Cold Spring Harbour Symp. Quant.Biol. 36, p.405.
21.McDonald,C.C. and Phillips,W.D.:1973, Biochem. 12, p.3170.
22.Keller,R.M. and Wuthrich,K.:1978, Biochim.Biophys.Acta 533,p.195.
23.Stellwagen,E. and Shulman,R.G.:1973, J.Mol.Biol. 75, p.683.
24.Patel,D.J. and Canuel,L.L.:1976,Proc.Natl.Acad.Sci.USA 73,p.1398.
25.Keller,R.M. *et al.*:1973, FEBS Lett. 36, p.151.
26.Cohen,J.S. *et al.*:1974, J.Biol.Chem. 249, p.1113.
27.Dobson,C.M. *et al.*:1975, FEBS Lett. 51, p.60.
28.Moore,G.R. and Williams,R.J.P.:1975, FEBS Lett. 53, p.334.
29.Campbell,I.D. *et al.*:1976, FEBS Lett. 70, p.96.
30.Keller,R.M. and Wuthrich,K.:1978, Biochim.Biophys.Acta 53,p.195.
31.Gupta,R.K. and Redfield,A.G.: 1970, Science 169, p. 1204.
32.Keller,R.M. and Wuthrich,K.:1978, Biochem.Biophys.Res.Commun. 83, p.1132.
33.Dickerson,R.E. *et al.*:1976, J.Mol.Biol. 100, p.473.
34.Dickerson,R.E. and Timkovich,R.:1975, "The Enzymes" (ed.Boyer, P.) 11, Academic Press, N.Y., p.397.
35.McDonald,C.C. *et al.*: 1974, Biochem. 13, p.1952.
36.Xavier,A.V. *et al.*:1978, Nature 275, p.245.
37.Takano,T. *et al.*:1977, J.Biol.Chem. 252, p.776.
38.Almassy,R.J. *et al.*:1978, Proc.Natl.Acad.Sci.USA 75, p.2674.
39.Senn,H. *et al.*:1980, Biochem.Biophys.Res.Commun. 92, p.1362.
40.MooreG.R. and Williams,R.J.P.:1977, FEBS Lett. 79, p.229.
41.Pettigrew,G.W. *et al.*:1975, Biochim.Biophys.Acta 430, p.197.
42.Smith,G.M.:1979, Biochem. 18, p.1628.
43.Moore,G.R. *et al.*:1980, Biochim.Biophys.Acta 590, p.261.
44.Campbell,I.D. *et al.*:1975, Proc.R.Soc.London, Ser.B189, p.503.
45.Wuthrich,K. and Wagner,G.:1975, FEBS Lett. 50, p.265.
46.Cave,A. *et al.*:1975, FEBS Lett. 65, p.190.
47.Williams,R.J.P. *et al.*:1977, "Biological Aspects of Inorganic Chemistry" (eds. Addison,A.W., Cullen,W.R., Dolphin,D. and James,B.R.) J.Wiley, N.Y., p.369.
48.Moore,G.R. *et al.*:1980, J.Inorg.Biochem. 12, p.1.
49.Moore,G.R. *et al.*:1980, J.Inorg.Biochem. 13, p.347.
50.Gupta,R.K. and Mildvan,A.S.:1978, Meths.Enz. 49, 322.
51.Kowalsky,A.:1965, Biochem. 4, p.2382.
52.Gupta,R.K. and Redfield,A.G.:1970, Biochem.Biophys.Res.Commun. 41, p.273.
53.Gupta,R.K. and Koenig,S.H.:1971, Biochem.Biophys.Res.Commun. 45, p.1134.
54.Moura,J.J.G. *et al.*:1977, FEBS Lett. 81, p.275.
55.Moura,I. *et al.*:1980, Cienc.Biol.(Portugal) 5, p.189.
56.Moura,I. *et al.*:1980, Cienc.Biol.(Portugal) 5, p.195.

SECTION D

EXAMPLE OF AN OXYGEN CARRIER

SUBSTITUTION AND ELECTRON TRANSFER PROCESSES IN HEMERYTHRIN

Patricia C. Harrington and Ralph G. Wilkins
Chemistry Department, New Mexico State University, Las Cruces, NM 88003

Barry B. Muhoberac and David C. Wharton
Biochemistry Department, The University of Texas Health Science Center, San Antonio, TX 78284

The kinetics and mechanisms of reactions at the binuclear iron site of hemerythrin are discussed. These include substitution, redox and configuration changes involving deoxy, oxy, semi-met and met forms of the protein. Emphasis will be on results with the octameric and monomeric forms of the protein from Themiste zostericola.

I. INTRODUCTION

Hemerythrin is an easily isolable iron-containing protein which transports and/or stores dioxygen in certain marine organisms. The protein has been found in an annelid, two brachiopods and two priapulids (1). However the sipunculids are the major source of hemerythrin which has been studied. One of the fascinations of the protein is in the different polymeric forms in which it can exist. All sipunculids examined have, in the coelomic fluid, erythrocytes loaded with the protein which in most species so far examined is octameric, but sometimes trimeric (2,3) and in one instance dimeric and tetrameric (4). From the retractor muscle of *Themiste zostericola*, the protein has been characterized as a monomer (5). The monomer (6) and the subunits of the trimer (7) and octamer (8) are remarkably similar in tertiary structure, having a M.W. of about 13,500 daltons and each unit containing one binuclear iron site. There is no porphyrin ring and the irons are coordinated only to amino acids, some of which, as well as probably an oxy group, form the bridging atoms (9). The extensive structural studies have been well reviewed (6,8,10, 11), see also contribution by L.C. Sieker and R. Stenkamp).

H. B. Dunford et al. (eds.), The Biological Chemistry of Iron, 145–160.

In the form of the protein which picks up O_2, both irons are in the oxidation state +2. The stoichiometry of dioxygen addition is 2Fe(II) : $1O_2$, and there is strong evidence that in the product both irons are +3, combined with a peroxide moiety (10). Both deoxy- and oxy-hemerythrin react with oxidants (e.g. $Fe(CN)_6^{3-}$) to give a new form of the protein, methemerythrin, in which a binuclear Fe(III) unit is present. This is no longer O_2-sensitive but combines with a number of univalent anions (12,13). In 1978 (14) evidence for a semi-reduced form, containing formally one iron +2 and one iron +3, was obtained from reduction of the met form by dithionite. One electron oxidation of deoxyhemerythrin gave, surprisingly, a semi-oxidized form with distinctive characteristics from those of the other mixed oxidation state (15,16). EPR spectroscopy at liquid He temperatures has been invaluable in characterizing the species and the nature of the changes which they undergo (17,18). Electronic spectroscopy has been used extensively for the measurement of rates (13-16, 18-20).

II. KINETICS AND MECHANISMS OF REACTIONS OF THE BINUCLEAR IRON SITE IN HEMERYTHRIN

This contribution is concerned with the coordination chemistry of the binuclear iron site in hemerythrin. The dynamic aspects will be emphasized, in context with other observations. The reactions with which we shall be concerned, and in fact which make up the major part of hemerythrin chemistry so far explored, are best discussed by using the scheme shown in Figure 1. It is assumed

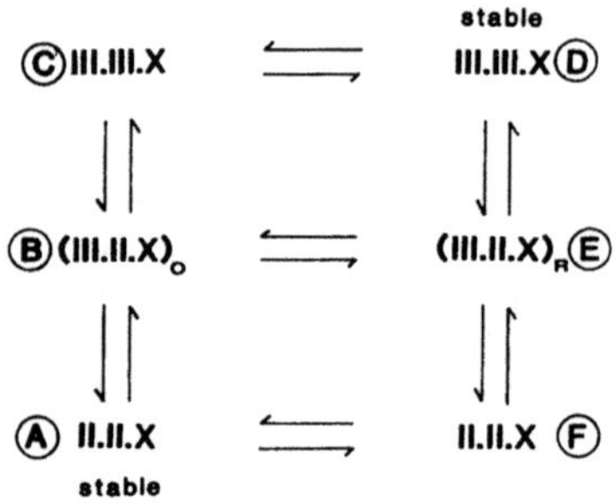

Figure 1. Scheme for reactions of the binuclear site in hemerythrin.

that there are two types of binuclear iron units, one associated with the thermodynamically stable form of deoxyhemerythrin, designated (A), and the other associated with the stable met (D). In the native protein X is assumed to be a replaceable H_2O or OH^- group or blank. Otherwise, X may represent an anion in (A) - (E) or the dioxygen molecule in (A). The mixed oxidation states are represented by $(III.II.X)_O$, Ⓑ and $(III.II.X)_R$ Ⓔ, these being forms produced respectively by one electron oxidation of stable II.II.X and one electron reduction of stable III.III.X. An unstable form of deoxyhemerythrin, represented Ⓕ and the two linked steps have not yet been characterized. All other species and steps have been observed with X = H_2O or OH^- but not all with X = anions. Some steps are observed only with the monomeric form of the protein. In some cases the distinction between the two types of structure may vanish. Rate, but not fundamental, differences are observed between octameric forms of *T.zostericola (T.z.)*, *P.gouldii (P.g.)* and *T.dyscritum (T.d.)*. The detailed mechanism of some of the steps with the octamer however is profoundly different from that of the monomer.

a) Substitution Reactions (one X-group replacing another)

Ⓐ, X = H_2O, OH^- $\leftrightarrow$ X = O_2. This important reaction, involving the oxygenation of deoxyhemerythrin, probably also involves an internal redox process since the product is best formulated as an $[Fe(III)]_2$-peroxo complex (10). The system contrasts sharply with hemoglobin in showing little sign of cooperativity (1). The kinetics have been studied by flow and temperature-jump techniques (20-23). Results for *T.z.* are shown in Table 1 (23). Those for *P.g.* are quite similar to octameric *T.z.* (20). The main features are (i)

Table 1. Rate data for formation (k_{on}) and dissociation (k_{off}) of oxyhemerythrin *(T.z.)* at 25°C, pH 8.2, I = 0.10M (23)

Species	$10^{-6}k_{on}$ $M^{-1}s^{-1}$	k_{off} s^{-1}	$\Delta H^{\neq}_{off}$	$\Delta S^{\neq}_{off}$	$10^{-5}K^a$ M^{-1}	ΔH^b	ΔS^b
$(T.z.)_8$	7.5(9.0)^c	82	19^d	+15^d	0.9(1.3)^b	-16	-30
$(T.z.)$	78(80)^c	315	17	+9	2.5(1.5)^b	-13	-20

[a]Value obtained by kinetics, k_{on}/k_{off}. [b]Value obtained by spectral analysis. [c]Recombination after laser photodissociation (25). [d]Ref. 22.

a single relaxation for octameric and monomeric protein (ii) good agreement in equilibrium constants K calculated from kinetic data and measured spectrally on equilibrated solutions; the values of K are slightly larger for the monomer than the octamer but the deoxymyohemerythrin-O_2 system is considerably more labile (larger k_{on} and k_{off} values) than the deoxyhemerythrin-O_2 equilibrium. (iii) the high rate constant and low value of $\Delta H^{\neq}$ (4 kcal.mole^{-1}) for the oxygenation of deoxymyohemerythrin strongly suggests that H_2O replacement is not involved, i.e. the process is additive rather than substitutional in nature (23).

The laser photolysis of oxyhemerythrin *(Sipunculus nudus)* in water and glycerol-water has been reported (24). The second-order rate constant for recombination of deoxy and O_2, after photodissociation, is $2.9 \times 10^7 M^{-1}s^{-1}$, quite close to that obtained by temperature-jump (21). We have examined (25) laser photolysis of oxyhemerythrin and oxymyohemerythrin and reformation rates following photodissociation (Figure 2) are in excellent agreement with those obtained by temperature-jump (Table 1). *If* the photodissociation

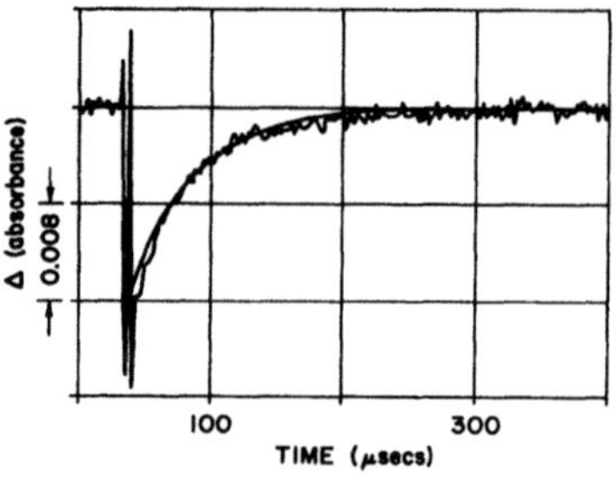

Figure 2. Laser photolysis of oxymyohemerythrin (~0.3mM) in air saturated solution. Monitoring wavelength 490 nm. Pulsing wavelength 500 nm. Dye Laser, 500 mJ, 400 ns pulse width.

(<100nsec.) produces an unstable form of deoxyhemerythrin then it is apparent that it reverts to the stable form prior to the oxygenation, i.e. $< 50\mu sec$. This assumes that the unstable form binds to O_2 with kinetic characteristics different from those of normal deoxyhemerythrin.

Ⓐ, X = H_2O, OH^- ↔ X = N_3^-, NCO^-. Only a *very small* spectral change accompanies the addition of these anions to deoxyhemerythrin. The interaction can however be easily seen in the rapid bleaching of oxyhemerythrin when N_3^- and NCO^- are added. The study of the kinetics of the bleaching yields values for the binding and rate constants for anation (26). The binding site is unknown but it is reasonable to suppose that it is close to, or at, that of the O_2 which it replaces.

Ⓑ or Ⓔ, X = H_2O, OH^- ↔ anions. Since the semi-met forms contain one Fe(III) for each active site it is not unexpected that they form anionic adducts. Quantitative studies with N_3^- show that they are very much less stable than those of the met but much more labile (larger k_{on} and k_{off} values), Table 2 (19). Anation is best effected by using 0.1M solutions of SCN^-, F^-, NCO^-,

Table 2. Reaction parameters for the formation (k_{on}) and dissociation (k_{off}) of met- and semi-methemerythrin adducts with N_3^- at 25°C, pH 8.2, I = 0.15M (19)

Species	k_{on} $M^{-1}s^{-1}$	k_{off} s^{-1}	K M^{-1}
met*(T.z.)*$_8$	1.4	$<10^{-4}$	$>10^4$
met*(T.z.)*	10 (14)[a]	$<10^{-3}$	$>10^4$
(semi-met)$_R$*(T.z.)*$_8$	33	~0.01	~4 x 10^3
(semi-met)$_O$*(T.z.)*$_8$	3.2 x 10^2	0.2	1.6 x 10^3

[a]Value for unstable metmyohemerythrin

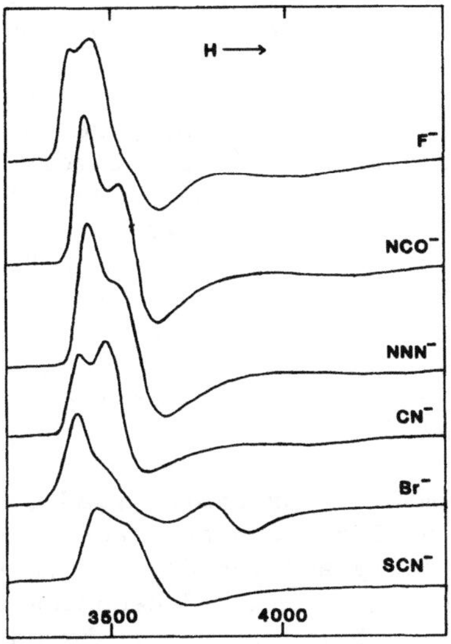

Figure 3. EPR of semi-met anion adducts. Conditions: (semi-methemerythrin)$_R$ *(T.z.)* 0.3mM treated with X^- (1.0M). Stored for ≥ 30 min. at 25° after adding X^- (this allows complete disproportionation of any uncomplexed semi-met). Instrument settings: microwave power 5mW; field modulation 3.2 gauss; receiver gain 2.5 x 10^3; microwave frequency 9.22MHz; time constant 0.35; field center 3500 gauss; field range ± 1000 gauss; scanning speed 8 min. Temp. 10°K.

Br^- and CN^-, or lower concentrations of N_3^- ion. The EPR spectra at liquid He temperatures are shown in Figure 3 (27). The form of the EPR spectrum for the Br^- adduct suggests that it might be a mixture. Generation of the N_3^- and SCN^- adducts from either semi-met form gave identical electronic and EPR spectra, but there were significant differences in the semi-met azide EPR from *P.g.* and *T.z.* monomer (g = 1.90, 1.81 and 1.49) and $(T.z.)_8$ (g = 1.94, 1.85 and 1.57) (17,27). Kinetic studies of (semi-metmyo) interactions with SCN^- are difficult to interpret (18).

Ⓓ, X = H_2O, OH^- ↔ anions. The ability of methemerythrin to combine with a number of univalent anions has been known for some years (12). The kinetics of formation and dissociation of the adducts have been measured and a comparison with the corresponding iron(III) complexes has been discussed (12,13). Data with N_3^- ion are shown in Table 2. It is now clear that N_3^- is bound to only one Fe in the binuclear unit (9) and any previous arguments based on bridging N_3^- must be abandoned. The reaction of SCN^- at pH 6.3 with metmyohemerythrin and methemerythrin *(T.z)* is very slightly biphasic. At pH = 8.2, the biphasic character is marked. The fast step is second order (k = 530 $M^{-1}s^{-1}$ and 26.4 $M^{-1}s^{-1}$, for monomer and octamer *T.z.)*. The slower step is first-order (k = 3.9×10^{-3} s^{-1} and $5 \times 10^{-3}s^{-1}$) independent of SCN^- concentration. The spectrum of the intermediate resembles quite closely that of the azide adduct. A (not convincing) explanation is that binding is initially through the N of SCN^-, and that this then undergoes linkage isomerism to finish as S-bound. Alternatively, the first-order transformation could be a conformational change following SCN^- binding to Fe(III), but it is difficult then to rationalize the *absence* of biphasic character to N_3^- binding at pH 8.3 (28).

b) Net Redox Reactions

Table 3 contains data for a number of different redox reactions.

Ⓐ ↔ Ⓑ Of a number of oxidants tried (19) only $Co(terpy)_2^{3+}$ and Fe(III) cyano complexes convert deoxyhemerythrin to $(semi\text{-}met)_O$ [X = H_2O or OH^-] cleanly and rapidly. With the others, parallel oxidation of the protein or marked decomposition of the oxidant in the presence of the protein occurred. By using a series of iron(III) complexes of the type $Fe(III)(CN)_5X^{n-}$ with redox potentials varying from 300 to 600mV it was possible to establish spectrally the potential for Ⓑ → Ⓐ as 310mV (octameric *T.z.*, pH 8.2 and 6.3) (29). The $(semi\text{-}met)_O$ product is characterized by its electronic and EPR spectra (15-17). These are identical whether generated from monomeric or octameric protein. $(Semi\text{-}met)_O$ is reduced to deoxyhemerythrin, rapidly by excess $Fe(II)(CN)_5thiourea^{3-}$ and dithionite and slowly, but completely, by reduced DCIP (29).

Table 3. Rate data for a number of net redox reactions of octameric and monomeric hemerythrin species from *T.z.* at 25°C, pH 8.2 and I = 0.15M

Transformation	Reactants		k(monomer) $M^{-1}s^{-1}$	k(octamer) $M^{-1}s^{-1}$
(A) ↔ (B)	deoxy	$Fe(CN)_6^{3-}$	3×10^6 [a]	10^5 [b]
	$(semi\text{-}met)_O$	$S_2O_4^{2-}$	-	4×10^5 [bc]
(A) → (D)	deoxy	H_2O_2	97 [de]	6.6 [dc]
(B) → (C)	$(semi\text{-}met)_O$	$Fe(CN)_6^{3-}$	5.7×10^2 [a]	1×10^2 [bf]
(B) → (D)	$(semi\text{-}met)_O$	$Fe(CN)_6^{3-}$	-	1.3×10^{-3} [ghi]
	$(semi\text{-}met)_O$	H_2O_2	-	27.5 [j]
(D) ↔ (E)	met	$S_2O_4^{2-}$	1.1×10^6 [ka]	4.7×10^5 [kl]
	met	e^-_{aq}	4.5×10^9 [1]	3.9×10^9 [1]
	met	red.MV		2.5×10^6 [1m]
	$(semi\text{-}met)_R$	$Fe(CN)_6^{3-}$	$\geq 6 \times 10^6$ [a]	4×10^5 [b]
	$(semi\text{-}met)_R$	H_2O_2	-	12.3 [j]
(D) → (A)	met	red.Mb	2.0 [no]	5.6 [no]
(E) → (A)	$(semi\text{-}met)_R$	red. Mb	-	10 [j]
	$(semi\text{-}met)_R$	$S_2O_4^{2-}$	2.5×10^{-4} [ag]	1.1×10^{-3} [i]

[a]Ref. 18. [b]Ref. 15,16. [c]Reaction of SO_2^- with *P.g.* at 8°C. [d]Reaction at pH = 9 (0.5M Tris). [e]Ref. 22. [f]Only observed with *P.g.* at pH = 6.3. [g]s^{-1}. [h]$\Delta H^{\neq}$ = 21.9kcal·mole^{-1}, $\Delta S^{\neq}$ = +1. [i]Ref. 19. [j]Ref. 29. [k]Reaction of SO_2^-. [l]Ref. 31. [m]*P.g.* [n]Ref. 34. [o]pH = 7.0.

(A), X = O_2 ↔ (B) Oxyhemerythrin is oxidized by iron(III) cyano complexes to $(semi\text{-}met)_O$. Detailed kinetic studies, using *T.z.* and *P.g.* octamer show that the oxidation by $Fe(CN)_6^{3-}$ occurs through the deoxy species (16). Since the oxidation potential for the $(semi\text{-}met)_O$/oxyhemerythrin couple is more positive than for the (semi-met)$_O$/deoxyhemerythrin couple, it is easier to establish with the oxy equilibria (rather than complete reaction). The value of E° for the former couple (460mV, *T.z.*) is consistent with that for the latter (310mV) (29). The (semi-met)$_O$ species produced by oxidation of oxyhemerythrin has a similar EPR spectrum and kinetic characteristics to that obtained from deoxyhemerythrin (16,27).

Ⓐ → Ⓓ. Deoxyhemerythrin reacts with H_2O_2 to give a methemerythrin form. This, and its azide adduct, have spectra somewhat different from normal methemerythrin (22). It is uncertain whether there is a (semi-met)$_O$ intermediate, since this reacts more rapidly with H_2O_2 (Table 3). Oxyhemerythrin is oxidized by H_2O_2, like $Fe(CN)_6{}^{3-}$, *via* the deoxy species (22).

Ⓐ, X = O_2, anion → Ⓓ, X = anion. In the absence of anions, oxyhemerythrin spontaneously autoxidizes to methemerythrin very slowly ($t_{1/2}$ ~4 hours, at pH = 7). The process is markedly catalyzed by a number of univalent anions with the result that oxyhemerythrin is converted into the corresponding metanion adduct (26). The reaction probably does not proceed *via* (semi-met)$_O$ since this + N_3^- + O_2 remains unchanged for hours. The following mechanism is suggested:

$$Fe(II)\cdot Fe(II)\cdot O_2 + X^- \overset{K}{\rightleftharpoons} Fe(II)\ Fe(II)\cdot X^- + O_2 \qquad [1]$$

$$\searrow k_B \qquad\qquad k_A \swarrow$$

$$Fe(III)\cdot Fe(III)\cdot X^- + O_2^{2-} \qquad [2]$$

with the relative contribution of k_A and k_B unknown. If k_B is considered a nucleophilic replacement of O_2^{2-} by X^-, the absence of a semi-met intermediate can be readily understood. A marked bleaching of oxyhemerythrin in [1] is observed only for X^- = N_3^- or NCO^- and for these the observed second-order autoxidation rate constant $k = k_B$ or Kk_A. For other anions studied (Cl^-, HCO_2^-, NO_2^-, and F^-) the value of K is unknown (26). An interesting correlation between the two types of kinetic (biphasic and monophasic above) and exchange behavior has been pointed out (30). Resonance Raman spectroscopy of the met products shows that reaction of oxyhemerythrin in the presence of N_3^-, CN^- or NCO^- leads to incorporation of ^{18}O from solvent water into the product, but no such incorporation occurs when autoxidation occurs in the presence of F^-, SCN^-, NO_2^- or $Fe(CN)_6^{3-}$. The ^{18}O incorporation is assigned to an iron-OH^-, H_2O or μ-oxo bridge stretching vibration (30).

Ⓑ ↔ Ⓒ. (Semi-met)$_O$ is oxidized directly by $Fe(CN)_6^{3-}$ to an unstable met form only with monomeric *T.z.* The product transforms to the stable met species (see Section IId), and is easily reduced by dithionite to deoxyhemerythrin (Ⓒ → Ⓐ) possibly *via* the species Ⓑ (18).

(B) → (D). Octameric $(semi\text{-}met)_O$ hemerythrin is oxidized by $Fe(CN)_6^{3-}$ and $Coterpy_2^{3+}$ to methemerythrin slowly via a disproportionation reaction (see Section IIc). The first-order rate constant is independent of the concentration of the oxidant (16). The oxidation of $(semi\text{-}met)_O$ by H_2O_2 is a direct second-order reaction (Table 3) but as with deoxy, the characteristics of the met product are unusual. It does *not* transform to the normal methemerythrin.

(B), $X = N_3^-$ → (D) $X = N_3^-$. $(Semi\text{-}met)_O\ N_3^-$ reacts with $Fe(CN)_6^{3-}$ in a biphasic manner at pH 6.3 (16). The first phase is a second-order oxidation with elimination of coordinated azide (k = (7 - 9) x $10^2 M^{-1}s^{-1}$). The second phase is the reaction of methemerythrin, resulting from the first step, with N_3^- ion. The oxidation by H_2O_2 is similar (first step, k = 0.7 - $1.0M^{-1}s^{-1}$ pH 8.2) with methemerythrin-azide as the final product (29). The reaction has not been studied with myohemerythin.

(D) ↔ (E). A number of reductants convert methemerythrin and metmyohemerythrin to $(semi\text{-}met)_R$ rapidly, before subsequent slow reduction to the deoxy form can occur. These include dithionite (14,18,19), reduced DCIP and reduced methylviologen and the radicals e_{aq}^- and CO_2^- generated by pulse radiolysis (31). Rate constants for some of these reductions are collected in Table 3. Oxidation potentials for met/$(semi\text{-}met)_R$ have been measured using DCIP (E° = 120mV at pH 8.2 (32)) The conversion (D) → (E) photochemically using a riboflavin/EDTA mixture (33) takes place much quicker than the complete reduction (Figure 4). Quite surprisingly, Cr^{2+} ion which we were hoping to

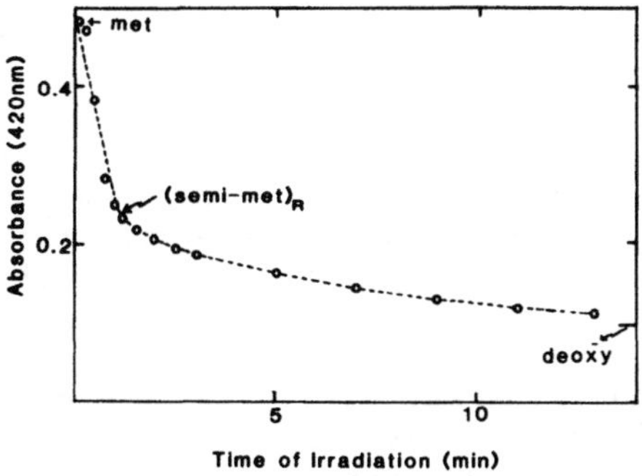

Figure 4. Absorbance vs. time of irradiation of methemerythrin (*(T.z.)* protein (0.26mM), riboflavin (4μM) and EDTA (4.9mM) irradiated at distance of 20 cm by a 300W projector lamp.

use as an inner-sphere reductant does *NOT* reduce, at pH = 6.3 over a long period, either methemerythrin or methemerythrin-N_3^-, both from *T.z.* (29). All reductants with *T.z.* and *P.g.* proteins give a (semi-met)$_R$ product with identical g values 1.95, 1.72 and 1.69 (17). That from *T.d.* has, however, an unusual appearance (Figure 5) (27). The oxidants $Fe(CN)_6^{3-}$, H_2O_2 and oxidized DCIP convert (semi-met)$_R$ directly into the met form and this is also observed with myohemerythrin (Table 3). Methemerythrin is converted slowly to deoxyhemerythrin by deoxymyoglobin ((D) → (A)) (Table 3) (34).

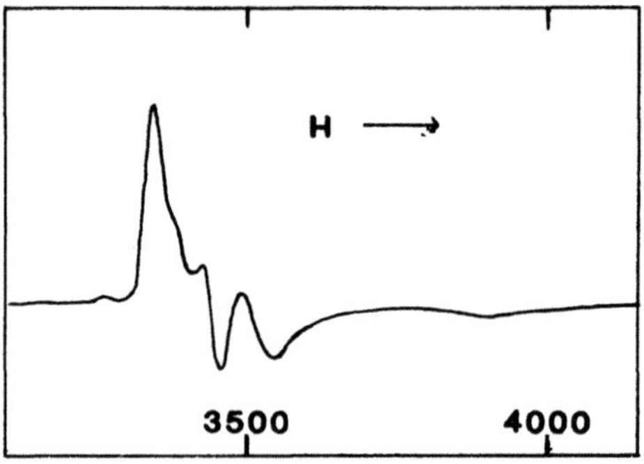

Figure 5. EPR of (semi-met)$_R$ 0.34mM, from *T.dyscritum.* Instrument settings and temperature as in Figure 3, except receiver gain was 1.0 x 10^3.

This is a uniphasic reaction and it is difficult then to understand why the likely intermediate (semi-met)$_R$ is reduced at a comparable rate (Table 3) and therefore its' incursion in the overall reaction not detected by deviations from uniphasicity.

(E), X = N_3^- → (D), X = N_3^-. The oxidation of the adduct formed from (semi-met)$_R$ and N_3^- with $Fe(CN)_6^{3-}$ and H_2O_2 have very similar characteristics and rates to those of the (semi-met)$_O$-N_3 adduct. Both azide adducts have very similar electronic and EPR spectra (see Section IIa) and these facts suggest that the azide adduct is *one* species.

(D), X = anion → (A). The conversion of methemerythrin to deoxyhemerythrin is markedly modified in the presence of anions which produce met-anionic adducts. Detailed kinetic studies show that reduction by dithionite proceeds *via* rate-determining dissociation of the anionic complex, e.g.

$$Fe(III)\cdot Fe(III)\cdot X^- \rightleftharpoons Fe(III)\cdot Fe(III) \xrightarrow{rapid} Fe(II)\cdot Fe(II) \quad [3]$$

The intermediate Fe(III)Fe(III) species cannot however be the stable met form (D) since it is reduced rapidly to deoxyhemerythrin with no sign of the slow (hours) tailing in (D) → (A) (13). The uniphasic, rapid and quantitative reduction of methemerythrin•SCN^- by dithionite is of value for the facile preparation of deoxyhemerythrin (13,30). Since the dissociation rate constant of met-azide adduct is so small ($< 10^{-4}s^{-1}$, Table 2), then it is very difficult to reduce with dithionite (days at pH ~7). However, we have found that reduction is complete within *90 minutes*, if the mixture of methemerythrin, (0.20mM) N_3^- (2.5mM) and $S_2O_4^{2-}$ (1.0mM) is subjected to illumination using a 500 watt tungsten lamp, *T.z.* octamer (29).

Ⓔ → Ⓐ. Dithionite, DCIP (reduced) and photochemical (Figure 4) reductions are all slow. With the octamer, they appear largely controlled by disproportionation (19). With the monomer, the kinetics of the slow reduction by $S_2O_4^{2-}$ cannot easily be interpreted (18). In none of these reductions is there a sign of an intermediate deoxy form Ⓕ; neither has direct oxidation of Ⓐ to Ⓔ been observed. Ⓔ, X = N_3^- ↔ Ⓐ. The reduction of the (semi-met)$_R$-N_3^- adduct by dithionite parallels closely that of the (semi-met)$_O$-N_3^- adduct (19) and suggests again that the two species may be the same.

c) Disproportionation Reactions

Ⓑ ↔ Ⓐ + Ⓓ. This reaction is observed only with the octameric forms of *T.z.* and *P.g.* It is an *intramolecular* redox reaction in which the electron transfer occurs over distances of 28-30Å with a rate constant around $10^{-3}s^{-1}$ at pH 6.3 - 8.2 (Table 4). It is easily monitored directly (spectrally) at 360 nm and other wave-lengths, and also by the addition of O_2 or SCN^- when the appearance of one of the products (deoxy or met respectively) can be followed (15,16,19). The amplitude of the EPR spectrum of (semi-met)$_O$, measured at liquid He temperatures, decreases on incubating at 25° (Figure 6) (17,27):

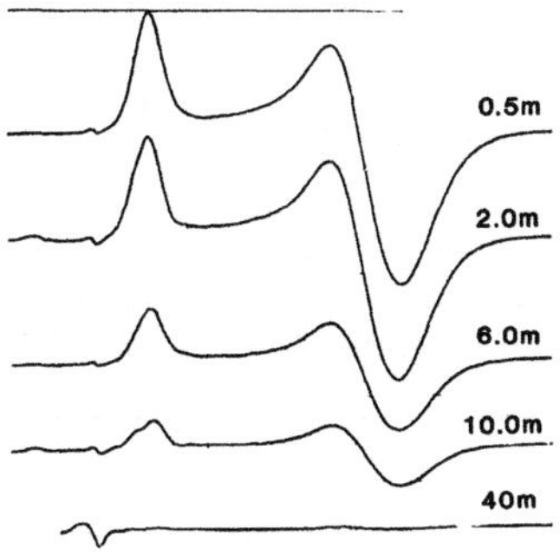

Figure 6. Loss of EPR signal of (semi-met)$_O$ (*T.z.*) at 25°, pH = 8.2. Instrument settings as in Figure 5. Semi-met prepared from 0.53mM deoxyhemerythrin and 0.54mM $Fe(CN)_6^{3-}$.

$$[Fe(III)Fe(II)]_8 \rightleftarrows [Fe(III)Fe(III)]_4[Fe(II)Fe(II)]_4 \quad [4]$$

The rate constants for disproportionation obtained by various monitoring methods are in very good agreement with each other. Disproportionation of (semi-met)$_O$ from *T.z.* and *T.d.* is complete at pH 8.2 and 25°C (19). Disproportionation of *P.g.* occurs only to the extent of 10 - 20%, a value most easily estimated from EPR measurements (19,27). The semi-met azide adduct is stable towards disproportionation for many hours (15,19).

Ⓔ ↔ Ⓐ + Ⓓ This reaction is a first-order intramolecular process with octameric protein from *T.z.*, *T.d.* and *P.g.* All rate constants are quite similar (19). Those for *T.z.* are given in Table 4. The disproportionation is a second-order intermolecular reversible reaction with monomeric protein (18):

Table 4. Rate data for disproportionation and comproportionation reactions of octameric and monomeric hemerythrin species from *T.z.* at 25°C, pH 8.2 and I = 0.15M

Reactants	k(monomer) $M^{-1}s^{-1}$	k(octamer) s^{-1}
(semi-met)$_O$ → ½met + ½deoxy	-	2.2×10^{-3} [ab]
(semi-met)$_R$ → ½met + ½deoxy	-	2.7×10^{-3} [ac]
2(semi-met)$_R$ → met + deoxy	0.9[d]	-
met + deoxy → 2(semi-met)$_R$	9.4[d]	-

[a]Ref. 19. [b]$\Delta H^{\neq} = 20.6$, $\Delta S^{\neq} = -1$. [c]$\Delta H^{\neq} = 22.9$, $\Delta S^{\neq} = +6$. [d]Ref. 18.

Table 5. Rate data for the conformational changes of (semi-metmyo)$_O$, (semi-metmyo)$_R$ and metmyohemerythrin at 25°C, pH = 8.2 and I = 0.15M (Monomeric *T.z.*) (18)

Transformation	k s^{-1}	$\Delta H^{\neq}$ kcal·mole^{-1}	$\Delta S^{\neq}$ e.u.
(semi-metmyo)$_O$ → (semi-metmyo)$_R$	1.0×10^{-2}	15.1	-17
(semi-metmyo)$_R$ → (semi-metmyo)$_O$	$\leq 10^{-3}$	-	-
*metmyo → metmyo	4.6×10^{-3}	16.8	-13

$$2(\text{semi-metmyo})_R \rightleftarrows \text{metmyohemerythrin} + \text{deoxyhemerythrin} \quad [5]$$

The decrease in the amplitude of the EPR signal due to (semi-metmyo)$_R$ after long standing indicates K = 0.19. The same equilibrium EPR spectrum can be generated by mixing equal and appropriate concentrations of metmyohemerythrin and deoxymyohemerythrin. Only with slightly aged protein are there small spectral changes accompanying reaction [5], see Figure 7, (29) and using these the rates

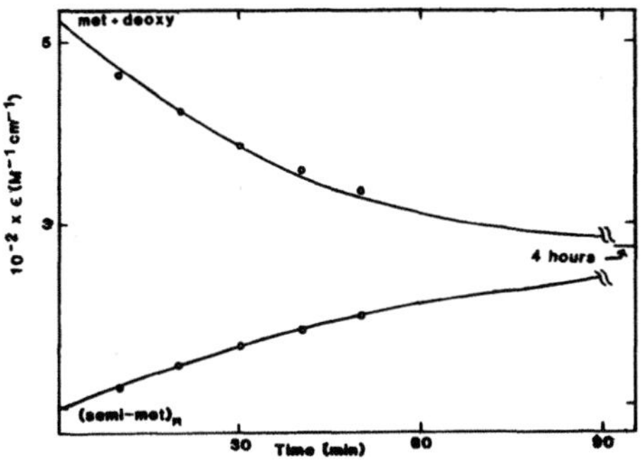

Figure 7. Disproportionation and comproportionation of (semi-metmyohemerythrin)$_R$. Total protein concentration in each direction = 57μM, at pH 8.2, I = 0.15 and 25°. Observation wavelength 500nm.

in both directions conform very well with second-order reversible kinetics (Table 4). Two EPR signals measured during the disproportionation-comproportionation reaction using *fresh* protein yielded values of k = 1.4$M^{-1}s^{-1}$ and k = 7.4$M^{-1}s^{-1}$ in good agreement with the spectal results. A discussion of the intramolecular disproportionation (19) and a comparison of the behavior of monomer and octamer has been given (18).

d) Conformational Changes

Changes from one type of binuclear site to another, without involving electron transfer, are substantiated only with myohemerythrin. No evidence yet for a type Ⓕ of deoxyhemerythrin has been obtained.

Ⓑ ↔ Ⓔ. The transformation

$$(\text{semi-metmyo})_O \rightleftarrows (\text{semi-metmyo})_R \quad [6]$$

is most easily characterized by EPR (18) and the rate most easily measured at 500nm. The (semi-metmyo)$_R$ subsequently disproportionates slowly (Section IIc). The EPR spectrum indicates that there is $\leq$ 10% of (semi-metmyo)$_O$ remaining after equilibration of [6]. On this basis, activation parameters are shown in Table 5.

Ⓒ ↔ Ⓓ. Addition of $Fe(CN)_6^{3-}$ to deoxymyohemerythrin (2:1 mole ratio) produces a species in which both irons are trivalent. This species, which has no EPR signal, changes quite quickly (Table 5) to a final product whose spectrum is identical to the product of the work-up of muscle protein (metmyohemerythrin) and very similar to methemerythrin from coelomic fluid protein. The change [7] occurs in one step, with an isosbestic point observed at 390nm (18). We are not able to

$$\text{unstable* metmyo Ⓒ} \rightleftarrows \text{stable metmyo Ⓓ} \qquad [7]$$

determine what percentage, if any, of Ⓒ remains after equilibration. It is interesting that the activation parameters for the Ⓑ → Ⓔ and Ⓒ → Ⓓ changes are remarkably similar, as might be anticipated if they reflect similar types of iron coordination changes in the two types of sites. There is evidence for a Ⓒ → Ⓓ type of conversion with *P.g.* at pH = 6.3, but the system requires further examination (16).

III. CONCLUSIONS

Hemerythrin represents a wonderful opportunity to study unique inorganic reactions at a binuclear iron site in a protein matrix. Since a number of oligomeric forms are presented by the protein, it is also possible to study the effect of tertiary structure on these reactions, particularly intermolecular and intramolecular electron transfer processes. By chance, a system has been found in which electron transfer rates and site separation are firmly established. The same type of binuclear iron site appears to exist in the important enzyme ribonucleotide reductase (35) and studies similar to those with hemerythrin appear possible. There are enough parallelisms between the behavior of hemerythrin and the respiratory globins to ensure that results for the former may be valuable for understanding the behavior of the (more important) globins.

IV. ACKNOWLEDGEMENT

The work of the authors is supported by NIH Grants.

REFERENCES

1. Bonaventura, J., Bonaventura, C., and Sullivan, B.: 1977, in *Oxygen and Physiological Function* (Jöbsis, F.F., ed.) Professional Information Library, Dallas, TX, pp. 177-220.
2. Liberatore, F.A., Truby, M.F., and Klippenstein, G.L.: 1974, *Arch. Biochem. Biophys. 160*, pp. 328-229.

3. Addison, A.W., and Bruce, R.E.: 1977, *Arch. Biochem. Biophys.* *183*, pp. 328-332.
4. Manwell, C.: 1977, *Comp. Biochem. Physiol. 58B*, pp. 331-338; Sieker, L.C., Bolles, L., Stenkamp, R.E., Jensen, L.H., and Appleby, C.A.: 1981, *J. Mol. Biol. 148*, pp. 493-494.
5. Klippenstein, G.L., Van Riper, D.A., and Oosterom, E.A.: 1972, *J. Biol. Chem. 247*, pp. 5959-5963.
6. Hendrickson, W.A.: 1978, *Naval Res. Rev. 31*, pp. 1-20.
7. Hendrickson, W.A., Smith, J.L.: Private Communication.
8. Stenkamp, R.E., and Jensen, L.H.: 1979, *Adv. Inorg. Biochem.* *1*, pp. 219-233.
9. Stenkamp, R.E., Sieker, L.C., Jensen, L.H., and Sanders-Loehr, J.: 1981, *Nature, 291*, pp. 263-264.
10. Kurtz, D.M., Jr., Shriver, D.F., and Klotz, I.M.: 1978, *Coordn. Chem. Rev., 24*, pp. 145-178.
11. Loehr, J.S., and Loehr, T.M.: 1979, *Adv. Inorg. Biochem., 1*, pp. 235-252.
12. Keresztes-Nagy, S., and Klotz, I.M.: 1965, *Biochemistry, 4*, pp. 919-931 ; Meloon, D.R., and Wilkins, R.G.: 1976, *Biochemistry, 15*, pp. 1284-1290.
13. Olivas, E., deWaal, D.J.A., and Wilkins, R.G.: 1979, *J. Inorg. Biochem. 11*, pp. 205-212.
14. Harrington, P.C., deWaal, D.J.A., and Wilkins, R.G.: 1978, *Arch. Biochem. Biophys. 191*, pp. 444-451.
15. Babcock, L.M., Bradić, Z., Harrington, P.C., Wilkins, R.G., and Yoneda, G.S.: 1980, *J. Amer. Chem. Soc., 102*, pp. 2849-2850.
16. Bradić, Z., Harrington, P. C., Wilkins, R.G., and Yoneda, G.: 1980, *Biochemistry, 19*, pp. 4149-4155.
17. Muhoberac, B.B., Wharton, D.C., Babcock, L.M., Harrington, P.C., and Wilkins, R.G.: 1980, *Biochem. Biophys. Acta, 626*, pp. 337-345.
18. Harrington, P.C., Muhoberac, B.B., Wharton, D.C., and Wilkins, R.G.: 1981, *Biochemistry, 20, pp.*
19. Harrington, P.C., and Wilkins, R.G.: 1981, *J. Amer. Chem. Soc., 103*, pp. 1550-1556.
20. deWaal, D.J.A., and Wilkins, R.G.: 1976, *J. Biol. Chem. 251*, pp. 2339-2343.
21. Bates, G., Brunori, M., Amiconi, G., Antonini, E., and Wyman, J: 1968, *Biochemistry, 7*, pp. 3016-3020.
22. Bradić, Z., Harrington, P.C., and Wilkins, R.G.: 1979 in *Oxygen: Biochemical and Clinical Aspects* (Caughey, W.S., ed.) Academic Press, New York, pp. 557-571.
23. Petrou, A.L., Armstrong, F.A., Sykes, A.G., Harrington, P.C., and Wilkins, R.G.: 1981, *Biochem. Biophys. Acta* .
24. Alberding, N., Lavalette, D., and Austin, R.H.: 1981, *Proc. Natl. Acad. Sci. USA, 78*, pp. 2307-2309.
25. Harrington, P.C., and Wilkins, R.G.: 1981, unpublished results. These experiments were carried out at the Center for Fast Kinetics Research, The University of Texas at Austin. The Center is partly supported by the Biotechnology Branch of the Division of Research Resources of NIH.

26. Bradić, Z., Conrad, R., and Wilkins, R.G.: 1977, *J. Biol. Chem.*, *252*, pp. 6069-6075.
27. Harrington, P.C., Muhoberac, B.B., Wharton, D.C., and Wilkins, R.G.: 1981, unpublished results.
28. Babcock, L.M., Harrington, P.C., and Wilkins, R.G., unpublished results.
29. Harrington, P.C., and Wilkins, R.G., unpublished results.
30. Freier, S.M., Duff, L.L., Shriver, D.F., and Klotz, I.M.: 1980, *Arch. Biochem. Biophys.* *205*, pp. 449-463.
31. Harrington, P.C., and Wilkins, R.G.: 1979, *J. Biol. Chem.*, *254*, pp. 7505-7508.
32. Armstrong, F.A., Harrington, P.C., and Wilkins, R.G.: 1980, unpublished results.
33. Massey, V., and Hemmerich, P.: 1977, *J. Biol. Chem.* *252*, pp. 5612-5614.
34. Bradić, Z., Harrington, P.C., and Wilkins, R.G.: 1979, *J. Biol. Chem.* *18*, pp. 889-893.
35. Sjöberg, B.M., Gröslund, A., Loehr, J.S., and Loehr, T.M.: 1980, *Biochem. Biophys. Res. Commun.* *94*, pp. 793-799.

THE ENVIRONMENT OF THE BINUCLEAR IRON COORDINATION COMPLEX IN METHEMERYTHRIN

L. C. Sieker, R. E. Stenkamp and L. H. Jensen

Departments of Biological Structure and Biochemistry, University of Washington, Seattle, Washington, 98195.

The three-dimensional model of the crystal structure of octameric methemerythrin has been refined to a point where features of the molecule which contribute to its biological activity can be discussed with some certainty. Attention is focused on the environment of the binuclear iron complex where the amino acid side-chains would be expected to affect the reversible oxygenation of the complex. Several aromatic residues are located in the region surrounding the iron complex and appear to interact with the iron ligands. Consideration is given not only to the O_2 binding site but also to residues or groups of residues closely associated with the iron ligands: His 25, His 54, His 73, His 77 and His 101.

An examination of these interactions by alternative techniques should contribute to a better understanding of the chemical functioning of the molecule.

INTRODUCTION

Hemerythrin is a non-heme iron metalloprotein found in the marine invertebrate phyla, sipunculidea, annelida, priapulida and brachiopoda, where the protein functions in oxygen transport and respiration. The pigment was discovered by Delle Chiaie in 1823 (1), and was named hemerythrin by Krukenberg, who also observed the reversible color change in the presence of oxygen in 1880 (2). Subsequently, the protein was shown to be an iron-containing globin with no associated heme (3,4). In the first thirty years of this century, the emphasis in hemerythrin

H. B. Dunford et al. (eds.), The Biological Chemistry of Iron, 161–175.

research shifted to physiological studies of this molecule and comparative studies with other oxygen-transport pigments (5,6,7). Florkin, in 1933, purified the protein to the point where he could obtain crystals, determine an iron-oxygen ratio, demonstrate the absence of a Bohr effect and observe the failure of carbon monoxide to bind to the protein (6).

For the last twenty-five years, studies of hemerythrin have been oriented toward characterizing the iron complex in the biologically functional forms of the protein. Chemical modification studies, sequence analyses, dissociation-equilibrium experiments, spectral studies (absorption, resonance Raman, NMR, Mossbauer), EXAFS, magnetic susceptibility measurements, and x-ray crystallographic studies have provided information concerning the structure and function of the molecule. This work has been extensively reviewed (8-16), and only a brief summary of the properties of the molecule will be given here.

Sipunculid hemerythrin is found in several different oligomeric forms. Examples of octameric, tetrameric, trimeric and dimeric forms have been reported although no more than two forms have been found in any one species. Myohemerythrin, a monomeric analog of the hemerythrin subunit, is found in the muscle of the sipunculid analogous to the case with myoglobin in higher organisms. Each hemerythrin subunit consists of a single polypeptide chain of 113 amino acids (MW 13500 daltons) and two iron atoms. No other prosthetic groups are found, the iron atoms being bound to the protein by amino acid side-chains. As in the case of hemoglobin and myoglobin, hemerythrin exists in deoxy, oxy and met forms.

The iron atoms are in the +2 oxidation state in the deoxy form and the +3 oxidation state in oxy and met forms. Magnetic susceptibility studies show substantial anti-ferromagnetic coupling between the iron atoms in oxy and met, indicating that the iron atoms are not widely separated. The antiferromagnetic coupling and comparison of spectroscopic results for the protein and model compounds have been interpreted in terms of a mu-oxo bridged binuclear iron complex. Several different met forms of the protein can be generated with different small molecules (usually anions) bound to the iron complex. Complexes with N_3^-, SCN^-, Cl^-, OH^-, or H_2O yield absorption spectra, which on the basis of their similarity with the spectrum for the oxy form, have been used as evidence for a description of the Fe-dioxygen complex as $Fe(III)_2$-O_2^{2-}. Further evidence for the peroxide nature of the bound O_2 is provided by resonance Raman spectroscopy.

Little information is available concerning the nature of the metal complex in deoxyhemerythrin. It is not yet clear whether the mu-oxo bridge is retained, since the reduced antiferromagnetic coupling could simply reflect the presence of Fe(II) atoms. Also, the lack of a pH change associated with the oxygenation reaction might indicate the retention of the mu-oxo bridge (11).

Neither the deoxy nor oxy forms have been successfully investigated by crystallographic analyses because of the difficulty of keeping the molecule in either state for the long period of data collection. Attempts to obtain stable crystals and data for these forms are in progress. The crystallographic studies of methemerythrin have had as one of the higher priorities the identification of the metal ligands and the determination of the structure of the binuclear iron complex. The ultimate goal of all these studies is to provide a structural basis for understanding the reversible oxygenation process and how this process is controlled.

GENERAL STRUCTURAL FEATURES

The molecule studied in this laboratory is the octameric form isolated from the coëlomic fluid of the sipunculid, *Themiste dyscritum* (17,18). Figure 1 is a diagram of the octameric packing of the subunits. The octamer is composed of two rings of four subunits each set back to back with two-fold axes at the corners and through the middle of each face of the oligomeric arrangement.

T. dyscritum hemerythrin crystallizes in the tetragonal space group P4 (40,000 reflections to 2.0 Å resolution, 4000 atoms per asymmetric unit), the molecular four-fold axes being coincident with the crystallographic four-fold axes. In this crystal form, the octamers show C_4 symmetry and not D_4 or 422 symmetry; the molecular two-fold axes are pseudo two-fold axes. However, *Phascolopsis gouldii* hemerythrin, studied by Hendrickson and collegues (19,20), crystallizes in space group P422 where the molecular two-fold axes are exact, coinciding with the crystallographic two-fold axes. Thus, the octamers display D_4 symmetry, insuring the equivalence of each subunit. The entire ensemble might not display this symmetry while carrying out its function, but for further discussion, the D_4 symmetry will be assumed.

Each polypeptide chain of 113 residues(118 for myohemerythrin) is folded into a structure of four nearly parallel alpha helices associated in a left twisted bundle as

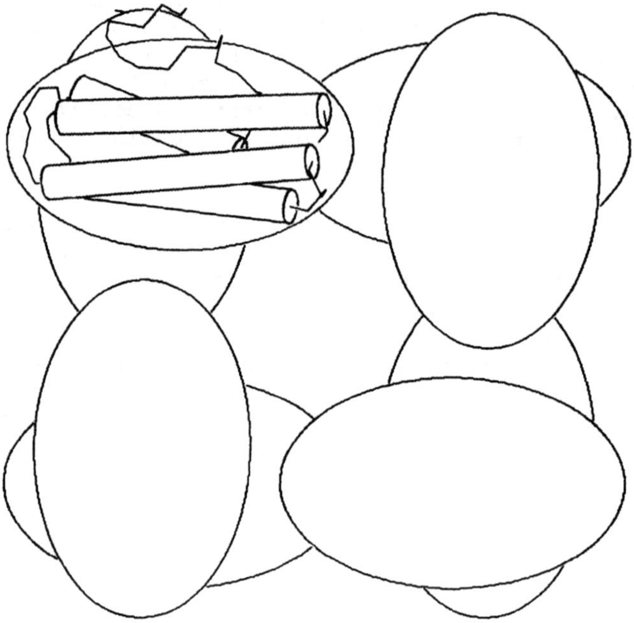

Figure 1. Schematic showing tertiary and secondary structure of hemerythrin.

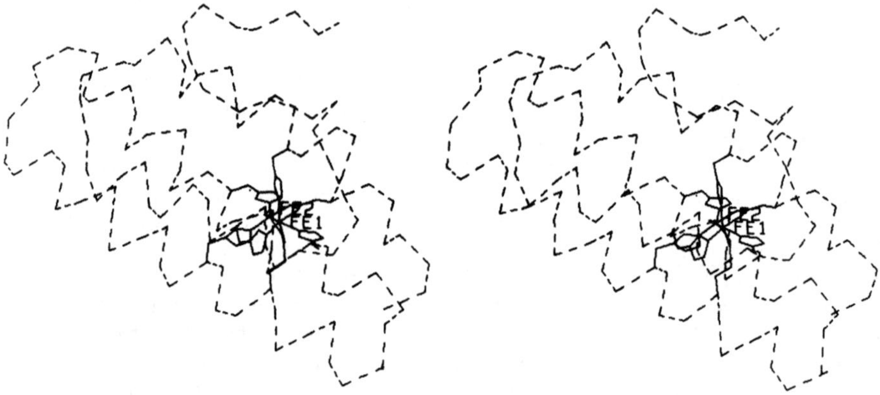

Figure 2a. Stereo diagram showing helices and iron complex.

Figure 2b. Alternate stereo diagram of helices and iron complex looking down the helical axes.

shown in Figures 2a and 2b which also indicate the location of the Fe complex relative to the alpha helices. This kind of helix packing is not unique to hemerythrin since cytochrome b_{562}, cytochrome c', the ferritin subunit and tobacco mosaic virus coat protein all display the same arrangement in spite of different primary structures and function (21). It is remarkable that several different proteins with very different sequences and biological functions have similar tertiary structures.

BINUCLEAR IRON COMPLEX

Several crystallographic models of the binuclear complex have been proposed over the past six years, but recent work has answered the major questions concerning the structure of the complex in methemerythrin (22). The two iron atoms are coordinated in each subunit by five histidines (25,54,73,77 and 101) and Glu 58 and Asp 106, where the carboxylates of the two acidic groups are shared as shown in Figures 3 and 4. The bridging oxygen atom, designated a hydroxide ion in previous studies on methydroxohemerythrin from this laboratory, coincides with the oxygen of the mu-oxo bridged models suggested by other studies (11,23).

The coordination of one iron (Fe1) is octahedral, with the six ligands being His 73, His 77, His 101, Glu 58, Asp 106 and the mu-oxygen atom. The second iron atom (Fe2), at least for the azide form of methemerythrin, is coordinated by His 25, His 54, azide, the shared Glu 58 and Asp 106, and the mu-oxygen bridge. The azide binding site is probably the binding site for the O_2 molecule in oxyhemerythrin, but it is not at all clear what this

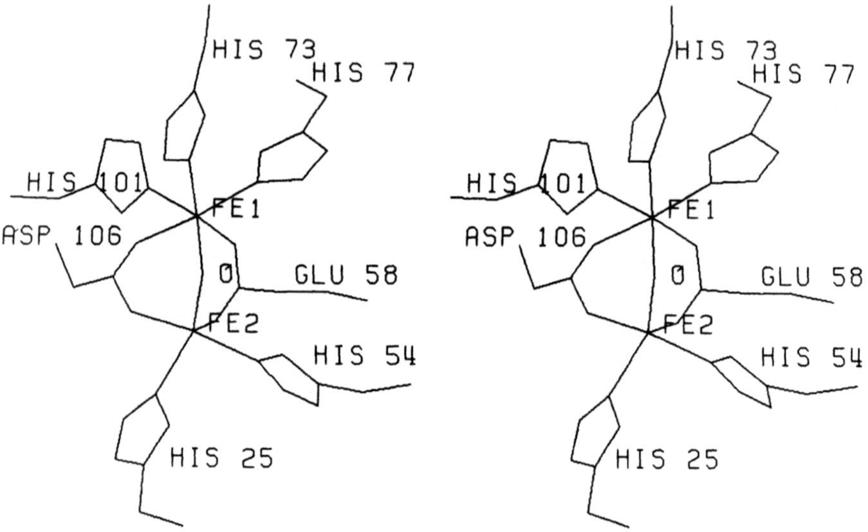

Figure 3. Stereo diagram of complex in methydroxohemerythrin.

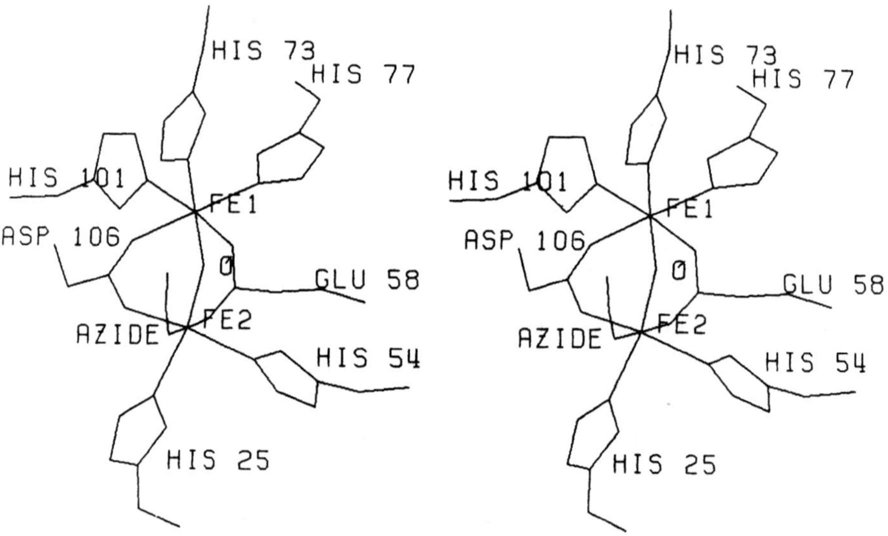

Figure 4. Stereo diagram of complex in metazidehemerythrin.

site contains in the other met (22) and deoxy hemerythrins, if the site even exists in deoxy. It appears that the Fe2 of methydroxohemerythrin is penta-coordinate with no small molecule ligand bound to the complex. However, there is still some question whether the form studied crystallographically is methydroxo or metaquo. There was no hint of the presence of a penta-coordinated iron or of the bridging carboxylate groups until the crystallographic results were obtained. To our knowledge, no model compound mimicking this active site has been synthesized.

A review by Ibers and Holm (24) has discussed the advantages offered by model compounds in chemically characterizing the metal complexes found in metalloproteins. Working with model complexes is certainly in keeping with the reductionist approach to chemical problems where detailed information concerning smaller molecules is much easier to obtain and is essential for understanding the more complex biological processes. Nevertheless, the protein portion of metalloproteins is important in modulating the properties of the metal complexes to give rise to their biological activity, and studies of complete metalloproteins will be necessary to achieve an adequate level of understanding.

THE ENVIRONMENT OF THE COMPLEX

Knowledge of the structure of the iron complex and its surroundings is essential for understanding how hemerythrin reversibly binds oxygen. The residues which are not directly coordinated to the iron atoms provide an environment for the binuclear iron complex, a possible pathway for entrance and egress of the O_2 molecule and possibly a subtle "poising" of the complex for reversible oxygenation. Although the interactions among these residues are necessary to maintain the three-dimensional structure of the polypeptide, they also serve to modulate the properties of the complex.

The general packing of aromatic and other hydrophobic residues in the interior of this molecule is in accord with what is observed for other globular proteins. The arrangement of hydrophobic residues internally and hydrophilic residues externally not only solubilizes the molecule in an aqueous medium, but also helps maintain its integrity as a globular unit. The polypeptide chain is arranged or folded with hydrogen bonds, coulombic interactions and van der Waals forces holding the chains in a relatively rigid cylindrical framework with a hydrophobic core.

A role of the hydrophobic core in hemerythrin is that of providing an environment for the iron-ligand complex which modulates the relative stability of the met, oxy and deoxy forms. The protein matrix provides specific interactions between residues causing the metal complex to display unique chemical properties. These "stereotaxic" associations of residues allow specific interactions to exist, persist or change in a concerted manner which gives protein molecules functional properties distinct from small molecules. While the internal core of protein structures are not truly static, neither are they as dynamical in nature as small molecules in solution.

These types of arrangements are a major feature of biological macromolecules. In the serine proteases, a hydrogen bond network is believed to give Ser 195 the unusual pK and activity it displays. The properties of the active site serine are induced by the "charge relay system" where the proposed polarization system arises from a buried aspartic acid hydrogen bonded to a histidine which in turn is hydrogen bonded to the active serine (25,26).

An example where a cofactor's properties are modified and stabilized to display new characteristics can be found in the various cytochromes. The heme pocket and involvement of the coordinating ligands with the iron porphyrin change the cofactor's properties, mediating electron transfer at a redox potential different from that normally found for metalloporphyrin systems. The interactions of the protein moiety with the metal porphyrin involve covalent, van der Waals and hydrogen bonding interactions (27,28).

Similar sets of interactions can be found in hemerythrin, with as yet unknown effects on the properties of the iron complex. One of the regions which should be considered as a potential modulator of the binuclear complex in hemerythrin is in the vicinity of His 25 shown in Figure 5. This is the region close to the complex which is on the inside of the protein and yet can be considered somewhat hydrophilic since a water molecule is hydrogen bonded to NE2 of His 25, NE1 of Trp 10, the carbonyl oxygen atom of Phe 14, and OD1 of Asp 22. Table 1 lists the hydrogen bond distances for this water.

Another region with several significant interactions is the O_2 binding pocket. The bridging oxygen atom is situated in the center of a shallow pocket defined by four histidine residues, two iron atoms and the bridging oxygen. The four histidine residues are His 25, His 54, His 77 and His 101 as shown in Figure 6. Table 2 is a list of the close contacts between His 77 and Phe 55. His 77 is in van der Waals contact with Phe 55 in a

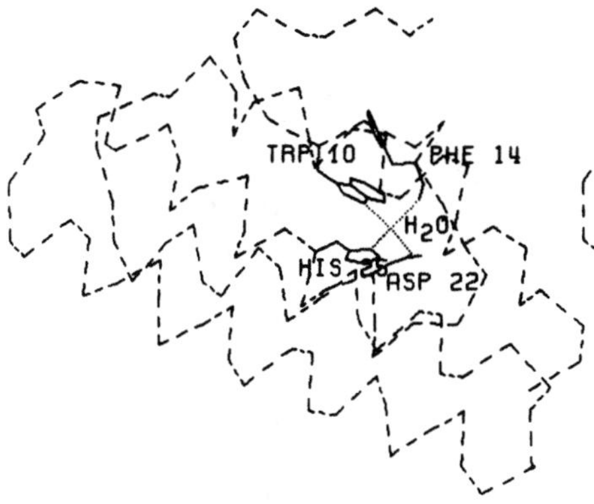

Figure 5. Stereo diagram showing vicinity of water 149.

Table 1. Hydrogen bonds to water 149.

residue	atom	residue	atom	distance	RMS deviation
Asp 22	OD1	Water 149	OW	2.59	--
Trp 10	NE1	Water 149	OW	3.19	0.22
His 25	ND1	Water 149	OW	2.66	0.07
Phe 14	O	Water 149	OW	2.61	0.08

RMS deviation based on four independent subunits.

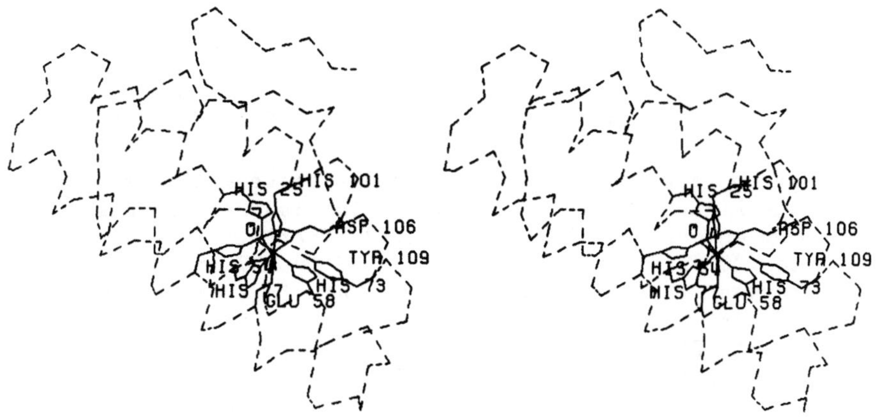

Figure 6. Stereo diagram showing complex and Tyr 109.

Table 2. Contacts between Phe 55 and His 77.

residue	atom	residue	atom	distance	RMS deviation
Phe 55	CB	His 77	ND1	3.51	0.10
Phe 55	CG	His 77	CG	3.51	0.03
Phe 55	CG	His 77	ND1	3.63	0.08
Phe 55	CG	His 77	CD2	3.76	0.09
Phe 55	CD1	His 77	CG	3.79	0.06
Phe 55	CD1	His 77	CE1	3.87	0.11
Phe 55	CD1	His 77	NE2	3.63	0.11
Phe 55	CD1	His 77	CD2	3.55	0.04
Phe 55	CE1	His 77	CD2	3.67	0.08
Phe 55	C	His 77	CD2	3.96	0.14
Phe 55	CD2	His 77	CG	3.65	0.06
Phe 55	CD2	His 77	CB	3.50	0.01

RMS deviation based on four independent subunits.

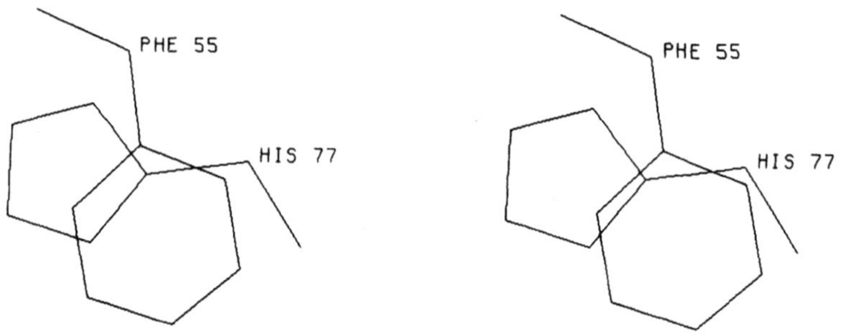

Figure 7. Stereo diagram of overlap between Phe 55 and His 77.

manner appropriate for pi-pi interactions between the aromatic groups, see Figures 7 and 9. Trp 97 is located between His 101 and Phe 80, as shown in Figures 8 and 9. There is interaction between the CG of His 101 and the ring of Trp 97 and good overlap between the latter and Phe 80. The rings of Phe 80 and Tyr 93 are coplanar, the edges of the rings being in van der Waals contact. Figure 8 also shows the stacking arrangement of Phe 80, Tyr 93, Trp 97 and His 101. These residues make up two adjacent sides of a possible O_2 entrance tunnel (29), and Table 3 summarizes the close approaches between them. His 25 appears to be in van der Waals contact with the CG1 and CD

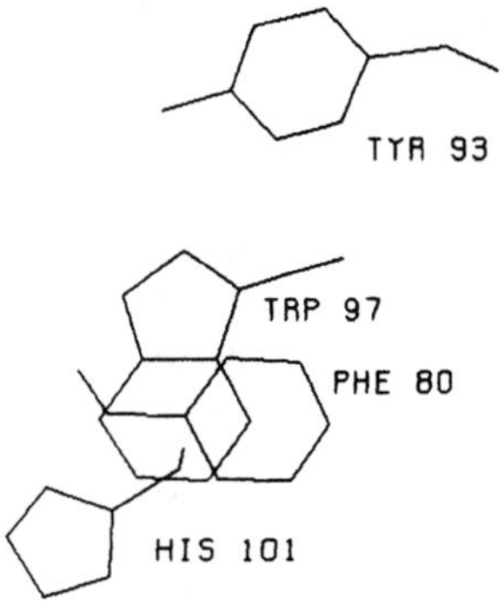

Figure 8. Stereo diagram showing overlaps with Trp 97.

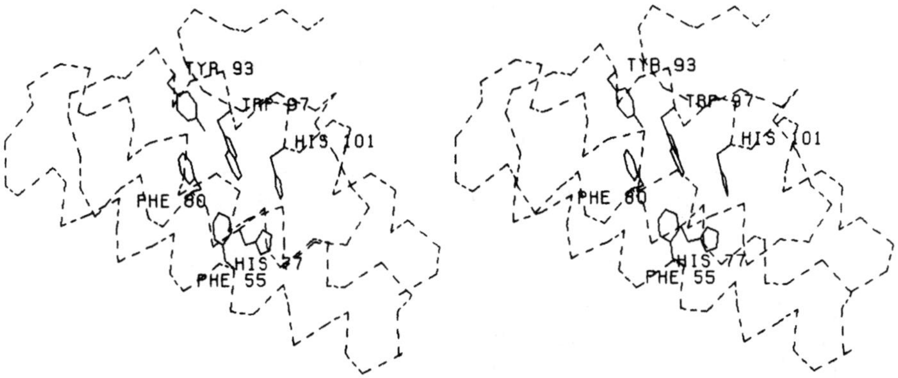

Figure 9. Stereo diagram of aromatic residues near complex.

groups of Ile 102 as shown in Table 4 completing the description of hydrophobic interactions for this region.

Two hydrophilic interactions close to this region involve an invariant Glu 24 carboxylate oxygen hydrogen bonded to ND1 of His 54 (2.81 Å ± 0.08) and the semi-invariant Gln 59 hydrogen bonded to ND1 of His 77 (2.74 Å ± 0.11).

Table 3. Close contacts between Phe 80, Tyr 93, Trp 97 and His 101.

residue	atom	residue	atom	distance	RMS deviation
Phe 80	CD1	Tyr 93	CD1	4.23	0.13
Phe 80	CE1	Tyr 93	CE1	4.34	0.14
Phe 80	CB	Trp 97	CE2	3.69	0.08
Phe 80	CB	Trp 97	CZ2	3.55	0.05
Phe 80	CB	Trp 97	CH2	3.77	0.07
Phe 80	CG	Trp 97	CD2	3.71	0.08
Phe 80	CG	Trp 97	CE2	3.82	0.10
Phe 80	CG	Trp 97	CE3	3.69	0.08
Phe 80	CG	Trp 97	CZ3	3.79	0.05
Phe 80	CD1	Trp 97	CD2	3.45	0.06
Phe 80	CD1	Trp 97	CE3	3.65	0.06
Phe 80	CD2	Trp 97	CZ3	3.70	0.08
Phe 80	CD2	Trp 97	CE3	3.84	0.11
Phe 80	CE1	Trp 97	CE3	3.78	0.08
His101	CB	Trp 97	CH2	3.68	0.09
His101	CB	Trp 97	CZ3	3.65	0.11
His101	CG	Trp 97	CH2	3.43	0.03

RMS deviation based on four independent subunits.

Table 4. Contacts between His 25 and Ile 102.

residue	atom	residue	atom	distance	RMS deviation
His 25	CG	Ile102	CD	3.78	0.13
His 25	ND1	Ile102	CG1	4.09	0.09
His 25	ND1	Ile102	CD	3.82	0.02
His 25	CE1	Ile102	CG1	3.93	0.08
His 25	CE1	Ile102	CD	4.13	0.04
His 25	NE2	Ile102	CG1	4.01	0.04
His 25	CE2	Ile102	CD	4.06	0.18

RMS deviation based on four independent subunits.

Figure 10 is a stereo diagram of the region about His 73 showing His 73, Tyr 109, Met 62, Ile 113, and Ile 105. Table 5 lists close approaches between side-chain atoms of these groups. The ring of His 73 is in van der Waals contact with SD and CE of Met 62 on one side and CD of Ile 105 on the other. Tyr 109 is in van der Waals contact edge-on with His 73. The CG1 and CD of Ile 113 are in van der Waals contact with the ring of Tyr 109.

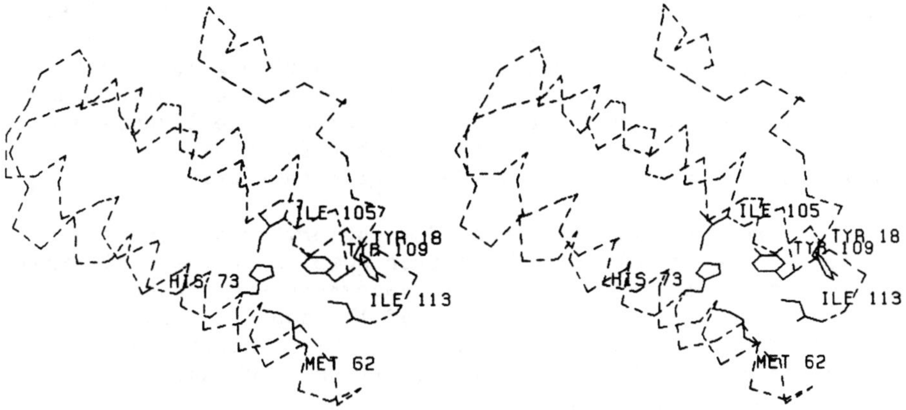

Figure 10. Stereo diagram showing vicinity of His 73.

Table 5. Contacts among residues near His 73.

residue	atom	residue	atom	distance	RMS deviation
Met 62	SD	His 73	CG	3.85	0.09
Met 62	SD	His 73	ND1	3.65	0.04
Met 62	CE	His 73	CG	3.64	0.08
Met 62	CE	His 73	ND1	3.84	0.21
Met 62	CE	His 73	ND2	3.85	0.21
Ile105	CD	His 73	CG	3.50	0.01*
Ile105	CD	His 73	ND1	3.41	0.20*
Ile105	CD	His 73	CE1	3.82	0.18*
Ile105	CD	His 73	CE2	4.16	0.05*
Ile105	CD	His 73	CD2	3.90	0.05*
Ile113	CG1	Tyr109	CG	3.89	0.07
Ile113	CG1	Tyr109	CD1	3.64	0.04
Ile113	CD	Tyr109	CG	3.56	0.04
Ile113	CD	Tyr109	CD1	3.48	0.03
Ile113	CD	Tyr109	CE1	3.88	0.09
Ile113	CD	Tyr109	CD2	3.90	0.11

Asterisked RMS deviations based on three independent subunits. Unasterisked RMS deviations based on four.

The final region to be considered here is on the opposite side of the iron atoms from the four histidine, bridging oxygen region. Except for the 2.84 Å distance between the hydroxyl atom of Tyr 109 and the OE2 of Glu 58 (which is sufficiently short to be considered a hydrogen bond), there is no strong interaction with the carboxylate ligands. Figure 6 is a stereo view of this region. Tyr 109 is located between the carboxylate ligands and below His 73 in a position to be available for possible interaction in the deoxy or oxy forms of hemerythrin.

The structure determination of methemerythrin alone can not explain all the properties of the iron complex or the biological function of the molecule. Nor is it necessary that the regions described here be the only portions of the subunit which are important for the molecule to function. This description has also not addressed the subunit interactions which may be important for causing the properties of the complex in the octamer to be different from myohemerythrin. Nevertheless, crystallographic studies can provide sufficient detail forming a firm basis for suggesting other experiments and correlating other experimental observations. In particular, NMR and other spectroscopic techniques can now be used to study further the interactions between the complex and surrounding side-chains described here. Additional crystallographic work directed toward determining the structures of the oxy and deoxy forms of hemerythrin, coupled with other experimental data, can be expected to provide a more complete description of the molecule's biological function of reversible oxygenation.

This work has been supported by grant GM-10828 from the National Institutes of Health and equipment grant PCM 76-20557 from the National Science Foundation.

REFERENCES

1. Delle Chiaie, S.: 1823, "Memorie sulla storia e notomia degli animali del Regno di Napoli," Stamp. Fernandes, Naples, Italy, p. 444.
2. Krukenberg, C.F.W.: 1880, Vergl. Physiol. Studien zu Tunis, Mentone u. Palermo, Abt. III. Reihe I, pp. 66-123.
3. Ray Lankester, E.: 1872, Proc. Roy. Soc. London, 21, pp. 70-81.
4. Andrews, E.A.: 1890, J. Johns Hopkins Univ., 4, pp. 389-430.
5. Marrian, G.F.: 1927, Brit. J. Exptl. Biol., 4, pp. 357-364.
6. Florkin, M.: 1933, Arch. intern. physiol., 36, pp. 247-328.
7. Roche, J.: 1935, "Essai sur la biochimie générale et comparée des pigments respiratoires," Masson, Paris.

8. Ghiretti, F.: 1962, in Oxygenases, Ed. O. Hayaishi, Academic Press, New York, pp. 517-553.
9. Klotz, I.M.: 1971, in Biological Macromolecules, Vol. 5, Eds. S.N. Timasheff and G.D. Fasman, M. Dekker, Inc., New York, pp. 55-103.
10. Llinas, M.: 1973, Structure and Bonding, 17, pp. 135-220.
11. Okamura, M.Y. and Klotz, I.M.: 1973, in Inorganic Biochemistry, Vol. 1., Ed. G.I. Eichorn, Elsevier, New York, pp. 320-343.
12. Klotz, I.M., Klippenstein, G.L. and Hendrickson, W.A.: 1976, Science, 23, pp. 335-344.
13. Kurtz, D.M. Jr.,Shriver, D.F. and Klotz, I.M.: 1977, Coord. Chem. Reviews, 24, pp. 145-178.
14. Stenkamp, R.E. and Jensen, L.H.: 1979, in Advances in Inorganic Biochemistry, Vol. 1., Eds. G.I. Eichorn and L.G. Marzilli, Elsevier, New York, pp. 219-233.
15. Loehr, J.S. and Loehr, T.M.: 1979, in Advances in Inorganic Biochemistry, Vol. 1., Eds. G.I. Eichorn and L.G. Marzilli, Elsevier, New York, pp. 235-252.
16. Klippenstein, G.L.: 1980, Amer. Zoologist, 20, pp. 39-51.
17. Loehr, J.S., Meyerhoff, K.N., Sieker, L.C. and Jensen, L.H.: 1975, J. Mol. Biol., 91, pp. 521-522.
18. Stenkamp, R.E., Sieker, L.C., Jensen, L.H. and Loehr, J.S.: 1976, J. Mol. Biol., 100, pp. 23-34.
19. Ward, K.B., Hendrickson, W.A. and Klippenstein, G.L.: 1975, Nature, 257, pp. 818-821.
20. Ward, K.B. and Hendrickson, W.A.: 1977, in Metal-Ligand Interactions in Organic Chemistry and Biochemistry, part 2, Eds. B. Pullman and N. Goldblum, D. Reidel Publishing Co., Dordrecht-Holland, pp. 173-186.
21. Weber, P.C. and Salemme, F.R.: 1980, Nature, 287, pp. 82-84.
22. Stenkamp, R.E., Sieker, L.C., Jensen, L.H. and Sanders-Loehr, J.: 1981, Nature, 291, pp. 263-264.
23. Hendrickson, W.A.: 1978, Naval Research reviews, 31, pp. 1-20.
24. Ibers, J.A. and Holm, R.H.: 1980, Science, 209, pp. 223-235.
25. Blow, D.M., Birktoft, J.J. and Hartley, B.S.: 1969, Nature, 221, pp. 337-340.
26. Birktoft, J.J., Kraut, J. and Freer, S.T.: 1976, Biochemistry, 15, pp. 4481-4485.
27. Salemme, F.R.: 1977, Ann. Rev. Biochem., 46, pp. 299-329.
28. Adman, E.T.: 1979, Biochem. Biophys. Acta, 549, pp. 107-144.
29. Stenkamp, R.E., Sieker, L.C. and Jensen, L.H.: 1978, J. Mol. Biol., 126, pp. 457-466.

SECTION E

IRON-SULFUR CLUSTERS AND ENZYMES

SIMPLE IRON-SULFUR PROTEINS: METHODOLOGY FOR ESTABLISHING THE TYPE OF CENTER

Isabel Moura and José J.G. Moura

Centro de Química Estrutural, Complexo Interdisciplinar I.S.T., Av. Rovisco Paris, 1000 Lisbon (Portugal) and Gray Freshwater Biological Institute, University of Minnesota, P.O. Box 100, Navarre, Minnesota, U.S.A. 55392

In this article we have put together available information on how to determine the structures of the different types of iron-sulfur centers. The discovery of some new basic structures especially the |3Fe-3S| core has brought a renewed interest into the field of iron-sulfur proteins.

1. INTRODUCTION

There are four well described iron-sulfur centers found in different enzymes and electron transfer proteins. The elucidation of these four basic structures has been particularly difficult due to the instability of their active centers, but structural determination efforts have provided four basic structures |Rb| (rubredoxin); |2Fe-2S|; $|3Fe-3S|_1$ and |4Fe-4S| centers (see Figure 1).

The simple constitution of the active center, iron and sulfur contrasts with the diversified role played by these proteins in key biological oxidation-reduction processes, such as carbon, hydrogen, sulfur and nitrogen metabolism (over a wide range of redox potentials, +350 mV in photosynthetic bacteria to -600 mV in chloroplasts) as described in Table 1.

Iron-sulfur proteins, which contain these basic structures are generally electron-transfer proteins mediating electron exchange between enzymatic systems. Table 2 describes the

H. B. Dunford et al. (eds.), The Biological Chemistry of Iron, 179–192.

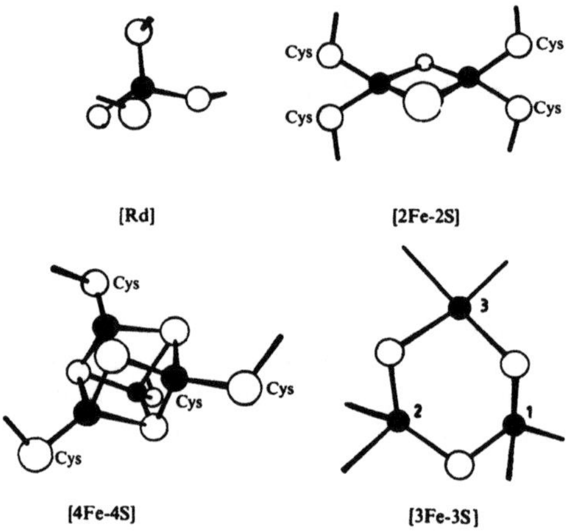

Figure 1. Simple iron-sulfur clusters. Abbreviation for rubredoxin is commonly Rb or Rd.

formal charges of the structures and the redox behavior as well as the way they may associate into more complex arrangements.

We will focus on the general properties of these centers giving more emphasis to those recently discovered which are less well characterized in the literature. We shall try to give information on how to identify the type of iron-sulfur center present in newly isolated proteins. Other more complex iron-sulfur structures present in nitrogenase (P clusters and molybdenum iron-sulfur cofactor) are reviewed in the following chapters.

Table 1. Fe-S Centers in Biological Relevant Reactions

Phosforoclastic reaction
Hydrogen metabolism and methane formation
Nitrogen fixation
Photosynthetic processes
Sulfate and sulfite reduction
Oxidation of xanthine and aldehydes
Krebs cycle
Respiratory chain

Table 2. Active Centers in Iron-Sulfur Proteins (a)

Simple	Complex
|Rb| (1Fe)	n|Fe-S|
|Rb| type (2Fe)	|Fe-S| + flavin
$\|2Fe\text{-}2S\|^{+2\,(+2,+1)}$	|Fe-S| + flavin + haem
$\|3Fe\text{-}3S\|^{+3\,(+3,+2)}$	|Fe-S| + haem
$\|4Fe\text{-}4S\|^{+2\,(+2,+3)}$	|Fe-S| + Mo
$\|4Fe\text{-}4S\|^{+2\,(+2,+1)}$	|Fe-S| + Mo + flavin
$2\|4Fe\text{-}4S\|^{+2\,(+2,+1)}$	|Fe-S| + Mo + Se
$\|3Fe\text{-}3S\|+\|4Fe\text{-}4S\|^{+2\,(+2,+3)}$	|Fe-S| + Fe-S-Mo nitrogenase cofactor
$\|3Fe\text{-}3S\|+\|4Fe\text{-}4S\|^{+2\,(+2,+1)}$	|Fe-S| + chlorophyll

(a) The formal charges are determined by adding the charges of the iron and the labile sulfur atoms of the core, without taking into account the charge of protein ligands. Possible redox transitions are indicated in the superscript parentheses. See also Table 4.

This paper relies to a considerable extent upon both Mössbauer and EPR spectral data. Those readers not familiar with these techniques are asked to consult the paper by B.H. Huynh elsewhere in this volume (Mössbauer) and, for example, to a paper by G. Palmer (EPR) in the volume edited by R. Wilkins and D. Darnell described in the preface.

2. PROPERTIES OF THE IRON-SULFUR CENTERS

2.1 One-Iron Center (no Labile Sulfur)

Rubredoxin is one of the simplest low molecular weight iron-sulfur proteins which can be isolated from both aerobic and anaerobic organisms. This protein can be easily recognized by its distinct optical spectrum (Table 3), with the center having only one iron atom. This atom is coordinated by four cysteinyl residues and therefore labile sulfur is absent.

Table 3. Characteristics of the different types of iron-sulfur centers
Table adapted from references (39,41,45)

	Rb		2Fe-2S		3Fe-3S		4Fe-4S		
	3+	2+	2+	1+	3+	2+	3+	2+	1+
Absorption spectrum mm(λ_{max},nm)	7.0(493) 8.4(376) 17.2(278)		5.0(325) 10.0(415) 9.6(455)		23.1(305) 15.7(405) 13.3(453 shoulder)	9.6(425 shoulder)	32.4(325) 20.0(375) 18.4(450)	15.2(400)	
EPR spectrum	9.4,4.3	no signal	no signal	2.02,1.93, 1.93	2.02(fairly isotropic)	no signal	2.12,2.04 2.04	no signal	2.06,1.92 1.88
Mössbauer spectrum									
δ(mm/s)	0.32	0.70	0.27	0.25,0.54	0.27	0.46,0.30	0.40,0.29	0.42	0.50,0.58
ΔE_Q(mm/s)	-0.50	-3.25	0.65	0.64,3.00	0.54	1.47,0.47	1.03,0.88	1.50,1.20, 1.10,0.66	1.32,1.89
Redox potential[(a)] (mV)	0 to -60		-250 to -450		-130 and -420		+350	-300 to -450	

(a) Rubredoxin type proteins have a narrow range of redox potentials when compared with |2Fe-2S| and |4Fe-4S| containing proteins. The redox potentials indicated for the |3Fe-3S| core referred to *D.gigas* FdII and *A.vinelandii* Fd.

Although the physiological role of rubredoxin has not yet been determined in anaerobes the structural studies performed on this class of proteins are well advanced. The X-ray structures of *C. pasteurianum* and *D. vulgaris* rubredoxins are known and show a nearly tetrahedral iron complex. All the iron-sulfur distances are similar (2.24 ± 0.01 A) with no shortened bond, as was first reported (1,2,3). Extended X-ray absorption fine structure studies (4) also confirmed that the average iron sulfur distance is the same as the average of the crystallographic distances in the protein.

The oxidized form of rubredoxin gives typical EPR spectra with signals at g = 4.3 and 9.4 (Table 3). Such a spectrum is characteristic of a S = 5/2 system (5) and the EPR signals are due to transitions in the middle and lower Kramer doublets with a high degree of rhombic distortion (E/D = 0.23). As the temperature is lowered the g = 4.3 signal decreases in magnitude as the middle Kramer doublet becomes depopulated, and the g = 9.4 signal from the ground state increases. The magnitude of the zero-field splitting (D) can be determined from the temperature dependence of the EPR signals.

Chemical models of the rubredoxin center shows similar properties (6,7,8). The X-ray studies of $|Fe(S_2\text{-o-xyl})_2|^-$ show a tetrahedral symmetry; there is no difference between the four iron sulfur distances as was shown also for the protein. The visible and near-infrared spectra are quite similar. Magnetic susceptibility measurements of $|Fe(S_2\text{-o-xyl})_2|^-$ and Rd_{ox} follow the simple behaviour of the Curie law with μ_{eff}/Fe of 5.8 - 5.9 μ_B in accordance with high spin iron (III). Mössbauer parameters of the iron(II) complexes (Rd_{red}) also indicate a very similar electronic structure. Both have temperature independent quadrupole splittings (3.1-3.3 mm/s), similar isomeric shifts (0.54-0.66 mm/s) and the same sign for the principal component of the electric gradient field.

Until now all of the spectroscopic methods applied to these systems show that the synthetic complexes are very good analogs of the Rd_{ox} and Rd_{red} centers.

We have isolated a protein named desulforedoxin from the sulfate reducing organism *Desulfovibrio gigas* (9) and shown that it represents a variation on the basic rubredoxin type of structure. It also has four cysteines per iron atom which can be displaced by four equivalents of a mercurial agent (10).

A comparison of the visible spectra of desulforedoxin and rubredoxin shows close similarities between their chromophores. Oxidized desulforedoxin shows a complex EPR spectrum with principal features around g = 7.7 and g = 5.7. These spectra were interpreted by using a spin hamiltonian for high-spin ferric ion ($S = 5/2$). Taking $D > 0$ and $E/D = 0.08$ the observed g-values can be quite well reproduced (10). The resonances at 7.7, 4.1 and 1.8 are assigned to the ground state doublet (±1/2) and the resonance at g = 5.7 to the middle doublet (±3/2).

The zero field parameters obtained for oxidized desulforedoxin ($D = 2.2 \pm 0.2$ cm^{-1} and $E/D = 0.08$) are indicative of geometrical differences between desulforedoxin and rubredoxin ($E/D = 0.23$) (11). The saturation fields calculated for desulforedoxin (-385 kG) (10) and rubredoxin (-410 kG) indicate more covalent bonding in the first protein. The sign of the quadrupole splitting in the reduced proteins have opposite signs showing that the orbital ground state of reduced desulforedoxin has a predominantly $d_{x^2-y^2}$ character ($\Delta E_Q > 0$ and $\eta > 0.5$) while reduced rubredoxin is predominantly d_{z^2} ($\Delta E_Q < 0$ and $\eta > 0.5$).

The differences observed in the electronic structures of the chromophore of desulforedoxin and rubredoxin were interpreted (10) as the result of different stereochemical constaints imposed by the polypeptide chain. Indeed the spacing of the four cysteine residues in sequences of the type Cys(6)-a-b-(Cys)-Gly(9) and Cys(39)-c-d Cys(42)-Gly as observed in rubredoxin are replaced by the sequences Cys(9)-x-y Cys(12)-Gly and Cys(28)-Cys(29)-Gly. This unusual arrangement may be responsible for the stereochemical constaints imposed by the polypeptide chain.

2.2 Two-Iron Center

|2Fe-2S| centers were elucidated by a combination of spectroscopic techniques (i.e., Mössbauer, EPR, ENDOR, NMR, magnetic susceptibility and various optical methods), before an X-ray determination confirmed the structure.

|2Fe-2S| centers can exist in two oxidation states (Table 4). In the oxidized state (+2) the center is diamagnetic ($S = 0$) and no EPR signal is observed. Upon one electron reduction (+1) the center becomes paramagnetic ($S = 1/2$) and a EPR signal at g = 1.94 can be observed (Table 3). Two types of EPR signals have been observed: rhombic (chloroplast type) with g-values around 2.05, 1.95, 1.89, and axial with g = 2.02 and 1.94 (hydroxylase type).

A model for the antiferromagnetic coupling was proposed by Gibson (12) which accounts for the EPR features. This model was confirmed by Mössbauer spectroscopy (13,14); in the +1 state the two iron sites have coupling constants of opposite sign. Recently an X-ray study on the ferredoxin from *Spirulina platensis* (15) has confirmed the structure proposed from other spectroscopic techniques. Model compounds of two iron centers (6) are

Table 4. Formal charges of iron-sulfur centers and possible redox transitions (a)

|Rb|	|2Fe-2S|	|3Fe-3S|
Fe^{3+} (EPR)[(b)]	Fe^{3+} - Fe^{3+}	$3Fe^{3+}$ (EPR)
↕	↕	↕
Fe^{2+}	Fe^{2+} - Fe^{3+} (EPR)	$2Fe^{+2.5}$ - $1Fe^{3+}$

|4Fe-4S|		
(+3)$3Fe^{3+}$ - $1Fe^{2+}$ (EPR)	$HiPIP_{ox}$	
↕	↕ +350 mV	
(+2)$2Fe^{3+}$ - $2Fe^{2+}$	$HiPIP_{nat}$	Fd_{nat}
↕	↕ ≃ -600 mV	↕ ≃ -400 mV[(c)]
(+1)$1Fe^{3+}$ - $3Fe^{2+}$ (EPR)	$HiPIP_{super\ red}$	Fd_{red}

(a) See footnote Table 2. For the |3Fe-3S| core the electron added upon reduction of the center is shared by two of the iron atoms. However the spin coupled model for the |3Fe-3S| core indicates that this structure is a covalently linked structure and should not be considered as a |2Fe-2S| cluster weakly attached to a third iron atom (30). (b) (EPR) indicates EPR active species. (c) Mid point potentials for the redox transitions are indicated in Table 3.

also good structural models of the biological centers.

2.3 Three-Iron Center

This new structure for iron-sulfur complexes has been a recent addition to the field of electron transfer proteins. This structure was firmly established in 1980 in the iron-sulfur protein isolated from *A. Vinelandii* (16) and in a ferredoxin from *D. gigas* (17). The *A. vinelandii* ferredoxin first isolated by Shetna (18) and Yoch (19) contains 7-8 iron atoms and exhibits EPR and absorption features similar to those of proteins with |4Fe-4S| centers. Sweeney *et al.* (20) and Yoch and Carithers (21) reported potentiometric titrations of the protein and suggested the presence of two |4Fe-4S| centers: one of the centers present in the native form of the protein gives an "isotropic" EPR signal. Another "isotropic" signal almost superimposable with the first one appears upon oxidation with ferricyanide. EPR quantitation gave approximately 1 spin/mol for the protein in the native state and 2 spin/mol in the ferricyanide treated protein. It was proposed that the two clusters were of the "HiPIP" type using the +3/+2 states (Table 4) with midpoint redox potentials 770 mV apart (+350 and -420 mV, were the measured redox potentials). In a further attempt to identify the clusters of ferredoxin I, Howard *et al.* (22) interpreted the extrusion experiments on the basis of two |4Fe-4S| centers, but Averill *et al.* (23) could not find evidence for more than one |4Fe-4S| center. Although EPR spectra strongly suggest the presence of |4Fe-4S| centers, the use of complementary techniques proved this assumption to be wrong. Indeed, preliminary X-ray results at 4.0 A (24) suggested the existence of two types of centers: one |4Fe-4S| and one |2Fe-2S|.

Meanwhile, Münck and co-workers studied the high and low potential centers of ferredoxin I using Mössbauer and EPR spectroscopy and suggested the presence of a high potential |4Fe-4S| center and a low potential center arranged in a novel type of iron-sulfur core, containing a three-nuclear iron-cluster (16). X-ray studies of *A. vinelandii* ferredoxin at 2.5 A (25) described this 3Fe center as a "distinctly planar structure that could not be modeled with |2Fe-2S| and |4Fe-4S| structures". The ligands to the |3Fe-3S| center are five cysteines and the sixth ligand, which may be a carboxylate group from glutamate or an oxygen provided by coordination of a water molecule (26).

The data analysis presented for the *A. vinelandii* ferredoxin (16) is complicated due to the presence of two types of centers in the same molecule. However, this center was shown to be present in Ferredoxin II isolated from *D. gigas*. This case represents a much simpler situation since the |3Fe-3S| center is the only core present. We will use this example as a prototype for proteins containing the three iron center; its properties are representative of this new type of core.

Isolated Fd II exhibits a farily isotropic EPR signal around g = 2 (27). This EPR spectrum is similar to those observed in other bacterial ferredoxins in the oxidized state (28) and has also been attributed in those cases to the +3 state of the |4Fe-4S| center (29). *D. gigas* Fd II is an example of how EPR techniques can be misleading in the identification of new iron-sulfur centers. Quantitation of the EPR signal accounts for 0.91-0.95 spin/3Fe atoms (17) and fully reduced Fd II gives a very weak g = 1.94 signal that accounts for only 0.02-0.03 spins/3Fe atom.

Mössbauer studies on oxidized Fd II show at 4.2 K a magnetic spectrum due to three distinct iron sites (17). At higher temperatures (77 K) the three components give the same Mössbauer spectra. The isomeric shifts and quadrupole splitting values are typical of high-spin Fe^{3+} with tetrahedral coordination by sulfur ligands (Table 3). The low temperature spectrum can be simulated by the spin Hamiltonian formalism for a system with S = 1/2 having different hyperfine coupling constants for the three subsites. A model for the spin-coupling between the three irons in the oxidized form of the three iron cluster (30) can account for the different hyperfine coupling constants for each iron.

Upon one electron reduction (E = -130 mV) an EPR silent state is observed and two distinct environments could be detected by Mössbauer spectroscopy at zero field (17). The two environments designated by doublet I and II have a ratio of 2:1. Doublet I is high-spin in character. The increase in δ and ΔE_Q of the iron atoms of doublet I points to a formal oxidation state of +2.5 for the two iron atoms. Application of a magnetic field to the reduced material leads to the conclusion that the irons belong to a spin coupled system with integer S > 0.

The low temperature magnetic circular dichromism (m.c.d.) of oxidized and reduced form of *D. gigas* Fd II has been measured in the range 300-800 nm (31). The main conclusions are:

1) The m.c.d. spectra of oxidized and reduced forms of Fd II are quite distinct from one another.

2) The oxidized state is different from the oxidized HiPIP state. The m.c.d. spectrum may be a useful fingerprint of the Fd II center as well as the well characterized Mössbauer properties (17). The m.c.d. spectrum may also be a good criterion in general to distinguish the 3Fe and 4Fe cores in higher oxidation states.

3) Both oxidation states are paramagnetic.

4) m.c.d. magnetization curve studies were fitted for the oxidized state to a ground state S = 1/2 with an isotropic g factor of 2.01. The curves were tentatively fitted for the reduced state with S = 2.

Magnetic susceptibility measurements performed by NMR show that upon reduction the magnetic moment of the cluster increases. (unpublished data).

Besides the two proteins that we have just described |3Fe-3S| clusters are now firmly established for aconitase (44), and appear to exist in *Thermus thermophilus* ferridoxin (Eckard Münck, personal communication) and *Methanosarcina barkeri* ferredoxin (unpublished data).

2.4 Four-Iron Centers

The |4Fe-4S| cluster has been one of the most studied structures by a wide variety of physical techniques.

The X-ray structure of oxidized *Peptococus aerogenes* ferredoxin (2x|4Fe-4S|) (32) and reduced HiPIP from *Chromatium* (33) are nearly identical. Based on this similarity and on other common properties Carter *et al.* proposed the "three state" hypothesis (Table 4). This hypothesis considers that the |4Fe-4S| clusters can be stabilized in three oxidation states, +3, +2, +1. Reduced HiPIP and oxidized *P. aerogenes* ferredoxin are in the +2 state. So, oxidized HiPIP is representative of a +3 state and reduced *P. aerogenes* ferredoxin of a +1 state.

The |4Fe-4S| center in the +2 state is EPR silent. Upon one electron reduction the center becomes EPR active (Table 3).

In the +3 state the center presents an EPR signal with g values bigger than 2 (35). In the +1 state which is typical of reduced bacterial ferredoxins, the |4Fe-4S| cluster presents a typical axial EPR signal with features around 2.07, 1.94 and 1.92 (the "g = 1.94" type signal). The oxidation-reduction transition which the cluster can undergo, can be denoted $|4Fe\text{-}4S|^{+2(+2,+3)}$ for HiPIP and $|4Fe\text{-}4S|^{+2(+2,+1)}$ for bacterial type ferredoxins using the IUB-IUPAC nomenclature (36) (Table 4). Further evidence for the "three state" hypothesis comes from the fact that native HiPIP (+2 state) can be further reduced to a +1 state in unfolding conditions (37) showing the typical g = 1.94 type spectrum with a redox potential estimated to be approx. -600 mV. However, there is no evidence at the present stage that clostridial ferredoxin can be superoxidized to the oxidation level +3 with ferricyanide (29) since this procedure leads to a conversion of |4Fe-4S| into |3Fe-3S| centers (38).

Holm and co-workers succeeded in designing model compounds which mimic the |4Fe-4S| cluster found in proteins (6) and explored the chemistry of this cluster. These chemical analogues proved to be of great importance for the determination of the role played by the polypeptide chain in controlling the physico-chemical properties of the inorganic iron-sulfur structures.

The development of techniques to displace or extrude the |4Fe-4S| center made by Orme-Johnson, Holm and co-workers is a useful way of analyzing the type of center present in a protein. The method involves the unfolding of the protein by an inorganic solvent under anaerobic conditions in the presence of a thiol which displaces the center. The displaced center is accepted by an apoferredoxin. The apoferredoxin can be a specific acceptor for |4Fe-4S| centers like *B. polymixa* ferredoxin. In refolding conditions the extruded core is recaptured by the added standard apoprotein giving a typical ferredoxin and the low temperature EPR can be quantitated (39,40).

3. METHODOLOGY FOR IDENTIFICATION OF TYPE OF IRON-SULFUR CENTER

The proteins containing iron-sulfur centers are generally brown or red in colour. The optical spectra are characteristic of the type of center (41) but in general they are broad and in complex systems they are difficult to resolve. The EPR spectra of the cores in different oxidation states are distinct (42); therefore EPR has been one of the most used methods for detecting this type of protein in complex systems (Table 3). However, the use of EPR as the sole technique can be misleading when applied to new situations as we have seen for the case of the |3Fe-3S| core. Due to these properties, iron-sulfur proteins are frequently isolated before the physiological role is determined. The analysis of the type and number of iron-sulfur centers is not an easy task. Iron and labile sulfide analysis can give wrong values due to protein interference and the uncertainty of the determination of molecular weights and proteins concentrations. Only the combination of different techniques can give a full characterization of the type of active center.

3.1 One-Iron Center

The rubredoxin type center presents a very characteristic UV/visible spectrum (see Table 2), and a very intense CD and MCD spectrum. The redox potential is close to 0 mV (43) and, in general, reduction with ascorbic acid can be accomplished. The EPR spectrum shows a signal at $g = 4.3$ that can be observed at 77 K and additional features at $g = 9.4$ below 20 K in the oxidized form ($S = 5/2$). No signal is observed in the reduced state ($S = 2$). The saturation field determined by Mössbauer spectroscopy in the +3 state is around 410 kG, which is a very low value and can be used for recognition of the center. Magnetic moments are 5.70 and 4.90 Bohr magnetons for the oxidized and reduced form, respectively.

3.2 Two-Iron Center

This center shows a characteristic UV/visible spectrum (Table 3). In general if no other center is present it is very easy to identify a |2Fe-2S| center by its absorption bands at 330, 420 and 465 nm. EPR spectra can be observed in +1 state with $g_{av.} < 2$ at 77 K. No signal is observed in the +2 state. In 80% DMSO and under reducing conditions an isotropic EPR signal is observed below 77 K. The Mössbauer spectrum at high temperatures in the +1 state shows two doublets in the ratio 1:1. At low temperatures the two sites show coupling constants of opposite sign. Extrusion and displacement reaction with o-xylyldithiol in

80% DMSO give characteristic products that can be transferred to apoadrenodoxin. The core transfer is specific for |2Fe-2S| centers.

3.3 Three-Iron Center

The EPR spectrum in the oxidized state is fairly isotropic centered around g = 2.02. There is no signal in the reduced state. A wide range of redox potentials have been determined (-420 mV for *A. vinelandii* Fd and -130 mV for *D. gigas* Fd II). Typical Mössbauer spectra in the reduced form at 4.2 K in zero field gave two doublets in the ratio 2:1 (Table 3). A substantial broadening at weak applied fields (600 G) might be a typical feature. Magnetic susceptibility measurements indicate that in going from the oxidized to the reduced form the center is converted from a state with S = 1/2 to another paramagnetic state, probably with S = 2. (This is a big difference from |4Fe-4S| or |2Fe-2S| centers where the change from a state with S = 1/2 (+3 or +1 in |4Fe-4S| or +1 in |2Fe-2S|) to the +2 oxidation state causes a decrease in the magnetic susceptibility. MCD spectra of both oxidized and reduced Fd II are very different from those of the four-iron sulfur cluster and should provide a useful criterion for distinguishing between three and four iron sulfur clusters (31).

3.4 Four-Iron Center

The UV/visible spectrum observed in the +2 state shows broad features around 400 and 300 nm similar to those observed for |3Fe-3S| centers. A |4Fe-4S| in the +3 state can be easily identified if no other center is present by its maxima near 325, 375 and 450 nm (Table 3). The CD spectrum is weak and therefore not easily identifiable. The EPR spectrum in 80% DMSO (reducing conditions) gives an axial type spectrum observable below 35 K. HiPIP and ferredoxin type centers may be easily distinguished based on EPR characteristics. In the +3 state $g_{av.} > 2$ and the +1 state $g_{av.} < 2$. So far all the HiPIP proteins studied have a very positive redox potential. Extrusion and displacement reactions with o-xylydithiol in 80% DMSO give characteristic products that can be transferred to apoferredoxin from *B. polymixa* a specific acceptor for |4Fe-4S| centers.

REFERENCES

1. Watenpaugh, K.D., Sieker, L.C., Herriott, J.R. and Jensen, L.H.: 1973, *Acta Crystal.* B29, pp. 943-956.
2. Adman, E.T., Sieker, L.C., Jensen, L.H., Bruschi, M. and LeGall, J.: 1977, *J. Mol. Biol.* 112, pp. 113-120.
3. Watenpaugh, K.D., Sieker, L.C. and Jensen, L.H.: 1979, *J. Mol. Biol.* 131, pp. 509-522.
4. Shulman, R.G., Eisenberger, P., Teo, B.K., Kincaid, B.M. and Brown, G.S.: 1978, *J. Mol. Biol.* 124, pp. 305-321.

5. Blumberg, W.E.: 1967 in "Magnetic Resonance in Biological Systems", Ehrenberg, A., Malmstrom, B.C., Vanngard, T., eds. Pergamon Press, Oxford, pp. 119-133.
6. Holm, R.H. and Ibers, J.A.: 1977, in "Iron Sulfur Proteins", Lovenberg, W., ed., Academic Press, New York, p. 205.
7. Kostikas, A., Petrouleas, V., Simopolous, A., Coucouvanis, D. and Holah, D.G.: 1976, *Chem. Phys. Lett.* 38, pp. 582-584.
8. Phillips, W.D., Poe, M., Weiher, J.F., McDonald, C.C. and Lovenberg, W.: 1970, *Nature* 227, pp. 574-577.
9. McCarthy, K.F. and Lovenberg, W.: 1970, *Biochem. Biophys. Res. Commun.* 40, pp. 1053-1057.
10. Moura, I., Huynh, B.H., Hausinger, R.P., LeGall, J., Xavier, A.V. and Münck, E.: 1980, *J. Biol. Chem.* 255, pp. 2493-2498.
11. Shulz, C. and Debrunner, P.G.: 1976, *J. Phys. Colloque* 37, suppl. 12, p. 153.
12. Gibson, J.F., Hall, D.O., Thornley, J.H.M. and Whatley, F.R.: 1966, *Proc. Natl. Acad. Sci. U.S.A.* 56, pp. 987-990.
13. Dunham, W.R., Bearden, A.J., Salmeen, I.T., Palmer, G., Sands, R.H., Orme-Johnson, W.H. and Beinert, H.: 1971, *Biochim. Biophys. Acta,* 253, pp. 134-152.
14. Münck, E., Debrunner, P.G., Tsibris, J.C.M. and Gunsalus, I.C.: 1972, *Biochemistry* 11, pp. 855-863.
15. Tsukihara, T.K., et al.: 1978, *Abstracts, Sixth International Biophysics Congress,* Kyoto, Japan.
16. Emptage, M.H., Kent, T.A., Huynh, B.H., Rawlings, J., Orme-Johnson, W.H. and Münck, E.: 1980, *J. Biol. Chem.* 255, pp. 1793-1796.
17. Huynh, B.H., Moura, J.J.G., Moura, I., Kent, T.A., LeGall, J., Xavier, A.V. and Münck, E.: 1980, *J. Biol. Chem.* 255, pp. 3242-3244.
18. Shetna, Y.T.: 1970, *Biochim. Biophys. Acta* 250, pp. 58-62.
19. Yoch, D.C. and Arnon, D.I.: 1972, *J. Biol. Chem.* 247, pp. 4514-4520.
20. Sweeney, W.V., Rabinowitz, J.C. and Yoch, D.C.: 1975, *J. Biol. Chem.,* 250, pp. 7842-7847.
21. Yoch, D.C. and Carithers, R.P.: 1978, *J. Bacteriol.* 136, pp. 822-824.
22. Howard, J.B., Lorsbach, T. and Que, L.: 1976, *Biochem. Biophys. Res. Comm.,* 70, pp. 582-588.
23. Averill, B.A., Bale, J.R. and Orme-Johnson, W.H.: 1978, *J. Am. Chem. Soc.,* 100, pp. 3034-3043.
24. Stout, C.D.: 1979, *Nature,* 279, p. 83.
25. Stout, C.D., Ghosh, D., Pattabhi, V. and Robbins, A.H.: 1980, *J. Biol. Chem.,* 255, pp. 1797-1800.
26. Stout, C.D.: 1981, in "Iron Sulfur Proteins", Spiro, T., ed., in press.
27. Cammack, R., Rao, K.K., Hall, D.O., Moura, J.J.G., Xavier, A.V., Bruschi, M., LeGall, J., Deville, A. and Gayda, J.-P.: 1977, *Biochim. Biophys. Acta,* 490, pp. 311-321.

28. Palmer, G., Sands, R.H. and Mortenson, L.E.: 1966, *Biochem. Biophys. Res. Commun.*, 23, pp. 357-362.
29. Sweeney, W.V., Bearden, A.J. and Rabinowitz, J.C.: 1974, *Biochem. Biophys. Res. Commun.*, 59, pp. 188-194.
30. Kent, T.A., et al.: 1980, *Proc. Natl. Acad. Sci. U.S.A.*, 77, p. 6574.
31. Thomson, A.J., et al.: 1981, *Biochim. Biophys. Acta*, in press.
32. Adman, E.T., Sieker, L.C. and Jensen, L.H.: 1976, *J. Biol. Chem.*, 251, pp. 3801-3806.
33. Carter, C.W. Jr.: 1977, *J. Biol. Chem.* 252, pp. 7802-7811.
34. Carter, C.W. Jr., Kraut, J., Freer, S.T., Alden, R.A., Sieker, L.C., Adman, E. and Jensen, L.H.: 1972, *Proc. Natl. Acad. Sci. U.S.A.*, 69, pp. 3526-3629.
35. Palmer, G., Brintzinger, H., Estabrook, R.W. and Sands, R.H.: 1967, in "Magnetic Resonance in Biological Systems", Ehrenberg, A., Malmstrom, B.G. and Vanngard, T., eds., Pergamon Press, Oxford, pp. 159-171.
36. IUB - IUPAC: Nomenclature of Iron-Sulfur Proteins - Recommendations: 1979, *Biochim. Biophys. Acta*, 549, pp. 101-105.
37. Bruschi, M., Moura, I., LeGall, J., Xavier, A.V. and Sieker, L.C.: 1979, *Biochem. Biophys. Res. Commun.*, 90, pp. 596-605.
38. Thomson, A.J., et al.: 1981, *Biochim. Biophys. Acta*, in press.
39. Averill, B.A. and Orme-Johnson, W.H.: 1978, in "Metal Ions in Biological Systems", Siegel, H., ed., vol. 7, New York, Marcel Decker, p. 127.
40. Orme-Johnson, W.H. and Holm, R.H.: 1978, in "Methods in Enzymology", Fleisher, S. and Packer, L., eds., Vol. III, New York, Academic Press, p. 631.
41. Hall, D.O., et al.: 1974, in "Iron in Biochemistry and Medicine", Jacobs, A. and Worwood, M., eds., New York Academic Press, p. 279.
42. Orme-Johnson, W.H. and Sands, R.H.: 1973, in "Iron Sulfur Proteins", Lovenberg, W., ed., Vol. 3, New York, Academic Press, p. 195.
43. Moura, I., Moura, J.J.G., Santos, M.H., Xavier, A.V. and LeGall, J.: 1979, *FEBS Lett.*, 107, pp. 419-421.
44. Kent, T.A., et al.: 1980, in "Symposium on Interaction Between Iron and Proteins in Oxygen and Electron Transport", Ho, C., ed., Elsevier/North Holland, Amsterdam, in press.
45. Xavier, A.V., Moura, J.J.G. and Moura, I.: 1981, *Structure and Bonding*, 43, pp. 187-213.

CATALYSIS BY HIGHLY ACTIVE 12Fe-12S CONTAINING HYDROGENASES

Hans J. Grande[a], Cees van Dijk[a], William R. Dunham[b] and Cees Veeger[a].

a) Department of Biochemistry, Agricultural University, Wageningen, The Netherlands

b) Department of Biophysics, University of Michigan, U.S.A.

Steady state kinetics has been used to investigate the mechanism for the production and consumption of hydrogen by M.elsdenii *hydrogenase. Similar kinetics has been found for* D.vulgaris *hydrogenase. It is proposed that both enzymes act according to a random mechanism. Electron spin resonance is used to gather evidence for some of the states following from the proposed mechanism. It is found that several* [Fe_4-S_4]*-type of clusters are present. In fact at least three types of ferredoxin clusters and one HIPIP cluster are needed to explain the EPR data. It is concluded that a super-reduced HIPIP plays a role in the 2-electron transfer step essential in the catalytic mechanism.*

1. INTRODUCTION

Hydrogenase is an enzyme which can produce and consume molecular hydrogen. Although the existence of hydrogenase had been demonstrated already in 1931 by Stephenson and Strickland [1], the purification of this kind of enzyme has taken a much longer time. Nowadays many purified preparations from a wide range of micro-organisms have been described. In Table 1 a survey of hydrogenases from different sources and their properties is given. It is apparent that these enzymes all contain Fe/S in different contents and in some cases also flavin. Some of the enzymes are oxygen stable when isolated, others are highly oxygen-sensitive and deactivate in minutes when exposed to air. Also there is a large difference in the hydrogen production activity.

In fact only three enzymes, from *M.elsdenii*, *C.pasteurianum* and from *D.vulgaris* (Hildenborough) are very active in producing hydrogen. Hydrogenase from *M.elsdenii* for example produces under optimum conditions (low pH, high salt concentration) about 2 l H_2 per minute per mg of protein. Since these three proteins have

H. B. Dunford et al. (eds.), The Biological Chemistry of Iron, 193–206.

Table 1. Catalytic, structural and photo-chemical properties of various hydrogenases.
++ very oxygen sensitive; + oxygen sensitive; - oxygen insensitive;
-- very oxygen insensitive.

Micro-organisms	*Megasphaera elsdenii*	*Clostridium pasteurianum*	*Desulfovibrio vulgaris* (Hildenborough)	*Desulfovibrio gigas*
Physiology	obligately anaerobic fermentation		obligately anaerobic, with with electron phosphorylation	
Localization	cytoplasm?	cytoplasm	periplasm	periplasm
Natural electron donor/acceptor	flavodoxin ferredoxin	ferredoxin	cytochrome c_3	cytochrome c_3
Molecular weight	50.000	60.500	50.000	89.500
Subunits	1x50.000	1x60.500	1x50.000	1x62.000 1x26.000
E_m(ox)($mM^{-1}cm^{-1}$, 400 nm)	46	36	46	46.5
Fe	12	12	12	12
S^{2-}	12	12	12	12
Flavin/mol	--	--	--	--
Structure of cluster	3x(4Fe-4S)	3x(4Fe-4S)	3x(4Fe-4S)	3x(4Fe-4S)
EPR oxidized	2.101, 2.052 2.005	2.099, 2.046 2.005	2.029-1.995	
EPR reduced	2.087-1.89	2.079-1.89	2.055-1.893	
Oxygen sensitivity	++	++	--	
V(μmol.H_2.min^{-1}.mg^{-1}) donor MV^+	7000, pH 8 140000, pH 5.5	4000, pH 8	10400, pH 8	91, pH 7.6
References	2,3	4-8	9	10

a. Observed at 50-77 K
b. Observed at <30 K

Table 1. Continued

Chromatium	*Chromatium vinosum*	*Rhodospirillum rubrum*	*Proteus mirabilis*	*Alcoligenes eutrophus*
	anaerobic	phototropic	faculta-tively an-aerobic	aerobic chemo-antrotrophic
membranes	soluble	membranes	membranes	membranes
unknown	unknown	ferredoxin I, II? cyt "c_3"?	unknown	NAD^+
98.000	55.000	66.000	205.000	200.000
2x50.000	1x55.000	1x66.000	2x63.000 2x33.000	1x68.000 1x60.000 2x29.000
14			106	60 (A_{420})
4 4 --	4 4 --	4 4 --	24 24 --	12 12 2
1x(4Fe-4S)	1x(4Fe-4S)	1x(4Fe-4S)		2x(4Fe-4S) 2x(2Fe-2S)
2.019, 2.003	1.94, 1.96 1.98, 2.003, 2.018, 2.03	2.023, 2.004	2.025, 2.018	2.02
2.2, 2.125, 2.06	no signal	no signal	no signal	2.04, 1.95[a] 2.04 - 1.86[b]
--	--	--	--	--
		33.3, pH 7		48
11	12	13, 14	15	16, 17, 18

several similarities (see Table 1) the question arises whether they have a similar organisation of the active site and electron donor groups. To answer this question we have studied the hydrogenases from *M.elsdenii* and *D.vulgaris* quite extensively. In this article we want to illustrate how Electron Spin Resonance can be used to identify Fe/S clusters in different redox states of the enzyme. Adams *et al.* concluded in a recent review [20] on hydrogenases that the common feature to all hydrogenases is a so-called high potential $[Fe_4-S_4]$-cluster (HIPIP). They suggested that because of the conformity this cluster might be the active center of all hydrogenases. We want to show that indeed in our enzymes the HIPIP cluster might be the active site, but that it does not function as an ordinary HIPIP. Furthermore we will show that the other clusters function as electron donor sites to this active site.

2. RESULTS

I. The kinetic behaviour of *M.elsdenii* has recently been published by our group [2,21]. It was shown in [2] that in the H_2-producing reaction no linear kinetics is observed. Especially at high concentrations of the artificial donor, reduced methyl viologen (MV^+), deviation of a linear relation in the reciprocal plot is obtained. We proposed that since there is no reason for cooperativity (evidence for this will be given later on) a random mechanism can explain the observed non-linearity.
The following reactions can occur:

(1) $E + H^+ \rightarrow EH^+$ $\qquad$ $E + e^- \rightarrow E^-$

(2) $EH^+ + e^- \rightarrow E + \frac{1}{2}H_2$ $\qquad$ $E^- + H^+ \rightarrow E + \frac{1}{2}H_2$

In a random mechanism both reactions operate and compete with each other:

(3)

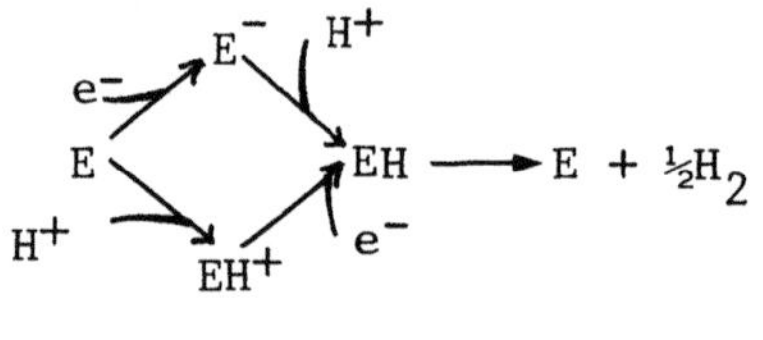

In these reactions e^- is transferred from a suitable electron donor. Reaction 3 is a simplified representation of the proposed mechanism [21].

In the meantime we studied the kinetic behaviour of the *D.vulgaris* (Hildenborough) hydrogenase and again non-linear kinetics is observed. A typical fitted curve to the experimental data is shown in Fig. 1. Erbes and Burris [19] stated that *C.pasteurianum* hydrogenase is linear with respect to its hydrogen production activity. Van Dijk [2] measured equilibrium constants for the dimerisation of MV^+ at different salt concentrations. In the simu-

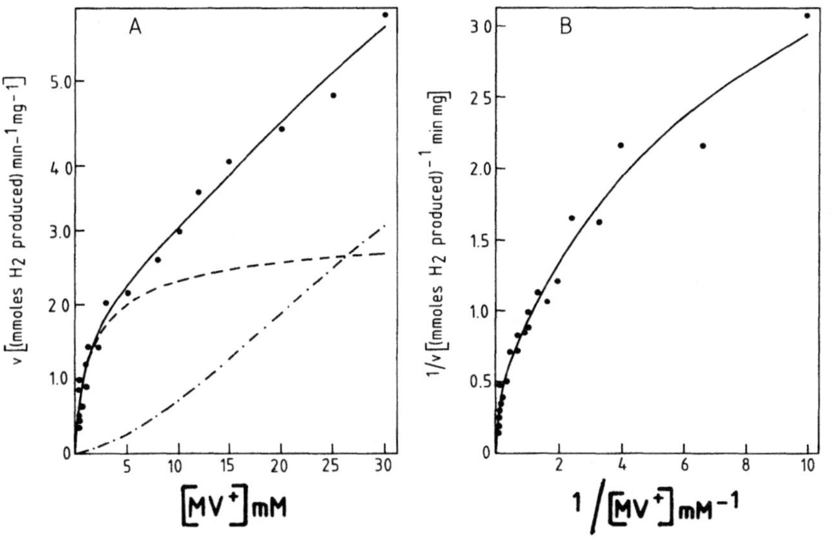

Fig. 1. *Direct Plot (A) and double-reciprocal plot (B) of the hydrogen production activity versus the methyl viologen concentration.*
--- Monomeric MV^+ contribution to the activity; -·- Dimeric MV^+ contribution to the activity; —— Total activity.

lated curves for *M.elsdenii* [2] and for *D.vulgaris* (Fig. 1) the dimer was taken into account as a separate substrate. At high concentrations of reduced methyl viologen it is shown in the review article of Adams *et al.* [20] that *C.pasteurianum* hydrogenase also shows curvature in the double reciprocal plot. For all three enzymes [2,19, this article] at low MV^+ concentrations the double reciprocal plot is linear which means that one path, the one in which H^+ is the first substrate, becomes much faster. Under those conditions the mechanism changes to compulsory and the kinetics is again linear.

Another explanation in literature for the observed nonlinearity in the kinetics could be cooperativity. The *Chromatium* hydrogenase data in [22] are treated as being apparent cooperative. In our opinion those data can be fitted also by the random model. Moreover no evidence is proposed for the apparent cooperativity. In case of *D.vulgaris* and *M.elsdenii* hydrogenase there is no evidence for di(poly)merisation of hydrogenase, which means that the cooperativity should be found inside the monomer enzyme itself. The only possibility then could be the presence of more $[Fe_4-S_4]$ clusters which cooperate. We will show EPR evidence that this is not the case.

Several other observations have been made with *M.elsdenii* hydrogenase. Since the kinetics of this enzyme has been studied ex-

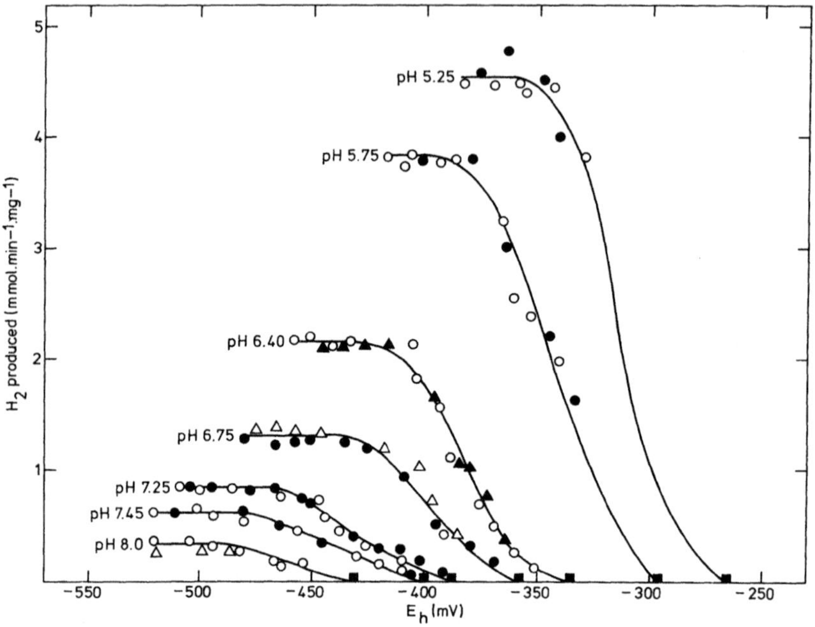

Fig. 2. *Effect of pH and redox potential on the hydrogen production activity with methyl viologen semiquinone* [*from ref. 21*]. The methyl viologen concentration is 0.3 mM. Activities were determined spectrophotometrically [23]. The different potentials were obtained by varying the relative dithionite/sulfite concentrations. The dithionite concentrations used: (o) 2 mM; (o) 2 mM; (Δ) 5 mM; (Δ) 10 mM; (□) end potentials. (Reproduced with permission from the European Journal of Biochemistry)

tensively we will discuss this enzyme and show wherever possible the similarities with *D. vulgaris* (Hildenborough) hydrogenase. In Fig. 2 the activities at different pH and different redox potentials are given. From those curves it is apparent that in all cases the activity of the enzyme follows a redox type of titration curve. That is: at low potential the activity is maximal, upon rising the potential (at constant pH) the activity stays the same, then decreases rapidly until no activity is left. When the curves are analysed, they show a n=2 type of activity - redox potential profile (that means that two-electrons are transferred in one step).

Other observations can be made from Fig. 2. The activity becomes higher at lower pH, while still a n=2 redox curve is followed. The shift of the midpoint potentials follows in fact the hydrogen electrode (e.g. at pH 8.0 the midpoint potential relative to the hydrogen electrode at pH=0 is 8 x -60 mV = -480 mV). This type of kinetic behaviour has some consequences for the mechanism, since the activity is related to the functioning of Fe/S clusters in electron transport. First of all, the kinetics do show a n=2 redox curve. The question is: where do those 2-electrons come from or go to? Secondly, how is it possible that the enzyme is even active at pH 5.5? If the activity of the enzyme is associated with the redox state of the clusters the n=2 profile means that there must be a catalytic centre which is able to accept two electrons at a time, while normally the Fe/S centres bound to protein are thought to be one-electron donors or acceptors. The fact that we have still activity at pH 5.5, means that a ferredoxin type of cluster can be reduced at -330 mV (5.5 times -60 mV) which is a very high potential for those type of clusters. It is clear that evidence is needed for a 2-electron accepting site. Recent results indicate that the midpoint potentials of the clusters are pH-dependent. In the second part of this article EPR results for both *M.elsdenii* and *D.vulgaris* hydrogenase are presented as evidence for the observations made above.

II. It has been known already for some time that Fe/S clusters can be observed by Electron Spin Resonance in different redox states depending on the type of cluster which is present. In Fig. 3 a typical ferredoxin type (dotted line) of spectrum is shown. The spectrum is of the partially reduced hydrogenase of *D.vulgaris*. Such a type of spectrum arises from a spin one half system, which has a non-spherical electron distribution. The electron in this case has different electron densities in different directions and thus resonates at different fields.

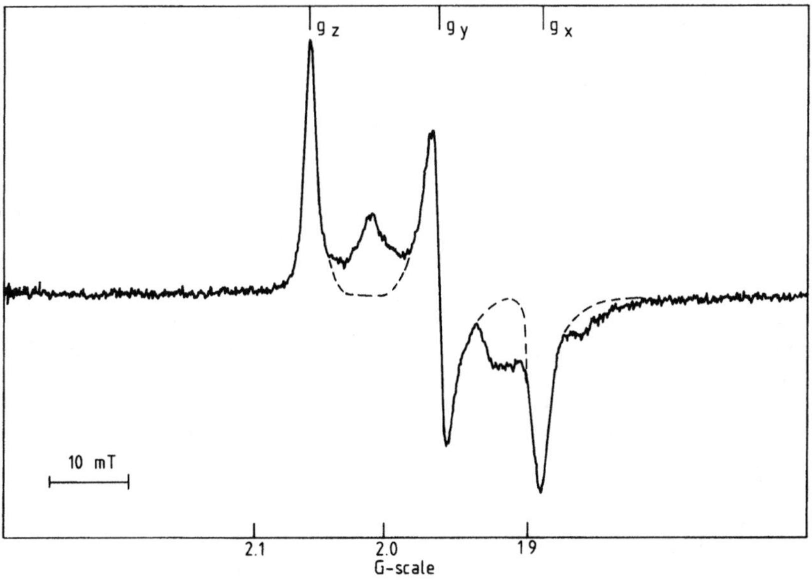

Fig. 3. *Partially reduced hydrogenase from* D.vulgaris. Enzyme concentration 90 μM, pH 8 in the presence of dithionite and sulfite. The spectrum was taken at a temperature of 10.8 ^{o}K; microwave power 0.05 mW. Modulation amplitude 1 mT. The G-scale on the lower axis is calculated from the measured values of both the sweep-width and microwave frequency.

Three main axes are defined as X, Y and Z as is shown in Fig. 3. The spectrum is taken at low temperature and therefore the molecules do not rotate and all directions relative to the external field are present. All those directions contribute to the spectrum but the main absorptions are along the main axes. To assign a value to positions in EPR the so-called g-value is used. The resonance condition is:

(4) $h\nu = g\beta B$

in which h is Plank's constant, ν the microwave frequency, β the Bohr magneton of the electron and B the magnetic field at which resonance takes place. Experimentally the value of ν is constant in an experiment and the magnetic field is swept. A position in the spectrum is therefore defined by its B-field. Since however the frequency ν is different from experiment to experiment $g = \frac{h\nu}{\beta B}$ is an absolute measure of position in an EPR experiment. The three positions in Fig. 3 at X, Y and Z can now be defined by their respective g-values g_x, g_y and g_z. In this way a comparison with other species can be made. In case of the hydrogenase the g-values are 1.891, 1.959 and 2.055 which are typical values observed for all ferredoxin type of proteins.

Ferredoxins are proteins which act in the cell as redox carriers. They contain, depending on the species, either a $[Fe_2-S_2]$ cluster (plant type) or $[Fe_4-S_4]$ cluster(s).

They are characterised by a low midpoint potential, around -420 mV relative to the hydrogen electrode. Upon reduction a spectrum which is similar to Fig. 3 is obtained. One can distinguish between spectra from a $[Fe_2-S_2]$ cluster and a $[Fe_4-S_4]$ cluster by studying the temperature dependence of the signals. Since in both cases ($[Fe_2-S_2]$ and $[Fe_4-S_4]$) the unpaired electron is more or less delocalised over all the Iron nuclei of the cluster, this electron is in a highly perturbed environment. For this reason low temperature measurements which freeze the environment are needed, otherwise one could not observe the electron at all. The unpaired electron of a $[Fe_2-S_2]$ cluster is less delocalised and therefore observable up to higher temperatures. Typically a $[Fe_4-S_4]$ is not observed anymore above 40^oK and a $[Fe_2-S_2]$ above $70-80^oK$. In case of *D.vulgaris* hydrogenase the observed spectrum of Fig. 3 disappears around 35^oK and is thus of the $[Fe_4-S_4]$-ferredoxin type.

Another type of the Fe/S-signals is shown in Fig. 4. This spectrum is of *M.elsdenii* hydrogenase. The calculated redox potential at low temperature is about -280 mV. This type of spectrum is described in the literature as a HIPIP-type of signal. Characteristic for such a signal is an average g-value above g=2 in contrast to ferredoxins. Another characterisation is that they are much less sensitive to temperature and can also be observed at a temperature around 60^oK. A HIPIP type of cluster is also a $[Fe_4-S_4]$ cluster, but with a different redox state.
HIPIP can be oxidized from the $[Fe_4-S_4]^{2+}$ state (two irons have a

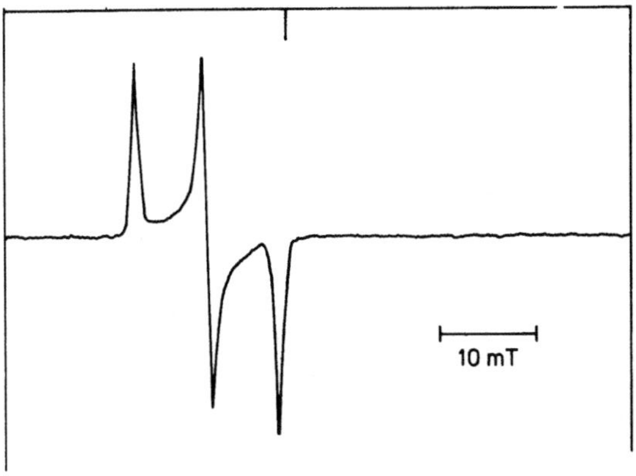

Fig. 4. *EPR spectrum of oxidized* M.elsdenii *hydrogenase [from ref. 3]*. Enzyme concentration 40 μM in 100 mM phosphate buffer pH 6, 1 mM $Na_2S_2O_4$ and 150 mM Na_2SO_3. The spectrum was taken at 9.7°K; microwave power 0.05 mW; Modulation amplitude 0.2 mT. The vertical line above the figure indicates the calculated (calibrated) g=2. (Reproduced with permission from the European Journal of Biochemistry)

formal charge of 2+ and two of 3+) to the $[Fe_4-S_4]^{3+}$ state. A ferredoxin cluster can be reduced from the $[Fe_4-S_4]^{2+}$ state to the $[Fe_4-S_4]^{+}$ state, in which case the formal charges of iron become 3 of 2+ and 1 of 3+. In general the ferredoxin-type of cluster can accept only one electron and the HIPIP can only donate one electron, although exceptions are becoming more apparent [e.g. C. Veeger *et al.*, this book].

Fig. 5 (upper curve) shows an X-band EPR spectrum of *D.vulgaris* hydrogenase which is reduced under hydrogen. Clearly the spectrum is very complex and several species are present. The question is how can one get information from such a spectrum.
First of all temperature can be varied and it is found that the intensity of almost the whole spectrum decreases rapidly with the exception of a signal around g = 2.03. From this it is concluded that the signal is mainly of the $[Fe_4-S_4]$ ferredoxin type while also a small amount of HIPIP type of spectrum is present. The problem is whether the spectrum has to be built up out of a sum of rhombic type of spectra like the one in Fig. 3 or that spin-spin interaction between two or more clusters can be the reason for its complexity. If two electrons are close enough they will interact and quite complex spectra can arise.

As an illustration of spin-spin interaction the $2[Fe_4-S_4]$ ferredoxin from *Micrococcus lactyliticus* can be chosen [24]. The lower spectrum of Fig. 6 shows the spectrum of this protein reduced with dithionite. The spectrum is clearly very much different from that of Fig. 3 and it is impossible to obtain such a spectrum by adding two spectra, like the one of Fig. 3, each with slightly different g-values.
In fact the spectrum arises due to interaction of the two spins, which have a coupling of about 350 MHz. The arrows indicate some very low intensity wings which also arise from the the interaction. The interaction of the spins is dependent both on the angle between the two spin distributions and the distance. Formally one can define a spin-spin interaction tensor D, which describes the

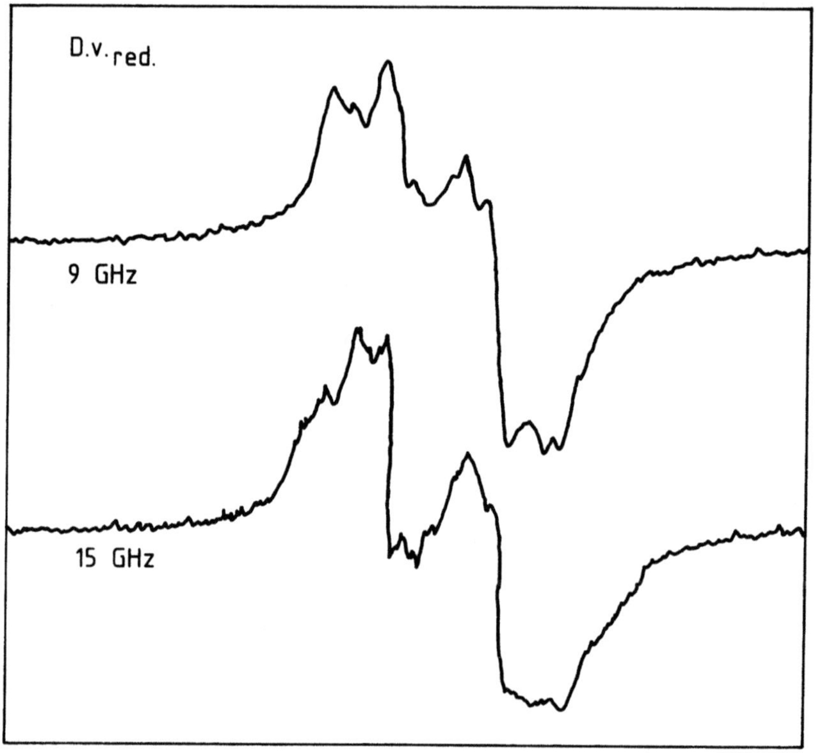

Fig. 5. *Reduced* D.vulgaris *hydrogenase at X- and P-band EPR.* Upper curve: microwave frequency 9.42 GHz; microwave power .51 mW; Modulation amplitude, 1 mT. Temperature 10.8°K. Lower curve: microwave frequency 14.97 GHz; microwave power 2.4 mW; Modulation amplitude 1 mT; Temperature 15°K. The spectra are given at the same g-scale.

interaction. This interaction is independent of the applied B-field. If the frequency ν of the spectrophotometer is changed, B has to be changed proportionally in order to contain the resonance condition (equation 4). Since the interaction is not dependent on B, the term $S_1.D.S_2$. (S_1 and S_2 are the spins of cluster 1 and 2) will not change with frequency. Changing the frequency thus becomes a tool for observing the presence of interaction. (Hyperfine interaction is not discussed here, but it will be also independent of B). If the interaction is relatively weak at 9 GHz it might be relatively big at 3 GHz, since the term gβB becomes three times smaller. Reversely a strong interaction might become less important if the frequency is raised. The upper curve of Fig. 6 shows this for *M. lactilyticus* ferredoxin. Changing the frequency from about 9 GHz to about 35 GHz changes the spectrum drastically. The wings (at the arrows in the upper curve) are much more intense and move

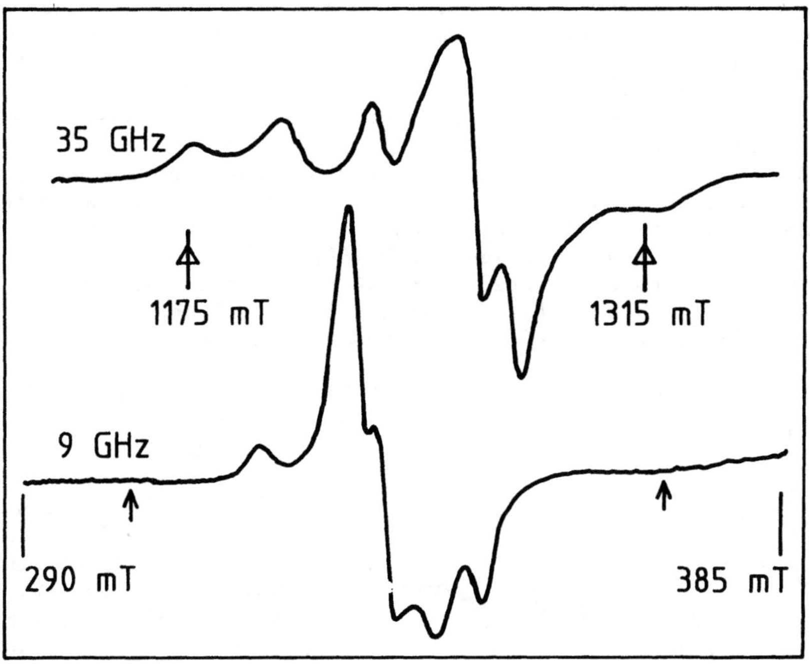

Fig. 6. *Fully reduced* Micrococcus lactilyticus *ferredoxin at X- and Q-band EPR [from ref. 24]*. Upper curve: microwave frequency 34.16 GHz; microwave power 8 mW; modulation amplitude .9 mT; Temperature, 20°K. Lower curve: microwave frequency 9.24 GHz; microwave power .003 mW; modulation amplitude 1.8 G; Temperature 10°K.

inside, which is in accord with the theory. The spectrum even at 35 GHz does not resemble a normal rhombic-type of spectrum like Fig. 3, which means that the interaction is still relatively strong. The value of 350 MHz follows both from simulation at different frequencies and ELDOR [24].

If we now consider again Fig. 5, which represents the reduced hydrogenase from *D.vulgaris*, we see that there is hardly any difference at the two frequencies. The lower curve which is a spectrum at 15 GHz is almost identical with the one at 9 GHz. The conclusion of this result is that apparently there is only a very small interaction between the electron spins in this case. The same is true for hydrogenase from *M.elsdenii* in the reduced state (not shown).

The spectrum observed in Fig. 5 therefore consists of a sum of rhombic-type spectra, like the ones in Figs. 3 and 4. It is however clear that the appearance of Fig. 5 is much more complex than a simple sum of two ferredoxin-like spectra. As stated in Table 1 3[Fe_4-S_4] clusters are found in case of both *M.elsdenii* and *D.vulgaris* enzyme. In case of *D.vulgaris* hydrogenase this is not only based on the EPR spectrum and the Fe/S content (Table 1) but also on cluster extrusion experiments (to be published). One of those three clusters however must be of the so-called HIPIP-type, since this cluster is also observed (Fig. 4). Also in case of *D.vulgaris* enzyme there is a clear HIPIP-type signal present under different circumstances. This apparent contradiction, more than two ferredoxin-type of spectra and a HIPIP type cluster needed, while only 3 clusters are present, can be solved by assuming that the high potential cluster can transfer in this case two electrons at a time. Thus oxidized HIPIP (EPR of Fig. 4) becomes super-reduced and has then a ferredoxin-type of spectrum ($[Fe_4-S_4]^{3+} \xrightarrow{2e^-} [Fe_4-S_4]^+$). This explains also why in Fig. 5 there is some HIPIP-type of spectrum left in the reduced enzyme.

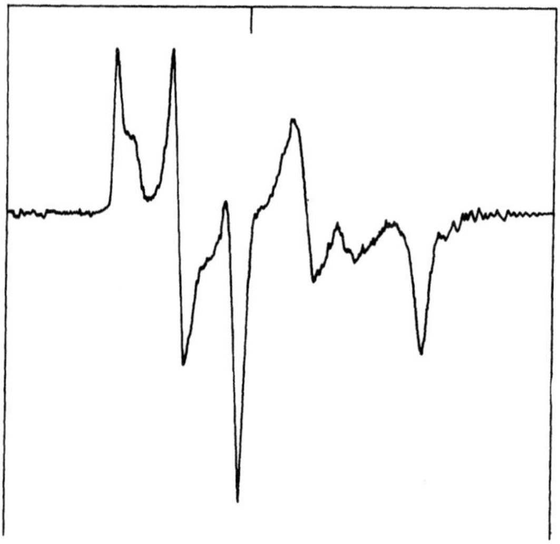

Fig. 7. *EPR spectrum of* M.elsdenii *hydrogenase in a partial reduced state* [*from ref.* 3]. Enzyme concentration 52 μM in 100 mM phosphate buffer, pH 7.6 with 45 mM $Na_2S_2O_4$. EPR settings microwave power, .05 mW; modulation amplitude, 1 mT; temperature 10°K. The sweep width is the same as in Fig. 4. The vertical line above the spectrum indicates g=2. (Reproduced with permission from the European Journal of Biochemistry)

Fig. 7 shows a spectrum of *M.elsdenii* which is clearly a mixture of a ferredoxin-type of cluster and of the HIPIP cluster (Fig. 4). This spectrum can now also be explained. Clearly the HIPIP-cluster after being super-reduced turns back to its $[Fe_4-S_4]^{3+}$ state and shows the HIPIP-type of spectrum again. Next to this there is one ferredoxin cluster still reduced. This can be either the super-reduced HIPIP cluster or one of the two other clusters, which has perhaps a slightly higher midpoint potential. Earlier [2] it was hard to explain how species with such a large difference in midpoint potential can occur at this potential.

3. CONCLUSIONS

In the first part the kinetic results and their implications were given. From the EPR results it became clear that there are several ferredoxin-type of clusters which do not interact. From this, cooperativity or apparent cooparativity between the two ferredoxin sites can be excluded.

The finding that there is a 2-electron accepting HIPIP makes this cluster a likely candidate for the n=2 redox step in the catalytic mechanism.

From the interpretation given it follows that at maximum 4 electrons can be present on the protein. In that case both the ferredoxin-clusters are reduced and the HIPIP-cluster is super-reduced. Under hydrogen or with dithionite one does not expect to find all four electrons. Reduction of the ferredoxin clusters has to take place first. If at an enzyme molecule 2 clusters are reduced they pass on their electrons to the HIPIP type of cluster, which becomes super-reduced. This centre in turn reacts rapidly with the H^+'s and turns into a HIPIP-centre again. At that moment the electrons leave as H_2 the enzyme. Therefore it will be difficult to observe more than 2 electrons on the protein. In fact integration of our signals give about 1.6 electron-equivalents both for the hydrogen- and dithionite-reduced enzyme. The observation that both hydrogen- and dithionite-reduced enzyme have about the same spectrum proves that also with dithionite the reduction of the clusters is at about the midpoint potential. Since those enzymes are highly active the dithionite is used to produce hydrogen and before freezing the enzyme is already at its hydrogen potential. Only reduction of the enzyme under a high hydrogen pressure will result in fully reduced enzyme. This type of experiment will be done by us in the near future.

With the results shown here it becomes possible to interpret the spectra of partially reduced hydrogenases. Since interaction is negligibly small at 15 GHz (Fig. 5) simulation of the spectra can now be undertaken. In analyzing spectra in different redox states (like e.g. Fig. 7) the composing spectra can be found and the amount of each species can be calculated. With those numbers at hand further proof can be found for the mechanism and about the preferential pathways in the complex model [21].

ACKNOWLEDGEMENTS

We thank Mr. B.J. Sachteleben and Mr. M.M. Bouwmans for drawing the figures and Mrs. J.C. Toppenberg-Fang for typing the manuscript. This investigation was supported by the Netherlands Foundation for Chemical Research (S.O.N.) with financial aid from the Netherlands Organisation for the Advancement of Pure Research (Z.W.O.), by Nato-grant 1854 and EEC-grant ESD/029/NL.

REFERENCES

1. Stephenson, M. and Stickland, L.H. (1931) Biochem.J. *25*, 205.
2. Van Dijk, C., Mayhew, S.G., Grande, H.J. and Veeger,C. (1979) Eur.J.Biochem. *102*, 317-330.
3. Van Dijk, C., Grande, H.J., Mayhew, S.G. and Veeger, C. (1980) Eur.J.Biochem. *107*, 251-261.

4. Chen, J.S. and Mortenson, L.E. (1974) Biochim.Biophys.Acta *371*, 382-389.
5. Erbes, L., Burris, R.H. and Orme-Johnson, W.H. (1975) Proc. Natl.Acad.Sci. USA, *72*, 4795-4799.
6. Gillum, W.O., Mortenson, L.E., Chen, J.S. and Holm, R.S. (1977) J.Amer.Chem.Soc. *99*, 584-595.
7. Chen, L.S., Mortenson, L.E. and Palmer, G. (1976) in "Iron and Copper Proteins" (Yasunobo, K.T., Mower, H.F. and Hayaishi, O., eds.) pp. 68-82, Plenum Press, New York.
8. Mortenson, L.E. and Chen, J.S. (1976) in " Microbiol Production and Utilization of Gases (Schlegel, H.G., Gottschalk, G. and Pfennig, N., eds.) pp. 97-108, E. Goltze, Göttingen.
9. Van der Westen, H.M., Mayhew, S.G. and Veeger, C. (1978) FEBS Lett. *86*, 122-126.
10. Hatchikian, E.C., Bruschi, M. and Le Gall, J. (1978) Biochem. Biophys.Res.Commun. *82*, 451-461.
11. Gitlitz, P.H. and Krasna, A.J. (1975) Biochemistry *14*, 2561-2568.
12. Van Heerikhuizen, H., Albracht, S.P.J., Slater, E.C. and van Rheenen, P.S. (1981) Biochim.Biophys.Acta *657*, 26-39.
13. Kakuno, T., Kaplan, N.O. and Kamen, M.D. (1977) Proc.Natl.Acad. Sci. USA, *74*, 861-863.
14. Adams, M.W.W. and Hall, D.O. (1979) Arch.Biochem.Biophys. *195*, 288-299.
15. Schoenmaker, G.S., Oltmann, L.F. and Stouthamer, A.H. (1979) Biochim.Biophys.Acta *567*, 511-521.
16. Schneider, K. and Schlegel, H.G. (1976) Biochim.Biophys.Acta *452*, 66-80.
17. Schneider, K., Cammack, R., Schlegel, H.G. and Hall, D.O. (1979) Biochim.Biophys.Acta *578*, 445-461.
18. Schink, B. and Schlegel, H.G. (1979) Biochim.Biophys.Acta *567*, 315-324.
19. Erbes, D.L. and Burris, R.H. (1978) Biochim.Biophys.Acta *525*, 45-54.
20. Adams, M.W.W., Mortenson, L.E. and Chen, J.S. (1981) Biochim. Biophys.Acta *594*, 105-176.
21. Van Dijk, C. and Veeger, C. (1981) Eur.J.Biochem. *114*, 209-219.
22. Llama, M.J., Serra, J.L., Rao, K.K. and Hall, D.O., (1981) Eur.J.Biochem. *114*, 89-96.
23. Van Dijk, C., Mayhew, S.G. and Veeger,C. (1981) Eur.J.Biochem. *114*, 201-207.
24. Mathews, R., Charlton, S., Sands, R.H. and Palmer, G. (1974) J. Biol. Chem. *249*, 4326-4328.

HYDROGENASES: PHYSIOLOGY, LOCATION AND RELEVANCE FOR SULFATE REDUCING AND METHANE FORMING BACTERIA

Jean LeGall and Harry D. Peck, Jr.

Biochemistry Department, University of Georgia, Athens, Georgia 30602 USA

ABSTRACT

Hydrogenases play an essential role in the process known as interspecific hydrogen transfer which occurs in many important complex fermentations such as methane formation from cellulose degradation. The best type of interaction involving molecular hydrogen transfer is obtained using sulfate reducing and methane forming bacteria. Current knowledge of the physiology of the bacteria gives some insight into the mechanism of this process. In particular, the regulation of electron flow between the reduction of sulfate and the reduction of protons can be explained at the molecular level by the existence of different electron transfer chains feeding two types of hydrogenases.

INTRODUCTION

Although hydrogenase activity has been known since the early thirties (1), its physiological importance has not been uncovered until the last decade. It has long been regarded as a biological curiosity rather than a primardial enzyme. The recent search for new energy sources has renewed interest in the study of the metabolism of combustible gases such as hydrogen and methane and, as a result, the importance of molecular hydrogen in the anaerobic decomposition of organic matter has been uncovered.

H. B. Dunford et al. (eds.), The Biological Chemistry of Iron, 207–222.

IMPORTANCE OF MOLECULAR HYDROGEN AS AN INTERMEDIARY FUEL IN THE CELLULOSE FERMENTATION PROCESS

In anaerobic digestors, as well as in natural environments, where sulfate, nitrate and other electron acceptors are sufficiently depleted, the overall fermentation of cellulose that leads to the formation of methane and CO_2 can be represented as follows:

Step 1 (cellulose degradation by cellulolytic bacteria):

$$\text{cellulose} \longrightarrow \text{cellobiose} + \text{glucose} + CO_2 + H_2$$

Step 2 (oxidation of sugars):

$$\left.\begin{matrix}\text{cellobiose}\\ \text{glucose}\end{matrix}\right. \longrightarrow \begin{matrix}\text{lactate}\\ \text{alcohols}\\ \text{fatty acids}\end{matrix} + \begin{matrix}\text{acetate}\\ H_2\\ CO_2\end{matrix}$$

Step 3 (oxidation of alcohols and organic acids):

$$\begin{matrix}\text{lactate}\\ \text{alcohols}\\ \text{fatty acids}\end{matrix} \longrightarrow \text{acetate} + CO_2 + H_2$$

Step 4 (removal of acetate and H_2):

$$4H_2 + HCO_3^- + H^+ \longrightarrow CH_4 + 3H_2O$$

$$CH_3COO^- + H^+ \longrightarrow CH_4 + CO_2$$

The process (steps 1 and 2) begins with the degradation of cellulose into lactate, alcohols and fatty acids with some acetate, H_2 and CO_2 by a mixture of bacteria containing cellulolytic organisms and sugar oxidizers (2). These fermentation products are converted into substrates for methanogenesis: acetate, H_2 and CO_2 (step 3). During the last step, these compounds are removed to give CH_4 and CO_2. The role of hydrogen appears only when the process is written step by step since the overall reaction can be written as follows:

$$2CHOH\ (\text{cellulose}) \longrightarrow CH_4 + CO_2$$

This process can be used by the organisms as a "thermodynamical trick" to pull reactions that are not favorable. For example, the oxidation of ethanol:

$$CH_3CH_2OH + 2H_2O \longrightarrow 2CH_3COOH + 4H_2 \quad (\Delta G'o = + 1.5\ Kcal)$$

is pulled by the constant removal of hydrogen by the methanogenic bacterium. Such a mechanism requires very specialized redox transfer chains allowing the rapid release and utilization of H_2 gas, together with a precise regulation of the electron flow either toward the reduction of sulfates or to the reduction of protons. The building of a molecular model will be possible only after a complete description of the redox proteins which are involved in the process will be made available, together with the understanding of their spacial localization within the cell and their relationship with the energy yielding mechanisms.

As far as their physiology is concerned, sulfate reducing bacteria (main genera: *Desulfovibrio, Desulfotomaculum*) are strict anaerobes that obtain energy for growth from the oxidation of a limited number of organic compounds and molecular hydrogen (8). The electrons are used for the reduction of sulfur compounds ($SO_4^=$, $SO_3^=$, S°) to the level of H_2S or for the reduction of protons to form molecular hydrogen. During the transfer of electrons, ATP is generated (9). Reviews on the metabolism and physiology of the bacteria have been published (8,10,11). Their complex electron transfer chains have attracted considerable research in the field of structure/function relationships of the redox proteins. In *Desulfovibrio*, numerous electron carriers have been identified and characterized: cytochromes c, b and d, non-heme iron proteins (ferredoxins, rubredoxins, desulforedoxin, hydrogenases), flavodoxins and a complex redox protein containing non-heme iron and molybdenum (12).

The methanogenic bacteria constitute a very heterogenous group of strict anaerobes (13) with the general capability of forming methane from a limited number of substrates. They can be divided into two broad groups, depending upon their physiology (14): the first comprises hydrogen utilizing bacteria which reduce CO_2 and formate to methane; the second group is more versatile and can form methane from H_2 and CO_2, methanol, methyl amines and acetate. A variety of electron carriers have also been isolated such as a deazoflavin F_{420} (15,16) which most probably functions *in vivo* in association with one or several proteins, a 3Fe-3S containing ferredoxin (17), and a cytochrome b, (18).

Thus, hydrogen plays an essential role in nature through an exchange between bacteria which is now known as the "interspecies hydrogen transfer mechanism" (3). This process is also fundamental in many natural ecosystems; attempts to give a correct interpretation of the methanogenic processes in lake sediments (4,5) lead to two interesting conclusions: sulfides apparently "inhibited" methanogenesis, and lactate appeared to be a common product of the decomposition of organic matter. It was later found that sulfate actually is used by the sulfate reducing bacteria and reduced to sulfide, its apparently inhibitory effect on methanogenesis being due to the preferential reduction of sulfate over the production of molecular hydrogen for interspecies hydrogen transfer by sulfate reducing bacteria (6). This concept was clearly defined (3) when it was shown that *Methanobacillus omelianski*, which metabolized ethanol and CO_2 to acetate and methane was, in fact, a mixed culture composed of a methanogenic bacterium and of another organism which is able to convert ethanol to acetate and hydrogen. The following scheme:

Organism S: $2CH_3CH_2OH + 2H_2O \longrightarrow 2CH_3COOH + 4H_2$

Organism H: $4H_2 + CO_2 \longrightarrow CH_4 + 2H_2O$

Total: $2CH_3CH_2OH + CO_2 \longrightarrow 2CH_3COOH + CH_4$

shows that the two organisms are linked through interspecies hydrogen transfer. The same type of relation exists between the sulfate reducing bacteria and the methanogenic bacteria. Here, the situation is more complex since sulfate can be preferentially used as the terminal acceptor so that hydrogen becomes available for methane formation only after the supply of sulfate is sufficiently depleted (7):

Sulfate Reducing Bacterium

$$2CH_3CH_2OHCOOH \longrightarrow 2CH_3COOH + 2CO_2 + 2H_2$$

$$2H_2 + SO_4^{=} \longrightarrow S^{=} + 4H_2O$$

Intracellular hydrogen oxidation

and, in the presence of a methanogenic bacterium and low concentration of sulfates:

Methanogenic Bacterium

$$2H_2 + CO_2 \longrightarrow CH_4 + 2H_2O \quad (\Delta G'o = 33.2 \text{ Kcal})$$

Intercellular hydrogen oxidation

HYDROGENASES FROM SULFATE REDUCING BACTERIA

Table I shows the different results concerning the localization of several hydrogenases which have been isolated so far:

TABLE I

Localization of hydrogenases in various strains of *Desulfovibrio*

Strain	Periplastim	Membrane	Cytoplasmic
D. vulgaris			
Hilden (19)	M_r 52,000	0	+
Myazaki (20)		Large M_r 180,000 Small M_r 89,000	No more than 10%
Marburg (21)	+	0	0
D. gigas (22)	M_r 89,000	0	+
D. desulfuricans			
Norway 4 (23)	about 30%	M_r 59,000	?
ATCC 7757 (24)	M_r 52,000	10-20%	?

Desulfovibrio hydrogenases can be either periplasmic (external), membrane bound, or cytoplasmic (internal). No homogeneity seems to exist even within the same species, as evidenced by the results concerning three different strains of *D. vulgaris*. It is not yet possible, from the existing data, to formulate a general pattern, if it exists, concerning the number and localization of the hydrogenases. However, three remarks can be made concerning the interpretation of these results: 1. The number and distribution of hydrogenases in *Desulfovibrio* could be a reflect of the extreme heterogeneity within this genus, a heterogeneity which is supported by the comparative study of the sequences of several families of redox proteins such as cytochrome c_3, ferredoxin and rubredoxin (25); 2. Since the selective elution of hydrogenase from the external surface of the membrane seems to be a complex process (19,24) and that simple breakage of the cell can yield a membrane bound enzyme, the above results could be due to artifacts and lead to an apparent heterogeneity; 3. It is somewhat difficult to obtain *Desulfovibrio* cells grown in standard conditions since scientists have a tendency to prefer their own medium and growth conditions with the result that the cells which are utilized for enzyme purification are in different physiological stages. Thus, standard conditions will have to be used before definite conclusions can be reached concerning the localization of hydrogenase, as far as the growth conditions and the enzyme extraction are concerned.

Some properties of three purified hydrogenases from *D. desulfuricans* Norway 4 (23), *D. vulgaris* Hildenborough (19) and *D. gigas* (22) are reported in Table II. Some other hydrogenase preparations have been reported from *D. vulgaris* Miazaki (20) and *D. desulfuricans* ATCC 7757 (24), however the three hydrogenases in Table II have been chosen because they represent extreme cases. The membrane bound hydrogenase from *D. desulfuricans* Norway 4 has a molecular weight of 58,000 daltons, a single polypeptide chain, contains 6 to 7 atoms of iron and labile sulfur and has a low specific activity (100 µmoles H_2 evolved/mn/mg protein); *D. gigas* periplasmic hydrogenase also has a similar low specific activity, is constituted by two polypeptide chains (62,000 and 26,000 daltons respectively) and possesses twelve irons and labile sulfur atoms per mole of enzyme. Finally, the periplasmic hydrogenase from *D. vulgaris* has a very high specific activity (10,000 µmoles H_2 evolved/mn/mg protein), a molecular weight of 52,000 daltons with a single polypeptide chain and twelve iron and sulfur atoms.

The common property of these enzymes seems to be their specific reactivity with cytochrome c_3 (apparent Km = 2 µM for the Norway 4 hydrogenase), a particularity which has been reported very early (26). It has also been shown that *D. gigas* hydrogenase (27) will reduce ferredoxin, flavodoxin and rubredoxin only in the presence of either cytochrome c_3 (Mr 13,000) or cytochrome c_3 (Mr 26,000). However, such results are to be taken with caution, in particular because of the reversible inactivation of *Desulfovibrio* hydrogenases by oxygen. Since cytochrome c_3 is very reactive toward O_2 (28), its role could be limited in this type of experiment to the removal of oxygen from the hydrogenase, thus reactivating the enzyme. Mass spectrometry with cytochrome c_3-free *D. gigas* hydrogenase preparations (23) have shown that the exchange reaction can be initiated by addition of either cytochrome c_3 or by the flucose/glucose oxidase system, or even after long flushing periods with oxygen free argon. The problem remains to know whether such a reactivated hydrogenase would transfer its electrons directly to the non-cytochromic electron carriers and, if so, with rates which are comparable to the ones which we found in the presence of cytochrome c_3. The absolute proof for cytochrome c_3 involvement

TABLE II

Properties of hydrogenases from three *Desulfovibrio* strains

	D. desulfuricans Norway 4	*D. vulgaris* Hildenborough	*D. gigas*
Natural electron acceptor	cytochrome c_3 (apparent Km = 2μM)	cytochrome c_3	cytochrome c_3
Molecular weight	58,000	50,000	89,000
Sub-units	1 x 58,000	1 x 50,000	1 x 62,000
Absorption band	around 400 nm	around 400 nm	around 400 nm
Fe and S/mole	6-7	12	12
Specific activity μmole H_2 ev/mn/mg	100	10,400	91
O_2 sensitivity	relatively stable	relatively stable	relatively stable

in a chain of reaction such as this:

$$\bar{e} \longrightarrow \text{ferredoxin} \longrightarrow \text{cytochrome } c_3 \longrightarrow \text{hydrogenase} \longrightarrow H_2$$

would be that hydrogen evolution would only occur in its presence even if ferredoxin would eventually be reduced by hydrogenase alone. It would mean that hydrogenase could pass electrons to the ferredoxin, or *vice versa*, but at a very slow rate not compatible with a physiological activity. In our present state of knowledge, the three following possibilities cannot be excluded:

$$H_2 \longleftrightarrow \text{hydrogenase} \longleftrightarrow \text{cytochrome } c_3 \longleftrightarrow \begin{matrix}\text{ferredoxin}\\ \text{flavodoxin}\\ \text{rubredoxin}\end{matrix} \longleftrightarrow \begin{matrix}\text{other}\\ \text{electron}\\ \text{carriers}\end{matrix}$$

or:

$$H_2 \longleftrightarrow \text{hydrogenase} \longleftrightarrow \begin{matrix}\text{cytochrome } c_3\\ \text{ferredoxin}\\ \text{flavodoxin}\\ \text{rubredoxin}\end{matrix} \longleftrightarrow \begin{matrix}\text{other}\\ \text{electron}\\ \text{carriers}\end{matrix}$$

or:

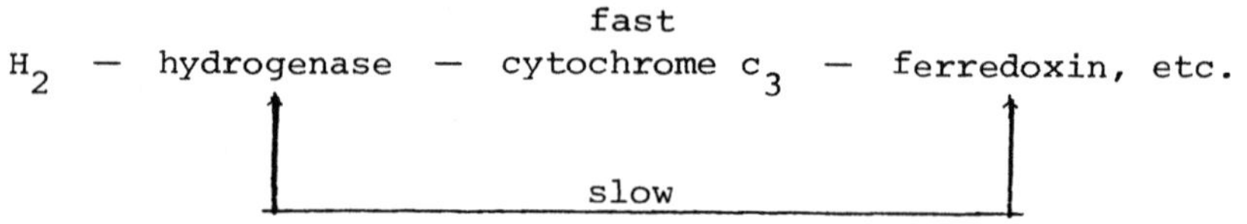

The fact that neither *M. elsdenii* (30), *C. pastorianum* (31) nor *Methanosarcina barkeri* (29) hydrogenases seem to be able to react with cytochrome c_3 is yet the best evidence for its specificity toward *Desulfovibrio* hydrogenases.

The first demonstration of the capability of non-photosynthetic anaerobes to perform oxidative phosphorylations was given by the use of *Desulfovibrio* whole cells (32), then cell-free extracts with hydrogen as the electron donor and sulfite (9) or fumarate (33) as the electron acceptor. From the results of sulfur reduction with hydrogen by inverted vesicles of *D. gigas*, a chemi-osmotic model has been proposed which involves vectorial electron transfer across the cell membrane (34):

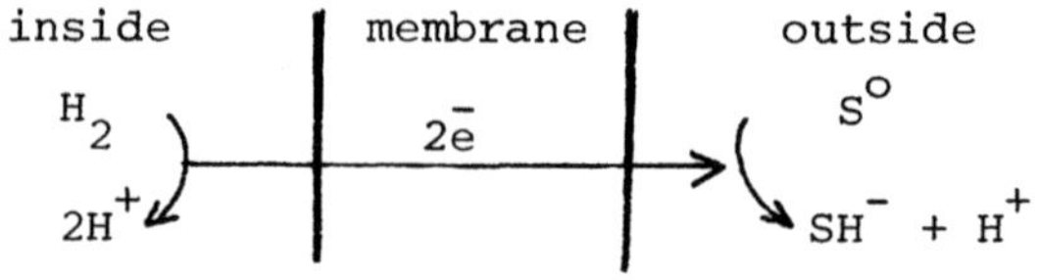

A similar mechanism has also been proposed to explain the results of the enzyme localization study with *D. vulgaris* Marburg grown on hydrogen and sulfate (21). The protons produced would be utilized for the production of ATP by means of ATPase (35). This mechanism for the generation of a proton gradient is different from the typical chemiosmotic model since it does not involve a typical Mitchell loop.

When it was first shown that some *Desulfovibrio* strains did accumulate hydrogen during growth on lactate plus sulfate media (36), it was felt necessary to let more precise data concerning the metabolic balances of the different strains. It was found that (37,38) while *D. gigas* and *D. africanus* NCIB 8310 have a metabolic balance corresponding to the classical lactate/sulfate reaction:

$$2CH_3CHOHCOO^- + SO_4^= \longrightarrow 2CH_3COO^- + 2CO_2 + S^= + 2H_2O$$

D. desulfuricans Norway 4 metabolized lactate according to the following reaction:

$$CH_3CHOHCOO^- + 0.375\ SO_4^= \longrightarrow CO_2 + 0.8\ CH_3COO^- + 0.375\ S^=$$
$$+ 0.1\ CH_3CH_2CH_2CH_2OH + 0.8\ H_2O$$

and *D. vulgaris* Hildenborough according to:

$$CH_3CHOHCOO^- + 0.35\ SO_4^= + 0.56\ H^+ \longrightarrow CO_2 + 0.98\ CH_3COO^-$$
$$+ 0.02\ CH_3CH_2OH + 0.16\ H_2S + 0.215\ HS^- + 0.5\ H_2O + 0.48\ H_2$$

showing that, in contrast to the common belief, some *Desulfovibrio* species can produce all the precursors for methane production by *Methanosarcina* (H_2, CO_2 and acetate) even in the presence of sulfate. Tsuji and Yagi (20) have reported a burst of hydrogen production during the early stages of growth of *D. vulgaris* Myazaki and a re-utilization of the hydrogen during later stages of the growth cycle. They have also reported the presence of two membrane bound hydrogenases, one of M.W. 70,000 and another of M.W. 180,000. They suggested that the lower molecular weight hydrogenase was involved in hydrogen utilization and the higher molecular weight enzyme was involved in hydrogen production. It was shown recently (39) that *D. vulgaris* Hildenborough grown in interspecific hydrogen transfer conditions in the presence of

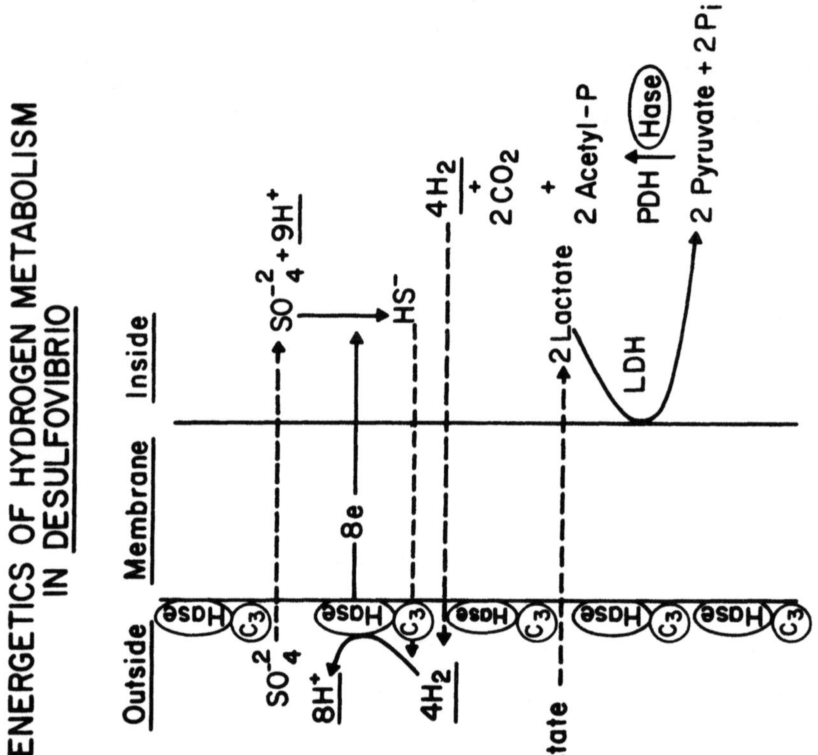

Figure 1. Hydrogen Cycling in Desulfovibrio Adapted from ref. 39

Methanobacterium bryantii had ten times more of an internal hydrogenase than when grown alone on a lactate-sulfate medium. This indicates that this internal hydrogenase, rather than the periplasmic one is involved in the formation of hydrogen from lactate or pyruvate. Proof for the presence of a cytoplasmic hydrogenase in *D. gigas* has also been given using the relative impermeability of the cytoplasmic membrane to oxidized benzyl viologen (40). Furthermore, it was shown that spheroplasts will not oxidize lactate in the presence of sulfate unless both pure periplasmic hydrogenase and cytochrome c_3 were added to the reaction. These results have prompted the authors to propose that molecular hydrogen is an intermediate in the reduction of sulfate by lactate. In the scheme proposed in Figure 1, molecular hydrogen, which is formed inside of the cell from the oxidation of lactate and pyruvate, diffuses freely through the membrane where it is then oxidized by the periplasmic hydrogenase + cytochrome c_3, the electrons being brought inside of the cell to reduce sulfate. This role of the external hydrogenase is similar to the role of hydrogenase in nitrogen fixation which has been proposed to act in order to prevent waste of reducing power (41). This new concept of a hydrogen cycling due to a spacial separation of the fermentation process (lactate) and pyruvate to hydrogen) and the respiratory process is very attractive. It gives a very satisfactory explanation for the capability of *Desulfovibrio* to grow either in the presence of sulfate or in its absence by interspecific hydrogen transfer. It remains to be seen if such a scheme will apply to all *Desulfovibrio* strains, in other words if the absence of internal hydrogenase which has been reported in some strains is due to the use of different growth conditions or to the fact that the bacteria reduce sulfate according to a different mechanism. Actually, the fact that no internal hydrogenase was found in *D. vulgaris* Marburg (21) is in favor of this scheme since the cells were grown on a hydrogen/sulfate medium where the internal hydrogenase was unnecessary. Purification of the internal hydrogenase will be awaited with much interest since its specificity toward the electron carriers could well be very different from the periplasmic or membrane bound enzymes. Also, it remains to be seen if the multiple electron carriers found in *Desulfovibrio* species have evolved to specialize some toward the fermentative reaction, some toward the respiratory chain. In this regard, the finding of multiple ferredoxins in *Desulfovibrio* could be of particular significance (42). The fact that different strains of *Desulfovibrio* placed in the same growth conditions do not have the same metabolic balance, in particular as far as their hydrogen metabolism is concerned, remains to be explained. It probably means that the hydrogen cycling scheme, whence proven to be able to provide for a general explanation for sulfate reduction in *Desulfovibrio*, will have to be refined to be adapted to each particular case.

HYDROGENASES FROM METHANE FORMING BACTERIA

Data on hydrogenases from methanogenic bacteria are still scarce, although the presence of hydrogenase activity in their extract has been known for a long time (43). The localization of hydrogenase activity is still an unsolved problem, however they appear to be internal since the activity with viologen dyes can be increased 80 fold after cell rupture (44). In a study of hydrogenase activity in crude extracts of *Methanobacterium thermoautotrophicum* (43), the presence of multiple bands of hydrogenase activity after electrophoresis on polyacrylamide gels has been reported. A high molecular weight hydrogenase (Mr 500,000) has been purified to 50% homogeneity (46) from the same organism, it has subunits of 26,000, 31,000 and 40,000 daltons. The enzyme was reported to contain iron and labile sulfur and to be distinct from a second hydrogenase. Another hydrogenase, again from *M. thermoautotrophicum* (47), was found to have a molecular weight of 185,000 daltons and to be composed of four subunits of two types with molecular weights of 64,000 and 40,000 daltons. The enzyme was reported to have a specific activity, in the presence of methyl viologen, of 1,000 μmoles H_2/mn/mg protein at 60°C. As for the hydrogenase from *Methanobacterium* strain G2R (44), it was rapidly inactivated by O_2, but this inactivation could be completely reversed by dithionite or the glucose/glucose oxidase system. A soluble hydrogenase has been partly purified from *Methanosarcina barkeri* (48), but no physico-chemical data are available on this enzyme apart from the fact that its optical spectrum suggest the presence of iron-sulfur centers.

Since the deazoflavin F_{420} has been shown to participate in the reduction of $NADP^+$ by extracts of *Methanobacterium ruminiantium* (49,50) in the presence of hydrogen and that it is also active in the formation of molecular hydrogen from formate in the same organism, it is of interest to test the activity of the purified hydrogenase toward this co-factor. The large hydrogenase from *M. thermoautotrophicum* (46) was found to be able to reduce it with an apparent Km of 19 μM. The second hydrogenase which has been reported by the same authors will not reduce it, although it reduces viologen dyes. This hydrogenase appears then to be similar to the 180,000 daltons enzyme which has also been found unable to react with F_{420} (47). However, these results have to be taken with some caution since it could well be that in crude extracts where $NADP^+$ reductase and formic hydrogenelyase activities were assayed, the reactive F_{420} is bound to a protein. The partly purified hydrogenase from *Methanosarcina barkeri* has been found to reduce the ferredoxin and the soluble cytochrome b from the same organism, but not F_{420} (48).

No substrate level phosphorylation has been found so far in methanogenic bacteria. It is then very tempting to postulate that all phosphorylations that occur in this organism are linked to electron transfer. ATPase activity has been shown to exist in *M. thermoautotrophicum* (51). ATP synthesis in whole cells was stimulated by an artificially created proton motive force as a shift of pH from 7.5 to 3.0 was found to be most effective in generating ATP. Because of the scarcity of data, it is still quite premature to present general models concerning the bioenergetics and electron transfer pathways of the methane bacteria. However, an attractive model has been presented for the generation of a proton gradient by vectorial electron transfer (52) that can be represented as follows:

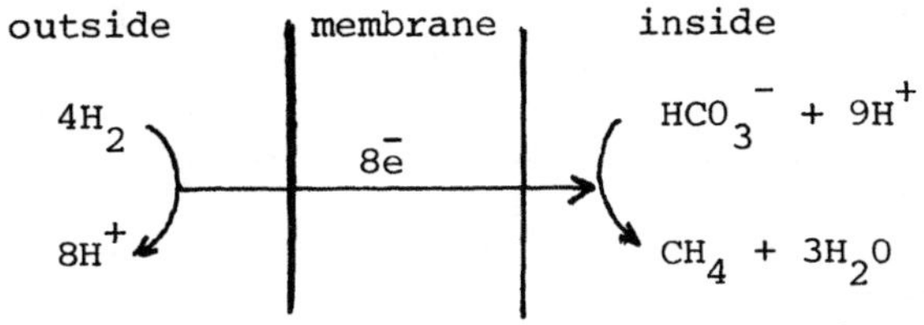

With such a model, the proton gradient could be equivalent to four molecules of ATP if $H^+/ATP = 2$. However, such an efficiency is not consistent with the low growth yield of methane bacteria (53).

CONCLUSION

When data on the influence of interspecies hydrogen transfer conditions (sulfate reducing bacteria + methanogenic bacteria) upon the hydrogenase contents of sulfate reducing bacteria begin to appear in the literature (39), our knowledge concerning the methane forming bacteria is still too insufficient to make a similar study significant concerning this group of bacteria. Much more work is needed on the proteins which are involved in the redox systems of these very specialized organisms. On another hand, if more data have been accumulated concerning the sulfate reducing bacteria, their electron transfer chains are still to be built up: in this case we have actually too many redox proteins that are needed for our simplified models.

ACKNOWLEDGEMENTS

The authors want to acknowledge grants from SERI (Contract No: XK-1-1182-1) from COMES (J.L.G.) and from DOE (Contract No: DEAS09-80 ER 10499) to H.D.P.

REFERENCES

1. Stephenson, M. and Stickland, L. H.: 1931, *Biochem. J.* 25, pp. 205-214.
2. Peck, H. D., Jr. and Odom, M.: 1981, in *Trends in the Biology of Fermentation for Fuels and Chemicals*, Plenum Press, in press.
3. Bryant, M. P., Wolin, E. A., Wolin, M. J. and Wolfe, R. S.: 1967, *Arch. Microbiol.* 59, pp. 20-31.
4. Cappenberg, T. E.: 1974, Antonie van Leeuvenhoek 40, pp. 285-296.
5. Cappenberg, T. E.: 1974, Antonie van Leeuvenhoek 40, pp. 297-312.
6. Winfrey, M. R. and Zeikus, J. G.: 1977, *Applied and Environ. Microbiol.* 33, pp. 275-286.
7. Bryant, M. P., Campbell, L. L., Reddy, C. A. and Crabill, M. R.: 1977, *J. Bacteriol.* 33, pp. 1162-1169.
8. LeGall, J. and Postgate, J. R.: 1973, in *Advances in Microbial Physiology* (A. H. Rose and D. W. Tempest, Eds.) Acad. Press, Inc., London, 10, pp. 81-133.
9. Peck, H. D., Jr.: 1966, *Biochem. Biophys. Res. Commun.* 22, pp. 112-118.
10. Peck, H. D., Jr.: 1974, in *Evolution of the Microbial World* (M. J. Carlile and J. J. Skehel, Eds.) Symp. Soc. Gen. Microbial., Cambridge Press Univ. 24, pp. 241-262.
11. LeGall, J., Bruschi, M. and Hatchikian, C. E.: 1975, Proc. Tenth FEBS Meeting (P. Desnuelle and A. M. Michelson, Eds.) North Holland, American Elsevier, 44, pp. 277-285.
12. LeGall, J., Moura, J. J. R., Peck, H. D., Jr. and Xavier, A. V.: 1981, in *Iron Sulfur Proteins*, Vol. IV, T. Spiro Ed., J. Wiley and Sons, New York, in press.
13. Balch, W. E., Fox, G. E., Magrum, L. J., Woese, G. R. and Wolfe, R. S.: 1979, *Microbial Rev.* 43, pp. 260-296.
14. Wolfe, R. S. and Higgins, I. J.: 1979, in *Microbial Biochemistry* (J. R. Quayl Ed.) MTP Press Ltd., Lancaster, England, pp. 267-283.
15. Cheeseman, P., Toms-Wood, A. and Wolfe, R. S.: 1975, *J. Bacteriol.* 121, pp. 192-203.
16. Eirich, L. D., Vogels, G. D. and Wolfe, R. S.: 1978, *Biochemistry* 17, pp. 4583-4593.
17. Moura, I., Moura, J. J. G., Boi, H. Huyind, Hausinger, R. P., Santon, H., LeGall, J. and Xavier, A. V.: unpublished results.
18. Kuhn, W., Fiebig, K., Walther, R. and Gottschalk, G.: 1979, *FEBS Letters* 105, pp. 271-274.
19. Van der Westen, H. M., Mayhew, S. G. and Veeger, C.: 1978, *FEBS Letters* 86, pp. 122-126.
20. Tsuji, K. and Yagi, T.: 1980, *Arch. Microbiol.* 125, pp. 35-42.
21. Badziong, W. and Thauer, R. K.: 1980, *Arch. Microbiol.* 125, pp. 167-178.

22. Hatchikian, E. C., Bruschi, M. and LeGall, J.: 1978, *Biochem. Biophys. Res. Commun.* 82, pp. 451-461.
23. Lalla-Maharajh, W. V., Hall, D. O., Rao, K. K. and LeGall, J.: unpublished results.
24. Glick, G. R., Martin, W. G. and Martin, S. M.: 1980, *Can. J. Microbiol.* 26, pp. 1214-1223.
25. Bruschi, M. and LeGall, J.: 1976, in *Iron and Copper Proteins*, Plenum Publish. Corp. (K. T. Yasunabu, M. F. Mower, and O. Hayaishi, Eds.) pp. 57-67.
26. Yagi, T., Hanya, M. and Tamuja, N.: 1968, *Biochim. Biophys. Acta* 153, pp. 692-698.
27. Bell, G. R., Lee, J. P., Peck, H. D., Jr. and LeGall, J.: 1978, *Biochimie* 60, pp. 315-320.
28. Postgate, J. R.: 1956, *J. Gen. Microbiol.* 14, pp. 545-572.
29. Berlier, Y., Fauque, G., Lespinat, P. and LeGall, J.: unpublished results.
30. Mayew, S. G., Van Dick, C. and Van der Westen, E. M.: 1972, in *Hydrogenases: Their Catalytic Activity, Structure and Function* (H. G. Schlegel and K. Schneider, Eds.) Goltze, K. G., Gottingen, pp. 125-136.
31. Mortenson, L. E. and LeGall, J.: unpublished results.
32. Peck, H. D., Jr.: 1960, *J. Biol. Chem.* 235, pp. 2734-2738.
33. Barton, L. L., LeGall, J. and Peck, H. D., Jr.: 1970, *Biochem. Biophys. Res. Commun.* 41, pp. 1036-1042.
34. Fauque, G. D., Barton, L. L. and LeGall, J.: 1980, in *Sulphur in Biology*, CIBA Foundation Symposium 72 (Excerpta Medica Ed.) Elsevier North-Holland, pp. 71-86.
35. Guarraia, L. J. and Peck, H. D., Jr.: 1971, *J. Bacteriol.* 106, pp. 890-896.
36. Hatchikian, E. C., Chaigneau, M. and LeGall, J.: 1976, in *Microbial Production and Utilization of Gases* (H. K. Schlegel, K. Gottschalk, and N. Pfennig, Eds.) Goltze, K. G., Gottingen, pp. 109-118.
37. Traore, A. S., Hatchikian, C. E., Belaich, J. P. and LeGall, J.: 1981, *J. Bacteriol.* 145, pp. 191-199.
38. Traore, A. S., Hatchikian, C. E., LeGall, J. and Belaich, J. P.: unpublished results.
39. Odom, J. M. and Peck, H. D., Jr.: 1981, *FEMS Microbial. Letters*, in press.
40. Jones, R. W. and Garland, P. B.: 1977, *Biochem. J.* 164, pp. 199.
41. Dixon, R. O. D.: 1972, *Arch. Microbiol.* 85, pp. 193-201.
42. Bruschi, M., Hatchikian, E. C., Golovleva, L. and LeGall, J.: 1977, *J. Bacteriol.* 129, pp. 30-38.
43. Wolin, E. A., Wolin, M. J. and Wolfe, R. S.: 1963, *J. Biol. Chem.* 238, pp. 2882-2893.
44. McKellar, R. C. and Spratt, G. D.: 1979, *J. Bacteriol.* 139, pp. 231-242.

45. Fuchs, G., Wall, J., Scherer, P. and Thauer, R. K.: 1978, in *Hydrogenases: Their Catalytic Activity, Structure and Function* (H. G. Schlegel and K. Schneider Eds.) Goltze, K. G., Gottingen, pp. 132-146.
46. Daniels, L., Fox, J., Jacobson, F., Orme-Johnson, W. H., and Walsh, C.: 1981, Abstracts VIIth Inter. Symp. on Flavins and Flavoproteins, Ann Arbor, MI, USA.
47. Mortenson, L. E., Peck, H. D., Jr. and LeGall, J.: unpublished results.
48. Moura, I., Moura, J. J. G., Santos, M. H., Xavier, A. V. and LeGall, J.: unpublished results.
49. Tzeng, S. F., Wolfe, R. S. and Bryant, M. P.: 1975, *J. Bacteriol.* 121, pp. 184-191.
50. Tzeng, S. F., Bryant, M. P. and Wolfe, R. S.: 1975, *J. Bacteriol.* 121, pp. 192-196.
51. Doddema, H. J., Hutten, T. J., Van der Drift, C. and Vogels, G. D.: 1978, *J. Bacteriol.* 136, pp. 19-26.
52. Thauer, R. K., Songarmann, K. and Decker, K.: 1977, *Bacteriol. Rev.* 41, pp. 100-180.
53. Taylor, G. T. and Pirt, S. J.: 1977, *Arch. Microbiol.* 113, pp. 17-22.

GENERATION, TRANSPORT AND TRANSFER OF LOW-POTENTIAL REDUCING EQUIVALENTS IN NITROGENASE CATALYSIS

Arnold Braaksma, Hans Grande, Huub Haaker, Colja Laane and Cees Veeger*
Department of Biochemistry Agricultural University, De Dreijen 11, 6703 BC Wageningen, The Netherlands.

1. INTRODUCTION

Studies with nitrogenase systems, isolated from a variety of microorganisms, revealed that biological N_2 fixation with the catalyst nitrogenase component I or MoFe-protein needs another protein, nitrogenase component II or Fe-protein, anaerobic conditions, energy in the form of ATP and a strong reductant. Despite the specific demand for a strong reducing agent and an anaerobic environment for the site of the nitrogenase system, N_2 fixation is observed in a variety of microorganisms, including obligate aerobes. There is little information about the absolute value of the redox potential necessary for nitrogenase activity. The midpoint potential of the Fe-protein of *C.pasteurianum* in the presence of MgATP or MgADP is approximately -400 mV. The MoFe-protein is completely reduced at -500 mV [1]. It has been demonstrated that the nitrogenase complexes isolated from *A.vinelandii* and *C.vinosum* show full activity at redox potentials below -500 mV and no significant activity at redox potentials higher than -430 mV [2-4]. It is clear that besides ATP the cell has to invest significant amounts of energy to generate the reducing power for N_2 fixation.

Several factors are known to influence the amount of energy used to fix N_2:

(a) From studies with isolated nitrogenase systems, it is known that the ratio of MoFe-protein to Fe-protein influences the amount of ATP hydrolysed per 2 electrons transferred to substrates of the MoFe-protein [5]. The temperature also influences this ratio [6]. At the moment no molecular explanation for these phenomena is available;

*Who presented this contribution

H. B. Dunford et al. (eds.), The Biological Chemistry of Iron, 223–240.

(b) During the reduction of N_2 to ammonia by nitrogenase, protons are reduced to H_2. Part of the energy lost by H_2 evolution is recaptured by H_2 oxidation. Hydrogen oxidation is coupled to ATP-formation in *A.vinelandii* [7] through an additional proton-translocating site;
(c) Aerobic N_2 fixing organisms have the ability to use energy sources more completely, but they have to face the effect of O_2 on N_2 fixation. Several mechanisms are known to protect N_2 fixation in aerobes, for example [8] respiratory protection, physical protection (slime production, a thick cell wall, etc.) and an O_2 buffering system such as leghaemoglobin. Despite all these protective mechanisms, the possibility cannot be excluded that the strong reducing agents generated for N_2 fixation react with O_2. The toxic superoxide anion O_2^- is formed and the energy of the reducing equivalents is lost as heat if O_2^- reacts with superoxide dismutase and H_2O_2 reacts with catalase.

This article will deal with some of the recent insights in terms of generation of low-potential reducing equivalents, transfer of reducing equivalents to nitrogenase and the mechanism of electron transport from the Fe-protein to the MoFe-protein as well as catalysis.

2. ELECTRON TRANSFER TO NITROGENASE

The only known electron carriers operating in the transfer of reducing equivalents to nitrogenase are ferredoxin and flavodoxin. In anaerobes like *C.pasteurianum* low-potential reducing equivalents are produced in the phosphoroclastic reaction and transported in the sequence:

$$\text{pyruvate} \rightarrow \begin{matrix}\text{ferredoxin}\\ \text{flavodoxin}\end{matrix} \rightarrow \text{nitrogenase} \rightarrow N_2$$

in which either ferredoxin or flavodoxin acts as electron carrier [9,10]. In aerobes however the situation is different. Beneman *et al.* [11] postulated a sequence:

$$\text{NADPH} \rightarrow \begin{matrix}\text{ferredoxin}\\ \text{flavodoxin}\end{matrix} \rightarrow \text{nitrogenase} \rightarrow N_2.$$

It is difficult to visualize how *A.vinelandii* flavodoxin, of which the hydroquinone-semiquinone couple E_m is -495 mV, can be reduced, since Haaker *et al.* [12] showed that within the cell the ratio NAD(P)H/NAD(P)$^+$ was around 2-3:1. In a critical review Haaker *et al.* [13] seriously questioned any role of reduced pyridine nucleotides in nitrogenase fixation. Even in *C.pasteurianum* rather low ratios reduced/oxidized pyridine nucleotides are found. Similarly no evidence could be found from published data for such a system in photosynthetic bacteria, cyanobacteria and aerobic nitrogen fixers.

In agreement with Yates [14], but in contrast with Beneman *et al.* [15], flavodoxin hydroquinone serves as one-electron donor for the nitrogenases of *A.vinelandii* and *R.leguminosarum* [3,16]. *A.vinelandii* flavodoxin hydroquinone is a more efficient electron donor than reduced ferredoxin I; its K_m is around 5 μM. In fact at infinite carrier concentration the activity with flavodoxin hydroquinone is around 1.5 times the activity with $Na_2S_2O_4$. The flavodoxins of *Megasphaera elsdenii* and *Desulfovibrio vulgaris* have a maximum velocity of about 30% and 15% respectively. The hydroquinone of *A.vinelandii* flavodoxin is extremely reactive

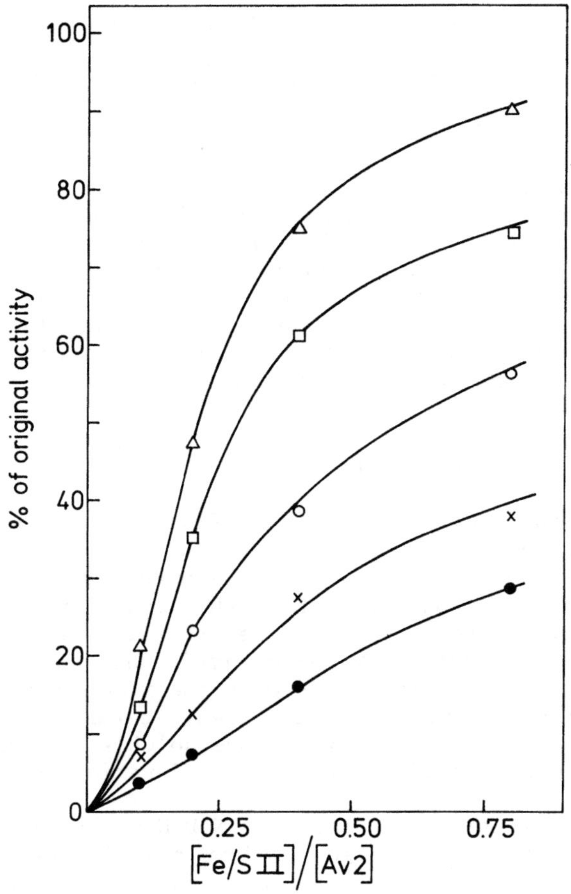

Fig. 1. Effect of $MgCl_2$ on the protection of Av_1 + Av_2 by the $[Fe_2-S_2]$-protein, [from ref. 16]. $MgCl_2$ concentrations ●-●, none; x-x, 0.5 mM; o-o 1.0 mM; □-□, 2.5 mM; Δ-Δ, 5.0 mM.

with O_2 and plays in our opinion an essential role in protecting nitrogenase against damage by O_2 [3,16]. The Fe-protein inactivates with a $t_{\frac{1}{2}}$ = 10 s; for the MoFe-protein $t_{\frac{1}{2}}$>15 min. It was found that *A.vinelandii* contains a small $[Fe_2-S_2]$-ferredoxin-type protein (M_r~26 000), which protects nitrogenase fully against O_2-damage [3,16]. In fact it was found (Fig. 1) that, only at 10 mM Mg^{2+}-ions, a three-component complex is formed of probably 6:6:6 stoichiometry and M_r~2 x 10^6. Interestingly, neither the Fe-protein (Av_2) nor the MoFe-protein (Av_1) alone were protected by the small $[Fe_2-S_2]$-protein, only the combination of the two proteins is stabilized. Furthermore, it could be shown [16] that this $[Fe_2-S_2]$-protein reacts rapidly with Av_2(red). This oxidative reaction, takes out electrons from Av_2, thus making it less sensitive towards O_2. This view is supported by studies of van der Westen *et al.* [17], demonstrating, that aeration of *D.vulgaris* cells prior to opening, leads to the isolation of an O_2-stable hydrogenase.

Scherings *et al.* [3] showed that the nitrogenase reaction comes to a stop when the redox potential of the donor exceeds -450 mV. In fact recent work [18, see below] shows that the switch off potential of the nitrogenase reaction under carefully controlled conditions of pH and ionic strength with different electron donors is around -400 mV, about 50 mV higher than the value reported earlier. In addition the lag in the activity [3], when the enzyme is not reduced prior to the assay, is always observed. It seems that *A.vinelandii* flavodoxin semiquinone specifically inhibits the nitrogenase, in contrast to *M.elsdenii* flavodoxin semiquinone.

3. GENERATION OF LOW-POTENTIAL ELECTRONS

An enzyme, which reduces *A.vinelandii* flavodoxin initially to the hydroquinone in the presence of NAD(P)H, was purified 200 times [16]. The specific activity of the pure FAD-containing protein is not very high, and varies between 6 and 30 nmoles.$min^{-1}.mg^{-1}$. In addition, this reductase reacts with the flavodoxins from *M.elsdenii*>*D.vulgaris*>*A.vinelandii* under anaerobic conditions, an order which corresponds with the redox potentials of the semiquinone-hydroquinone couples. Even catalyzing initially the formation of flavodoxin hydroquinone the soluble enzyme does not meet the requirements needed for electron donation to nitrogenase.

Although it is unlikely that the free NADPH-flavodoxin reductase is involved in the generation of low-potential reducing equivalents for nitrogenase, a role of this enzyme, when bound to the bacterial membrane, cannot be ruled out. It can be visualized [16,19] that by a process of reversed electron-flow, driven by the energized membrane or by its proton gradient, flavodoxin hydroquinone is formed.

The involvement of the proton motive force or the electrochemical proton gradient ($\Delta\tilde{\mu}_{H^+}$) of the cytoplasmic membrane as driving force in electron donation to nitrogen fixation in aerobes has been demonstrated by Haaker *et al.* [12,20,21]. It is clear from these studies that a decline in the energized state of the *A.vinelandii* membrane, under conditions without any effect on the intracellular NAD(P)H/NAD(P)$^+$ and ATP/ADP ratio's, leads to full inhibition of the nitrogen fixation activity. The major effect induced by the membrane was proposed to be on the generation of low-potential reducing equivalents, mainly flavodoxin hydroquinone [3,21]. The proton motive force across the cytoplasmic membrane, generated by respiration, drives the thermodynamically unfavourable formation of flavodoxin hydroquinone. This driving force, the so-called electrochemical proton gradient ($\Delta\tilde{\mu}_{H^+}$) is composed of an electrical gradient or membrane potential ($\Delta\Psi$) and a concentration gradient (ΔpH) across the membrane. It was shown, by manipulating $\Delta\Psi$ and ΔpH by the addition of ionophores like valinomycin and nigericin or by addition of NH_4^+ or tetraphenyl phosphonum bromide, that the value of $\Delta\Psi$ determines the level of nitrogen fixation in *A.vinelandii*, in *Rhizobium leguminosarum* bacteroids and in the photosynthetic bacterium *Rhodopseudomonas sphaeroides* [19,22,23].

Fig. 2 shows that a minimum value of the membrane potential is needed for nitrogenase fixation; beyond -80 mV nitrogenase is completely inactive. A more negative value of $\Delta\Psi$ below -80 mV

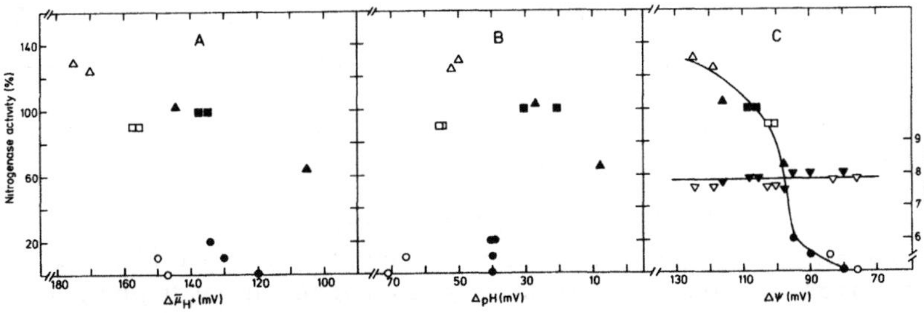

Fig. 2. Relationship between nitrogenase activity and $\Delta\tilde{\mu}_{H^+}$ (A), ΔpH (B), $\Delta\Psi$ (C) in bacteroids of *R.leguminosarum*. External pH 7.4 (dark symbols); external pH 6.7 (open symbols). (□,■) Without addition; (Δ-▲) with nigericin; (o,●) with valinomycin; (∇,▼) internal pH, [from ref. 22]. (Reproduced with permission from FEBS Letters)

leads to an enhanced nitrogenase activity. A 40% enhancement of activity over the normal activity was observed in the presence of nigericin, which can increase the membrane potential under certain conditions.

It has been firmly established, that the process of nitrogen fixation by free-living, bacteria is inhibited by ammonia in two ways; firstly by rapid switch-off (short-term effect), secondly by repression of N_2-ase synthesis (long term effect). In isolated bacteroids of *R.leguminosarum* on the other hand, nitrogen fixation is not affected. Flow dialysis experiments with intact *A.vinelandii*, either grown on succinate or on glucose, show, that N_2-ase is inhibited by the uptake of ammonium by the cells without influence on the ATP/ADP ratio [22]. Table I shows that uptake of ammonium lowers the energized state of the cell-membrane.

Table I. Effect of ammonium chloride on $\Delta\tilde{\mu}_{H^+}$, $\Delta\Psi$, ΔpH, internal pH, nitrogenase activity and the ATP/ADP ratio in succinate-grown *A.vinelandii* as measured by flow dialysis [from ref. 22].(Reproduced with permission from FEBS Letters)

ammonium chloride (mM)	$\Delta\tilde{\mu}_{H^+}$ (mV)	$\Delta\Psi$ (mV)	ΔpH (mV)	internal pH	nitrogenase activity (% remaining)	ATP/ADP ratio
--	132	106	26	8.05	100	1.8
0.06	133	95	38	8.25	40	1.9
0.10	128	90	38	8.25	10	2.0
0.20	120	82	38	8.25	0	3.0
0.50	105	75	26	8.05	0	2.5

The determining factor of the N_2-ase activity is again the membrane potential; a threshold value of at least -80 mV is also required here.

Laane *et al.* [22] demonstrated that ammonium is actively excreted by the bacteroid cell in response to respiration; in fact bacteroids excrete ammonium only in response to a pH-gradient.

The lipophylic tetraphenylphosphonium cation (PhP^+), when used in micromolar concentration as a $\Delta\Psi$-probe has no uncoupling effect; at higher concentrations it can decline $\Delta\Psi$. Table II shows that at a concentration of 0.4 mM PhP^+, nitrogenase activity is completely inhibited in *A.vinelandii*, in *R.leguminosarum* bacteroids and in *Rps.sphaeroides*, without any influence on the ATP/ADP ratio. At PhP^+ concentrations >0.4 mM a decline in ATP/ADP is observed in all cases. The inhibitory effect on PhP^+ on nitrogenase activity is however not observed in cell-free extracts; the inhibition of the nitrogenase activity is thus induced by an effect of this compound on the bacterial membrane.

These results demonstrate that membrane potential-driven nitrogen fixation is operating in a number of different aerobic

Table II. Effect of Ph_4P^+ on nitrogenase activity and ATP/ADP ratio in *A.vinelandii*, in *R.leguminosarum* bacteroids and in *Rps.sphaeroides* [from ref. 19].

organism	Ph_4P^+ (mM)	N_2ase act (%)	ATP/ADP
A.vinelandii	0	100	2.2
	0.4	0	2.2
R.leguminosarum	0	100	4.0
bacteroids	0.4	5	4.0
Rps.sphaeroides	0	100	3.9
	0.4	0	4.1

organisms. In fact, Rowell *et al.* [24] recently showed that the membrane potential is the driving force behind nitrogen fixation in the cyanobacterium *Anabaena variabilis*.

The actual mechanism of this process is still obscure. Several mechanisms have been proposed e.g. proton gradient reduction of flavodoxin [11], reversed electron flow [13] and membrane potential induced conformational transitions in flavodoxin [25]. Until now our trials to reconstitute the system have been unsuccessful.

4. ELECTRON TRANSFER AND CATALYSIS

One of the striking observations made by Scherings *et al.* [3] was the strong inhibition of the nitrogenase reaction above a redox potential of -460 mV, when *A.vinelandii* flavodoxin was used, an observation in accord with data in the literature [2,4]. On the other hand, when the flavodoxins of *M.elsdenii* and *D.vulgaris* were used, the reaction seemed to proceed at potential beyond -460 mV. These results are not in agreement with observations of Zumft *et al.* [1], who observed for the Fe-protein of *C.pasteurianum* a midpoint potential of -290 mV at pH 7.5. When MgADP or MgATP was bound to the protein the midpoint potential shifted to -400 mV. In all cases Zumft *et al.* [1] found that the titrations followed a n=1 redox profile.

Fig. 3 shows the results of the effect of the redox potential of the redox couple dithionite/sulfite on the nitrogenase activity. Conditions as described by Mayhew and van Dijk *et al.* [26,27] were followed. Not $S_2O_4^{2-}$ but its dissociation product, $SO_2^{\cdot-}$ is the actual electron donor for nitrogenase [6,28]. In all experiments the concentration $S_2O_4^{2-}$ was kept constant at 5 mM. With a dissociation constant for dithionite of 1.4 nM, it can be

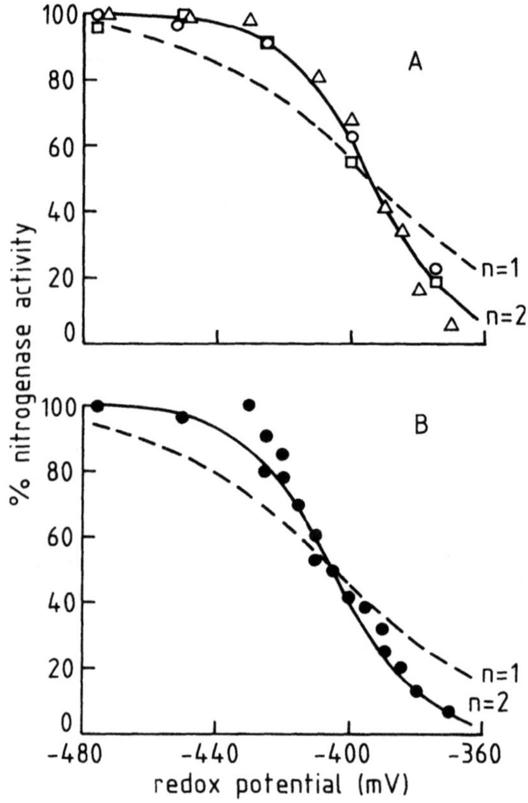

Fig. 3. The effect of the applied redox potential on the activity of different nitrogenase complexes. Redox potentials were set with $Na_2S_2O_4$ (5 mM) and varying amounts of Na_2SO_3 [from ref. 18]. A: o–o, $[Av_1]$ = 0.33 μM, $[Av_2]$ = 0.33 μM; □–□, $[Av_1]$ = 0.77 μM, $[Av_2]$ = 0.96 μM; Δ–Δ, $[Av_1]$ = 0.11 μM, $[Av_2]$ = 2.9 μM; B: ●–●, 0.25 mg/ml three component nitrogenase complex. The dotted lines are theoretical lines for a one-electron transition, the solid lines are the theoretical lines for a two-electron transition with E_m = -394 mV (A) or E_m = -405 mV (B). (Reproduced with permission from the European Journal of Biochemistry).

calculated that the concentration of $SO_2^{\bullet -}$ = 2.6 μM. The redox potential was increased by adding the oxidation product of $SO_2^{\bullet -}$,

Na_2SO_3 [26,27]. It is clear that the nitrogenase reaction is independent of the redox potential up to potentials of -440 mV. At higher potentials the nitrogenase activity declines and at potentials above -350 mV hardly any nitrogenase activity is detectable. It was ascertained that at low potentials a concentration of 2.6 μM $SO_2^{\bullet -}$ (5 mM $S_2O_4^{2-}$) is saturating in the nitrogenase reaction. Since during variation of the Na_2SO_3 concentration those of $SO_2^{\bullet -}$ and Av_2 are not changed it can be assumed that the rate of reduction of oxidized Av_2 by $SO_2^{\bullet -}$ is the same at high and at low potentials, although assuming equilibrium between the $SO_2^{\bullet -}/SO_3^{2-}$ and $Av_2(ox)/Av_2(red)$ couples, at higher redox potentials Av_2 becomes more oxidized.

In order to verify that the drop in nitrogenase activity at higher potentials is not due to non-equilibrium between Av_2 and the electron donor, we increased the absolute and relative concentrations of Av_2. The activities of three different reconstituted complexes (Fig. 3A) show the same dependence on the applied redox potential. It is important to notice that these reconstituted complexes have different specific activities. In a 1:1 complex the amount of Av_2 is thought to limit the overall activity. But in the 1:26 complex the flux of electrons from reductant via Av_2 to Av_1 is thought to be maximal. In all cases the same dependence of activity on applied redox potential is found. A similar result (Fig. 3B) was obtained with the O_2-stable three component nitrogenase complex, from *A.vinelandii* [2,3]. The activity versus the redox potential profile of the experimental points of Fig. 3 fits a theoretical curve for a 2-electron redox reaction much better than that for a 1-electron redox reaction.

To ascertain that decrease in nitrogenase activity at higher potentials is not due to an ionic strength effect due to the increasing concentrations of Na_2SO_3, we used substrate amounts of photochemically reduced low-potential electron carriers as electron donor by nitrogenase. A typical oxidation curve of *A.vinelandii* flavodoxin hydroquinone by nitrogenase is shown Fig. 4, showing that this flavodoxin hydroquinone is not oxidized beyond -460 mV. Similar curves were obtained at higher *A.vinelandii* flavodoxin concentrations, up to 500 μM. With a midpoint potential of -495 mV, it can be calculated that at a starting concentration of 500 μM, the flavodoxin hydroquinone concentration at -460 mV is 100 μM. However under continuous illumination flavodoxin concentrations around 10 μM give maximum activity [3]. Thus we can conclude that *A.vinelandii* flavodoxin semiquinone acts as inhibitor and that it is therefore not possible to use this flavodoxin as an electron donor and as indicator to study the redox potential dependence of the nitrogenase reaction.

Results with methyl viologen semiquinone and *M.elsdenii* flavodoxin hydroquinone oxidation by nitrogenase are shown in Fig. 4. *M.elsdenii* flavodoxin hydroquinone and methyl viologen semiquinone as electron donor give similar activity - redox potential

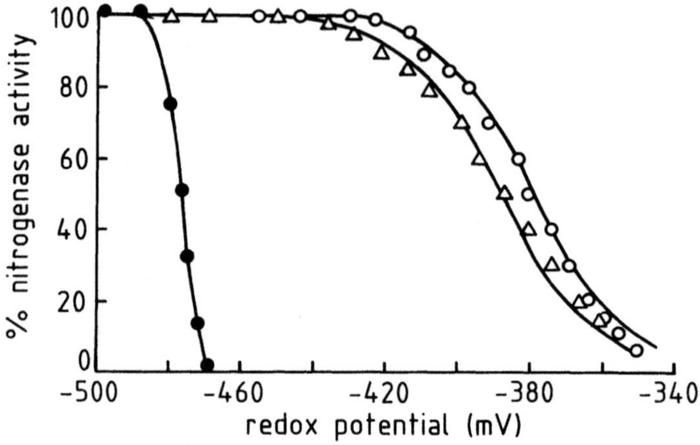

Fig. 4. Dependence of the nitrogenase activity on the redox potential of *A.vinelandii* flavodoxin, *M.elsdenii* flavodoxin and methyl viologen [from ref. 18]. Three component nitrogenase complex (0.25 mg/ml) was used. ●-●, 50 μM *A.vinelandii* flavodoxin; Δ-Δ, 72 μM methyl viologen, (E_m = -388 mV); o-o, 137 μM *M.elsdenii* flavodoxin (E_m = -379 mV). (Reproduced with permission from the European Journal of Biochemistry)

curves as $Na_2S_2O_4$ (Fig. 3). Again the best fit to the experimental points is a theoretical curve for a 2-electron redox reaction.

From the results of Figs. 3 and 4 it cannot be derived which of the two nitrogenase components is responsible for this redox behaviour. We performed redox titrations of Av_2 with the redox couple SO_2^{-}/SO_3^{2-} and measured the oxidation-reduction state of the $[Fe_4-S_4]$ cluster by EPR spectroscopy. The EPR spectra of Av_2, differ considerably from the data published [18], Table III.

We used the height of the g_y peak as a measure of the amount of Av_2 in the reduced state. Fig. 5 shows the influence of the applied redox potential on the height of this signal. Three important redox features of Av_2 are given. The experimental points of the redox titrations of Av_2 and Av_2 + MgATP fit better a theoretical curve for a 2-electron redox reaction than a n=1 curve. In contrast, the experimental points of the Av_2 + MgADP curve fit better a n=1 redox behaviour than a n=2 one. The midpoint potential of Av_2 without adenine nucleotides is considerably more negative than Cp_2 and the shift in midpoint potential by MgATP is considerably less. A further difference is that with Av_2 the midpoint potential in the presence of MgADP is more negative than Cp_2 + MgADP.

Fig. 6A shows the oxidation reduction state of Av_2 + Mg^{2+} in the presence and absence of ATP, as measured by EPR spectroscopy

Table III. g-values of nitrogenase Fe-proteins [from ref.18]. (Reproduced with permission from the European Journal of Biochemistry)

Fe protein	g_x	g_y	g_z	n-value	midpoint potential (mV)	references
Av_2(Mg)	1.864	1.941	2.048	2	-392(pH 7.0)	[18]
Av_2(MgADP)	1.874	1.947	2.051	1	-473(pH 7.0)	[18]
Av_2(MgATP)	1.895-1.90	1.940	2.045	2	-438(pH 7.0)	[18]
Ac_2(Mg)	1.87	1.94	2.05	-	-	[35]
Kp_2(Mg)	1.865	1.942	2.053	-	-	[36]
Kp_2(MgATP)	-	1.929	2.036	-	-	[36]
Cp_2(Mg)	1.87	1.94	2.06	1	-294(pH 7.5)	[1,29]
Cp_2(MgATP)	-	1.93	2.04	1	-400(pH 7.5)	[1,29]
Cp_2(MgADP)	-	1.93	2.04	1	-380(pH 7.5)	[1,29]

and the activity of the nitrogenase at different redox potentials. It is clear that the activity of nitrogenase follows the oxidation reduction state of Av_2 without adenine nucleotides bound; both curves can be fitted to a theoretical 2-electron oxidation reduction reaction.

Our results indicate quite convincingly, that Av_2, considered to be a $[Fe_4-S_4]$-cluster containing protein, shows a redox behaviour characteristic for a two-electron transferring protein. This behaviour is found in the "oxidized" protein, as isolated, as well as in its MgATP complex. But not in its MgADP complex, which shows a one-electron transfer redox behaviour. Our data contrast studies with the Fe-protein from *C.pasteurianum*, where a one-electron redox transfer was found, irrespective of the binding of Mg-adenine nucleotides [1,29]. Furthermore Ljones and Burris [5] concluded from physiological oxidation experiments of Cp_2 by Cp_1, that a one-electron transfer takes place.

On the other hand Watt and Bulen [2] could not decide whether the nitrogenase activity-redox potential curve of the three-component complex of *A.vinelandii* followed a n=1 or n=2 pattern. Evans and Albrecht [4] concluded that crude nitrogenase from *C.vinosum* follows a n=2 pattern in its activity curve. Thorneley *et al.* [20] observed in rapid kinetic studies with dye-oxidized Ac_2, that two electrons could be accomodated. The uptake of the

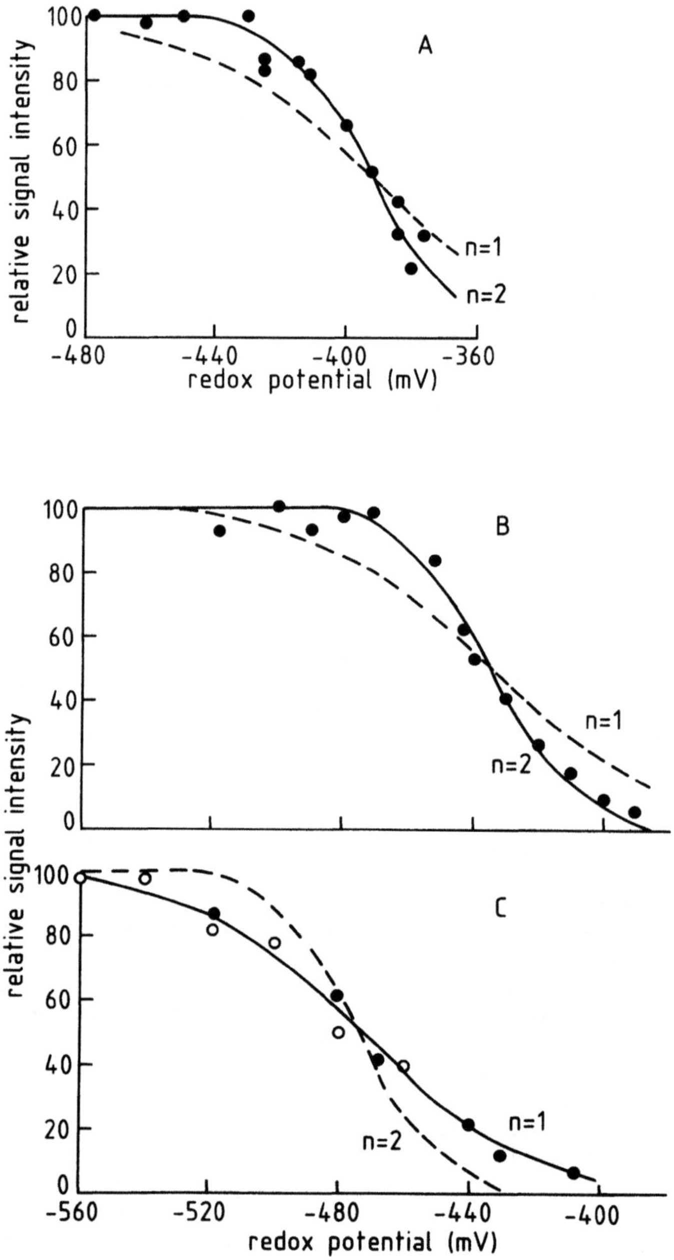
A
relative signal intensity
100
80
60
40
20
0
-480
-440
-400
-360
redox potential (mV)
n=1
n=2
B
C
relative signal intensity
100
80
60
40
20
0
-560
-520
-480
-440
-400
redox potential (mV)
n=1
n=2

Fig. 5. Dependence of the electron paramagnetic resonance signal of Av_2 in the presence or absence of Mg^{2+}-adenine nucleotides on the redox potential [from ref. 18]. The redox potential was set with $Na_2S_2O_4$ and Na_2SO_3. A: •-•, 56 μM Av_2, 7.5 mM $MgCl_2$ in 112 mM Hepes, pH 7.0. The dotted line is the theoretical curve for a one-electron transition, the solid line for a two-electron transition with an E_m = -392 mV; B: •-•, 31.5 μM Av_2, 7.5 mM $MgCl_2$ in 112 mM Hepes, pH 7.0; 5 mM ATP. The dotted line is the theoretical curve for a one-electron transition, the solid line for a two-electron transition with E_m = -435 mV; C: •-•, 56 μM Av_2, 7.5 mM $MgCl_2$, 5 mM ADP, in 112 mM Hepes pH 7.0; o-o, 21.5 μM Av_2, 7.5 mM $MgCl_2$, 5 mM ADP, 112 mM Hepes pH 7.8. The solid line is the theoretical curve for a one-electron transition, the dotted line for a two-electron transition with an E_m = -473 mV.
(Reproduced with permission from the European Journal of Biochemistry)

first, fast reacting electron is associated with the EPR-signal; the second electron-uptake was, alike the first electron-uptake, spectrally visible. Thorneley *et al.* [28] concluded that this second electron-uptake process is associated with the presence of inactive protein. Later Lowe [30] revised this view in suggesting that in Kp_2 another rapidly-relaxing paramagnetic centre is present, in view of the broad anisotropic EPR-spectra observed.

Our data on the activity-redox potential dependence, coinciding closely with the EPR titration of free Av_2 (Fig. 5A) oppose some of the current views expressed in kinetic models [28,31,32]. Thorneley *et al.* [28,32] suggested that the rate-limiting step is the dissociation of oxidized Fe-protein from the MoFe-protein.

It has been proposed [32,33] that the steps prior to the nitrogenase reaction proceed in a random mechanism. In other words free, reduced Fe-protein reacts with either the MoFe-protein followed by ATP-binding on the complex or by binding ATP followed by reaction of the ATP-Fe-protein complex with the MoFe-protein. All binding and complex formation proceeds in rapid equilibrium. Thereafter irreversible electron transfer takes place from the Fe-protein to the MoFe-protein with subsequent hydrolysis of ATP to ADP and phosphate and the final dissociation of the oxidized Fe-protein from the oxidized MoFe-protein is thought to be rate-limiting.

The activity-redox profiles coinciding fully with the EPR-redox behaviour of the free Fe-protein contrasts a random mechanism and requires a sequential mechanism as described in Scheme I.

The scheme gives the basic steps, as minimum hypothesis, necessary to explain the phenomena observed. In the theoretical considerations for an ordered mechanism in terms of the relations with the activity-redox profiles, the following facts and assumptions were taken into account.

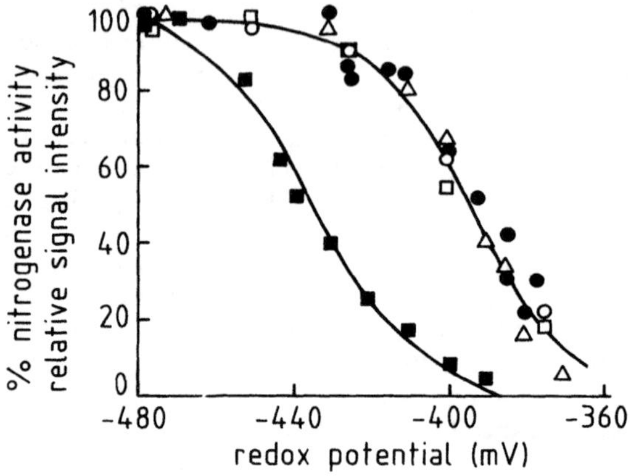

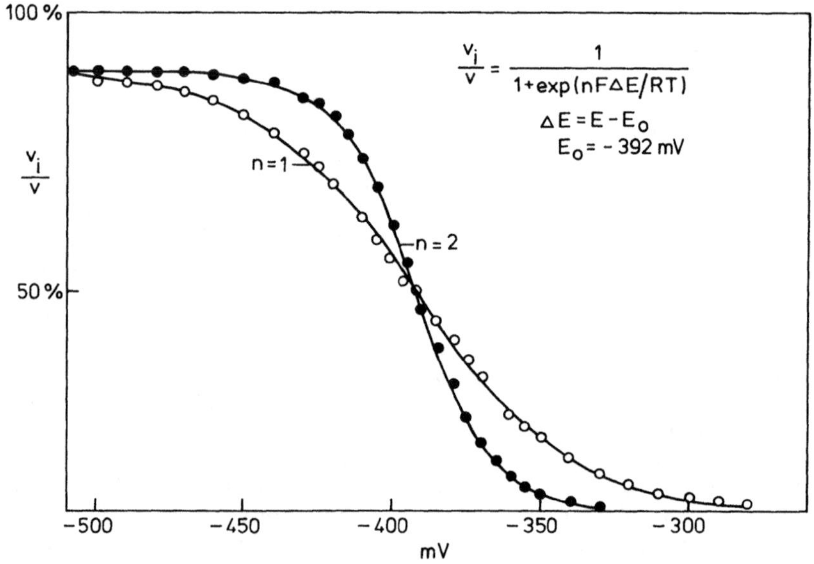

Fig. 6. Dependence of the redox state of Av_2 and of the nitrogenase activity on the redox potential [from ref. 18]. A: o-o, activity of a 1:1 nitrogenase complex; □-□, activity of a 1:14 nitrogenase complex; Δ-Δ, activity of a 1:26 nitrogenase complex; ●-●, relative g = 1.941 signal intensity y of Av_2 + $MgCl_2$ at the redox potentials indicated; ■-■, relative g = 1.940 signal intensity of Av_2 + MgATP at the redox potentials indicated. B: expected activity - redox profile from equation 1 of the proposed kinetic model, assuming a midpoint potential of the nitrogenase reaction of -392 mV. (Reproduced with permission from the European Journal of Biochemistry)

1. The concentration of the Av_2(red)-Av_1(ox) complex determines the rate of the reaction, because of the 1000-fold higher affinity of the Fe-protein for the MoFe-protein than for MgATP.
2. Av_2(red) and Av_2(ox) have similar association constants for Av_1(ox) as well as for MgATP.
3. The transfer of electrons in catalysis is irreversible and somewhere in the whole sequence of events the rate-limiting step of the reaction occurs.
4. All equilibrium steps are fast compared with the rate of catalysis.

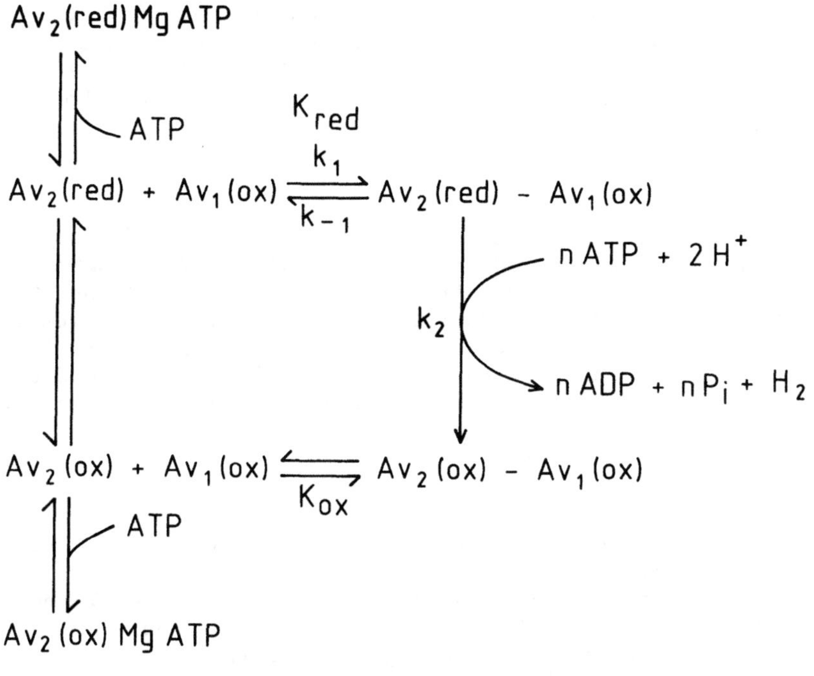

It can be shown [18], that in view of these considerations the activity-redox relation is described by equation (1), which follows the observed activity-redox potentials profile (Fig. 6B).

$$\frac{v_i}{v} = \frac{1}{1 + \exp(nF\Delta E/RT)} \qquad (1)$$

In this equation $\Delta E = E - E_o$ in which E is the applied redox potential; v_i is the inhibited activity; v is the maximum activity.

Our results clearly indicate that under redox potentials limitation catalysis proceeds by 2-electron transfer, not necessarily simultaneously, from Av_2 to Av_1, a view supported by the redox pattern of the Av_2-MgADP complex (n=1). The redox behaviour of the Av_2-MgADP complex also explains the strong inhibition by ADP of the nitrogenase reaction. Rather the absence of the second electron prevents catalysis than competition with bound ATP, as proposed [34].

ACKNOWLEDGEMENTS

We thank Mr. B.J. Sachteleben and Mr. M.M. Boumans for drawing the figures and Mrs. J.C. Toppenberg-Fang for typing the manuscript. This investigation was supported by the Netherlands Foundation for Chemical Research (S.O.N.) with financial aid from the Netherlands Organization for the Advancement for Pure Research (Z.W.O.).

ABBREVIATIONS

The MoFe- and Fe-proteins of the nitrogenase of *Azotobacter vinelandii*, *Azotobacter chroococcum*, *Clostridium pasteurianum* and *Klebsiella pneumoniae* are referred to as respectively Av_1 and Av_2, Ac_1 and Ac_2, Cp_1 and Cp_2 and Kp_1 and Kp_2; N_2-ase, nitrogenase; EPR, electron paramagnetic resonance; Hepes, N-2-hydroxyethylpiperazine-N'-2-ethane sulfonic acid; Tricine, N-tris(hydroxymethyl)methyl glycine; deazaflavine, 3,20-dimethyl-5-deazaisoalloxazine.

REFERENCES

1. Zumft, W.G., Mortenson, L.E. and Palmer, G. (1974) Eur.J. Biochem. *46*, 525-535.
2. Watt, G.D. and Bulen, W.A. (1976) "Proc.1st.Intern.Symp. on Nitrogen Fixation" (W.E. Newton and C.J. Nijman, eds.) Washington State Univ. Press pp. 248-256.
3. Scherings, G., Haaker, H. and Veeger, C. (1977) Eur.J.Biochem. *77*, 621-630.
4. Evans, M.C.W. and Albrecht, S.L. (1974) Biochem.Biophys.Res. Commun. *61*, 1187-1192.
5. Ljones, T. and Burris, R.H. (1972) Biochim.Biophys.Acta *275*, 93-101.
6. Watt, G.D. and Burns, A. (1977) Biochemistry *16*, 264-270.
7. Laane, C., Haaker, H. and Veeger, C. (1979) Eur.J.Biochem. *87*, 147-153.
8. Yates, M.G. and Jones, C.W. (1974) Adv.Microbiol.Physiol. *11*, 97-135.

9. Mortenson, L.E. (1964) Proc.Natl.Acad.Sci.USA *52*, 272-279.
10. Knight, E. and Hardy, R.W.F. (1966) J.Biol.Chem. *241*, 2752-2756.
11. Beneman, J.R., Yoch, D.C., Valentine, R.C. and Arnon, D.I. (1971) Biochim.Biophys.Acta *226*, 205-212.
12. Haaker, H., de Kok, A. and Veeger, C. (1974) Biochem.Biophys. Acta *357*, 344-357.
13. Haaker, H., Laane, C., and Veeger, C. (1980) "Nitrogen Fixation" (W.D.P. Stewart and J.R. Gallon, eds.) Academic Press, London, New York, pp. 113-138.
14. Yates, M.G., (1972) FEBS Lett. *27*, 63-67.
15. Beneman, J.R., Yoch, D.C., Valentine, R.C. and Arnon, D.I. (1969) Proc.Natl.Acad.Sci.USA *64*, 1079-1086.
16. Veeger, C., Laane, C., Scherings, G., Matz, L., Haaker, H. and van Zeeland-Wolbers, L. (1980) "Nitrogen Fixation" (W.E. Newton and W.H. Orme-Johnson, eds.), University Park Press, Baltimore, Volume I, p. 111-137.
17. Van der Westen, H.M., Mayhew, S.G. and Veeger, C. (1980) FEMS Lett. *7*, 35-39.
18. Braaksma, A., Haaker, H., Grande, H. and Veeger, C. (1981), Eur.J.Biochem., submitted for publication.
19. Veeger, C., Haaker, H. and Laane, C. (1981) "Current perspectives in Nitrogen Fixation" (A.H. Gibson and W.H. Newton, eds.), Elsevier-North-Holland, Amsterdam, pp. 101-104.
20. Haaker, H. and Veeger, C. (1976) Eur.J.Biochem. *63*, 499-507.
21. Haaker, H. and Veeger, C. (1977) Eur.J.Biochem. *77*, 1-10.
22. Laane, C., Krone, W., Konings, W.N., Haaker. H. and Veeger, C. (1979) FEBS Lett. *103*, 328-332.
23. Laane, C., Krone, W., Konings, W.N., Haaker, H. and Veeger, C. (1980) Eur.J.Biochem. *103*, 39-46.
24. Rowell, P., Reed, R.H., Hawkesford, M.J., Ernst, A., Diez, J. and Stewart, W.D.P. (1981) "Current Perspectives in Nitrogen Fixation" (A.H. Gibson and W.H. Newton, eds.) Elsevier-North Holland, Amsterdam, pp. 186-189.
25. Laane, C. (1980) Ph.D. Thesis, Agricultural University, Wageningen.
26. Mayhew, S.G. (1978) Eur.J.Biochem. *85*, 535-547.
27. Van Dijk, C., Mayhew, S.G. and Veeger, C. (1981) Eur.J.Biochem. *111*, 201-207.
28. Thorneley, R.N.F., Yates, M.G. and Lowe, D.J. (1976) Biochem. J. *155*, 173-144.
29. Zumft, W.G., Palmer, G. and Mortenson, L.E. (1973) Biochim. Biophys.Acta *292*, 413-421.
30. Lowe, D.J. (1978) Biochem.J. *175*, 955-957.
31. Thorneley, R.N.F. (1981) "Current Perspectives in Nitrogen Fixation" (A.H. Gibson and W.E. Newton, eds.), Elsevier-North Holland, p. 360.
32. Mortenson, L.E. and Thorneley, R.N.F. (1979) Ann.Rev.Biochem. *48*, 387-418.

33. Hageman, R.V., Orme-Johnson, W.H. and Burris, R.H. (1980) Biochemistry *19*, 2333-2342.
34. Thorneley, R.N.F. and Cornish-Bowden, H. (1977) Biochem.J. *165*, 255-262.
35. Yates, M.G. and Planqué, K. (1975) Eur.J.Biochem. *60*, 467-476.
36. Smith, B.E., Lowe, D.L. and Bray, R.C. (1973) Biochem.J. *135*, 331-341.

MÖSSBAUER AND EPR EVIDENCE ON THE PROSTHETIC GROUPS OF THE MoFe PROTEIN

B. H. Huynh,[a] E. Münck,[b] and W. H. Orme-Johnson[c]

a. Department of Physics, Emory University, Atlanta, GA 30322 U.S.A.
b. Gray Freshwater Biological Institute, University of Minnesota, Navarre, MN 55392 U.S.A.
c. Department of Chemistry, Massachusetts Institute of Technology, Cambridge, MA 02139 U.S.A.

Spectroscopic evidence strongly suggests the existence of two types of redox centers in the MoFe proteins, designated as the M center and the P cluster. Each MoFe protein was shown to contain two M centers and four P clusters, accounting for 28 of the 30 iron atoms in the molecule. The M center is a spin-coupled complex which contains most probably six irons and one molybdenum atom. It resides in the cofactor and can be stabilized in three oxidation states. The native state M^N *has a spin* $S = 3/2$ *and its ground doublet is EPR active. A one electron oxidation process transforms* M^N *into* M^{ox} *which is diamagnetic. Under nitrogen fixing conditions, the state* M^N *is reduced into state* M^R*. State* M^R *is not discussed in this manuscript but has been shown to have integer spin* $S > 0$ *(1). The P cluster contains four spin-coupled iron atoms, most probably bridged by labile sulfur. Two oxidation states,* P^{ox} *and* P^N*, have been characterized. The native state* P^N *is diamagnetic. The oxidized state* P^{ox} *has a half integer spin* $S \geq 3/2$*, and is characterized by a negative zero field parameter D. The nature of the remaining two irons is still nebulous and further investigations are warranted.*

1. INTRODUCTION

The nitrogenase enzyme complex catalyzes the six-electron reduction of molecular nitrogen into ammonia, allowing atmospheric nitrogen to enter the biosphere. This enzyme

H. B. Dunford et al. (eds.), The Biological Chemistry of Iron, 241–258.

complex consists of two protein components: component I and component II (2,3). Component II is also designated as the iron protein or nitrogenase reductase. It has a molecular weight of approximately 60,000 and comprises two polypeptide chains and one [4Fe-4S] cluster. Component I is also known as the molybdenum-iron (MoFe) protein, or simply, nitrogenase. It consists of four subunits and one or two cofactors and has a molecular weight on the order of 220,000. Each molecule of the MoFe protein contains approximately 30 iron atoms, comparable amounts of acid-labile sulfur and approximately 2 Mo atoms.

For the past few years, we have been using EPR and Mössbauer spectroscopy to study the MoFe protein from *Clostridium pasteurianum* and from *Azotobacter vinelandii* (1,4). Our studies reveal the presence of two types of spin-coupled iron clusters in the MoFe protein: two M centers (each center contains most likely six iron atoms), and four P clusters (each cluster contains four iron atoms). In what follows, we will discuss briefly some basic features of the ^{57}Fe Mössbauer spectroscopy and outline the relationship between EPR and Mössbauer measurements. In section 3, we discuss the EPR and Mössbauer evidence. Other physical data in support of the proposed model will also be mentioned.

2. ^{57}Fe MÖSSBAUER SPECTROSCOPY

2.1 Some Fundamental Aspects

Mössbauer spectroscopy is a recoiless nuclear resonance technique. A Mössbauer spectrometer consists basically of two parts: a nuclear γ-ray source and a recording system which records the transmission of the resonant γ-rays through an absorber of interest. For the study of iron proteins, the source is the 14.4 kev γ-rays emitted by the excited ^{57}Fe nuclei which are created by the radioactive decay of ^{57}Co. The Mössbauer spectrum is a record of the resonant transitions between the ground and the first excited states of the ^{57}Fe nuclei in the absorber. The ^{57}Co is generally diffused into a suitable host matrix so that the emitted γ-beam is monochromatic. The nuclear energy levels of the absorber, however, are split and shifted by the hyperfine interactions between the nucleus and the surrounding electrons. The hyperfine splitting are typically less than 10^{-6} eV and the source radiation may be brought into resonance with the absorber by moving the source relative to the absorber. A velocity of 1 mm/s will Doppler shift the 14.4 kev γ-rays by 4.8×10^{-8} eV. Below, we will discuss two types of hyperfine interactions that are important for the understanding of Mössbauer spectrum.

Magnetic hyperfine interaction. The ^{57}Fe nucleus possesses a magnetic moment which will interact with a local effective magnetic field $\vec{H}_{eff}$. The interaction can be written as

$$\hat{H}_M = -g_n \beta_n \vec{I} \cdot \vec{H}_{eff} \tag{1}$$

where β_n is the nuclear magneton and I is the nuclear spin (I = 1/2 and 3/2, respectively, for the ground and the excited states of the ^{57}Fe nucleus). The nuclear gyromagnetic ratio g_n has the values 0.1806 and -0.1033 for the ground and the excited states, respectively (5). The effective field is the sum of the external applied field ($\vec{H}_{app}$) and the internal field ($\vec{H}_{int}$) created by the unpaired electrons.

$$\vec{H}_{eff} = \vec{H}_{app} + \vec{H}_{int}$$

$$= \vec{H}_{app} - \frac{1}{g_n \beta_n} \tilde{A} \cdot \vec{S} \tag{2}$$

where $\tilde{A}$ is the magnetic hyperfine coupling tensor and $\vec{S}$ is the electronic spin. For iron complexes, the zero-field splitting of the electronic ground multiplet and the electron Zeeman interaction are much larger than the magnetic hyperfine interaction. In other words, the electron and nuclear spins are effectively decoupled. The operator $\vec{S}$ in equation 2 may therefore be substituted by the expectation value $\langle \vec{S} \rangle$, i.e., $\vec{H}_{int} = -[1/(g_n \beta_n)] \tilde{A} \cdot \langle \vec{S} \rangle$. The internal field for iron is in the order of a few hundreds of kG. Consequently, for small $\vec{H}_{app}$, the effective field approximates $\vec{H}_{int}$.

The effect of the magnetic hyperfine interaction completely removes the spin degeneracy of the nuclear states, and creates six allowed Mössbauer transitions as indicated in Fig. 1. The corresponding Mössbauer spectrum will exhibit six absorption lines and is termed magnetic Mössbauer spectrum. The intensities for the absorption lines depend on the angle θ between $\vec{H}_{eff}$ and the observed γ-beam and can be calculated from standard radiation theory (6). The results are:

$$\text{Intensity } (\pm 1/2 \rightarrow \pm 3/2) \propto 3(1 + \cos^2\theta)/2$$

$$\text{Intensity } (\pm 1/2 \rightarrow \mp 1/2) \propto (1 + \cos^2\theta)/2 \tag{3}$$

$$\text{Intensity } (\pm 1/2 \rightarrow \pm 1/2) \propto 2 \sin^2\theta$$

The idealized magnetic Mössbauer spectra corresponding to

three typical situations are shown in Fig. 1.

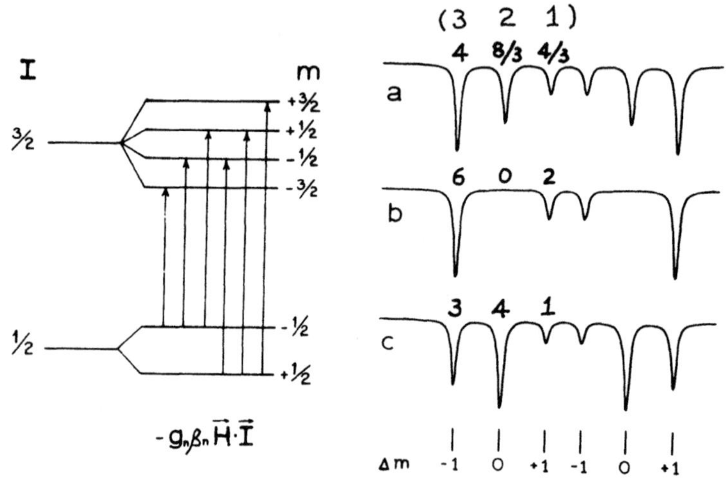

Figure 1. Energy level diagram of ^{57}Fe nucleus under the effect of magnetic interaction. The allowed Mössbauer transitions are indicated with vertical arrows. Also shown on the right are the magnetic Mössbauer spectra resulting from three situations: samples with randomly oriented $\vec{H}_{int}$ (a), with $\vec{H}_{int}$ parallel (b), or perpendicular (c) to the observed 14.4 kev γ-beam. The numbers on top of the spectra indicate the relative absorption intensities.

Electronic Quadrupole Interaction. The excited state of the ^{57}Fe nucleus has a spin of 3/2 and possesses an electric quadrupole moment Q. An electric field gradient (EFG) created by the surrounding electronic charges may interact with the quadrupole moment and split the nuclear excited state into two levels separated in energy by

$$\Delta E_Q = \frac{eQV_{z'z'}}{2}\,(1 + \frac{1}{3}\,\eta^2)^{1/2} \qquad (4)$$

where $V_{i'i'}$'s are the principal components of the EFG tensor and $\eta = (V_{x'x'} - V_{y'y'})/V_{z'z'}$. Conventionally, the principal axes are chosen so that $|V_{x'x'}| \leq |V_{y'y'}| \leq |V_{z'z'}|$. The ground state of the ^{57}Fe nucleus has no quadrupole moment and is not perturbed by the EFG. In the absence of a local magnetic field, only two resonant transitions are possible. The corresponding Mössbauer spectrum exhibits two absorption lines. Fig. 2 shows the nuclear energy level scheme and the resultant Mössbauer spectrum. Such a spectrum is termed a

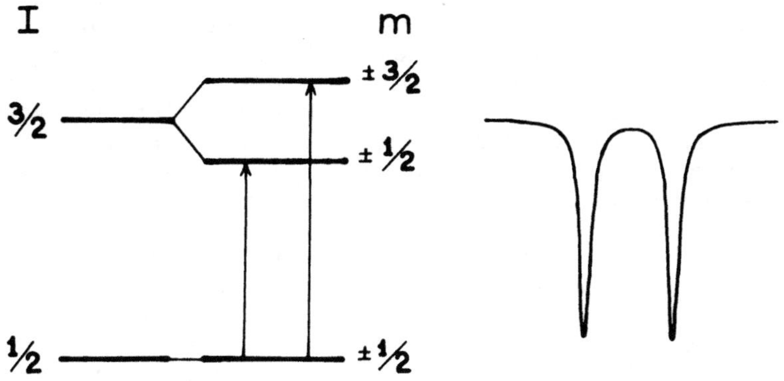

Figure 2. Energy level diagram of ^{57}Fe nucleus showing the effect of quadrupole interaction. The excited state is split into two levels. The allowed Mössbauer transitions are indicated with vertical arrows. A typical quadrupole doublet is shown on the right.

quadrupole doublet. The absorption lines of a quadrupole doublet should have equal intensities for absorbers with randomly oriented molecules. In most cases, a quadrupole doublet is symmetric with respect to the centroid of the spectrum.

For clarification, a comment on isomer shift is in order here. The excited ^{57}Fe nucleus is about 0.1% smaller in radius than the ground state nucleus causing the nuclear excitation energy to depend on the electron density at the nucleus (7). Since the iron environment of the absorber is different from that of the source which results in different electron densities at the nuclei, the centroid of the absorption spectrum is shifted away from the zero velocity. This energy shift is termed isomer shift. To allow isomer shift to be measured independently of the nature of the source matrix, it is now conventionally adopted the centroid of the room temperature absorption spectrum of metallic iron to be the zero energy. Isomer shift is usually treated as an empirical parameter and is particularly useful in determining the iron oxidation states of related series of materials.

2.2 Correlation with EPR

Paramagnetic complexes may be classified into two broad categories: those with an even number of electrons and those with an odd number. The system spin is integer for the even-number-electron system and is half-integer for the odd-number-electron system (Kramers' system). For integer-spin system, the spin-orbit coupling completely lifts the spin degeneracy of the ground multiplet resulting in non-degenerate electronic levels (except in cases of unusually high symmetry). Since singlet state can have no permanent magnetic moment, i.e., $\langle S\rangle = 0$, it produces no $\vec{H}_{int}$ at the nucleus. Consequently, *iron complexes with even number of electrons yield only quadrupole doublet in the absence of an external field.* The same complexes in general yield no EPR signal since the zero-field splittings within the ground multiplet are larger than the applied microwave frequency (typically, $h\nu \sim 0.3\ cm^{-1}$).

However, the situation changes for Kramers' system. According to Kramers' theorem half-integer spin systems will always be at least doubly degenerate in the absence of a magnetic field. EPR transitions can generally be induced within each Kramers' doublet. The expectation value $\langle S\rangle$ is not equal to zero. Consequently, *iron complexes with odd number of electrons should yield magnetic Mössbauer spectrum.* In small applied field, the expectation value $\langle S\rangle$ is given by

$$\langle S_i\rangle = g_i/4,\ i = x,y,z \tag{5}$$

where g_i's are the observed EPR g-values and x,y,z denote the principal axes of the $\tilde{g}$-tensor. The effective field produced by the Kramers' doublet can be written as,

$$\vec{H}_{eff} \sim \vec{H}_{int} = \frac{A}{4g_0}(\ell_x g_x^{\ 2}\ \hat{x} + \ell_y g_y^{\ 2}\ \hat{y} + \ell_z g_z^{\ 2}\ \hat{z}) \tag{6}$$

where $g_0 = \sqrt{\ell_x^{\ 2} g_x^{\ 2} + \ell_y^{\ 2} g_y^{\ 2} + \ell_z^{\ 2} g_z^{\ 2}}$ and ℓ_x, ℓ_y and ℓ_z are the direction cosines of $\vec{H}_{app}$. (To simplify our discussion, we have assumed $\tilde{A}$ to be isotropic). To discuss the relationship between the Mössbauer and EPR data, let us first consider a system with isotropic g-values, i.e., $g_x = g_y = g_z$. From equation 6 we notice that $\vec{H}_{eff}$ is always parallel to $\vec{H}_{app}$ in such an isotropic system regardless of the molecular orientation. Thus by applying a field of a few hundred Gauss parallel and perpendicular to the observed γ-beam, Mössbauer spectra with patterns as indicated in Fig. 1b and 1c, will be observed, respectively. In case that the g_i's are different but comparable to each other, the direction of $\vec{H}_{int}$ can still be effectively oriented by $\vec{H}_{app}$. Field-direction-dependent Mössbauer spectra will also be observed. Since Kramers' doublets with isotropic, or comparable g-components are EPR active, the observation of field-direction-dependent Mössbauer spectra as shown in Fig. 1b and 1c indicates an EPR active Kramers' doublet, or vice versa.

Let us now consider the case of an extremely anisotropic system with $g_x \sim g_y \sim 0$ and $g_z >> g_x$ and g_y. In such a situation, equation 6 indicates that $\vec{H}_{eff}$ is parallel to the z-axis. In other words, $\vec{H}_{eff}$ is fixed in direction with respect to the molecule regardless of the direction of $\vec{H}_{app}$. For a randomly oriented absorber, θ will be averaged over all possible values. The magnetic Mössbauer spectrum will have the pattern as shown in Fig. 1a and is independent of the direction of $\vec{H}_{app}$. Also, for a powder sample, Aasa and Vanngard (8) have shown that the EPR intensity at a field corresponding to g_z is given by

$$\text{Intensity (z)} \propto \frac{g_x^2 + g_y^2}{\sqrt{(1 - \frac{g_x^2}{g_z^2})(1 - \frac{g_y^2}{g_z^2})}} \qquad (7)$$

If $g_x \sim g_y \sim 0$, the EPR intensity is small and may be impossible to observe. Taken together we arrive at the conclusion that a magnetic Mössbauer spectrum with a pattern that is independent of the direction of $\vec{H}_{app}$ implies an "EPR-silent" Kramers' doublet.

In summary, for a system with an even number of electrons, EPR is generally not applicable. In the absence of an applied field, the associated iron site(sites) should yield only quadrupole doublet(doublets). For a system with odd number of electrons, the associated iron site(sites) should yield magnetic Mössbauer spectrum(spectra). In the presence of a small magnetic field two different types of Kramers' doublets can be distinguished. One, an "EPR-silent" doublet with two g-values close to zero, will have a field-direction-independent Mössbauer spectrum. The other type, an EPR-active doublet, will have a field-direction-dependent Mössbauer spectrum.

A factor having a major impact on Mössbauer spectrum is the electronic relaxation rate. The above discussion is valid only when the relaxation rate is slow compared with the nuclear precession frequency ($\sim 10^7$Hz). Slow electronic relaxation is common for biological compounds at temperatures below 4.2 K.

3. MÖSSBAUER AND EPR STUDIES

The MoFe protein is very sensitive to oxygen, and stringent anaerobic conditions are required during the purification procedure. The protein as isolated shows EPR resonances at g = 4.3, 3.7 and 2.0 (9). These signals result from the $S_z = \pm 1/2$ doublet of an S = 3/2 system. The observed g-values can be reproduced very well using an S = 3/2 spin Hamiltonian with E/D = 0.055 (10). Temperature studies of the EPR signal indicate that the zero field parameter D is approximately +5.5 cm^{-1}. Spin quantitation of the EPR signal yielded about 1 spin/Mo atom. Taken together with the metal analysis of 2 Mo per molecule, the data indicate that the MoFe protein contains two S = 3/2 centers. Under nitrogen fixing conditions, i.e., when MgATP, a proper reductant, and the Fe protein are added to the MoFe protein under N_2 atmosphere, the S = 3/2 EPR signal decreases to about 10% of its maximum intensity. This observation suggests that the S = 3/2 center may play an important role in the nitrogen fixing process.

A reason that we ventured into the studies of MoFe proteins using Mössbauer spectroscopy was to answer the following important question: Does the S = 3/2 signal result from an iron containing complex or from molybdenum? From the discussion presented in the previous section, we now know that if the signal is resulted from an iron complex we should observe at low temperature a field-direction-dependent magnetic Mössbauer spectrum. If no such spectrum is observed, the signal must associate with molybdenum.

Figure 3A shows a Mössbauer spectrum of the native MoFe protein from *C. pasteurianum* recorded at 4.2 K in a parallel field of 600 G. The spectrum is a superposition of four components, three quadrupole doublets plus a paramagnetic spectrum. The three quadrupole doublets are designated as spectral components D, Fe^{2+} and S as indicated in Fig. 3. The paramagnetic spectrum is termed component M^N. Since the EPR spectrum of the Mössbauer sample showed no other resonance except the characteristic S = 3/2 signals, component M^N must be associated with the S = 3/2 center. We will show in the following that the field direction dependence of spectrum M^N is in agreement with the observed g-values.

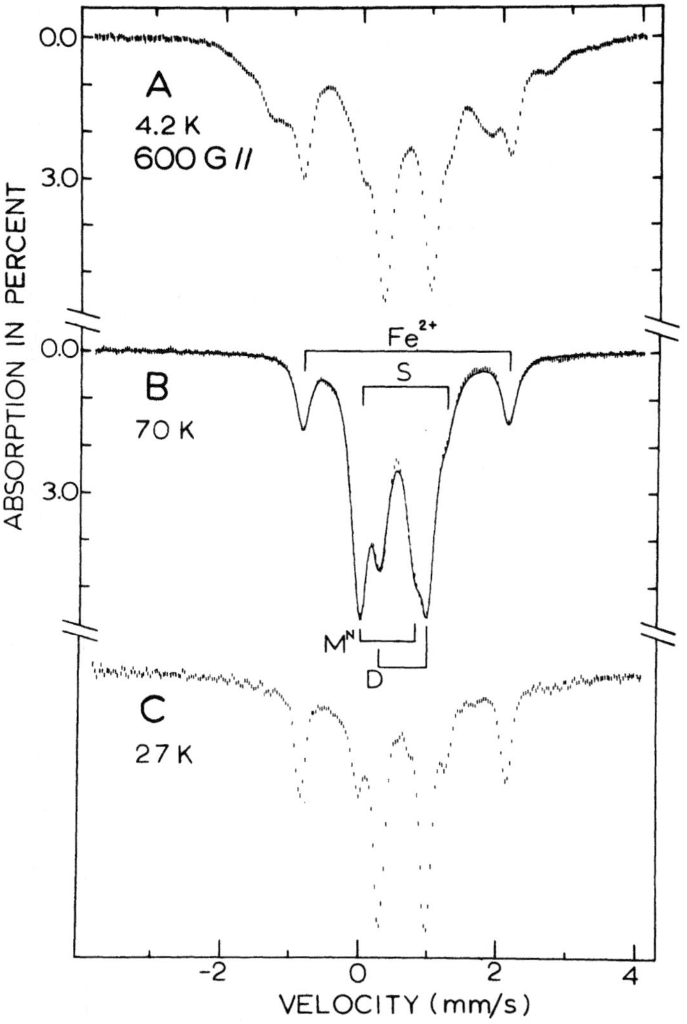

Figure 3. Mössbauer spectra of ^{57}*Fe-enriched native MoFe protein from* C. pasteurianum *W5 recorded in a parallel field of 600 G at (A) 4.2 K and (B) 70 K. The solid line in (B) is the result of fitting four quadrupole doublets to the data. The spectrum in (C) is a Fourier deconvoluted spectrum with the linewidth contribution of the* ^{57}*Co source being removed.*

Table I. *Mössbauer parameters for MoFe proteins from* Clostridium pasteurianum *and from* Azotobacter vinelandii.

Spectral Component	*C. pasteurianum*[a]			*A. vinelandii*[b]		
	ΔE_Q (mm/s)	δ (mm/s)	Absorption (%)	ΔE_Q (mm/s)	δ (mm/s)	Absorption (%)
Fe^{2+}	3.00 (2)[c]	0.64 (2)	12.8 (5)	3.02 (2)	0.69 (2)	13.0 (5)
D	0.70 (2)	0.64 (2)	39 (2)	0.81 (2)	0.64 (2)	42 (2)
S	1.37 (5)	0.64 (5)	∿6	1.37 (4)	0.64 (4)	∿6
M^N	0.8	0.41	42 (2)	0.76 (3)	0.40 (3)	40 (2)

(a) Ref. 1, (b) Refs. 1 and 10, (c) Values in parentheses give uncertainties in units of the least significant digit. If no uncertainties are quoted, the parameters are average values of unresolved subcomponents.

At 70 K the spin relaxation rate of the S = 3/2 center is fast, and component M^N collapses into a quadrupole doublet as indicated in Fig. 3B. In order to obtain reliable results from fitting such a complex spectrum, it is essential to reduce the number of adjustable parameters. Therefore, the following procedure was used to estimate the absorption line positions prior to the fitting. A spectrum was recorded at 27 K, a temperature at which the relaxation rate of the M center is in the intermediate region. As a result the spectral component M^N is broad and featureless. Then a Fourier deconvolution method (11) was used to remove the linewidth contribution of the ^{57}Co source. The resulting spectrum is shown in Fig. 3C. The quadrupole doublets of components D, Fe^{2+} and S are clearly discernable and their line positions are very well defined. The line positions of the collapsed M^N quadrupole pair at 70 K are estimated from a difference spectrum of Figures 3A and 3B. During these data manipulations, a limit on the line-width of each absorption line was also acquired. With all the positions predetermined and a crude limit for each line-width, the least-squares fit of the 70 K data could be performed with little ambiguity. The solid line shown in Figure 3B is the result of such a fit. The Mössbauer parameters and the absorption percentage of each component is listed in Table I. A similar method was used to analyze the Mössbauer spectra of the MoFe protein from *A. vinelandii* (12). The results are also listed in Table I. The published data of the MoFe protein from *K. pneumoniae* shows similar results (13).

3.1 The M Center

Iron quantitation (14,15) and Mössbauer data (4) have shown that each MoFe protein molecule contains a total of (30 $\pm$ 2) iron atoms. The relative absorption areas listed in Table I indicate that *each MoFe protein molecule contains approximately 12 iron atoms associated with component D, 4 with component* Fe^{2+}*, 2 with component S and 12 with component* M^N. Quantitation of the EPR signal indicates that there are two M centers per molecule. It follows then that each M center contains *six* iron atoms.

In order to study the spectrum of M^N in detail, the contributions of components D, Fe^{2+} and S were removed from the raw data using the parameters listed in Table I. Figure 4 shows such prepared spectra of M^N from *A. vinelandii* recorded at 4.2 K with parallel and transverse applied fields of 600 G. Close inspection of the spectrum in Figure 4 reveals an absorption line at + 2.7 mm/s and a shoulder at + 3.1 mm/s. The absorption intensities of these features observed in

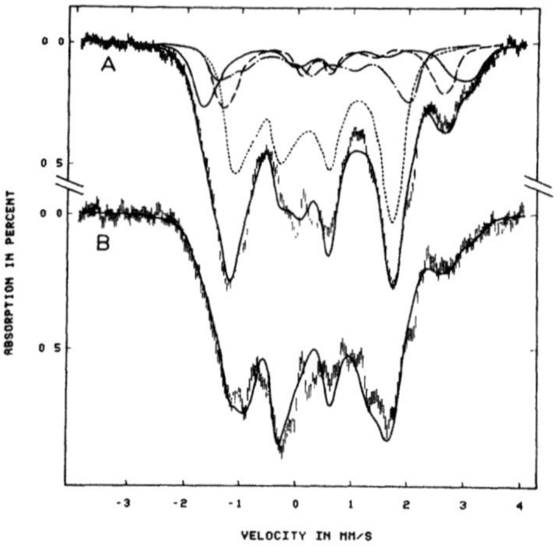

Figure 4. Mössbauer spectra of the M center from A. vinelandii *recorded at 4.2 K with 600 G fields applied parallel (A) and perpendicular (B) to the γ-beam. The spectra were obtained from the raw data by subtracting the contributions of components D,* Fe^{2+} *and S. The solid lines are theoretical simulations. Subcomponents are also shown in (A). For details of the theoretical analysis, readers are referred to reference 12.*

transverse field (Figure 4B) are about half of those observed in parallel field (Figure 4A). This field direction dependence suggests that the peak at + 2.7 mm/s and the shoulder at + 3.1 mm/s are the right-most absorption lines of two subcomponent spectra. (See discussion in section 2.2 and Figs. 1b and 1c). Similar features are also observed for the *C. pasteurianum* and *K. pneumoniae* MoFe proteins (1,13). A computer simulation revealed that about 30-35% of the total absorption of M^N belongs to these two subcomponents. This observation again suggested that the M center contains *six* iron atoms, which correlates well with the above-mentioned spin quantitation.

A detailed analysis of the spectra of M^N taken over a wide range of applied fields has been given by Huynh *et al.* (12) The following assumptions have been made to simplify the data analyses: (1) The Mössbauer spectrum of a spin-coupled iron cluster is the superposition of spectra resulting from each individual iron site. All sites share a common electronic

Hamiltonian but with different local A-tensors and EFG-tensors. (2) The A-tensors are isotropic. (3) The EFG-tensors and the electronic g-tensor share a common principal axis system. For details of the analysis, interested readers are referred to reference 12. The major conclusions are that the spectra of M^N can be adequately decomposed into four spectral components representing *six* irons and can be explained by means of a S = 3/2 spin Hamiltonian with E/D = 0.055. The solid lines in Fig. 4 are the results of such a decomposition. The value $(6 \pm 1.5)\,cm^{-1}$ obtained for D from analyzing the high field data is in reasonable agreement with that obtained from the EPR data.

In 1977, Shah and Brill (14) have successfully extracted an iron- and molybdenum-containing factor (designated as iron-molybdenum cofactor) from the MoFe proteins. An EPR and Mössbauer study of the cofactor (16) revealed yet another evidence for the M-center to be a six-iron spin-coupled cluster. The cofactors extracted from *A. vinelandii*, *C. pasteurianum* and *K. pneumoniae* were shown to exhibit the characteristic S = 3/2 EPR and Mössbauer spectra, firmly establishing that the M-center resides in the cofactor. Metal analysis determined 8 Fe per Mo atom for the cofactor materials. Spin quantitation of the cofactor EPR signal indicated one S = 3/2 center per Mo atom. In other words, the cofactor materials were shown to contain 8 Fe atoms per each S = 3/2 center. However, the Mössbauer data revealed that 20-25% of the iron atoms in the cofactor yielded a quadrupole doublet and therefore should not belong to the EPR active center. Consequently, the data again suggested that the S = 3/2 center should be a six-iron cluster.

The evidence for the Mo atom being also part of the M center was provided from the Mo EXAFS studies on the MoFe proteins and on the cofactor materials (17,18). It was shown that the nearest neighbors to Mo were Fe and S atoms. An Mo-Fe distance of ∿2.7 Å was reported. Recently, an ENDOR study (19) on the MoFe proteins enriched with ^{95}Mo has identified a Mo resonance within the S = 3/2 signal, demonstrating that the Mo is part of the M center. Combining all the spectroscopic evidence, we reach the conclusion that the M center is a six-Fe one-Mo spin-coupled cluster.

3.2 The P-Cluster

We will now direct our attention to components Fe^{2+} and D. Of particular interest is the parameters obtained for component Fe^{2+} (see Table I). Its quadrupole splitting (3 mm/s) and isomer shift (0.64 - 0.70 mm/s) are typical of tetrahedral-sulfur-coordinated, high-spin ferrous ions. However, Mössbauer data taken in strong applied fields indicate that component Fe^{2+} is diamagnetic (S = 0). Therefore, component Fe^{2+} must represent an iron site which resides in a spin-coupled diamagnetic cluster. In addition high field studies show that component D is also diamagnetic. Another interesting result is that components D and Fe^{2+} occur in the ratio of 3 to 1 for both MoFe proteins from *A. vinelandii* and from *C. pasteurianum*. A similar result was also obtained for *K. pneumoniae* (13). These observations have led to an intriguing speculation that the D and Fe^{2+} irons may be sites of a 4 Fe cluster. This idea was supported by studying the thionine oxidized proteins.

Upon addition of thionine, the M center is oxidized and becomes EPR silent. Zimmerman *et al.* (4) titrated the MoFe protein from *A. vinelandii* with thionine while monitoring the EPR signal of the M center. The results are shown in Fig. 5.

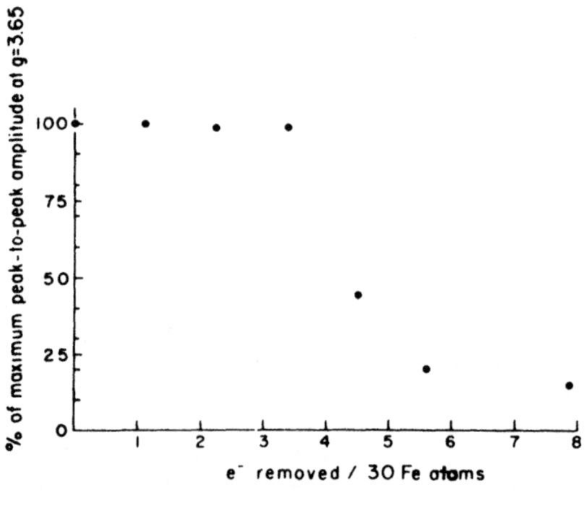

Figure 5. Oxidative titration of the MoFe protein from A. vinelandii. *The protein was oxidized with graded amounts of thionine. The ordinate represents the peak to peak amplitude of the EPR signal at g = 3.65 region. The abscissa represents the oxidizing equivalents of thionine added per g atom Fe present in the sample.*

The data indicated two distinct phases in the titration. As the first four equivalent electrons were removed, the EPR signal of the M center was unaffected. Upon removal of two additional equivalent electrons, the EPR signal decreased to its minimum intensity (about 10% of the maximum). Figs. 6A-C show Mössbauer spectra of MoFe proteins from *A. vinelandii* in three different oxidation states: (A) native, (B) 2.5-electron-oxidized and (C) fully oxidized proteins. For comparison, the spectrum of a fully oxidized MoFe protein from *C. pasteurianum* is also shown in Fig. 6D. The spectra were recorded at 4.2 K with a field of 600 G applied perpendicular to the γ-beam. The spectrum in Fig. 6C shows essentially two types of spectra: a central doublet (labelled M^{ox}) and a complex magnetic spectrum (labelled M1). A comparison of Figs. 6A and 6B indicates that the spectrum M1 appears in the early stage of the thionine titration and its appearance is accompanied by the disappearance of components Fe^{2+} and D. The decrease of intensity of component Fe^{2+} and D can be totally accounted for by the increase of the M1 intensity. Consequently, component M1 was identified as the oxidized form of components Fe^{2+} and D. More importantly, components Fe^{2+} and D were found to be oxidized concomitantly (4,20). Component M^{ox} appears in the later stage of the titration and was identified as the oxidized M component.

The above mentioned observations show that the removal of *four* electrons oxidized both components Fe^{2+} and D, which represent *sixteen* iron atoms per molecule. This strongly suggests that components Fe^{2+} and D must be organized into four 4Fe clusters. There are two possible ways to partition components Fe^{2+} and D: (1) two different types of clusters, one cluster containing four Fe^{2+}-type irons and the other three containing four D-type irons; (2) four identical clusters with each comprising one Fe^{2+}-type and three D-type irons. For case (1) to be true one has to postulate similar redox potentials for the two different types of clusters in order to explain the observed parallel oxidation behavior of components Fe^{2+} and D. Moreover, the electronic properties of component M1 discussed below are rather unique. To have such a unique electronic system occurring in two different types of clusters within one protein is rather unlikely. Therefore, case (2) was considered more probable and the cluster was named the P cluster.

In the absence of an applied field, component M1 exhibits a magnetic pattern revealing an electronic system of half-integer spin. In weak applied field (< 0.1 T) spectrum M1 shows no field direction dependence, an indication that the magnetic

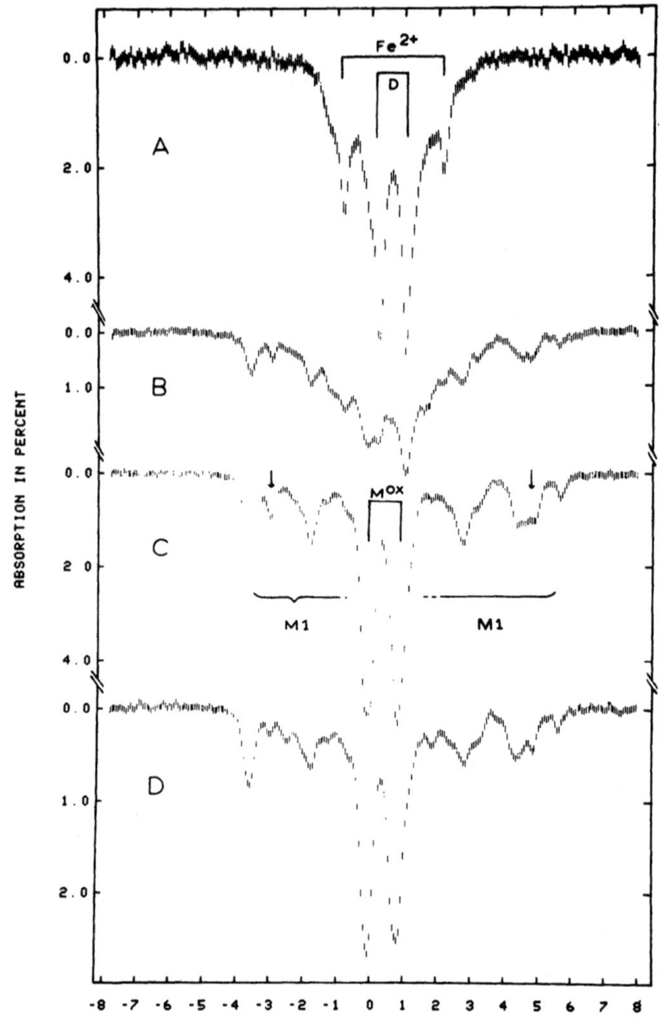

Figure 6 (A-C) Mössbauer spectra of MoFe proteins from A. vinelandii *in three oxidation states: (A) native, (B) 2.5 electron oxidized and (C) fully oxidized proteins. (D) Mössbauer spectrum of fully oxidized MoFe protein from* C. pasteurianum. *All spectra were recorded at 4.2 K with a field of 600 G applied perpendicular to the γ-beam.*

spectrum M1 results from a Kramers doublet with an extremely anisotropic g-tensor ($g_x \sim g_y << g_z$). Such a ground doublet would be "EPR-silent", in agreement with experiment, and can result from a half-integer spin system with $S \geq 3/2$, $D < 0$, and $E/D \sim 0$. A recent MCD study (21) on oxidized MoFe protein from *K. pneumoniae* is in support of this finding and suggests a spin value of either 5/2 or 7/2 for the oxidized P cluster. Cluster extrusion experiment (22) has also confirmed the existence of four [4Fe-4S] clusters in the MoFe proteins.

3.3 Component S

Little information has been acquired for component S because the spectrum of MoFe protein is complex and component S has relatively weak absorption (∿ 6% of the total iron absorption). However, component S is believed not to be an impurity but an essential part of the MoFe protein for the following reasons. Component S is present in equal amounts in the MoFe proteins from *A. vinelandii* and *C. pasteurianum*. The *K. pneumoniae* MoFe protein also appears to contain comparable amounts of component S. For all the proteins studied, component S has practically identical Mössbauer parameters (ΔE_Q = 1.37 mm/s, $\delta \simeq$ 0.65 mm/s). These values are atypical of commonly observed adventitious iron.

Acknowledgement

The authors acknowledge with gratitude the support by the National Science Foundation and the National Institutes of Health.

Figures and Table are reproduced with permission from Elsevier-North Holland Biomedical Press.

REFERENCES

1. Huynh, B. H., Henzl, M. T., Christner, J. A., Zimmermann, R., Orme-Johnson, W. H., and Münck, E.: 1980, *Biochim. Biophys. Acta*, 623, p. 124.

2. Orme-Johnson, W. H., and Davis, L. C., in *"Iron-Sulfur Proteins,"* Vol. III, Ed. Lovenberg, W., Academic Press, New York, 1977, p. 16.

3. Mortenson, L. E., and Thorneley, R. N. F.: 1979, *Ann. Rev. Biochem.*, 48, p. 387.

4. Zimmermann, R., Münck, E., Brill, W. J., Shah, V. K., Henzl, M. T., Rawlings, J., and Orme-Johnson, W. H.: 1978, *Biochim. Biophys. Acta*, 537, p. 185.

5. Violet, C. E., and Pipkorn, D. N.: 1971, *J. Appl. Phys.*, 42, p. 4339.

6. Frauenfelder, H., Nagle, D. E., Taylor, R. D., Cochran, D. R. F., and Visscher, W. M.: 1962, *Phys. Rev.*, 126, p. 1065.

7. Walker, R., Wertheim, G. K., and Jaccarino, V.: 1961, *Phys. Rev. Letters*, 6, p. 98.

8. Aasa, R., and Vanngard, T.: 1975, *J. Magn. Resonance*, 19, p. 308.

9. Davis, L. C., Shah, V. K., Brill, W. J., and Orme-Johnson, W. H.: 1972, *Biochim. Biophys. Acta*, 256, p. 512.

10. Münck, E., Rhodes, H., Orme-Johnson, W. H., Davis, L. C., Brill, W. J., and Shah, V. K.: 1975, *Biochim. Biophys. Acta*, 400, p. 32.

11. Dibar-Ure, M. Celia, and Flinn, P. A. in "*Mössbauer Effect Methodology*," Vol. 7, Ed. Gruverman, I. J., Plenum Press, New York, 1971, p. 245.

12. Huynh, B. H., Münck, E., and Orme-Johnson, W. H.: 1979, *Biochim. Biophys. Acta*, 576, p. 192.

13. Smith, B. E., and Lang, G.: 1974, *Biochem. J.*, 137, p. 169.

14. Shah, V. K., and Brill, W. J.: 1977, *Proc. Natl. Acad. Sci. USA*, 74, p. 3249.

15. Postgate, J., in "*Mechanism of Oxidative Enzymes*," eds. Ondanza, R., Edmundson, D., and Singer, T. P., Plenum Press, New York, 1978, p. 173.

16. Rawlings, J., Shah, V. K., Chisnell, J. R., Brill, W. J., Zimmermann, R., Münck, E. and Orme-Johnson, W. H.: 1978, *J. Biol. Chem.*, 253, p. 1001.

17. Cramer, S. P., Hodgson, K. O., Gillum, W. O., and Mortenson, L. E.: 1978, *J. Amer. Chem. Soc.*, 100, p. 3398.

18. Cramer, S. P., Gillum, W. O., Hodgson, K. O., Mortenson, L. E., Stiefel, E. I., Chisnell, J. R., Brill, W. J., and Shah, V. K.: 1978, *J. Amer. Chem. Soc.*, 100, p. 3814.

19. Hoffman, B. M., Roberts, J. E., and Orme-Johnson, W. H., to be published in *J. Amer. Chem. Soc.*

20. Smith, B. E., O'Donnell, M. J., Lang, G. and Spartalian, K.: 1980, *Biochem. J.*, 191, p. 449.

21. Johnson, M. K., Thomson, A. J., Robinson, A. E., and Smith, B. E., to be published in *Biochim. Biophys. Acta*, November issue, 1981.

22. Kurtz, D. M., Jr. McMillan, R. S., Burgess, B. K., Mortenson, L. E., and Holm, R. H.: 1979, *Proc. Natl. Acad. Sci. USA*, 76, p. 4986.

SECTION F

HEME MODEL SYSTEMS

MAGNETIC COMPLEXITIES IN PORPHINATOIRON(III) COMPLEXES

W. Robert Scheidt

Department of Chemistry
University of Notre Dame
Notre Dame, Indiana, U.S.A.

Crystallographic and other physical characterization of three magnetically complex porphinatoiron(III) derivatives are described. These systems are perchlorato-meso-tetraphenylporphinatoiron(III), bis (3-chloropyridine)octaethylporphinatoiron(III) perchlorate, and complexes having isothiocyanate and pyridine as the axial ligands. The systems provide experimental verification that relatively minor changes of phase, physical state, temperature, and porphinato ligand can have profound effects on the spin state of iron. Indeed, one system exhibits all three possible (low-, high-, and intermediate-spin) states.

1. INTRODUCTION

The iron atom of heme in a hemoprotein has several well-defined spin states. In addition, there are number of subtle magnetic properties observed for the hemoproteins. These include temperature dependent spin equilibria with certain axial ligands, quantitative differences in the spin equilibria for heme proteins having parity of axial ligands, changes in the spin properties on binding substrates even when the substrate does not bind at iron, and the possibility that a hemoprotein has different ground states depending on the purification of the protein. Such observations raise the question of how the protein matrix can effect such changes.

Studies with isolated heme complexes have shown that the spin state depends primarily on the type and number of axial ligands. [1] This matter has been discussed by Reed in this

H. B. Dunford et al. (eds.), The Biological Chemistry of Iron, 261–272.

meeting. [2] Briefly, for six-coordinate iron(III) species, strong-field axial ligands lead to low-spin (S = 1/2) complexes. Decreasing the axial ligand field strength leads first to high-spin (S = 5/2) complexes and in turn to intermediate (S = 3/2) or quantum admixed intermediate-spin (S = 3/2, 5/2) complexes.

Porphinato complexes of each spin state have characteristic geometric features. Thus, given the spin state of an iron porphyrin complex, its structure can be predicted with a high degree of metrical accuracy. Conversely, the structure of a porphinatoiron complex can be regarded as the final arbiter of its spin state. With the foregoing as background, the structural and other physical characterization of some porphinatoiron-(III) complexes which we regard as magnetically complex are presented. In some cases, the work is incomplete and the final description may differ from that presented here.

2. CHARACTERISTIC GEOMETRIES OF PORPHINATOIRON(III) COMPLEXES

In this section we describe structures that can be regarded as the prototypical structure of each spin state. [3] Low-spin six-coordinate species are well known and a number of structures of such species have been determined. A typical complex is that of $[Fe(TPP)(CN)_2]^-$ which has an in-plane iron atom, [4] an average $Fe-N_p$ bond distance 2.000 Å and relatively short axial bond distances. Exchanging one cyanide ligand for a pyridine to yield the relatively rare complex having nonequivalent axial ligands does not lead to significant structural changes. [5]

A number of six-coordinate high-spin complexes with two identical weaker field ligands have now been characterized. Two examples are $[Fe(TPP)(H_2O)_2]ClO_4$[6] and $[Fe(TPP)(TMSO)_2]ClO_4$.[17] These species are characterized by an in-plane iron atom. The larger high-spin iron atom is accommodated by a radial expansion of the porphinato core by about 0.05 Å, with $Fe-N_p$ distances of 2.045 Å, and longer axial bond distances.

Characterized six-coordinate intermediate-spin complexes are relatively rare and are best exemplified by the structure of polymeric $[Fe(TPP)(C(CN)_3)]_n$.[8] This spin state is characterized by short $Fe-N_p$ distances of 2.00 Å and relatively long axial bond distances. Mossbauer spectra of the intermediate-spin complexes have a quite distinctive quadrupole splitting constant of around 3 ms/s.

Five-coordinate iron(III) porphyrinates have a single axial anionic ligand and most are high spin. A typical example is Fe(TPP)(I).[9] The characteristic features of the class are long $Fe-N_p$ distances of 2.065 Å and a displacement of the iron atom by about 0.5 Å from the porphinato plane. These two features are the result of the large high-spin iron(III) atom. Large radius anions such as perchlorate, hexafluorophosphate, and hexafluoroantinomate yield intermediate spin complexes. These are probably best regarded as a quantum admixed (S = 3/2, 5/2) state. The structure of Fe(TPP)($OClO_3$) [10] shows significant structural differences compared to the high-spin complexes; namely, a decreased iron atom displacement of 0.27 Å and much shorter Fe-N_p bond distances.

Gouterman and Scheidt [11] have carried out iterative extended Huckel calculations of three six-coordinate Fe(P)L_2 complexes where L = CN^- (low spin), L = 3-chloropyridine (variable spin) and L = H_2O (high spin) in both high- and low-spin geometries. The Huckel calculations were used to establish changes in d orbital energies with changes in structure. The d orbital energies for the low-spin complex (L = CN) change very little with the change from low-spin to high-spin geometry. However, for the other ligands the energy gap between (d_{xy}, d_{xz}, d_{yz}) and d_{z^2} decreases sharply with a change from low-spin to high-spin geometry. The energy difference between (d_{xy}, d_{xz}, d_{yz}) and $d_{x^2-y^2}$ shows little change with the change in geometry. We have interpreted this result as suggesting that an iron porphyrinate can have, *with ligands that do not yield unambiguously low-spin complexes,* all three possible spin state lying close together in energy.

3. MAGNETICALLY COMPLEX SYSTEMS.

3.1. Perchloratoiron(III) Complexes.

All information concerning perchloratoiron(III) complexes is available in the literature. In addition to the *meso*-tetraphenylporphinato structure [10], an octaethylporphinato derivative [12] [13] has been reported. Structural studies show π donor-acceptor interactions for the porpinato complexes in the solid state. The *meso*-tetraphenylporphinato complex has a π interaction with a *m*-xylene solvate molecule and the octaethylporphinato derivative self-complexes. This type of solid state interaction is perhaps the explanation for the variation in the solid state EPR spectra of Fe(TPP)($OClO_3$). [10] Various crystalline modifications give an axial spectrum with g ~ 5 and 2; the exact spectrum is dependent on the particular aromatic solvate molecule incorporated into the lattice. EPR spectra of Fe(TPP)($OClO_3$) in frozen toluene solution suggest that mixtures of both high- and quantum admixed intermediate-spin complexes appear to be present in the solution phase. Mossbauer measurements in frozen toluene solution show two quadrupole split doublets, in distinct contrast to the single quadrupole doublet in the solid state, again suggesting the presence of two distinct species at low temperature in solution. NMR spectra of Fe(TPP)-($OClO_3$) show substantial solvent dependence and have been interpreted [14] in terms of aromatic solvents yielding solutions of higher spin than chlorinated solvents.

Taken together, the solution experiments suggest that the ligand field strength of coordinated perchlorate is at the spin crossover point so that very subtle changes in the environment control the spin state. However, only a single component (admixed intermediate spin) appears to be favored in the solid state.

3.2 Bis(3-chloropyridine)octaethylporphinatoiron(III) Perchlorate.

Buchler, Hill and co-workers [15] reported that a number of bis(substituted-pyridine) complexes of octaethylporphinato-iron(III) showed temperature dependent magnetic moments that could be interpreted in terms of a spin equilibrium between the high- and low-spin states. The bis(3-chloropyridine) complex showed the largest temperature dependence, with a magnetic moment at room temperature that corresponded to an approximate 50:50 mixture of spin states. At 77 K, the moment had decreased to nearly that expected for a low-spin state. Consequently, this system appeared to be most advantageous for crystallographic study at two temperatures. Crystals of the perchlorate salt, [Fe(OEP)(3-Cl_{py})$_2$]ClO_4, were prepared and examined by X-ray diffraction methods. The crystals initially obtained were found to crystallize in the triclinic system, with one molecule in the unit cell. This fact assumes subsequent importance since it requires that every molecule has an identical crystalline environment.

Crystallographic intensity data were collected at 98 K and 293 K; the results at 98 K are presented first. The structure found for the molecule at 98 K are clearly consistent with that expected for low-spin six-coordinate complex. [16] The average Fe-N_p distance is 1.994 (6) Å, well within the range observed for low-spin iron(III) porphyrinates. The axial Fe-N(py) distance is 2.031 Å, comparable to the 2.075 Å value observed for low-spin [Fe(TPP)(CN)(py)] [5]. A significant structural feature for subsequent discussion is the orientation of the 3-chloropyridine ligand plane with respect to the porphinato plane; the angle between the 3-Clpy plane and an equatorial coordinate plane is 41°; the most favorable value to reduce non-bonded interactions between the ligand and the core atoms is 45°. The results are shown in Figure 1.

The diffraction data collected for [Fe(OEP)(3-Clpy)$_2$]ClO_4 at 293 K were initially interpreted in terms of the 'average' structure of the thermal mixture of spin states. This 'average' structure had an axial Fe-N(py) distance of 2.194 Å and an average Fe-N_p distance of 2.014 (4) Å. The Fe-N_p distance is about midway between the value expected for low-spin and high-spin complexes. There thus appears to be some core expansion, consistent with the presence of a high-spin state at 293 K.

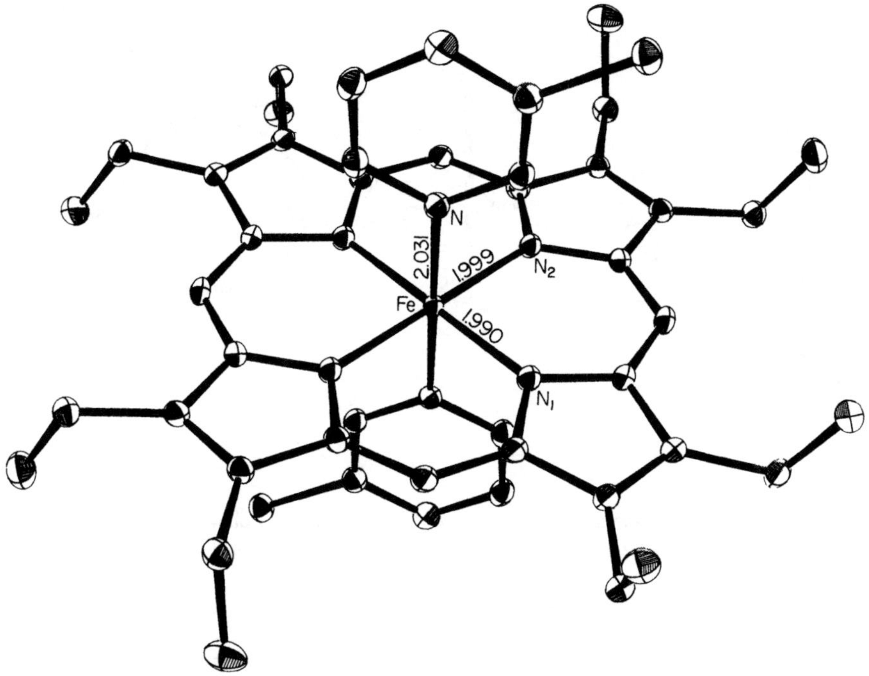

Figure 1. The molecular structure of the triclinic phase of $[Fe(OEP)(3\text{-}Clpy)_2]ClO_4$ *at 98 K.*

We have successfully carried out a 'crystallographic resolution' of the two spin isomers of $[Fe(OEP)(3\text{-}Clpy)_2]ClO_4$. The results are displayed in Figure 2. There are two axial ligand positions at 2.043 and 2.316 Å from the iron(III) atom. These positions are associated with the low- and high-spin forms of the molecule. Resolution of the small but expected real differences in the porphinato core atoms was not attempted. The total structural change for $[Fe(OEP)(3\text{-}Clpy)_2]ClO_4$ in the two spin states can be described as an 0.27 Å increase in the axial Fe-N(py) bonds and a probable increase of ~0.05 Å in the equatorial Fe-N_p bonds.

This reasonably consistent state of affairs was shattered

when a second crystalline phase of $[Fe(OEP)(3\text{-}Clpy)_2]ClO_4$ was accidentally obtained. Although the preparation of this phase has been repeated, the conditions for its preparation are still somewhat mysterious. Crystallographic examination of this phase showed that it belonged to the monoclinic system, with two

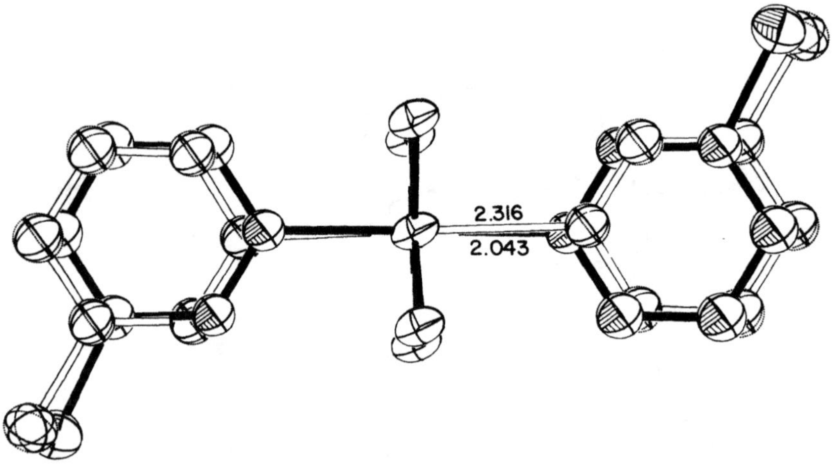

Figure 2. Crystallographic resolution of the low- and high-spin positions of the axial ligands of $[Fe(OEP)(3\text{-}Clpy)_2]ClO_4$ *at 293 K.*

crystallographically independent molecules. The two phases will be henceforth denoted as triclinic and monoclinic. *Both phases contain only* $[Fe(OEP)(3\text{-}Clpy)_2]ClO_4$ *molecules in the unit cell.* The initially disconcerting feature of the two phases is their apparent real differences in magnetic properties. This is illustrated in Figure 3.

Crystallographic studies of the monoclinic phase were undertaken to unravel these observations. [17] As noted, there are two crystallographically independent molecules. The structure of one of the molecules is shown in Figure 4. The structure of the other molecule is effectively identical to the first. The structure of the monoclinic phase is clearly different than the triclinic structure previously described. The orientation of all four pyridine rings is quite similar and different from that of the triclinic form. The planes of the 3-Clpy ligands form angles of 5-10° with an equatorial coordinate plane. This structural feature would cause severe nonbonded contacts between pyridine atoms and core atoms for Fe-N(py) bond distances smaller than those observed. It thus appears that this structural feature will resist a transformation from the observed state to

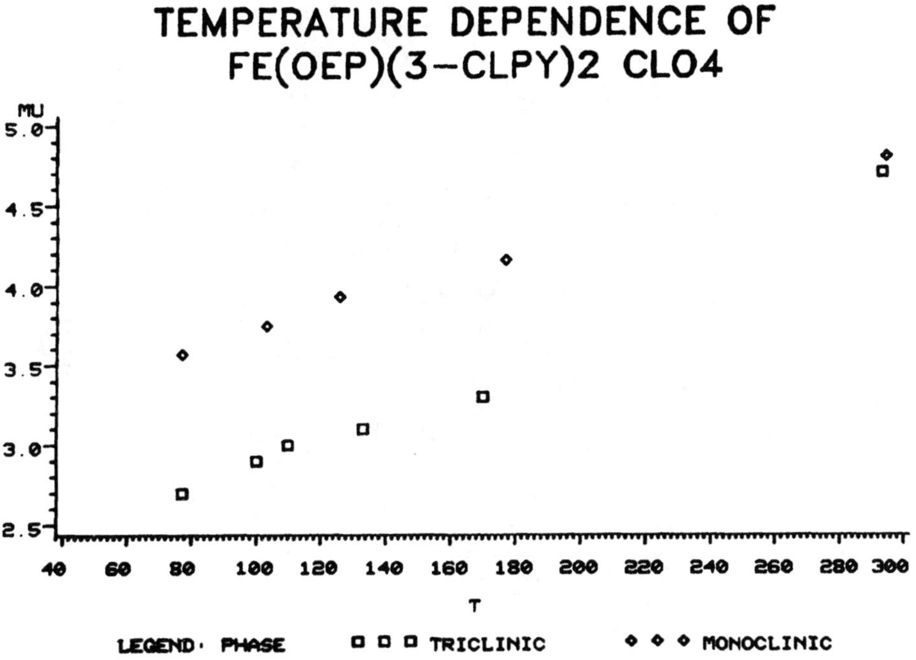

Figure 3. Comparison of the solid state magnetic moments of the two phases of [Fe(OEP)(3-Clpy)$_2$]ClO$_4$.

a low-spin state. The average Fe-N_p bond distance of 2.005 (6) Å and the axial Fe-N(py) bond distance of 2.310 (17) Å appears to us to be consistent only with the assignment of an intermediate-spin or quantum admixed intermediate spin state.

Additional evidence for this conclusion comes from Mossbauer studies on the two phases, undertaken in collaboration with G. Lang. Variable temperature studies in the absence and presence of an applied magnetic field have been performed, although the results have not been completely analyzed in detail. The triclinic phase has one doublet at all temperatures; the approximate values of the isomer shift and the quadrupole splitting constant are 0.25 and 2.1 mm/s (room temperature). The monoclinic phase also has one doublet; the isomer shift and quadrupole splitting constant are 0.4 and 2.7 mm/s at 77 K. In the presence of an applied 6 tesla field (4.2 K), the triclinic phase yields a spectrum consistent with a low-spin species. The monoclinic phase shows a spectrum similar to that of Fe(TPP)(OClO$_3$), [10] consistent with an admixed intermediate spin state. We thus conclude that the two phases have differing ground states that

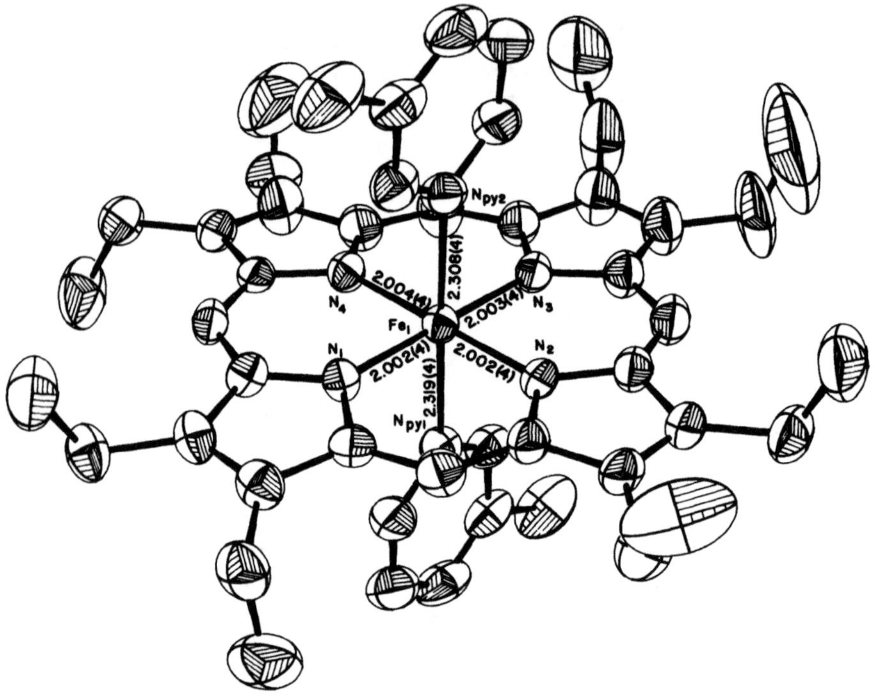

Figure 4. The molecular structure of one of the two crystallographically independent molecules of monoclinic [Fe(OEP)-(3-Clpy)$_2$]ClO$_4$.

are separated by only the small differences in energy consistent with packing energies.

3.3. Isothiocyanato(pyridine)iron(III) Complexes.

Our initial investigations of six-coordinate porphinatoiron(III) complexes having NCS$^-$ as a ligand were undertaken because of unexpected results in the structure of thiocyanato ligated hemoglobin. [18] We initially prepared a *meso*-tetraphenylporphyrin derivative, Fe(TPP)(NCS)(py). [19] The structure of this molecule is shown in Figure 5. The partly bent N-coordinated isothiocyanato ligand appears to be the result of crystal packing interactions. Most of the structural features of the molecule are quite in accord with the structure expected for a low-spin molecule. The magnetic moment for the complex in the solid state is however somewhat higher than expected (~3.5 μ_B at room temperature and does not decrease as much as would be expected with decreasing temperature). In CHCl$_3$-pyridine solution,

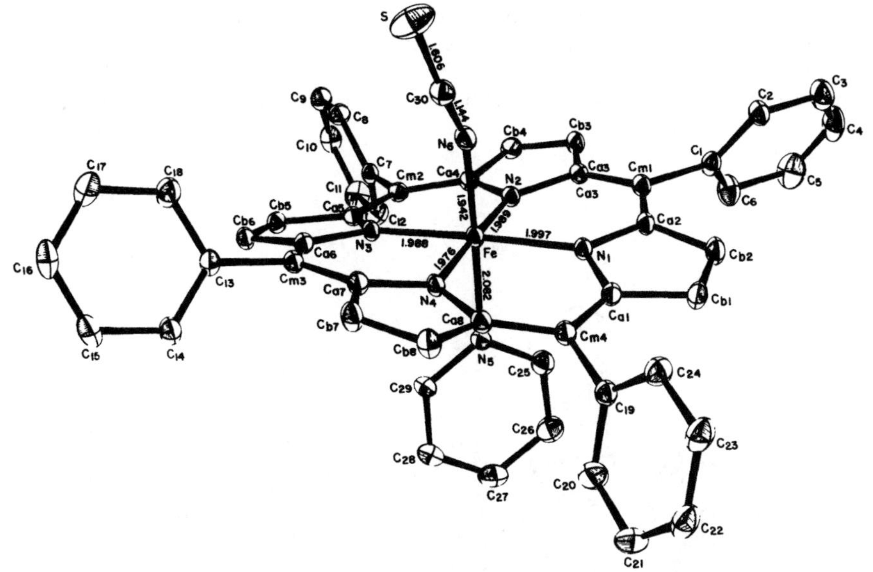

Figure 5. The molecular structure of low-spin Fe(TPP)(NCS)(py).

the moment decreases sharply with temperature. Typical values are 4.6 μ_B at 310 K and 2.1 μ_B at 262 K. Preliminary NMR spectra taken as a function of pyridine concentration and varying temperatures are interpreted in terms of an equilibrium involving high-spin five-coordinate Fe(TPP)(NCS) and six-coordinate Fe(TPP)(NCS)(py). The six-coordinate molecule is in a thermal spin equilibrium. Thus the axial ligand combination of isothiocyanate and pyridine brings this porphinatoiron(III) complex close to or at the spin crossover point.

Since we have observed that the TPP analog of [Fe(OEP)-(3-Clpy)$_2$]ClO_4 is low spin, we thought that the Fe(OEP)(NCS)(py) molecule could have a higher spin state. Accordingly, Fe(OEP)-(NCS)(py) was prepared. The spin state is indeed higher than that of Fe(TPP)(NCS)(py) and, in fact, Fe(OEP)(NCS)(py) has a moment corresponding to five unpaired electrons at all temperatures down to 77 K.

This complex appears to be the first isolated high-spin six-coordinate porphinatoiron(III) complex having two nonequivalent axial ligands. The molecule is closely related to a number of ferric hemoproteins in the high-spin state and its structure was highly interesting to us. We have determined the molecular structure of the molecule at room temperature. [20] The structure is illustrated in Figure 6. There are a number of remarkable differences between the OEP and TPP structures, many of which are expected. The increase in the average Fe-N_p bond distance from 1.988 to 2.046 Å, the increase in the Fe-N(NCS) bond distance from 1.942 to 2.031 Å are expected changes resulting

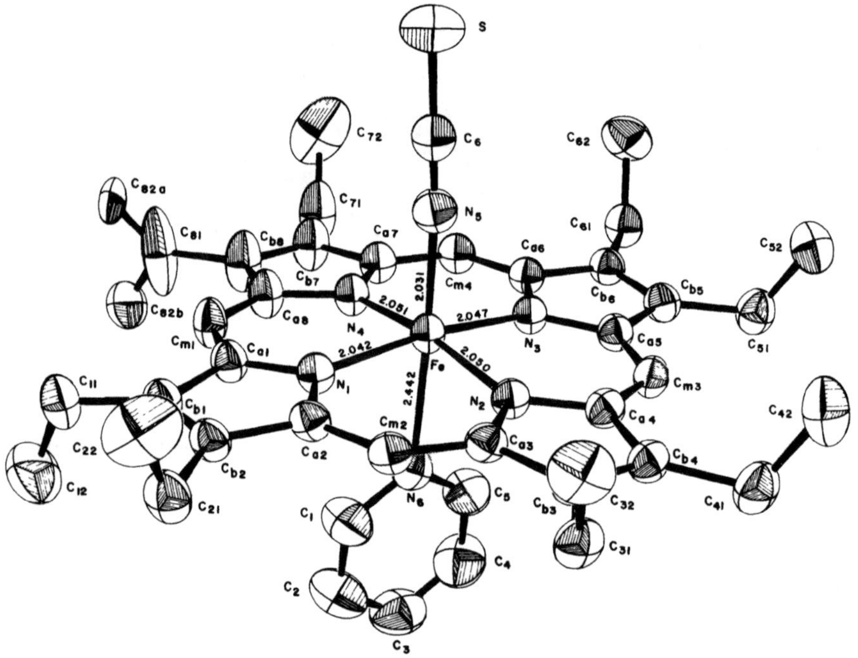

Figure 6. The molecular structure of high-spin Fe(OEP)(NCS)(py).

from the change in spin state. One structural feature that could not be anticipated with precision was the increase in the displacement of the iron atom out of the porphinato plane. The displacement of the iron atom is dependent, at least in part, on the relative importance of the bonding of the two dissimilar axial ligands. In Fe(OEP)(NCS)(py), the iron is displaced 0.23 Å from the mean plane of the core towards the isothiocyanato ligand. In the TPP complex, this displacement is only 0.05 Å. The large displacement of the iron atom in Fe(OEP)(NCS)(py) is the probable cause for the extremely long Fe-N(py) distance of 2.44 Å. A final structural feature of note is the doming of the porphinato core in Fe(OEP)(NCS)(py) with the pyrrole units of the ligand following the iron atom out of plane.

4. SUMMARY

Relatively minor changes can lead to different spin states for the same effective porphinatoiron(III) complex. These changes include a) state (solution vs. solid), b) phase (different crystalline modifications), c) temperature, and d) variation in porphinato ligand with identical axial ligands. Parallel observations of complex magnetic behavior for hemoprotein derivatives must be regarded seriously and indeed it is to be expected that many more such examples from hemoprotein derivatives will be found.

5. ACKNOWLEDGEMENTS

The collaboration of Profs. C.A. Reed, G. Lang, M. Gouterman, and K. Hatano is warmly acknowledged. The support of the National Institutes of Health (Grant HL-15627) for this research is appreciated.

REFERENCES

1. Scheidt, W.R., Reed, C.A.: 1981, *Chem. Rev.* 81, in press.
2. Reed, C.A. this volume.
3. Abbreviations used in this paper are: TPP, the dianion of *meso*-tetraphenylporphyrin; OEP, the dianion of octaethylporphyrin, py, pyridine; 3-Clpy, 3-chloropyridine; TMSO, tetramethylene sulfoxide; N_p, porphinato nitrogen atom.
4. Scheidt, W.R., Haller, K.J., Hatano, K.: 1980, *J. Am. Chem. Soc.* 102, pp. 3017-3021.
5. Scheidt, W.R., Haller, K.J., Lee, Y.J., Luandilok, W., Hatano, K., to be published.
6. Scheidt, W.R., Cohen, I.A., Kastner, M.E.: 1979, *Biochemistry* 18, pp. 3546-3552.
7. Mashiko, T., Kastner, M.E., Spartalian, K., Scheidt, W.R., Reed, C.A.: 1978, *J. Am. Chem. Soc.* 100, pp. 6354-6362.
8. Summerville, D.A., Cohen, I.A., Hatano, K., Scheidt, W.R.: 1978, *Inorg. Chem.* 17, pp. 2906-2910.
9. Hatano, K., Scheidt, W.R.: 1979, *Inorg. Chem.* 18, pp. 877-899.
10. Reed, C.A., Mashiko, T., Bentley, S.P., Kastner, M.E., Scheidt, W.R., Spartalian, K., Lang, G.: 1979, *J. Am. Chem. Soc.* 101, pp. 2948-2958.
11. Scheidt, W.R., Gouterman, M. Review article in "The Chemical Physics of Biologically Important Inorganic Chromophores" (Lever, A.B.P.; Gray, H.B., Eds.) Addison-Wesley, Reading, MA, in press.
12. Masuda, H., Taga, T., Osaki, K., Sugimoto, H., Yoshida, A.I., Ogoshi, H.: 1980, *Inorg. Chem.* 19, pp. 950-955.

13. Dolphin, D.H., Sams, J.R., Tsin, T.B.: 1977, *Inorg. Chem.* 16, pp. 711-713.
14. Goff, H., Shimomura, E.: 1980, *J. Am. Chem. Soc.* 102, pp. 31-37.
15. Hill, H.A.O., Skyte, P.D., Buchler, J.W., Lueken,H., Tonn, M., Gregson, A.K., Pellizer, G.: 1979, *J. Chem. Soc. Chem. Commun.*, pp. 151-152.
16. Scheidt, W.R., Geiger, D.K., Haller, K.J., submitted for publication.
17. Scheidt, W.R., Geiger, D.K., to be submitted for publication.
18. Korszun, Z.R., Moffat, K.: 1981, *J. Mol. Biol.* 145, pp. 815-824.
19. Scheidt, W.R., Lee, Y.J., Geiger, D.K., Hatano, K., to be submitted for publication.
20. Scheidt, W.R., Taylor, K.M., Hatano, K., to be submitted for publication.

THE EVALUATION OF STABILIZATION ENERGIES (EMPIRICAL RESONANCE ENERGIES) FOR BENZENE, PORPHINE AND [18]ANNULENE FROM THERMOCHEMICAL DATA, AND FROM AB INITIO CALCULATIONS.

Philip George

Biology Department, University of Pennsylvania, Philadelphia, Pennsylvania 19104

Charles W. Bock and Mendel Trachtman, Chemistry Department, Philadelphia College of Textiles and Science, Philadelphia, Pennsylvania 19144

ABSTRACT

The methodology involved in evaluating an empirical resonance energy for an aromatic molecule from thermochemical data is discussed with respect to benzene, nitrogen heterocycles and especially porphine - emphasizing that it is not a unique property of the aromatic molecule per se but a reaction energy. ΔH^o_f *data currently available suggest that the cyclic polyene, [18]annulene, is a poor model for porphine, and that the N-substitutions and additional bridging structural elements in porphine result in a much enhanced resonance energy relative to the polyene.*

INTRODUCTION

Some of the most important reactions of iron in biological systems are those of the hemoproteins. In this coordination chemistry environment the iron is bonded both to the protein and to the four nitrogen atoms of the porphyrin. Among the many significant properties of this macrocyclic ring system is the stabilization due to conjugation (1). In the classical Kekulé structure one of the individual rings is a pyrrole, two are of the pyrrolenine type, and the other is related to

H. B. Dunford et al. (eds.), The Biological Chemistry of Iron, 273–281.

maleimide. It is best regarded, not just as a "tetrapyrrole derivative", but as a six-equivalent higher oxidation state of the true tetrapyrrole, porphyrinogen, in which four pyrrole rings are simply joined together via $-CH_2-$ groups to give the corresponding macrocyclic structure. The oxidation gives rise to greater conjugation which can be treated as a delocalized 18-electron system in a 16-, 17- or 18-membered ring depending on the participation of electrons from the nitrogens[1]. Hence it could be inferred that porphine would have a resonance energy substantially greater than that of the four isolated pyrrole rings of porphyrinogen.

This paper deals with the methodology involved in calculating a resonance energy, RE, from thermochemical data and from *ab initio* calculations, and with the magnitude of the value for porphine compared to that for the structurally related cyclic polyene [18]annulene (3). ΔH_f^o values have been taken from Cox and Pilcher (4). A computer analysed revision of the experimental data has been published more recently (5). However the changes in the values of ΔH_f^o are very minor. All energy values are quoted in kcal mol^{-1}: 1 kcal mol^{-1} = 4.184 Joules mol^{-1}.

METHODOLOGY

A resonance energy evaluated from thermochemical data, whether as the difference between hydrogenation heats for the aromatic molecule and some mono-olefin, or with the use of bond energy tables, is in fact the energy of a reaction involving the aromatic molecule and other molecules whose bonding energy serves as reference. Hence rigorous comparison is possible between experiment and a value calculated *ab initio* as the difference between the total molecular energies of reactant and product species.

There is an important distinction to be drawn between this kind of resonance energy and a "theoretical" resonance energy, the magnitude of which depends on an hypothetical structure with idealized bonding of some sort, and which cannot therefore be compared with experiment. Many theoretical resonance energy calculations have been published in recent years (6).

With the recognition that an empirical resonance energy is not a property of the aromatic molecule *per se*, but a reaction energy, it follows that there is no unique value. This is demonstrated in Table I in the case of benzene. Reactions A1, 2 and 3 give RE values corresponding to bond energy calculations in which E(C-H) and E(C-C) are based on ΔH_f^o for methane and ethane, and E(C=C) on ΔH_f^o for ethylene, propene and trans-2-butene respectively. Reaction B is that proposed by Dewar and

Table I. Values of the empirical resonance energy (RE) for benzene calculated as $\Delta H°$ for various reactions from $\Delta H_f°$ data (kcal mol^{-1}).

REACTION	RE
Reactions underlying the bond energy calculations[a]	
1. Isodesmic bond separation (7)	
$C_6H_6 + 6CH_4 \longrightarrow 3CH_3\text{-}CH_3 + 3CH_2\text{=}CH_2$	64.2 ± 1.2
2. $C_6H_6 + 3CH_4 \longrightarrow 3CH_3\text{-}CH\text{=}CH_2$	48.5 ± 0.8
3. $C_6H_6 + 3CH_3\text{-}CH_3 \longrightarrow$ 3 trans $CH_3\text{-}CH\text{=}CH\text{-}CH_3$	31.9 ± 1.0
Reduction by ethane (8)	
$C_6H_6 + 3CH_3\text{-}CH_3 \longrightarrow C_6H_{12}$ (cyclohexane) + $3CH_2\text{=}CH_2$	48.8 ± 0.9
Difference between hydrogenation heats	
$C_6H_6 + 2C_6H_{12}$ (cyclohexane) $\longrightarrow 3C_6H_{10}$ (cyclohexene)	32.0 ± 1.0[b]
Homodesmotic group separation (9,10,11)	
1. $C_6H_6 + 3CH_2\text{=}CH_2 \longrightarrow$ 3 trans $CH_2\text{=}CH\text{-}CH\text{=}CH_2$	21.2 ± 0.9
2. $C_6H_6 + 3CH_2\text{=}CH_2 \longrightarrow$ 3 90° $CH_2\text{=}CH\text{-}CH\text{=}CH_2$	42.7 ± 3.0

a. The reaction which underlies a bond energy calculation can be established by the following steps - (i) express E(C-H), E(C-C) as functions of $\Delta H°_{atomiz.}$ for methane, ethane and the olefin, (ii) express $\Delta H°_{atomiz.}$ for benzene as a function of the bond energies plus RE, (iii) substitute the bond energies from (i) in (ii). RE then appears as a sum and difference of $\Delta H°_{atomiz.}$ terms which correspond to $\Delta H°$ for a particular reaction.

b. Corrected for the conventional ring strain energy, CRSE, of cyclohexene, i.e. 1.32 kcal mol^{-1}(4).

Schmeising (8). Reaction C is equivalent to the difference between the hydrogenation of benzene and three cyclohexenes (12). Reactions D1 and 2 are those proposed by George *et al* (9,10).

The reactions differ in the extent to which structural elements are matched in reactants and product species - ranging from minimal matching in terms of the number of C-H, C-C and C=C bonds, i.e. *isodesmic* reactions (7) to far more extensive matching with equal numbers of each type of carbon-carbon bond, $C_{sp^3}\text{-}C_{sp^3}$, $C_{sp^3}\text{-}C_{sp^2}$, $C_{sp^2}\text{-}C_{sp^2}$, $C_{sp^2}\text{=}C_{sp^2}$, etc. in reactants and products, and equal numbers of carbon atoms of each type, C_{sp^3}, C_{sp^2}, etc., with zero, one, two and three hydrogen atoms attached, i.e. *homodesmotic* reactions(9). The greater the matching the less the RE value contains extraneous energy contributions arising from changes in hybridization and C-H bonding (nearest neighbor effects), and hence homodesmotic reactions are to be preferred [13]. By utilizing the rotomer of 1,3-butadiene in which the C=C planes are 90° with respect to each other, the contribution arising from whatever π_x delocalization there might be in trans-1,3-butadiene as the reference molecule can be taken into account (11). But hyperconjugation in the 90° rotomer would restore some of the π_x delocalization energy that is lost(14).

Pauling and Sherman(15) did not specify which olefin they employed in their bond energy calculation of RE for benzene that gave 37.3 kcal mol^{-1}. But analysis shows that the change in bonding in reactions A3 and C is identical, which is probably the reason for the close agreement between their value and the 36.0 kcal mol^{-1} obtained by Kistiakowsky et al(12). If ethylene (16) or propene, are used to calculate E(C=C) far different values are obtained, see Table I A1 and 2.

AROMATIC NITROGEN-HETEROCYCLES: PORPHINE

Reactions similar to those for benzene in Table I can be set up for aromatic nitrogen-heterocycles, but to do this further reference molecules are needed. With reactions B and C these would be partly and/or fully hydrogenated products of the aromatic structure, and hence specific for each heterocycle. With reactions A1 and D1 on the other hand the same reference molecules supply the necessary nitrogen-carbon and nitrogen-hydrogen bonding elements in each case. Taking pyridine, pyrrole, and porphine as examples, these molecules are NH_3, CH_3-NH_2 and $CH_2=NH$ (formaldimine) for reaction A1, and $CH_2=CH-N=CH_2$ (2-aza-1,3-butadiene) and $CH_2=CH-NH-CH=CH_2$ (divinylamine) for reaction D1. The individual equations are:

Pyridine

$$\text{A1}\quad C_5H_5N + 5CH_4 + NH_3 \rightarrow 2CH_3-CH_3 + CH_3-NH_2 + 2CH_2=CH_2 + CH_2=NH$$

$$\text{D1}\quad C_5H_5N + 3CH_2=CH_2 \rightarrow 2CH_2=CH-CH=CH_2 + CH_2=CH-N=CH_2$$

Pyrrole

$$\text{A1}\quad C_4H_5N + 4CH_4 + NH_3 \rightarrow CH_3-CH_3 + 2CH_3-NH_2 + 2CH_2=CH_2$$

D1 $C_4H_5N + 2CH_2{=}CH_2 \longrightarrow CH_2{=}CH{-}CH{=}CH_2 + CH_2{=}CH{-}NH{-}CH{=}CH_2$

Porphine

A1 $C_{20}H_{14}N_4 + 28CH_4 + 4NH_3 \longrightarrow 11CH_3{-}CH_3 + 6CH_3{-}NH_2 + 9CH_2{=}CH_2 + 2CH_2{=}NH$

D1 $C_{20}H_{14}N_4 + 19CH_2{=}CH_2 \longrightarrow 11CH_2{=}CH{-}CH{=}CH_2 + 2CH_2{=}CH{-}N{=}CH_2 + 2CH_2{=}CH{-}NH{-}CH{=}CH_2$

In the case of pyridine and pyrrole the D1 reactions are homodesmotic, like that for benzene, but with porphine there is a residual mismatch. C-atoms bonded to two other C-atoms and an N-atom in porphine are not matched in the reference molecules. Appropriately substituted reference molecules, like the isobutene or divinylethylene in the reactions for polybenzenoid hydrocarbons (13) would therefore be needed to meet the homodesmotic criteria in this particular case. ΔH_f^o values are not yet available for one or more of the key reference molecules required for the direct calculation of the resonance energy as ΔH_r^o for any of the above reactions, i.e. $CH_2{=}NH$ for reaction A1, the partly and/or fully hydrogenated products for reactions B and C, and $CH_2{=}CH{-}N{=}CH_2$ and $(CH_2{=}CH)_2NH$ for reaction D1. Hence at present, calculations can only be made indirectly using values of bond energy terms derived from ΔH_f^o for more complicated structures that contain the same structural elements. The values reported in the literature for many aromatic nitrogen-heterocycles have been determined in this way (2,15).

As mentioned in the Introduction, one might expect the resonance energy of porphine to be at least four-times that for an isolated pyrrole ring because of the greater conjugation. Taking 21.9 kcal mol^{-1} for the pyrrole ring[2], the minimum value for porphine would thus be about 85 kcal mol^{-1}. There is, however, a very striking discrepancy between values based on combustion measurements reported in the literature (17,18). One set of measurements on eleven porphyrins with a variety of substituent groups gives values for the porphine skeletal structure ranging from 78 to 179 kcal mol^{-1}, with a mean of 149 $\pm$ 9 kcal mol^{-1}, whereas another set of measurements on porphine itself and the tetraethyl and tetraphenyl derivatives gives much larger values ranging from 450 to 500 kcal mol^{-1}, compared to the Kistiakowsky reaction value for benzene of 36 kcal mol^{-1} (2).

A consideration of the feasibility of various porphyrin reactions in which ring fission occurs, or in which hydrogenation breaks the conjugation, favors the lower value. In addition, an evaluation based upon data for a porphine/phlorine[3] couple (19) leads to 103 kcal mol^{-1} (2). But new experimental studies are obviously needed to establish a reliable value.

Empirical resonance energies are readily calculated *ab initio* as the difference between the total molecular energies of reactant and product species. To compare such values with those derived either directly or indirectly from ΔH^{o}_{f} at 298 K requires a knowledge of the heat capacity increments, $(H^{o}_{298}-H^{o}_{0})$, and zero-point energies. But *in toto* these contributions to ΔH^{o}_{298} are small, and significantly less in proportion for the homodesmotic group separation reaction than for the isodesmic bond separation on account of the greater matching of structural elements (10).

As with all *ab initio* calculations of reaction energies it is essential that the geometries employed be self-consistent. For a rigorous calculation optimized geometries are required becuase only in this way is the reaction energy evaluated from the bottom of the potential wells. But for a molecule as large as porphine geometry optimization is scarcely feasible at present. Experimental geometries are the next best choice. This would introduce some uncertainty because these geometries give energies a little more positive than those at the potential minima, although there is always the possibility that the increments would cancel to a certain extent.

A total molecular energy for porphine has been calculated by Almlof using a (7,3) basis set contracted to single zeta, with the experimental geometry slightly adjusted to obtain perfect D_{4h} symmetry for the skeletal structure (20). To calculate the resonance energy from reaction A1 total molecular energies would also be needed for CH_4, NH_3 CH_3-CH_3, CH_3-NH_2, CH_2=CH_2 and CH_2=NH using the same basis set. Such a calculation would be quite feasible because experimental geometries are available for all these species including CH_2=NH (21) which is not the case for the reference molecules, CH_2=CH-N=CH_2 and (CH_2=CH)NH, needed for reaction D1.

Some idea of the reliability of such a calculation can be gained from results for benzene. Using experimental geometries and the STO-3G basis set (7), which is similar in quality to the contracted (7,3) basis set employed by Almlof, the resonance energy calculated from reaction A1 and corrected to 298 K is 75.03 kcal mol^{-1} compared to the experimental value of 64.16 kcal mol^{-1}. With optimized geometries the value is 73.38 kcal mol^{-1} (22,23). Using the more extended, split valence-shell 4-31G basis set and experimental geometries (7)[4] better agreement is obtained, namely 68.23 kcal mol^{-1}. *Ab initio* calculations at this level thus tend to overestimate the value for benzene, but they would nevertheless be a useful guide as to whether a "low" or "high" value is more likely to be correct for porphine.

PORPHINE VIS-A-VIS [18]ANNULENE

Finally, the structural similarity between [18]annulene and porphine raises the interesting question as to whether the substitution and bridging groups in the latter bring about a marked change in the resonance energy. The most reliable ΔH_f^o for [18]annulene, + 124 ± 5.5 kcal mol^{-1}, is based on the enthalpy change for the thermal rearrangement.

[18]annulene ⟶ 1,2-benzo-1,3,7-cyclooctatriene + benzene

ΔH_f^o (benzene), and ΔH_f^o for the cyclooctatriene estimated using group increments (24). This leads to a value of RE very close to that for benzene if the underlying reaction is A3, but very different values are obtained for the other reactions, see Table II. With A1 [18]annulene has a large resonance energy, well over 100 kcal mol^{-1}, about twice that for benzene, whereas with D1 the resonance energy is very small indeed.

Table II. A comparison of empirical resonance energies for benzene and [18]annulene, evaluated using ΔH_f^o data, in kcal mol^{-1}.

Reaction Type	RE: benzene	RE: [18]annulene
A1	64.2 ± 1.2	127.9 ± 8.7
A2	48.5 ± 0.8	80.9 ± 7.6
A3	31.9 ± 1.0	31.3 ± 8.2
D1	21.2 ± 0.9	- 1.1 ± 7.8

In comparing the resonance energy of porphine with [18]annulene this feature has to be borne in mind. On the basis of reaction A3, the minimum value for porphine can be put at 85 kcal mol^{-1}, see above, and a value not much more than 100 kcal mol^{-1} would appear to be a reasonable estimate. The corresponding value for [18]annulene is about 30 kcal mol^{-1}. This difference is probably large enough to warrant the conclusion that the [18]annulene structure is a rather poor "model" for porphine, and that the substitutions and additional structural elements in porphine result in a much enhanced resonance energy.

STO-3G calculations on [18]annulene with partial geometry optimization - namely assuming either D_{6h} or D_{3h} symmetry, C-H set at 1.08Å, and <CCH set to bisect <CCC - find the D_{3h} alternating bond length structure to be the more stable, contrary to X-ray results (25). Nevertheless, bearing in mind that the total molecular energy is underestimated in this calculation, which will lead to an underestimate of the resonance energy (in a

positive sense), it is interesting to note that ΔE_T for reaction A1 is 99 kcal mol^{-1} compared to the STO-3G geometry optimized value for benzene of 70 kcal mol^{-1} (22,23), and ΔE_T for reaction D1 is -14 kcal mol^{-1} compared to the benzene value of 33 kcal mol^{-1} (22,23,26) i.e. the calculated values show the same trends as the experimental.

NOTES

1. See figures 1a,d:3a,d: and 4a,b of reference (2).

2. The average of the values for pyrrole and 2,5-dimethyl pyrrole, 21.57 and 22.25 kcal mol^{-1} respectively (2) compared to the value of 36.0 for benzene.

3. Phlorine is an isomer of dihydroporphine in which the conjugation is broken by a $-CH_2-$ group in the macrocyclic ring, see fig. 4d(2).

4. Komornicki, A., kindly provided the total molecular energy for benzene, -230.37755 a.u.

REFERENCES

1. George, P., 1973, *Ann. N. Y. Acad. Sci.*, 206, pp. 84-95.
2. George, P., 1975, *Chem. Rev.*, 75, pp. 85-111.
3. Badger, G. M., 1969, "Aromatic Character and Aromaticity", Cambridge University Press, London.
4. Cox, J. D. and Pilcher, G. 1970, "Thermochemistry of Organic and Organometallic Compounds", Academic Press, New York, N.Y.
5. Pedley, J. B., and Rylance, J., 1977, "Sussex-NPL. Computer Analysed Thermochemical Data: Organic and Organometallic Compounds", Univ. of Sussex, England.
6. See, for example, the following papers and references therein:
   a. Dewar, M. J. S., and de Llano, C., 1969, *J. Am. Chem. Soc.*, 91, pp. 789-795.
   b. Hess, Jr., B. A., and Schaad, L. J., 1971, *J. Am. Chem. Soc.*, 93, pp. 2413-2416.
   c. Hess, Jr., B. S. and Schaad, L. J., 1976, *J. Theo. Biol.*, 59, pp 237-240.
   d. Gutman, I., Milun, M., and Trinajstic, N., 1977, *J. Am. Chem. Soc.*, 99, pp. 1692-1704.
   e. Haddon, R. C., 1979, *J. Am. Chem. Soc.*, 101, pp. 1722-1728.
7. Hehre, W. J. Ditchfield, R. Radom, L, and Pople, J. A., 1970, *J. Am. Chem. Soc.*, 92, pp. 4796-4801.
8. Dewar, M. J. S., and Schmeising, H. N., 1959, *Tetrahedron*, 5, pp. 166-178.
9. George, P., Trachtman, M., Bock, C. W., and Brett, A. M., 1975, *Theoret. Chim. Acta, Berl.*, 38, pp. 121-129.
10. George, P., Trachtman, M., Bock, C. W., and Brett, A. M.,

1977, *J. Chem. Soc., London, Perkin Trans II,* pp. 1036-1047.
11. George, P. Trachtman, M. Bock, C. W., and Brett, A. M., 1976, *Tetrahedron,* 32, pp. 1357-1362.
12. Kistiakowsky, G. B., Ruhoff, J. R., Smith, H. A., and Vaughan, W. E., 1936, *J. Am. Chem. Soc.,* 58, pp. 146-153.
13. George, P., Trachtman, M., Bock, C. W., and Brett, A. M., 1976, *J. Chem. Soc., Perkin Trans. II,* pp. 1222-1227.
14. Mulliken, R. S., 1959, *Tetrahedron,* 6, pp. 68-87.
15. Pauling, L., and Sherman, J., 1933, *J. Chem. Phys.,* 1, pp. 606-617.
16. Glockler, G., 1953, *J. Chem. Phys.,* 21, pp. 1249-1254.
17. Stern, A., and Klebs, G., 1933, *Ann. Chem.,* 505, pp. 295-306.
18. Longo, F. R., Finarelli, J. D., Schmalzbach, E., and Adler, A. D., 1970, *J. Phys. Chem.,* 74, pp. 3296-3297.
19. Wilson, G. S., and Neri, B. N., 1973, *Ann. N. Y. Acad. Sci.,* 206, pp. 568-578.
20. Almlof, J., 1974, *Int. J. Quant. Chem., VIII,* pp. 915-924.
21. Pearson, Jr., R., and Lovas, F. J., 1977, *J. Chem. Phys.,* 66, pp. 4149-4156.
22. Radom, L., Lathan, W. A., Hehre, W. J., and Pople, J. A., 1971, *J. Am. Chem. Soc.,* 93, pp. 5339-5342.
23. Radom, L., and Schaefer III, H. F., 1977, *J. Am. Chem. Soc.,* 99, pp. 7522-7526.
24. Oth, J. F. M., Bunzli, J.C., and de Zelicourt, Y. de J., 1974, *Helv. Chim. Acta,* 57, pp. 2276-2288.
25. Haddon, R. C., 1980, *Chem. Phys. Lett.,* 70, pp. 210-212.
26. Haddon, R. C., and Starnes, W. H., 1978, *Advan. Chem. Sci.,* 169, pp. 333-362.

MODELS FOR PEROXIDASE AND CYTOCHROME P-450 ENZYMES

David Dolphin

Department of Chemistry
The University of British Columbia
Vancouver, B.C., Canada V6T 1Y6

The ways in which the enzymes catalase and horseradish peroxidase are oxidized by hydrogen peroxide to porphyrin π-cation radicals are described and compared to the intermediates involved in the activation of dioxygen by the enzyme cytochrome P-450. In the latter case a similar π-cation radical of an iron(IV) oxo-complex is examined as the probable intermediate in the oxidation of unactivated carbon-hydrogen bonds.

The redox chemistry of oxygen is extensive (Fig.1) and many of the derivatives of oxygen are highly reactive and toxic to living systems. Nevertheless, nature encounters and frequently uses many of the intermediates shown in figure 1. In the case of the highly reactive species such as singlet-oxygen $^1O_2$ and the hydroxyl radical their "indiscriminant" reactivity is used for the ultimate destruction of invading organisms and xenobiotics in the mammalian polymorphonuclear leukocyte (1) where myeloperoxidase, a heme protein, plays a major role (2). Similarly, nature uses both heme and non-heme iron proteins in the manipulation of other oxygen derivatives. Thus the four-electron reduction of dioxygen to water is controlled by the heme protein cytochrome oxidase (3) which in turn receives its electrons from the respiratory electron transport chain which includes *inter alia* the heme proteins cytochromes *b* and *c* (4).

H. B. Dunford et al. (eds.), The Biological Chemistry of Iron, 283–294.

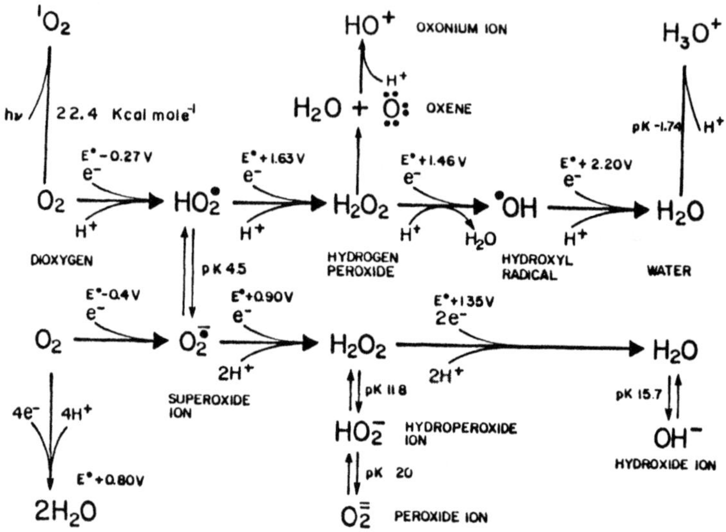

Figure 1. Some oxygen chemistry.

Hydrogen peroxide is a powerful oxidizing agent and if left unattended in a living cell would soon create havoc. Nature has devised at least three ways to handle this toxic yet necessary molecule. The heme protein catalase will, at essentially diffusion controlled rates, decompose H_2O_2 to oxygen and water (5). On the other hand, peroxidases use the oxidizing power of peroxide to bring about selective one-electron oxidations of organic substrates, usually phenols and anilines (6). Finally, and as we shall see below in some detail, hydrogen peroxide can be generated <u>in situ</u> by cytochrome P-450 by the coordination of dioxygen to ferrous P-450 followed by a one-electron reduction.

We originally became interested in the catalases and peroxidases as a result of studies on the reversible one-electron electrochemical oxidations of metalloporphyrins (7). Even when the metalloporphyrin contained a non-oxidizable metal such as Zn(II) or Mg(II) the system still underwent a reversible one-electron oxidation with the site of oxidation being the HOMO of the porphyrin (8). Thus a one-electron oxidation generates the corresponding porphyrin π-cation radical (P^{+}·). When a redox active transition metal is coordinated to a porphyrin both the metal and the porphyrin macrocycle can under go redox chemistry. Thus in the case of a Co(II) porphyrin the first one-electron electrochemical oxidation causes the metal to be oxidized with small perturbations in the optical spectrum (9). However, a second one-electron oxidation leaves the Co(III) unchanged and oxidizes the porphyrin to the π-cation radical. The second oxidation is associated with dramatic changes in the optical spectrum (Fig. 2) and since the electrochemical oxidations were carried out in the presence of tetra-n-propylammonium perchlorate the oxidized porphyrin is the perchlorate salt $[Co(III)P^{+\cdot}]^{2+}\ 2ClO_4^-$ (**1**).

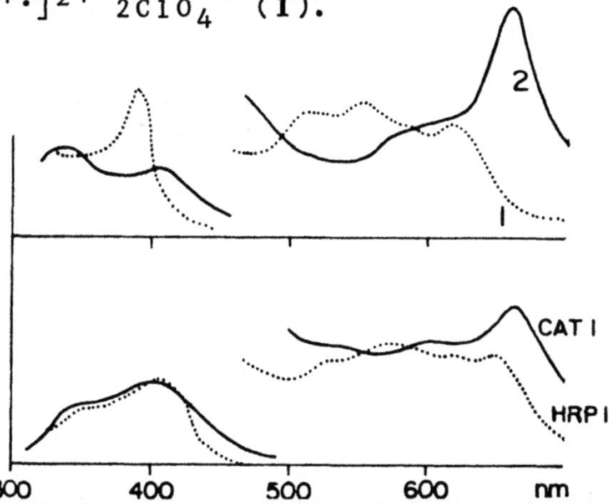

Figure 2. Optical spectra of the cobalt complexes 1 and 2 and the optical spectra of CAT I and HRP I.

Similar oxidations can be performed using bromine as the oxidant and first the metal and then the porphyrin macrocycle undergo one-electron oxidations to finally give the complex $[Co(III)P^{+\cdot}]^{2+}2Br^-$ (**2**). While dramatic changes occur in the optical spectra during this chemical oxidation the spectra of the final porphyrin π-cation radicals (**1** and **2**) differ as a function of their counter anions. That the spectra are indeed influenced by the counter ion is readily demonstrated by treating **2** with $AgClO_4$ which converts it to **1**. During this process the optical spectrum of **2** also changes to that of **1** (9). The question must then be asked as to how the two π-cation radicals of the same metalloporphyrin can have such different optical spectra?

The two highest-filled π-molecular orbitals of porphyrins are essentially degenerate and have a_{1u} and a_{2u} symmetries. Removal of a single electron from either of these orbitals will generate a free radical having either $^2A_{1u}$ or $^2A_{2u}$ ground states. The choice of which ground state is more stable in a given situation depends upon subtle criteria such as the nature of the fifth and sixth ligands coordinated to the metalloporphyrin (10) as exemplified above with **1** and **2**.

Of special interest was our observation that the optical spectra of the π-cation radicals **1** and **2** paralleled those of catalase and horseradish peroxidase in their highest, enzymatically active, oxidation states known as compounds I. Compounds I of both catalase (Cat I) and peroxidases (as exemplified by horseradish peroxidase(HRP I)) are generated when the resting ferric heme proteins undergo a two-electron, hydrogen peroxide mediated, oxidation.

The congruence of the optical spectra of **1** and **2** with Cat I and HRP I (Fig. 2) allowed us to describe Cat I and HRP I as porphyrin π-cation radicals(9). Such that the overall redox chemistry of both classes of enzymes is as follows:

$$Fe(III)P \longrightarrow Fe(IV)P \longrightarrow Fe(IV)P^{+\cdot}$$

From their optical spectra Cat I clearly exhibits the $^2A_{1u}$ and HRP I the $^2A_{2u}$ ground states. Since our original description of these high valent π-cation radicals there has been much controversy as to the accuracy of the descriptions of these electronic states. Nevertheless, the evidence is overwhelmingly in favor of the π-cation radical formation (11). In addition, the detection using ENDOR techniques by Hoffman's group of both the porphyrin π-cation radical (12) and of an oxygen coordinated to the iron(IV) of HRP I provides (13) conclusive proof for our original suggestion. At the same time it allows for HRP I (and by analogy Cat I, for bovine catalase) to be described as $O{=}Fe(IV)P^{+\cdot}$.

Since we know that catalase is coordinated by a proximal phenoxide (14) and horseradish peroxidase by a proximal imidazole ligand (15) the differences in their two ground states ($^2A_{1u}$ and $^2A_{2u}$) can be attributed to this difference in proximal ligation.

We have reconstituted horseradish peroxidase with a variety of synthetic hemes in order to explore changes in the electronic character of the compounds I of these reconstituted proteins as a function of the reconstituting heme (16). In the case of the deuteroheme (**3**) reconstituted protein the highest

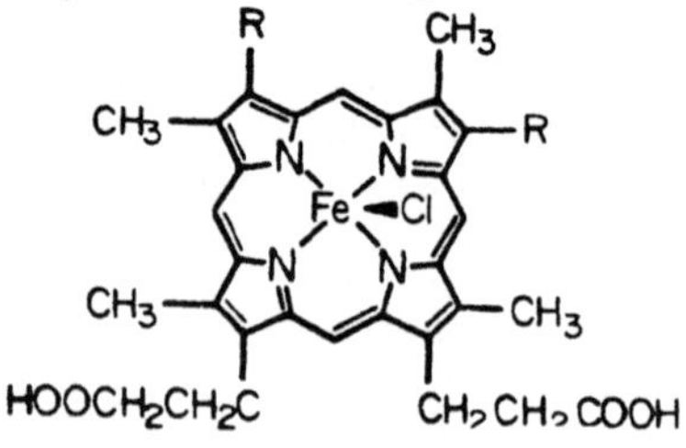

3 R = H
4 R = Et

oxidation state takes on the $^2A_{1u}$ ground state of catalase. However, since deuterohemin horseradish peroxidase possesses no catalase activity (it behaves like a peroxidase) the structure of the apoprotein is concluded to play the major role in determining the reactivity of HRP I toward hydrogen donors (16). Mesohemin (4) reconstituted horseradish peroxidase on the other hand exhibits an optical spectrum for its compound I which is half way between that typical of the $^2A_{1u}$ and $^2A_{2u}$ ground states. Since the reconstituted protein is homogeneous with respect towards porphyrin binding (17) we conclude that the electronic ground state is best described as a quantum mechanically mixed ground state. The symmetries of the a_{1u} and a_{2u} ground states allows for such quantum mechanical mixing (18).

We now turn our attention to the cytochromes P-450. These enzymes are widely distributed monooxygenases which activate molecular oxygen (19). Two principal classes of oxidations are controlled by P-450 namely the epoxidation of both aromatic and aliphatic double bonds and the oxidation of unactivated C-H bonds to give the corresponding alcohol (C-OH). In both cases the other oxygen atom of the dioxygen ends up as water.

The catalytic cycle for cytochrome P-450 is shown in figure 3. When functioning normally the

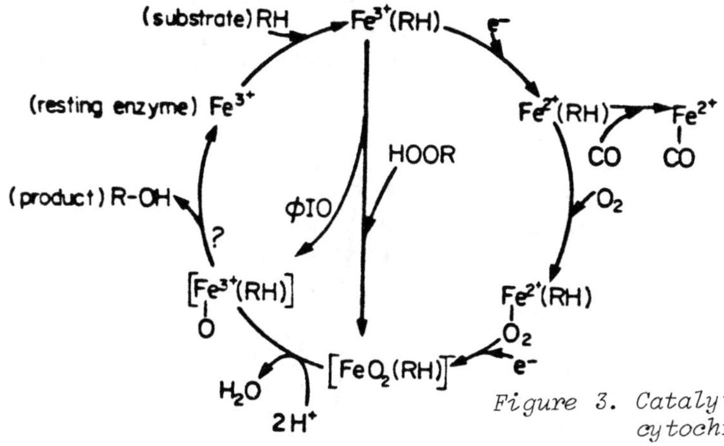

Figure 3. Catalytic cycle of cytochrome P-450

resting ferric hemoprotein binds substrate (RH) which generally causes a low to high spin conversion of the ferric iron. The high spin iron is then reduced, at about 0.0V to ferrous P-450 which like ferrous hemoglobin and myoglobin binds dioxygen. A further one-electron reduction of oxy P-450 occurs to give an intermediate which is formally a ferric porphyrin coordinated to hydrogen peroxide. This formalism has some reality since the dioxygen and two one-electron reductions can be shunted and the enzyme made to function by using peroxidatic oxidants such as alkyl hydrogenperoxides or peracids. No enzymatic intermediates have so far been observed beyond the "ferric-peroxide" complex, but we shall consider below some experimentally based speculations on the nature of the further intermediates. But first let us look at what is known about cytochrome P-450.

Cytochromes P-450 are widely distributed in nature and perform a great variety of oxidations based upon the general schemes given above. The majority of mammalian P-450's are membrane bound which makes studying them difficult (20). Fortunately a microbial enzyme which metabolizes camphor has been isolated and crystallized (21). The work that we shall describe below was carried out on this particular enzyme ($P\text{-}450_{cam}$) but the results can be generalized to the whole class of these enzymes.

Cytochrome P-450 is a protoheme-containing enzyme like hemoglobin, myoglobin, b-type cytochromes, peroxidase and catalase. However, unlike most ferrous-heme proteins which form CO complexes that absorb at ~420 nm, P-450 forms a CO complex which absorbs at 450 nm in its electronic spectrum (hence its name (22)). The major electronic differences that a CO complex of a protoheme containing enzyme experience can result only from the nature of the proximal ligand. Using model systems (Fig. 4) we were able to mimic the low energy Soret band at ~450 nm only when a mercaptide ligand (RS^-) was coordinated to a ferrous-CO complex (23) and further studies confirm that the ferrous state of P-450 is indeed coordinated by a cysteinyl anion when bound to CO. As shown in figure 4 not only did a band at ~450 nm appear in our model system but a second high energy band at ~360 nm was observed. Since our first

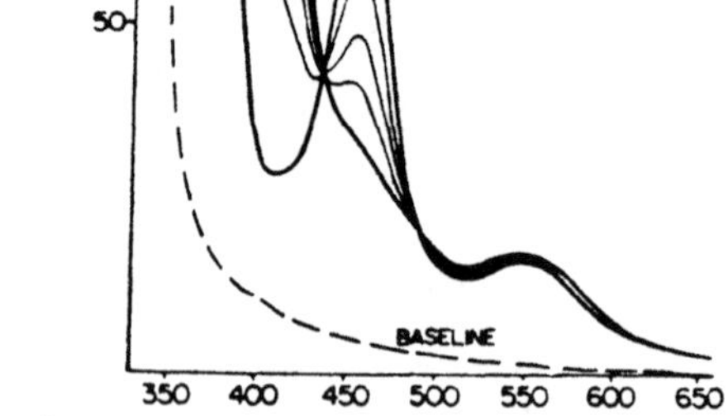

Figure 4. Formation of protoheme mercaptide-CO complex.

observation of this split-Soret band the phenomena has been observed in P-450 itself (24). The splitting of the normal $\pi-\pi^*$ transition occurs due to an interaction between the $\pi-\pi^*$ and a sulfur p — π^* charge transfer (facilitated by the high charge density at sulfur) which gives rise to two new transitions. Spectra which exhibit such split-Soret bands are known as hyper spectra and are characteristic, inter alia, of iron porphyrins coordinated by ligands of high charge density (24).

At first sight oxy-P450 resembles oxymyolobin since both exhibit Soret absorptions at ~418 nm. If oxy-P450 were coordinated by a thiolate ligand one would anticipate that it should show a hyper electronic spectrum. Yet its apparent similarity to oxymyoglobin suggests otherwise. If oxy-P450 is not coordinated by RS^- and, as we know, CO-P450 is coordinated by RS^- then the conversion of oxy-P450 to its CO complex should invoke the loss of one proton/mole. We have found that there is no net proton change in such a process and conclude that like CO-P450 oxy-P450 is indeed coordinated by a thiolate anion (25). This is confirmed by a closer inspection of the electronic spectrum of oxy-P450 which does indeed show a hyper spectrum (Fig. 5).

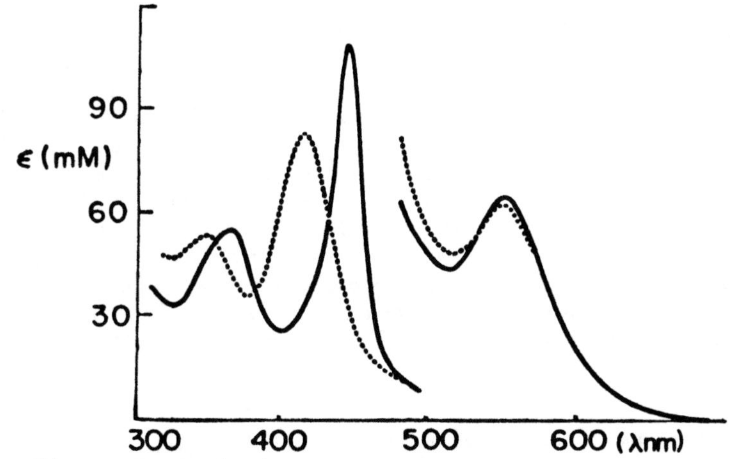

Figure 5. Optical spectra of carbonylated (——) and oxygenated (.....) ferrous P-450$_{cam}$.

It is merely coincidental that the low energy partner of the split Soret of oxy-P450 occurs at the same position as the Soret band of oxymyoblobin.

What happens when oxy-P450 is reduced by one electron? Shunting the enzyme by peroxide-like oxidants suggests that peroxide may indeed be found bound to iron. Addition of an electron to the ferric superoxide formulation (26) for oxyferrous porphyrins allow the visualization of such a process

$$Fe(II)O_2 \longrightarrow Fe(III)O_2^- \xrightarrow{e-} Fe(III)O_2^{2-}$$

We have explored this reaction using model systems and find that the one-electron reduction of oxygenated ferrous octaethylporphyrin does indeed give the corresponding ferric peroxide complex but that the peroxide is apparently bound sideways in a η^2-fashion (27). Whether or not these structural features prove to obtain for the enzyme must await further experimentation.

Whatever the structure of the "ferric-peroxide complex" in the functioning enzyme at some stage in the enzymatic process the O-O bond must break. Presently the evidence is strongly in favor of O-O cleavage before substrate oxidation. Such cleavage could be envisaged as double protonation followed by loss of water, or as suggested by recent observations (28) an initial formation of an intermediate peracid with subsequent loss of carboxylate to bring about O-O bond cleavage. In either case the breaking of the O-O bond will generate a species which is formally a ferric-oxene complex. The original observation of Ullrich (29) and the subsequent reactions of ferric porphyrins with iodosyl benzene as detailed by Groves (30) adds support to the concept of an iron complex bound to a single oxygen atom-containing species as the powerful oxidizing agent of P-450. The role of the thiolate anion in breaking the O-O bond and more importantly in stabilizing the Fe-O bond of an oxene complex has been discussed by Ullrich in this monograph. A ferric oxene complex is but one of several extreme resonance structure (Fig. 6) that can be written for the active oxidizing agent of P-450.

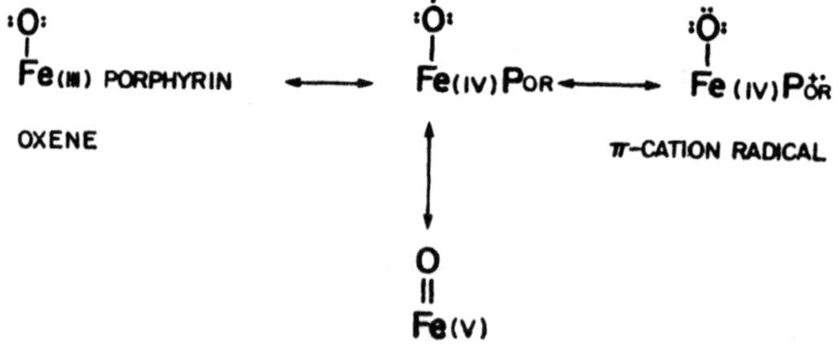

Figure 6. Resonance structures of an oxy-Fe(V) system.

A familiar electronic structure that can be written is the iron(IV) porphyrin π-cation radical that we have seen earlier describes the compounds I of catalase and horseradish peroxidase. Groves has recently isolated and characterized such an iron(IV) porphyrin π-cation radical from the reactions of a ferric porphyrin and iodosyl benzene (31). This appears to complete the circle and to suggest that those hemeproteins which control and utilize high oxidation states of oxygen have a common electronic configuration namely that of the iron(IV) porphyrin π-cation radical. That each of these high valent oxidation states perform different functions must be attributed both to the proximal coordination and active site environment provided by the protein.

Finally the mechanisms by which the active oxidizing agent oxidizes organic substrate is considered. We shall limit the discussion here to the oxidation of unactivated C-H bonds since this whole area is covered in greater depth in the next chapter by Loew. At first sight one might imagine that possible similarities between an iron-oxene and carbenes or nitrenes would suggest a direct insertion of the oxygen atom into the C-H bond to give the alcohol. Nevertheless, the elegant work of Groves (32) has shown that at least in the oxidation of tetradeuteronorbornane a substrate intermediate is an obligatory intermediate.

If the breaking of a substrate C-H bond is a

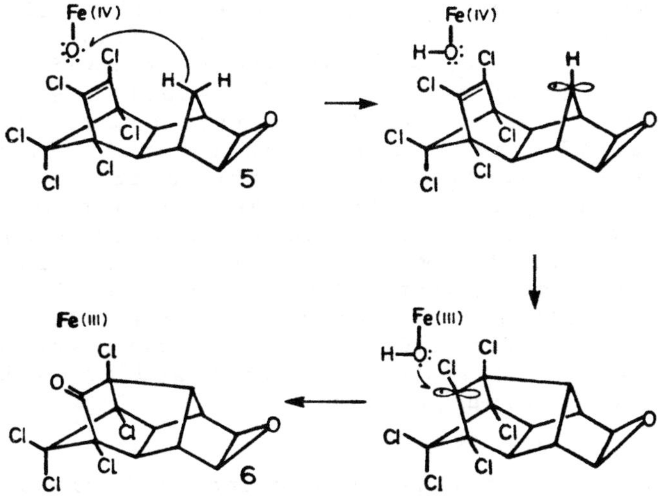

Figure 7. Proposed mechanism for the P-450 mediated oxidation of Dieldrin.

necessary step in the enzymatic reaction then the lowest energy intermediate is likely to be a substrate "free radical" since the generation of ionic intermediates, in what must be a strongly hydrophobic active site, would be energetically unfavorable. Indeed, the isotope effects and regio- and stereoselectivity of P-450 oxidations all argue for substrate radical intermediates. A further example of this is our own interpretatin of the microsomal metabolism of the insecticide Dieldrin (5) (33). Formation of the product (6) from the pentacyclic starting material requires the intervention of an intermediate and the chemistry outlined in figure 7 is best accommodated by a radical intermediate.

The formation and hydroxylation of such a substrate free radical is outlined in figure 8 where

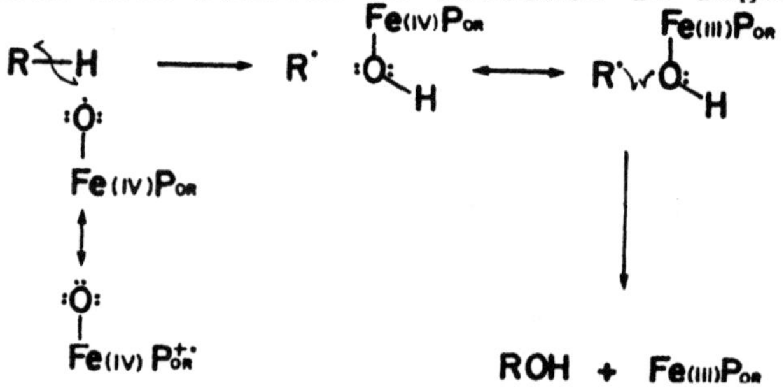

Figure 8. Proposed mechanism for the P-450 hydroxylations.

the abstraction of a hydrogen atom from substrate (R-H) by the active oxidizing agent of P-450 would give the substrate radical and an iron(IV) porphyrin complexed by hydroxide. Subsequent transfer of an iron bound hydroxy radical to the substrate radical could generate the oxidized substrate and the resting enzyme (Fe(III)) ready for another catalytic cycle. While these speculations on the details of the oxidative steps require more experimental support the intermediacy of both an oxene and hydroxyl radical (each stabilized by coordination to an iron porphyrin) would complete the picture for the oxygen chemistry outlined in figure 1 since all of the possible oxygen intermediates shown there would have been utilized by nature and stabilized in each case by the use of a heme protein.

Acknowledgment

The writing of this review and much of our work described here was supported by the United States National Institutes of Health (AM 17989).

References

1. Babior, B.: 1978, New England J. Med. 298, 721.
2. Harrison, J.E. and Schultz, J.: 1976, 251 1371.
3. Malstrom, B.G.: 1980 in "Metal Ion Activation of Dioxygen" ed. Spiro, T.G., Wiley Interscience, New York, pp. 181-207.
4. Dolphin, D., ed. "The Porphyrins" 1979, Academic Press, New York, Vol. VII.
5. Schonbaum, G.R. and Chance, C. in "The Enzymes" ed. Boyer P.D.: 1976, Academic Press, New York, Vol. XII, pp. 363-408.
6. Hewson, W.D. and Hager, L.P.: in "The Porphyrins" ed Dolphin, D. 1979, Academic Press, New York, Vol. VII, pp.295-332.
7. Fajer, J., Borg, D.C., Forman, A., Dolphin, D. and Felton,R.H.: 1970, J. Am. Chem. Soc. 92, 3451.
8. Dolphin, D., Muljiani, Z., Rousseau, K., Borg, D.C., Fajer,J. and Felton, R.H.:1973, Annals N.Y. Acad. Sci. 206, 177.
9. Dolphin, D., Forman, A., Borg, D.C., Fajer, J. and Felton, R.H: 1971, Proc. Natl. Acad. Sci. U.S.A. 6, 614.
10. Dolphin, D. and Felton, R.H.: 1974, Accounts Chem. Res. 7, 26.
11. Dolphin, D.: 1981, Israel J. Chem.21, 67.
12. Roberts, J.E., Hoffman, B.M., Rutter, R. and Hager, L.P.: 1981, J. Biol. Chem. 256, 2118.
13. Roberts, J.E., Hoffman, B.M., Rutter, R. and Hager, L.P.: J.Am. Chem. Soc., in press.
14. Rossmann, M., Tanaka, N. and Reid, T. III.: this volume.
15. Yonetani, T., Yamamoto, H., Erman, J.E., Leigh, J.S., Jr. and Reed, G.H.: 1972, J. Biol. Chem. 247, 2447.
16. DiNello, R.K. and Dolphin, D.: 1981, J. Biol. Chem. 256, 6903.
17. La Mar, G.N., personal communication.
18. Hanson, L.K., personal communication.
19. Coon, M.J. and White, R.E. in "Metal Ion Activation of Dioxygen" ed. Spiro, T.G., 1980, Wiley Interscience, New York, pp. 73-123.
20. Griffin, B.W., Peterson, J.A. and Estabrook, R.W. in "The Porphyrins" ed. Dolphin, D., 1979, Academic Press, New York, Vol. VII, pp. 333-375.
21. Wagner, G. and Gunsalus,I. C.: this volume.
22. Omura, T. and Sato, R.: 1964, J. Biol. Chem., 239, 2370.
23. Chang, C.K. and Dolphin, D.: 1975, J. Am. Chem. Soc., 97, 5948.

4. Hanson, L.K., Eaton, W.A., Sligar, S.G., Gunsalus, I.C., Gouterman, M. and Connel, C.R.: 1976, J. Am. Chem. Soc. 98, 2672.
25. Dolphin, D., James, B.R. and Welborn, H.C.: 1980, J. Mol.Cat., 7, 201.
26. Makinen, M.W. in "Biochemical and Clinical Aspects of Oxygen" ed. Caughey, W.S., 1974, Academic Press, New York, p. 143.
27. Welborn, C.H., Dolphin, D. and James, B.R.: 1981, J. Am.Chem. Soc., 103, 2869.
28. Sligar, S.G., Kennedy, K.A. and Pearson, D.C.: 1980, Proc.Natl. Acad. Sci. U.S.A., 77, 1240.
29. Lichtenberger, F., Nastainczyk, W. and Ullrich, V.: 1976, Biochem. Biophys. Res. Commun., 70, 939.
30. Groves, J.T. in "Metal Ion Activation of Dioxygen" ed. Spiro, T.G., 1980, Wiley Interscience, New York, pp. 125-162.
31. Groves, J.T., Haushatter, R.C., Nakamura, M., Nemo, T.E. and Evans, B.J.: 1981, J. Am. Chem. Soc., 103, 2884.
32. Groves, J.T., McClusky, G.A., White, R.E. and Coon, M.J.: 1978, Biochem. Biophys. Res.Commun., 81, 154.
33. Dolphin, D., Addison, A.W., Cairns, M., DiNello, R.K., Farrell, N.P., James, B.R., Paulson, D.R. and Welborn, C.:1979, Int. J. Quantum Chem., XVI, 311.

STRUCTURE AND SPECTRA OF STABLE AND TRANSIENT STATES AND MECHANISMS OF OXIDATION OF MODEL CYTOCHROME P-450

Gilda Loew, Dale Spangler and Andrew Pudzianowski

SRI International, Life Sciences Division, Menlo Park, California

1. INTRODUCTION

The cytochrome P450's are a class of heme proteins involved in a wide variety of metabolic transformations. However, the diverse reactions comprising P450 oxidative metabolism have a unifying characteristic: the enzyme uses one molecule of oxygen and two reducing equivalents [DOH_2] usually from NADPH, to produce one molecule of water and the oxidized substrate [ROH]. The two fundamental reactions are hydroxylation and epoxidation (1-2).

The P450's are ubiquitous and are usually integrated into a membrane together with the electron transport system that provides the reducing equivalence. To reach the enzymatically active state, P450 has a five-step reaction cycle which has been extensively characterized from the known behavior of the soluble P450 camphor enzyme (3-11).

In the first four steps of the normal enzymatic cycle, oxygen binds to the substrate bound once-reduced enzyme giving a stable oxy complex. If carbon monoxide is supplied at this point, the complex shows an absorption at 450 nm, part of a split or hyperchromic Soret absorption band, the characteristic band for which the enzyme was named (12-14). There is now evidence that the oxy-P450 complex also has a split Soret that is slightly blue-shifted from the carbonyl complex (14-20). Any change in the native enzyme that results in the restoration of a single intense Soret transition also results in deactivation (12-14).

The first four stages of the soluble cytochrome P450 camphor

H. B. Dunford et al. (eds.), The Biological Chemistry of Iron, 295–312.

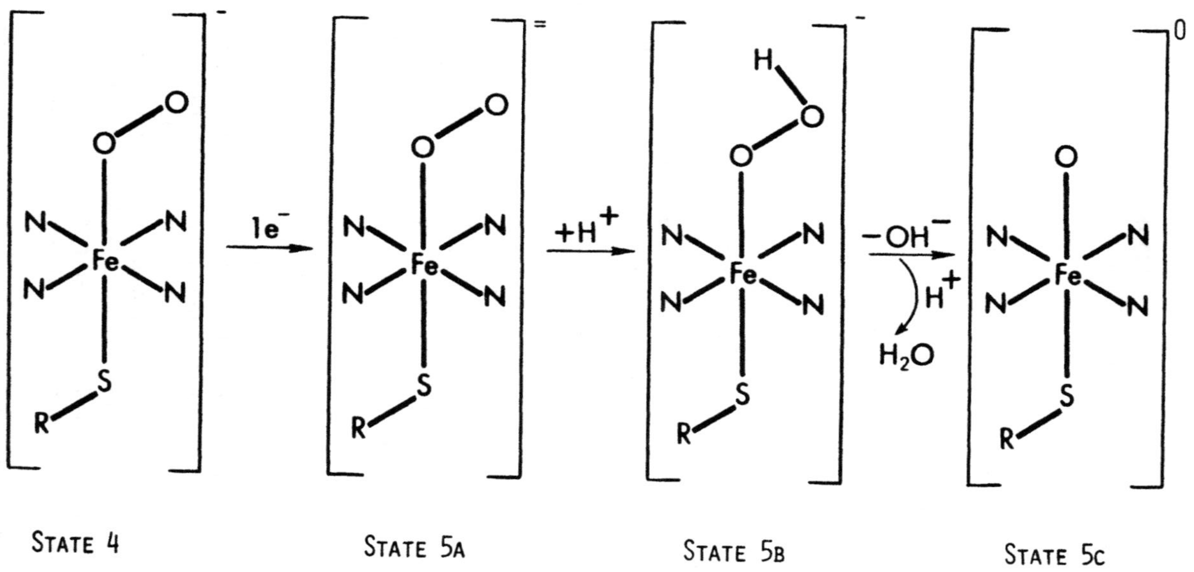

FIGURE 1: Postulated steps in formation of ferryl active state (5c) upon second electron reduction of oxy ferrous P450

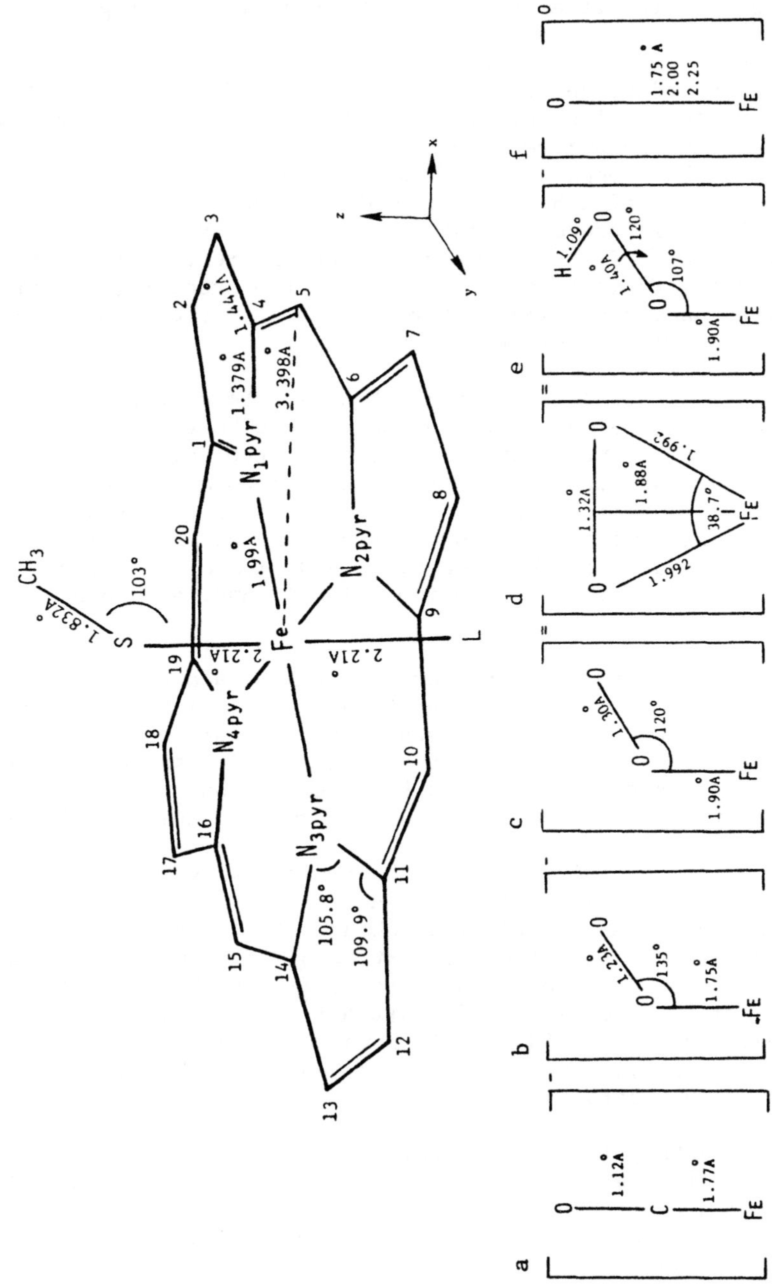

FIGURE 2: Model used for enzyme active site of cytochrome P450's

enzyme are all stable enough to have been isolated, and their spectroscopic properties have been extensively investigated.

The biologically active state of the enzyme is formed by one-electron reduction of the ferrous oxy-P450 complex; whereupon one oxygen atom is incorporated into the substrate, one into H_2O formation, and the enzyme returns to its resting state. This series of rather complex transformations occurs so rapidly that no form of the enzyme has been isolated beyond the oxy-ferrous P450 complex. Inferences about the nature of the fifth, or activated, state of the enzyme have come indirectly from experimental and theoretical investigations of the products, kinetics, and thermodynamics of oxidative reactions of intact protein and model systems with specific substrates.

As a result of these various studies, the currently accepted model of the biologically active state is an "oxenoid" or ferryl model in which a single oxygen is bound to the heme iron and reacts with substrates as something similar to a neutral oxygen atom or electrophilic oxene. Possible steps in the transformation from the stable oxy complex to a ferryl complex are indicated in Figure 1: second electron reduction; addition of a proton to the dioxygen ligand forming a peroxide anion ligand; and loss of OH^- by this ligand.

Support for this pathway comes partly by analogy with the known behavior of another heme protein, horseradish peroxidase (HRP). HRP is known to form a stable ferryl complex, called compound I (HRP-I), when it reacts with H_2O_2 in its resting state. Similarly, the cytochrome P450's have been shown to react in their resting state with organic peroxides, forming a complex like (5b).

Assuming a ferryl model for the biologically active state, there are still a number of important aspects of P450 aliphatic hydroxylation and epoxidations which are still unresolved. Among these are: (1) the extent of concertedness of Fe-O bond breaking and the transfer of O to the substrate; (2) whether the reactions involve radical or closed shell oxygen species: (3) whether the oxygen addition in epoxidation is bond centered or atom centered; (4) whether aliphatic hydroxylation is best described as C-H bond insertion or H abstraction.

Experimental efforts addressing these questions have used both model heme systems and intact cytochrome P450 proteins and have centered on product identifications, isotope effects on rates of reaction, and measurements of retention of configuration in products (21-29).

In the three-part study reported here we have used the techniques of theoretical chemistry to: (1) calculate the electronic structure and spectra of the stable oxy and carbonyl ferrous P450 complexes; (2) characterize plausible transient intermediates involved in enzymatic activity; and (3) elucidate mechanisms of hydroxylation and epoxidation of aliphatic hydrocarbons of these enzymes.

2. METHOD AND MODEL SYSTEMS USED

The techniques of theoretical chemistry embodied in large-scale computer programs have advanced to the point where they can be used to characterize the iron-coordination complexes forming the active sites of the cytochrome P450's and to elucidate mechanisms of biochemical transformation of substrates by these enzymes.

The model employed for the active site of the P450 enzyme as shown in Figure 2 was a planar porphyrin ring with D_{4h} symmetry and two axial ligands, one of which was methyl mercaptide. Geometries for the porphyrin ring (32-33), vinyl groups (37) mercaptide ligand (33-36) and second axial ligand (38) were chosen from structural features of relevant molecules. Figure 2 summarizes the geometries used for the stable carbonyl (a) and oxy (b) complexes, for the twice-reduced oxy complex (c,d), for the peroxide anion complex (e), and for the ferryl complex (f).

For the ferryl complex calculations were performed at three Fe-O bond lengths, r = 1.75, 2.00, and 2.25 Å in order to monitor changes in electron and spin distribution with lengthening of the Fe-O bond in the absence of substrate.

The calculations of model P450 complexes were performed using an INDO program (39-43) which includes parameterization for first-row transition metals and extensive configuration interaction, geometry optimization and unrestricted Hartree-Fock capabilities. This program has been successfully applied to describe the ground state and/or spectral properties of $[FeCl_4]^-$, $[CoCl_4]^{2-}$, $[CuCl_4]^{2-}$ (41,42), ferrocene (43), model oxyhemoglobin complexes (44), and model oxy and carbonyl P450 systems (45).

This program includes options for both theoretical and spectral (INDO/S) evaluation of the two-center repulsion integrals, the latter employing an empirical Weiss-Mataga-Nishimoto formula (46,47). The spectral parameterization was chosen to calculate electronic structures and electromagnetic properties of the model P450 complexes. The theoretical parameterization is appropriate for examining different geometries. Calculations on open shell systems were performed using the unrestricted Hartree-Fock formalism, and a method for spin-purification (42) was involved in an attempt to reduce the expectation value of S^2. Reported spin densities are after spin-purification.

Characterization of chemical reaction pathways by theoretical methods has advanced to the point where it is now possible to obtain a precise description of key species along a reaction pathway, i.e., transition states and intermediates as well as reactants and products, and to calculate enthalpies and entropies of activation and of reaction. To characterize hydroxylations and epoxidations by addition of an electrophilic oxygen species to aliphatic hydrocarbons, we have used a semiempirical molecular orbital method called MNDO (48) (Modified Neglect of Diatomic Overlap) in conjunction with rigorous analytic techniques for locating and characterizing all stationary point species, including transition states along the reaction surface (49-51).

However, such programs cannot yet be used to describe the reaction of an entire ferryl heme complex with substrates. Thus, we have begun characterizing mechanisms of P450 oxidations using the simplest systems that could serve as models for aliphatic hydrocarbon epoxidation and hydroxylations, i.e., triplet and singlet oxygen atom as the enzymatically active species, and methane, ethylene and propene as model substrates.

3. RESULTS AND DISCUSSION

3.1 Optical Spectra and Ground State Properties of Model Oxy and Carbonyl Ferrous Cytochrome P450 Complexes

Calculated ground state properties for the oxy and carbonyl model systems yielded values of ΔE_Q = -2.46 and 0.45 mm/sec respectively, in excellent agreement with observed values of -2.15 and 0.32 (30). Thus both the anomalously large negative electric quadrupole splitting (ΔE_Q) observed in Mossbauer resonance spectroscopy for the oxy-P450 complex (30) and the very small quadrupole splitting observed for the carbonyl-P450 complex (30) could be accounted for by the description of the ground state obtained by this method.

The origin of the difference in behavior between the oxy- and carbonyl-P450 complexes arises from the fact that an electron is transferred to the O_2 ligand from the iron $d\pi$ orbital which does not occur in the CO complex. This electron transfer results in a iron configuration resembling low-spin d^5 in the oxy complex and a nearly spherically symmetric d^6 low spin iron configuration in the carbonyl complex, although both are formally ferrous diamagnetic complexes.

The optical spectra of the model oxy- and carbonyl-P450 complexes were determined by performing single and double con-

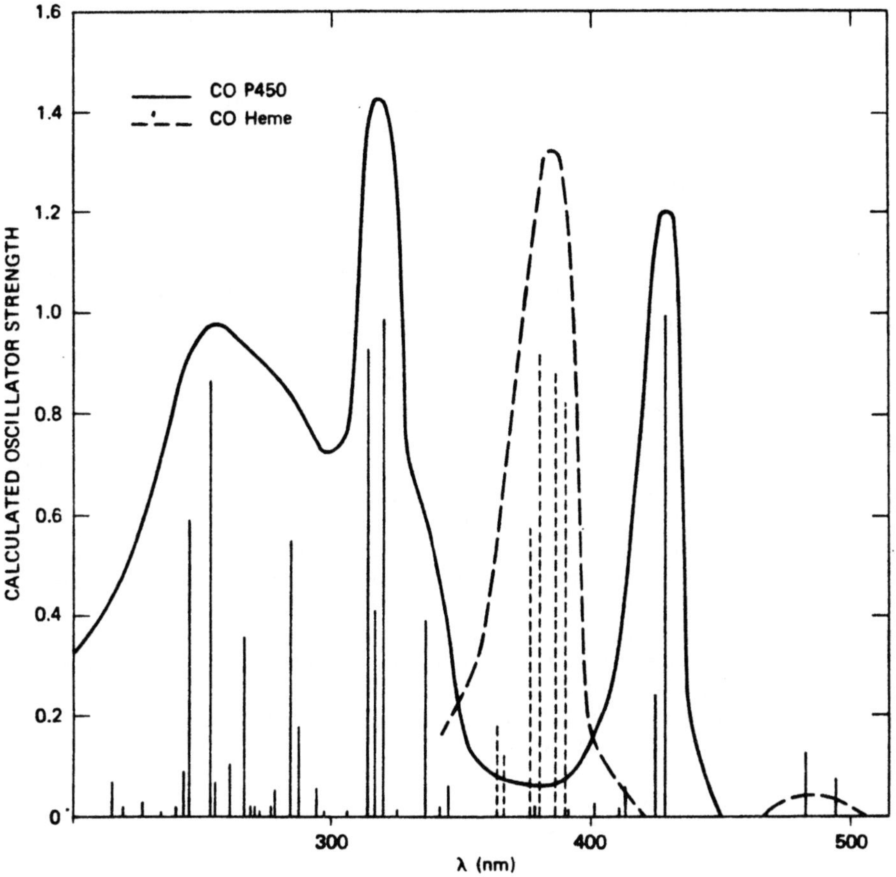

Figure 3: Calculated electronic transitions for model carbonyl-P450 (____) and carbonyl-heme (- - - -) complexes.

figuration interaction calculations on the respective SCF-level molecular orbitals. The calculated electronic spectra (oscillator strengths and frequencies) for model carbonyl-P450 and for a model carbonyl-heme (52) with an imidazole replacing the mercaptide ligand are shown in Figure 3. The calculated electronic spectra for model oxy-P450 and model carbonyl-P450 are presented in Figure 4, along with the experimental spectra for the two complexes. In Figure 3 and 4, envelopes have been drawn around the calculated oscillator strengths for convenience in visualization of the spectra. In Figure 3, we show that the Soret band is split for model carbonyl-P450, with calculated maxima at $\lambda \approx 430$ and 325 nm, while there is only one very intense band in the model carbonylheme complex at $\lambda \approx 380$ nm. From Figure 4, we see that both oxy and carbonyl complexes have split Soret bands and that the Soret bands observed for the model oxy-P450 complex are blue-shifted relative to those for the carbonyl-P450 complex. These results are in excellent agreement with the experimentally observed spectra (12-20, 31). Further verification of our calculated spectra would come from single crystal polarized absorption spectra, since the oxy- but not the carbonyl-P450 complex has a number of z-polarized transitions in the oxy-P450 complex with the major such peak occurring at $\lambda \simeq 210$ nm.

The split Soret band in the model carbonyl-P450 complexes arises because the a_{2u} (π) porphyrin-type orbital in the P450 complex has a significant contribution from the sulfur p_z basis function in the P450 complex, and the next four lower-energy orbitals have significant contributions from the *s* basis functions, whereas no such phenomenon occurs for the heme complex.

These calculations of the optical spectra of model complexes of P450 strongly imply that the mercaptide ligand has a key role in causing split Soret bands. While the nature of the second axial ligand modulates the details of the optical spectra, all P450 complexes with intact mercaptide bonds to the iron should have this anomalous spectral behavior. Since such behavior is linked to enzymatic activity, inhibition of activity appears to involve interference with the integrity of the iron bond to a mercaptide moiety of a nearby amino acid residue in the protein. Thus a link between structure, spectra, and function of cytochrome P450 is suggested by this work.

3.2 Electronic Structure and Spin Densities of Postulated Transient Active States

One-electron reduction of the already reduced oxy ferrous P450 complex leads to a dianionic radical (5a, Fig. 1). Such a radical has numerous possible spin and electron states and it is also possible that the geometry of the O_2 ligand could change from a bent, end-on to a triangular form with the two oxygens equally coordinated to the iron.

For the purpose of comparing the relative stability and electronic structure of the two geometries of the O_2 ligand, we have calculated the relative energies of a doublet state of each geometry. In these calculations, we have placed vinyl groups on the porphyrin ring at the C_3 and C_8 positions (Fig. 2) in which they are found in protoporphyrin IX.

The main effects of the second-electron reduction are as follows: if the oxy ferrous complex is described as a Fe(II) (d^6)-O_2 ferrous-dioxygen complex, the added electron and spin density go on to the oxygen forming an O_2^- radical. The bent,

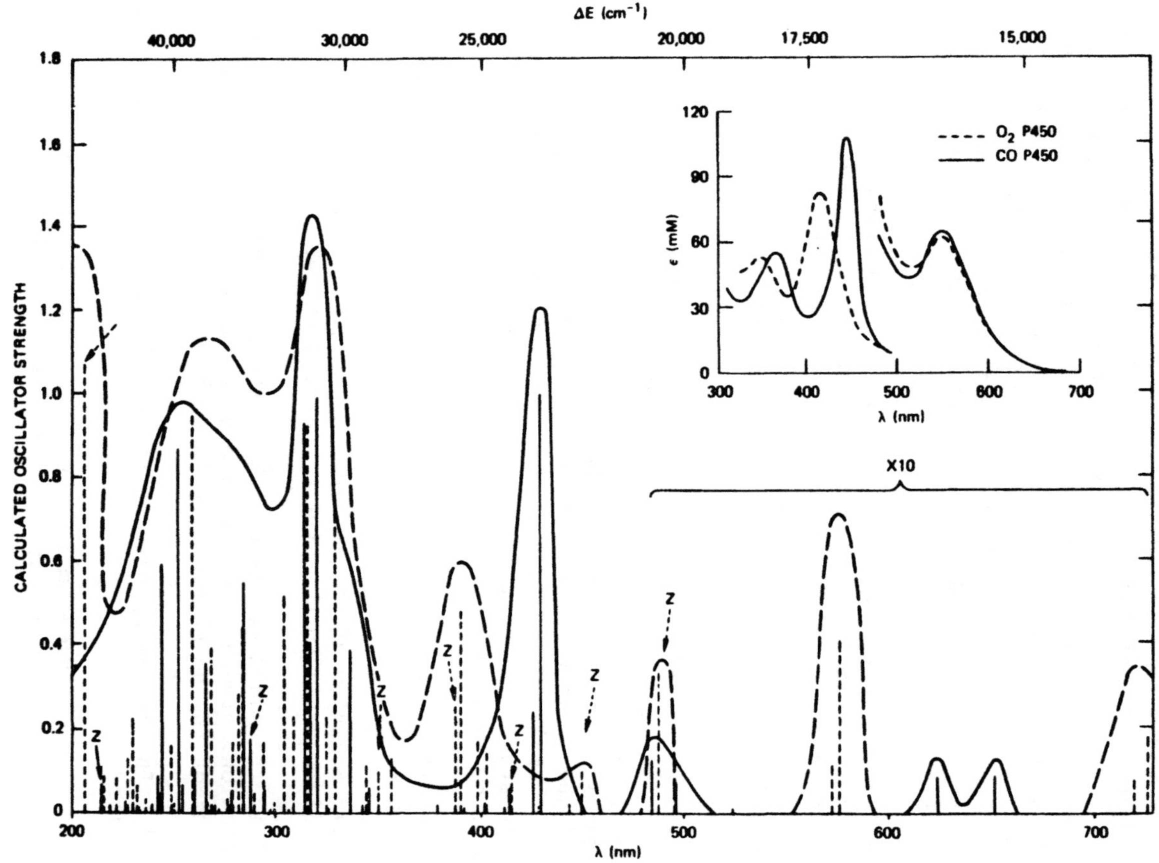

FIGURE 4: Calculated electronic transitions for model oxy-P450 (- - -) and carbonyl-P450 (——). The experimental spectra of the two complexes are shown in the insert.

end-on O_2^- ligand geometry is more stable than the triangular structure by 33 kcal/mol. The destabilization of the dicoordinated triangular structure appears to have two major causes: the close contact between the oxygen atoms and the pyrrole ring which is only 1.88 Å and the weak Fe-O bonds formed, with calculated bond overlap densities of only 0.03. While this complex may not have a doublet ground state, it is unlikely that the relative energies of the two geometries would change for states of higher multiplicity, since the steric constraint of the triangular structure would be constant for all spin states. Thus our results suggest that the O_2 stays in a bent, end-on configuration when the oxy P450 complex is reduced in the last step of the enzymatic cycle.

The electron and spin distribution of the three reactive intermediates were further characterized considering only this bent, end-on geometry for state 5a (Fig. 2,c); the geometry in Figure 3e for state 5b; and the three geometries in Fig. 2f for the ferryl complex. Spectroscopic parameters which are best suited for calculation of spin and charge distributions were used.

With these parameters a description of the diamagnetic oxy-P450 complex was obtained with an electron distribution corresponding to a ($Fe^{III}d^5$-O_2^-) configuration, but with no unpaired spin on either the iron or the dioxygen. Such a description yielded the large negative quadrupole splitting observed for oxy-P450 complexes. Figure 5 summarizes the resulting electronic and spin states calculated for the three reactive intermediates formed from this complex.

The effect of one-electron reduction on this oxy ferrous P450 complex can be summarized as follows:

(1) Addition of an electron causes a major change in spin from a diamagnetic to a sextet or quartet state.

(2) The added electron goes on the porphyrin ring, making it a dianion. There is at least one and more likely two unpaired electrons on the porphyrin ring in $e_g\pi^*$ orbitals.

(3) The system appears to favor a Fe(IV S = 1) configuration on the iron.

(4) The dioxygen remains an O_2^- moiety with little change in electron density and has a radical character (S = 1/2) in a sextet state.

For the peroxide anion complex (State 5b, Figure 1) formed by addition of a proton to the twice-reduced oxy complex, only a quartet state was characterized. Protonation of the dianion

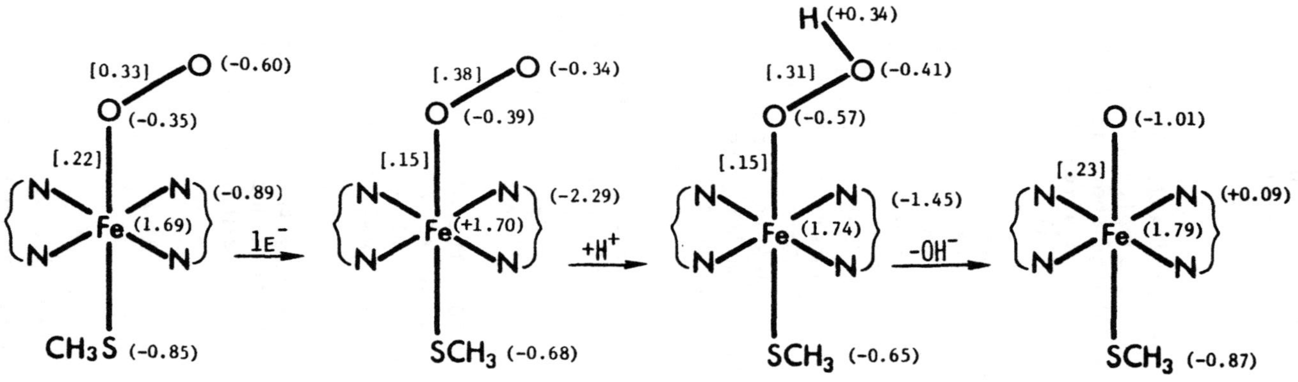

Fe(III d^5 S=0)O_2^-(S=0)por(S=0) ⟶ Fe(IV S=1)O_2^-(S=½) por(egπS=1) ⟶ Fe(IV S=1)O_2^-(S=0)por(egπS=½) ⟶ Fe(IV S≅ 1)O_2^-(S≅0) por(egπS=½)

$\langle S^2 \rangle = 0$ $\langle S^2 \rangle = 9.27$ $\langle S^2 \rangle = 4.64$ $\langle S^2 \rangle = 3.83$

FIGURE 5: Electronic structure and spin states of postulated intermediates in formation of ferryl active state by reduction of oxy ferrous P450 weakens the O-O bond making it more easily cleaved, but the Fe-O bond density remains essentially unchanged. The peroxide anion ligand has a net charge of -0.64, similar to that of O_2 before protonation, but the porphyrin ring charge is reduced from -2.29 to -1.45.

Loss of OH^- from the peroxide anion ligand leads to formation of the ferryl-type active state complex (State 5c, Figure 1). As shown in Figure 7, in this complex there is full negative charge on the oxygen and the Fe-O bond is strengthened, partly because it is also shortened (1.90 → 1.75 Å)

As already discussed, the postulated ferryl P450 complex (P450-I) is an analog of HRP-I, the stable compound two oxidizing equivalents above the resting state, formed by the direct reaction of horseradish peroxidase and H_2O_2. The difference between the two active sites is that HRP-I has an imidazole axial ligand, while P450-I has a mercaptide ligand. The stable HRP-I has been extensively characterized (53-60, 62, 63). In previous work (61), we have calculated the electronic structure, quadrupole splitting, and spin density for a number of electronic states of a model HRP-I complex. Fe(IV S = 1) porphyrin π cation radical states were found to be the lowest energy states and gave calculated properties in agreement with experiment.

Results obtained for P450-I are very similar to those for HRP-I. Without assignment, the state obtained for P450-I is a Fe(IV S = 1) por (a_{1u} S = 1/2) radical, with the same electronic structure and spin distribution as HRP-I. These salient characteristics may be summarized as follows:

(1) In both HRP and P450-I the electron and spin distribution on the iron correspond to a Fe(IV S = 1) $(d_{xy})^2(d_{xz})^1(d_{yz})^1$ configuration with a spin density of 0.83 e in the d_{yz} and d_{xz} orbitals. There is some spin delocalization from the iron to the oxygen.

(2) In both HRP-I and P450-I, there is nearly a full negative charge on the oxygen and about 0.3 unpaired spin density in the $O\pi$ orbitals. Thus before Fe-O bond cleavage the oxygen resembles an O^- radical.

(3) The a_{1u} cation radicals have very similar spin distributions in both HRP-I and P450-I.

(4) The active site of HRP-I has a formal +1 charge, while that of P450-I is neutral. This difference in overall charge is manifest in a large positive charge on the porphyrin ring of HRP-I.

In order to further characterize the nature of the oxygen atom as the Fe-O bond cleaves, we have calculated the electronic structure and spin distribution of P450-I at two longer Fe-O distances, r = 2.00 Å and 2.25 Å, with the constraint of maintaining an S = 3/2 state. With this constraint, as the $Fe(d_{z}2)-p_{z}(O)$ σ bond breaks, the O atom maintains a large negative charge and develops unpaired spin density in the p_z orbital. In this simplified system, without including the reattachment of the imidazole as a sixth ligand, or the simultaneous attack on the substrate, cleavage of the Fe-O bond appears to lead to a Fe(III S = 3/2) por π (S = 1/2) state of the active site with an O^- radical the leaving species.

3.3 Mechanisms of P450 Hydroxylation and Epoxidation of Aliphatic Hydrocarbons

In initial mechanistic studies modeling P450 aliphatic hydrocarbon hydroxylations and epoxidations we have used a simple triplet and singlet oxygen atom (64-66) (already cleaved from the Fe) as the electrophile adding to substrates. Since the substrates themselves are nonpolar and bind to hydrophobic sites, our working assumption is that electronic aspects of an oxene model can be elucidated by a "gas-phase" theoretical treatment of the interactions between atomic oxygen and model substrates. As model substrates, propene was selected as a prototype unsymmetrical alkene, and methane and ethylene because their reactions with $(^3P)O$ are well documented in the literature and can be used to gauge the performance of the methods used.

Reactants, transition states, intermediates and products were characterized. Enthalpies and entropies of reaction and activation and relative rates of reaction were calculated for hydroxylations of methane, ethylene, and the allylic and vinylic sites of propene, and for epoxidation of ethylene and propene. In general, calculated transition state geometries were accurate as reflected by the good agreement with known entropies of activation, while activation enthalpies were higher than experiment. Experimental values of enthalpies were used to calibrate results for similar systems. The main results obtained from these studies are summarized below.

(1) Aliphatic hydroxylation by $(^3P)O$ proceeds by H atom abstraction in a two-step reaction proceeding through a radical ·CR intermediate. (2) Aliphatic epoxidation by $(^3P)O$ proceeds by asymmetric addition to the double bond in a two-step reaction involving a $\cdot R_2C-CR_2O\cdot$ radical intermediate. (3) Aliphatic hydroxylation by $(^1S)O$ is a concerted reaction proceeding by partial H abstraction and reattachment of the leaving OH group to the carbon atom. (4) Aliphatic epoxidation by $(^1S)O$ is a concerted bond centered reaction. (5) The specificity of propene

oxidation by $(^3P)O$ is: C_1epox > C_2epox > allylic hydrox > 2 propenyl hydrox ≥ 1 propenyl hydrox (6). The specificity of propene oxidation by $(^1S)O$ is: expoxidation > allyl hydroxylation; but all reactions are highly exothermic indicating little specificity.

The results obtained from this study were compared with experimental results for ethylene (68) and styrene (67) oxidations by P450 and cyclohexene oxidation by a model P450 system (22). The combined results are most consistent with a radical oxygen mechanism and appear to corroborate the detailed two-step mechanism we have described for hydroxylation and epoxidation. In particular, secondary isotope effects on the rate of epoxidation of styrenes by P450 indicate a two-step asymmetric reaction for epoxidation. Further evidence for a two-step reaction is provided by the nature of the adduct identified (N_{pyr}-CH_2-CH_2OH) in suicide inactivation of cytochrome P450 by ethylene. Such an adduct can readily be formed by the biradical, $\cdot CH_2$-$CH_2O\cdot$, we have identified in the two-step epoxidation of ethylene. Oxidation of cyclohexene by a model P450 heme system indicates more epoxide formation than allyl alcohol, in agreement with predicted selectivity.

While the mechanisms presented have been obtained using a simple oxene model, it is possible to combine them with the description obtained for the ferryl active state, P450-I, to propose reasonable mechanisms for oxidation of aliphatic hydrocarbons by P450 with the oxygen atom still attached to the heme iron in the ferryl active state.

As shown in Figure 6a, epoxidation could occur by the asymmetric attachment of the oxygen to the carbon atom of a double bond forming a radical ligand stabilized by the porphyrin ring. This radical is then in a position either to attack the pyrrole nitrogen as it does in suicide substrates such as ethylene, or to form an epoxide which is first a ligand on the iron and is then displaced by the nearby imidazole ligand to form the product plus the resting state of the enzyme.

As shown in Figure 6b, hydroxylation by the ferryl oxygen, as with the free oxene atom, could also proceed by H atom abstraction to form radical intermediates which recombine to form an alcohol. The alcohol could be a ligand on the iron until it is displaced by the more strongly binding imidazole group.

Work is now in progress in our laboratory to characterize the resting state and to further investigate the proposed mechanisms for epoxidation and hydroxylation of aliphatic hydrocarbons by intact ferryl P450-I complex.

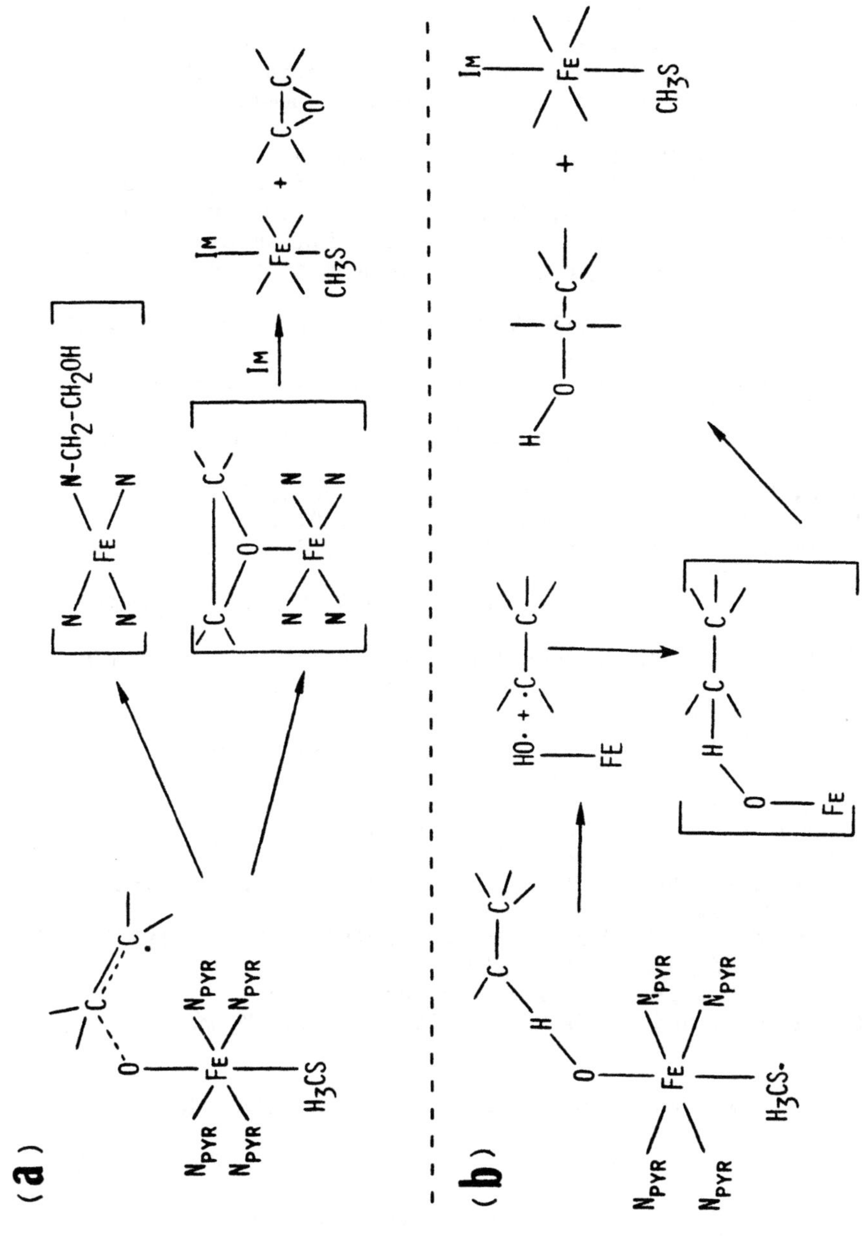

FIGURE 6: (a) Proposed mechanism of epoxidation by active ferryl state of cytochrome P450.
(b) Proposed mechanism of hydroxylation by active ferryl state of cytochrome P450.

REFERENCES

1. Gillette, J.R.: 1966, *Adv. Pharmacol.* 4, p. 219.
2. Levin, W., Wood, A.W., Lu, A.Y.H., Ryan, D., West, S., Cooney, A.H., Thakker, D.R., Yagi, H. and Jerina, D.M.: 1977, In "Drug Metabolism Concepts", Jerina, D.M., ed.: ACS Symposium Series no. 44, American Chemical Society. Washington, D.C. p. 99.
3. Estabrook, R.W., Matsubara, T., Mason, J.I., Werringloer, J. and Baron, J.: 1973, *Drug Metab. and Disp.* 1, p. 98.
4. Gunsalus, I.C., Meeks, J.R., Lipscomb, J.D., Debrunner, P., and Muenck, E.: 1974, In "Molecular Mechanisms of Oxygen Activation". Hayaishi, O., ed. Academic Press, New York, p. 559.
5. Collman, J.P., Sorrell, T.N. and Hoffman, B.M.: 1975, *J. Am. Chem. Soc.* 97, p. 913.
6. Koch, S., Tang, S.C., Holm, R.M. and Frankel, R.B.: 1975, *J. Am. Chem. Soc.* 97, p. 914.
7. Papaefthymiou, G. and Frankel, R.B.: 1975, *Bull. Amer. Phys. Soc.* 20, p. 425.
8. Koch, S., Tang, S.C., Holm, R.H., Frankel, R.B. and Ibers, U.A.: 1975, *J. Am. Chem. Soc.* 97, p. 916.
9. Collman, J.P. and Sorell, T.N.: 1975, *J. Am. Chem. Soc.* 97, p. 4133.
10. Peisach, J. and Blumberg, W.E.: 1970, *Proc. Nat. Acad. Sci.* 67, p. 172.
11. Tsai, R., Yu, C.A., Gunsalus, I.C., Peisach, J., Blumberg, W., Orme-Johnson, W.H. and Beinart, H.: 1970, *Proc. Nat. Acad. Sci.* 66, p. 1157.
12. Hanson, L.K., Eaton, W.A., Sligar, S.G., Gunsalus, I.C., Gouterman, M. and Connell, C.R.: 1976, *J. Am. Chem. Soc.* 98, p. 2672.
13. Hanson, L.K., Sligar, S.G. and Gunsalus, I.C.: 1977, *Croatica Chem. Acta* 49 (2), p. 237.
14. Ullrich, V., Ruf, H.H. and Wende, P.: 1977, *Croatica Chem. Acta* 49 (2), p. 213.
15. Peterson, J.A., Ishimura, Y. and Griffin, B.W.: 1972, *Arch. Biochem. Biophys.* 149, p. 197.
16. Dolphin, D., private communication.
17. Chang, C.K. and Dolphin, D.: 1976, *J. Am. Chem. Soc.* 98, p. 1607.
18. Reference 9.
19. Chang, C.K. and Dolphin, D.: 1976, *Proc. Nat. Acad. Sci.* 73 (10), p. 3338.
20. Stern, J.O. and Peisach, J.: 1974, *J. Biol. Chem.* 249 (23), p. 7495.
21. Groves, J.T., McClusky, G.A., White, R.E. and Coon, M.J.: 1978, *Biochem. Biophys. Res. Commun.* 81, p. 154.
22. Groves, J.T., Nemo, T.E. and Myers, R.S.: 1979, *J. Am. Chem. Soc.* 101, p. 1032.

23. DePierre, J.W. and Ernster, L.: 1978, *Biochim. Biophys. Acta* 473, p. 149.
24. Jerina, D.M. and Daly, J.W.: 1974, *Science* 185, p. 573.
25. Borgen, A., Darvey, H., Castagnoli, N., Crocker, T.T., Rasmussen, R.E. and Wang, I.Y.: 1973, *J. Med. Chem.* 16, p. 502.
26. Huberman, E., Mager, R. and Sachs, L.: 1976, *Nature* 264, p. 360.
27. Miller, J.A. and Miller, E.C.: 1969, In "Jerusalem Symposia on Quantum Chemistry and Biochemistry", Bergman, E. and Pullman, B., eds., D. Reidel Pub. Dordrecht-Holland, 1, pp. 237-257.
28. Miller, J.A.: 1970, *Cancer Res.* 30, p. 559.
29. Daly, J.: 1971, in "Handbuch der Experimentellen Pharmakologie" XXVII/2. Brodie, B.B. and Gillette, J.R., eds., Springer Verlag. Berlin, Heidelberg, New York, p. 285.
30. Sharrock, M., Debrunner, P.G., Schulz, C., Lipscomb, J.D., Marshall, V. and Gunsalus, I.C.: 1976, *Biochim. Biophys. Acta* 420, p. 8.
31. Churg, A.K. and Makinen, M.W.: 1978, *J. Chem. Phys.* 68 (4), p. 1913.
32. Little, R.G., Dymock, K.R. and Ibers, J.A.: 1975, *J. Am. Chem. Soc.* 97, p. 4532.
33. Tang, S.C., Koch, S., Papaefthymiou, G.C., Foner, S., Frankel, R.B., Ibers, J.A. and Holm, R.H.: 1976, *J. Am. Chem. Soc.* 98, p. 2414.
34. Collman, J.P., Sorrell, T.N., Hodgson, K.O., Kulshrestha, A.K. and Strouse, C.E.: 1977, *J. Am. Chem. Soc.* 99, p. 5180.
35. Averill, B.A., Herskovitz, T., Holm, R.H. and Ibers, J.A.: 1973, *J. Am. Chem. Soc.* 95, p. 3523.
36. Cramer, S.P., Dawson, J.H., Hodgson, K.O. and Hager, L.P.: 1978, *J. Am. Chem. Soc.* 100, p. 7282.
37. Kuchitsu, K., Fukuyama, T. and Morino, Y.: 1968, *J. Mol. Structure* 1, p. 463.
38. Collman, J.P., Brauman, J.E., Halbert, T.R. and Suslick, K.S.: 1976, *Proc. Nat. Acad. Sci.* 73, p. 3333.
39. Ridley, J. and Zerner, M.: 1973, *Theor. Chim. Acta* 32, p. 111.
40. Ridley, J.E. and Zerner, M.C.: 1976, *Theor. Chim. Acta* 42, p. 223.
41. Bacon, A.D.: 1976, Ph.D. thesis, University of Guelph, Canada.
42. Bacon, A.D. and Zerner, M.C.: 1979, *Theor. Chim. Acta* 53, p. 21.
43. Zerner, M.C., Loew, G.H., Kirchner, R.F. and Mueller-Westerhoff, U.T.: 1980, *J. Am. Chem. Soc.* 102, p. 589.
44. Herman, Z.S. and Loew, G.H.: 1980, *J. Am. Chem. Soc.* 102, p. 1815.
45. Rohmer, M.M. and Loew, G.H.: 1979, *Int. J. Quant. Chen, QBS* 6, p. 32.
46. Weiss, K., unpublished

47. Mataga, N. and Nishimoto, K.: 1957, *Z. Physik. Chem. (Frankfurt am Main)* 13, p. 140.
48. Dewar, M.J.S. and Thiel, W.: 1977, *J. Am. Chem. Soc.* 99, p. 4899.
49. Flanigan, M.C., Komornicki, A. and McIver, J.W., Jr.: 1977, In "Modern Theoretical Chemistry", Segal, G.A. ed., Plenum Press, New York, 8, p. 1.
50. Murrell, J.N. and Laidler, K.J.: 1968, *Trans. Faraday Soc.* 64, p. 371.
51. McIver, J.W., Jr. and Komornicki, A.: 1972, *J. Am. Chem. Soc.* 94, p. 2625.
52. Herman, Z.S., Loew, G.H. and Rohmer, M.M.: 1980, *Int. J. Quant. Chem. QBS* 7, p. 137.
53. Dunford, H.B. and Stillman, J.S.: 1976, *Coord. Chem. Rev.* 19, p. 187.
54. Theorell, H. and Ehrenberg, A.: 1952, *Arch. Biochem. Biophys.* 41, p. 442.
55. Schulz, C.E., Devaney, P.W., Winkler, H., Debrunner, P.G., Doan, N., Chiang, R., Rutter, R. and Hager, L.P.: 1979, *FEBS Lett.* 103, p. 102.
56. Maeda, Y. and Morita, Y.: 1967, *Biochem. Biophys. Res. Commun.* 29, p. 680.
57. Moss, T.H., Ehrenberg, A. and Bearden, A.J.: 1969, *Biochemistry* 8, p. 4159.
58. Harami, T., Maeda, Y., Morita, Y., Trautwein, A. and Gonser, U.: 1977, *J. Chem. Phys.* 67 (3), p. 1164.
59. Dolphin, D., Forman, A., Borg, D.C., Fajer, J. and Felton, R.H.: 1971, *Proc. Nat. Acad. Sci.* 68, p. 614.
60. Morishima, I. and Ogawa, S.: 1978, *J. Am. Chem. Soc.* 100, p. 7125; 1978. *Biochem. Biophys. Res. Commun.* 83, p. 946; 1978. *Biochemistry* 17, p. 4384.
61. Loew, G.H. and Herman, Z.S.: 1980, *J. Am. Chem. Soc.* 102, p. 6173.
62. DiNello, R.K. and Dolphin, D.: 1979, *Biochem. Biophys. Res. Commun.* 86, p. 190.
63. La Mar, G.N. and De Ropp, J.S.: 1980, *J. Am. Chem. Soc.* 102, p. 395.
64. Pudzianowski, A.T. and Loew, G.H.: 1980, *J. Am. Chem. Soc.* 102, p. 5443.
65. Pudzianowski, A.T. and Loew, G.H. Submitted for publication in *J. Am. Chem. Soc.*, 1981.
66. Pack, G.R. and Loew, G.H.: 1979, *Int. J. Quant. Chem. QBS* 6, p. 381.
67. Hanzlik, R.P. and Shearer, G.O.: 1978, *Biochem. Pharmacol.* 27, p. 1441.
68. Ortiz de Montellano, P.R., Kunze, K.L. and Mico, B.A.: 1980, *Mol. Pharmacol.* 18, p. 602.

HEMES OF HYDROPORPHYRINS

C. K. CHANG

Department of Chemistry
Michigan State University
East Lansing, Michigan 48824

ABSTRACT Iron chlorins and isobacteriochlorins have been synthesized and studied to provide information on the functional roles of heme d and siroheme in enzymes which mediate multielectron reductions of nitrite, sulfite, and oxygen. Coordination complexes of ligands such as CO, NO, dioxygen, and nitrite were characterized by physicochemical means. It has been found that although the coordination behavior of heme d and siroheme is not significantly different from that of iron porphyrin, their redox properties are very different from normal hemes. In particular, the facile ring cation radical formation observed with siroheme analogues suggests that redox reactions involving the isobacteriochlorin ring may participate in the biological function of siroheme.

INTRODUCTION

Heme-like prosthetic groups derived from hydroporphyrins exist in a variety of redox enzymes. The presence of a green cytochrome in several microorganisms was discovered more than fifty years ago (1). It was originally designated cytochrome $\underline{a}_2$ by Keilin (2). Barrett first isolated a green heme from Aerobacter aerogenes and a number of other bacteria and showed that this so-called heme $\underline{a}_2$ is an iron complex of a chlorin structually related to protoporphyrin—which is completely different from the common a-type heme present in cytochrome oxidase (3). Iron chlorins are now classified as heme d. The main function of heme d cytochromes in many bacteria appears to be that of catalyzing the reduction of either nitrite or molecular oxygen. For example, the well-studied pseudomonas cytochrome oxidase from Pseudomonas aeruginosa contains two heme c and two heme d moieties and catalyzes the electron transfer from cytochrome $\underline{c}_{551}$ or azurin to dioxygen. The same enzyme also reduces nitrite when oxygen is not present; in such case, nitrous oxide and dinitrogen are produced (4-6). Other

H. B. Dunford et al. (eds.), The Biological Chemistry of Iron, 313–334.

cd-type nitrite reductases have been found in *Paracoccus denitrificans* (7), *Thiobacillus denitrificans* (8), *Alicaligenes faecalis* (9), and *Pseudomonas perfectomarinus* (10).

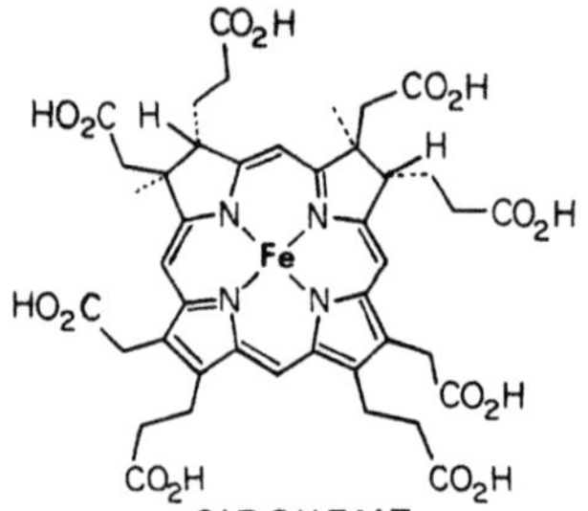

SIROHEME

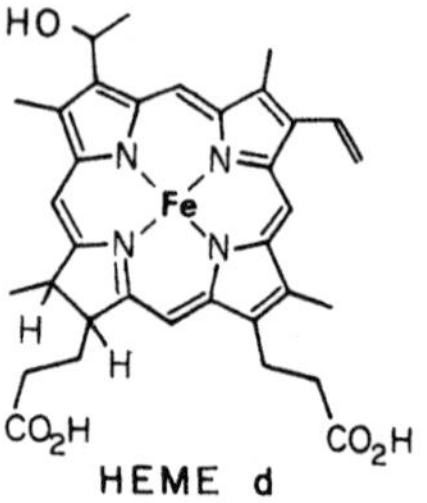

HEME d

(tentative structure)

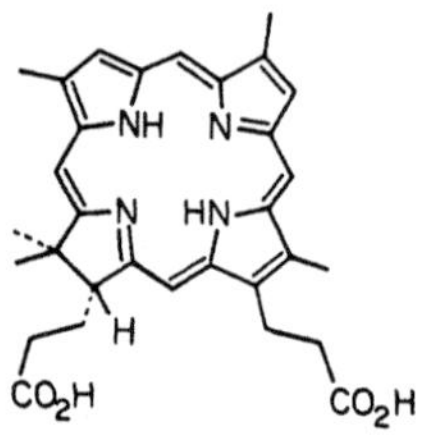

BONELLIN

In contrast to the small number of denitrifying bacteria, the assimilatory mechanism of reducing nitrite to ammonia is much widely distributed in plants and algae. This conversion is catalyzed by the enzyme ferredoxin:nitrite oxidoreductase which has recently been shown to contain an iron isobacteriochlorin, i.e., porphyrin in which two adjacent pyrrole rings are reduced (11-13).This heme is named *siroheme* since the same prosthetic group is also found in sulfite reductase which catalyzes the six-electron reduction of sulfite to sulfide (14-18). A substantial evidence now available indicates that in both sulfite and nitrite reductases siroheme is the substrate binding and activating site. In green plants, the reduction of nitrite is light driven with photosynthetically reduced pyridine dinucleotide serving as electron donor to the active enzyme :

Chlorophyll → NADPH → flavoprotein → siroheme → nitrite

The demetallated siroheme, called sirohydrochlorin, has been shown to be an intermediate in the vitamin B_{12} biosynthesis (19-21). The structure of sirohydrochlorin is unusual in that the pyrrole ring reduction is caused by methane addition rather than by hydrogenation of double bond. This averts the facile dehydrogenation reaction common in hydroporphyrins (22). Recently, a green catalase isolated from the nitrite reducing fungus *Neurospora crassa* has also been shown to contain an iron chlorin which, unlike heme *d*, cannot be dehydrogenated (23). Although presently there is no evidence confirming the presence of alkyl groups in this iron chlorin, another naturally occurring chlorin, bonellin, does contain a methyl group on the reduced pyrrole ring. (Bonellin plays a fascinating role in differentiating sex in the echurian worm *Bonellia viridis* (24-25).)

Considering the seemingly obligatory role of heme *d* and siroheme in enzymes responsible for multielectron redox reactions involving nitrite and sulfite, one is led to wonder whether there is a chemical or functional reason for their presence in the reductases. Are they better suited to serve as a prosthetic group in these redox reactions than the ubiquitous porphyrin hemes? To answer such questions detailed information regarding the redox and coordination properties of iron hydroporphyrins is needed. Beginning in

1979, we as well as others have examined iron chlorins and isobacteriochlorins in order to identify the special competence unique of these hemes in mediating the multielectron redox reactions. In this paper current knowledge on the chemistry of these hemes is reviewed.

SYNTHETIC MODELS

Dihydro- and tetrahydroporphyrins can be prepared from porphyrins by various reduction methods. Eisner (26) first showed that the sodium amylate procedure originally used by Fischer for preparing chlorins (27) can also provide isobacteriochlorin. Recently, Stoltzenberg et al. (28) have improved this method and obtained a 75% yield of octaethylisobacteriochlorin (OEiBC) from iron octaethylporphyrin (FeOEP). Oester and Whitlock (29) demonstrated that the diimide reduction of zinc tetraphenylporphyrin (ZnTPP) gives predominantly the isobacteriochlorin (ZnTPiBC) while the same reduction of metal free TPP leads to the bacteriochlorin (TPBC). An X-ray study (30) of ZnTPiBC(pyridine) complex showed that the diimide reduction adds hydrogen cis to the pyrrole rings while NMR studies of free base OEiBC suggest trans addition (31) during the sodium amylate reduction. All of these hydroporphyrins undergo oxidative dehydrogenations in the presence of oxidants.

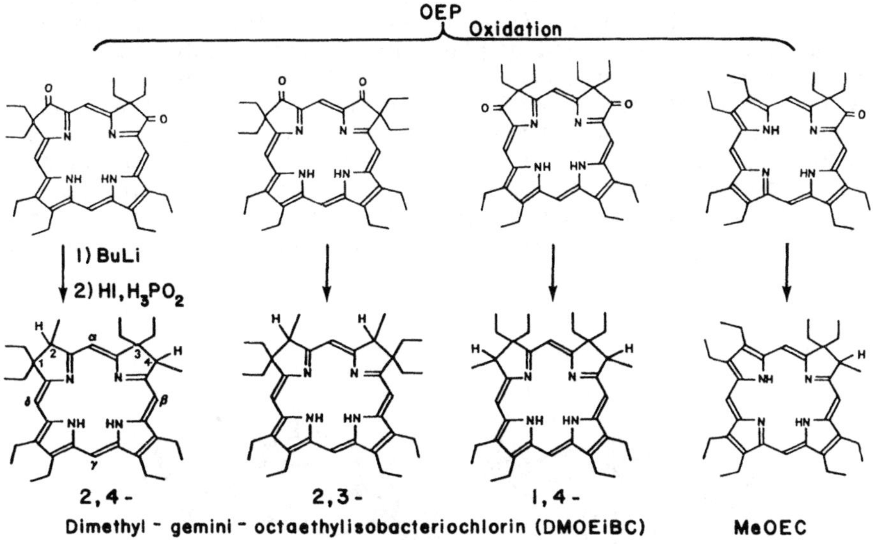

Scheme I

To synthesize true siroheme analogues, we have developed a synthesis by which methylated isobacteriochlorins can be prepared (32). Our synthesis begins with the readily available OEP and accomplishes in 3 steps the three isomeric dimethyl isobacteriochlorins (Scheme 1). This synthesis also produced an alkylated chlorin, like bonellin. Our method can further be adapted to prepare functional side chains, for example, an imidazole "tailed" chlorin can be synthesized according to Scheme 2 (33). Total synthesis of methylated iBC's and chlorins have also appeared recently (34-36) but it is questionable that these rather laborious syntheses could become a practical route to siroheme analogues.

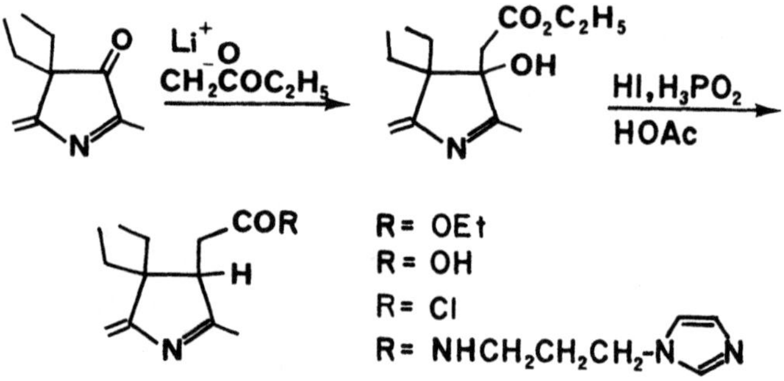

Scheme 2

The synthetic DMOEiBC's exhibit visible absorption peaks and key NMR chemical shifts nearly identical to those of sirohydrochlorin. The X-ray structure of the synthetic 2,3-DMOEiBC shows that the two reduced pyrrole rings are slightly twisted out-of-plane (Fig. 1), presumably because of the steric crowdedness of the substituents (37). An interesting feature of the free base isobacteriochlorin is that it has several distinct valence tautomeric structures. Although only one form gives an uninterrupted 18 pi-electron loop (shown in Scheme 1), X-ray density map has revealed that the two inner protons are equally distributed among the 4 pyrrole nitrogens (37). Variable temperature studies of spectral and luminescence properties of DMOEiBC in solution further revealed a dynamic equilibrium of these forms (32).

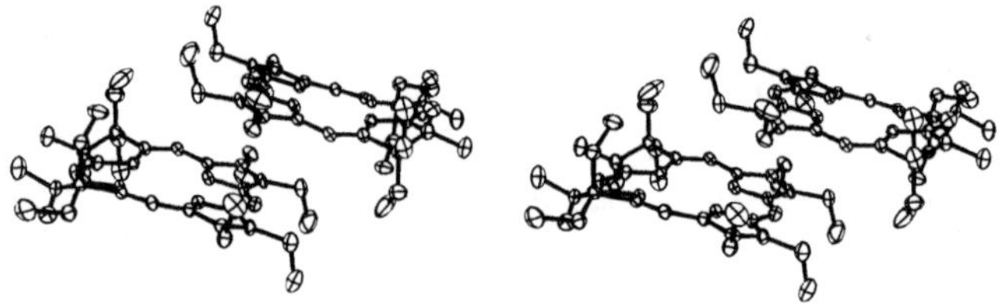

Figure 1. Stereoplot of the configuration of 2,3-DMOEiBC.

Metal insertions to chlorins and iBC have been accomplished using standard procedures, albeit the iBC pyrrole protons are less acidic thus more rigorous reaction conditions may be required. Metallochlorins and iBC, particularly when in solution, are not stable. Care should be exercised to avoid prolonged exposure to light and oxygen since even the alkylated hydroporphyrins are susceptible to nucleophilic attack on the photooxidized cation radical ring. These redox reactions will be discussed later. Here first, we would like to compare the coordination properties of iron porphyrin, chlorin, and iBC.

COORDINATION COMPLEXES OF IRON HYDROPORPHYRINS

Iron chlorin and iBC undergo almost all the coordination reactions established by iron porphyrins. Typical reactions of ferrous and ferric hemes are summarized in the following scheme:

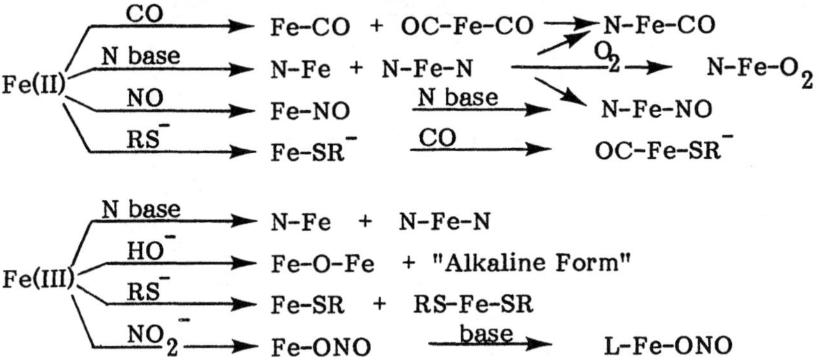

To provide a comparison of the affinities of hemes towards ligands, we have undertaken a detailed study of the binding of carbon monoxide, imidazoles, nitric oxide, nitrite ion, as well as oxygen to the three heme systems (33). It can be shown that although there is no marked difference in the type of reactions the 3 hemes can go, subtle differences in their equilibrium constants and kinetic rates do exist.

<u>CO - Binding</u> The coordination of CO to iron porphyrins has been extensively studied. Depending on the state of axial ligancy of the heme iron, two general reaction courses can take place. In non-coordinating solvents a "4-coordinate" bare heme binds successively two CO molecules while in the presence of either weak or strong ligands, only one CO can be bound (38-40). Ferrous chlorins and iBC bind CO in the same manner; the absorption spectra of the corresponding complexes are shown in Fig. 3 and 4. Equilibrium constants and the applicable kinetic rates are summarized in Table 1. The structures of "5-coordinate" chelated chlorin and iBC are shown in Figure 2.

As seen in Table 1, the equilibrium constants, indicated by the half-saturation pressures ($P_{\frac{1}{2}}$), of the bare hemes clearly indicate that the CO affinities increase in the direction: porphyrin < chlorin < iBC. The rate constants for the addition of the first CO to the heme only increase marginally from porphyrin to iBC, therefore the slower CO dissociation rate (k) seems to be responsible for the enhanced CO affinity in iBC. In the chelated heme series, the variations in equilibrium constants as well as kinetic rates are relatively smaller, even so, the iBC heme binds CO about twice as good compared to porphyrin heme.

TABLE 1. CO Binding Constants and Kinetic Rates

4-coordinate	$Fe \underset{k}{\overset{k'}{\rightleftharpoons}} Fe\text{-}CO \rightleftharpoons OC\text{-}Fe\text{-}CO$			
	$P_{\frac{1}{2}}(CO)$	$P_{\frac{1}{2}}(CO)_2$	$k'(M^{-1}s^{-1})$	$k(s^{-1})$
Etio heme	2.7torr	550torr	4.9×10^{8}	1.3×10^{4}
FeMeOEC	1.6	450	7.6×10^{8}	1.1×10^{4}
FeDMOEiBC	0.3	80	12.5×10^{8}	3.8×10^{3}

Titrations and kinetic rates were measured in dry toluene.
1 torr of CO = 1×10^{-5}M, 20-23°C.

5-coordinate	$Im\text{-}Fe \underset{l}{\overset{l'}{\rightleftharpoons}} Im\text{-}Fe\text{-}CO$		
	$P_{\frac{1}{2}}$	$l'(M^{-1}s^{-1})$	$l(s^{-1})$
Chelated Mesoheme	0.0013torr	1.1×10^{7}	0.019
Im-Chl	0.00093	2.0×10^{7}	0.025
Im-iBC	0.00065	3×10^{7}	0.03

Measured in 2% MTAB, 1 torr of CO = 1.35×10^{-6}M.

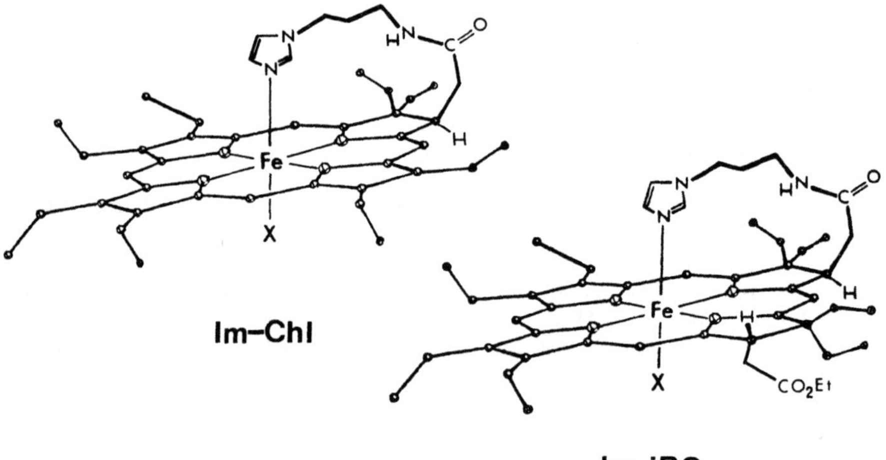

Im–Chl

Im–iBC

Figure 2. Structures of Chelated chlorin and isobacteriochlorin.

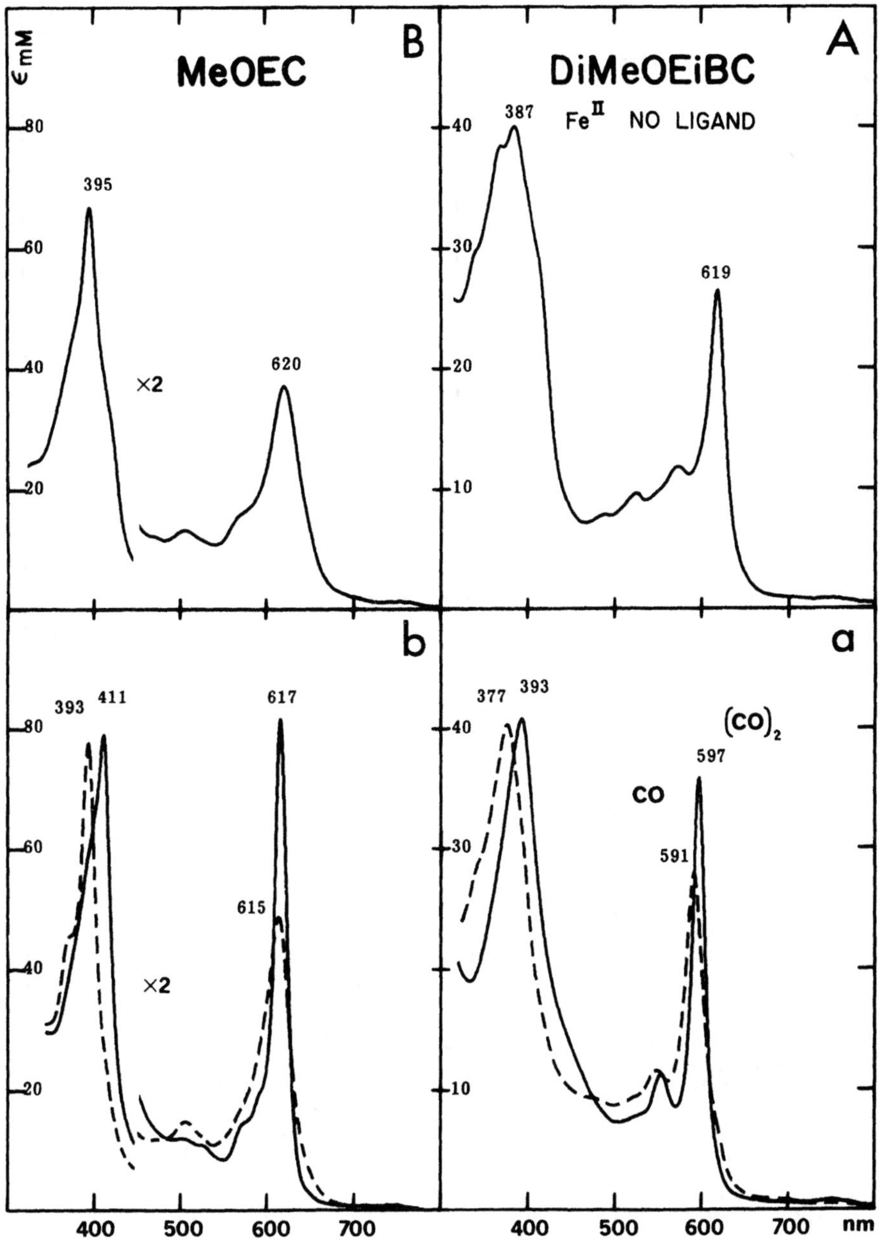

Figure 3. Visible absorption spectra of uncoordinate ferrous heme of DMOEiBC (A) and MeOEC (B) in toluene; their mono-CO adducts (a,b - - -); and bis-CO complexes (a,b ——). The bare hemes were generated using a benzophenone-assisted photoreduction (33).

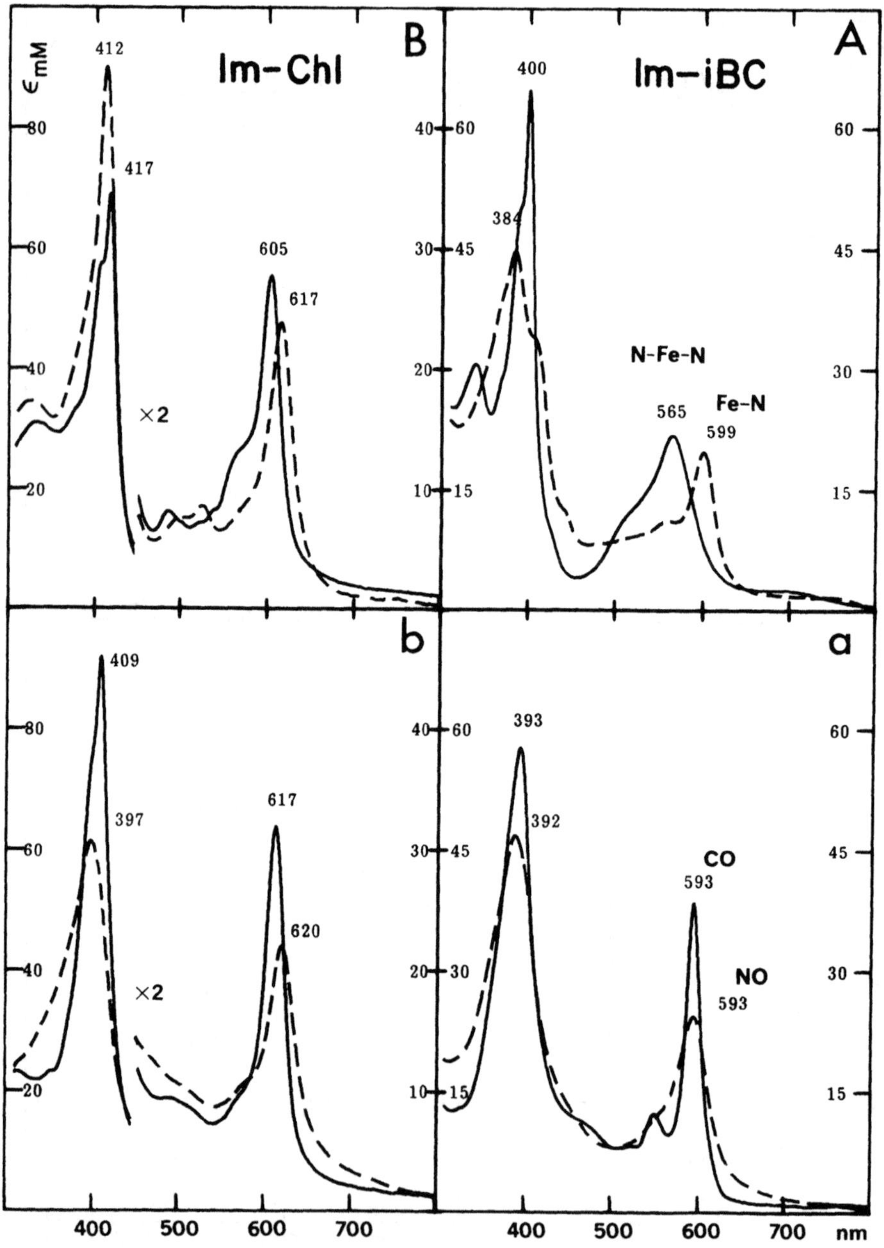

Figure 4. Visible absorption spectra of 5-coordinate chelated ferrous chlorin and iBC (structures shown in Fig. 2; A,B - - -); 6-coordinate hemochromes by addition of N-methylimidazole (A,B ——); Im-Fe-CO (a,b ——); and Im-Fe-NO complexes (a,b - - -), in toluene.

The variation in CO affinity observed here can be interpreted in terms of the cis-Effect (41). We have demonstrated in related studies using substituted porphyrins that electron withdrawing groups on the ring tends to increase $P_{\frac{1}{2}}$ by increasing the dissociation rate (42). This indicates that decreasing electron density in the iron d orbitals would lessen the Fe-CO π-backbonding and effectively weaken the Fe to CO bonding. This cis-electronic effect is most pronounced in the bare hemes where there are no other axial ligands to perturb the heme electron density and level off the ring contribution. Based on these observations, we conclude that the electron donating ability of the iron among the 3 heme systems follows the order: iBC>chlorin>porphyrin. The same trend can be seen also in oxygen binding.

O_2 - Binding Because of the unstability of oxyheme complex, oxygen binding to unprotected hemes is only observable by using low temperature

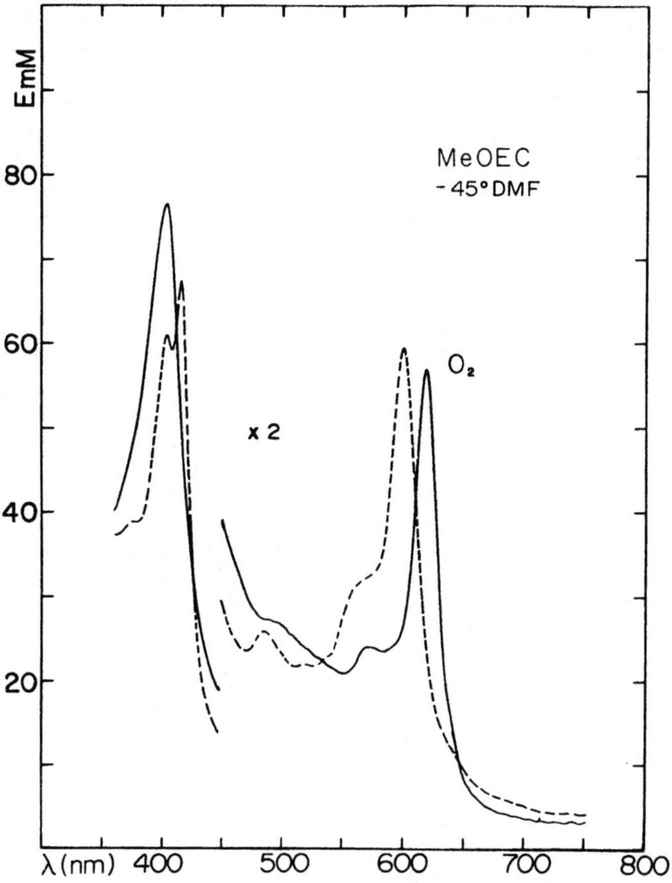

Figure 5. Absorption spectra of Fe(II)MeOEC in DMF with excess N-Methylimidazole at -45° C (- - -); after exposure to O_2 (——). The oxyheme has a half-life of 10 min.

techniques or by incorporating the heme into myoglobin or hemoglobin. Detailed thermodynamic and kinetic studies of chelated hemes as well as reconstituted hemoglobins are well-documented (39,43,44). In comparison with porphyrin hemes, the largest difference of chlorin and iBC hemes is not their rate or affinity for oxygen, but is the extremely fast autoxidation rate. With iron deuterochlorin - reconstituted equine myoglobin, for example, the oxy complex has a formation constant and kinetic on and off rates almost identical to those of deuteroheme - reconstituted Mb but the chlorin - Mb turned into the met form within 30 min at room temperature (42). In contrast, deuteroheme-Mb showed no detectable oxidation even after several hours under identical conditions. Similar autoxidation has also been observed with an imidazole hemochrome at low temperature (fig. 5). With iBC, the oxidation is so rapid that we have not been able to measure the oxygen binding parameters even at $-45^{o}C$. The oxidation rate is related to the electron donor power of the heme iron. Both theoretical calculations and experimental evidence have suggested that the $Fe-O_2$ bonding has an extensive charge transfer from iron to oxygen. Furthermore, oxidation of the oxyheme by releasing the superoxide ion has been demonstrated in several occassions (45). Whether the oxy form of chlorin and iBC - Mb is oxidized by superoxide dissociation is being investigated. The fact that electron deficient heme is more resistent to autoxidation (46) provides further support that autoxidation of ferrous iron is accelerated by electron rich macrocycles. In this context, FeiBC has the most electron donating ability among the 3 hemes.

Imidazole - Binding Ferrous hemes bind two pyridine or two imidazole ligands to form hexacoordinate hemochromes (47). The effect of ring saturation on the two formation constants is seen in Table 2. With porphyrin hemes, K_2 is much larger than K_1, therefore the 5-coordinate heme can not accumulate in solution. As the porphyrin ring becomes saturated, K_1 becomes larger whereas K_2 is diminished. Thus, it is possible to prepare a pentacoordinate iBC complex by having the correct amount of base in solution. The absorption spectra of both 5-coordinate and 6-coordinate complexes are shown in Figure 4. The decrease in K_2 is again believed to be related to the π-donor ability of the heme iron, which generates repulsion with the π-electron rich imidazole (Fig. 6). Similar arguments can be used to account for the failure of thiolate anion to form 6-coordinate complex with any of the hemes (48,49).

TABLE 2. Binding Constants for Imidazole Hemes

$$Fe^{II} \underset{}{\overset{K_1}{\rightleftharpoons}} Fe\text{-}Im \underset{}{\overset{K_2}{\rightleftharpoons}} Im\text{-}Fe\text{-}Im$$

	$K_1(M^{-1})$	$K_2(M^{-1})$
Deuteroheme*	$5x10^3$	$7x10^4$
FeMeOEC	$2.5x10^4$	$6x10^4$
FeDMOEiBC	$3.5x10^4$	$2x10^3$

Titrations were measured in toluene using N-methylimidazole.
* see White et al., J. Am. Chem. Soc. **101**, 2443 (1979).

Figure 6. Overlapping of π-orbitals in bis-imidazole hemochromes.

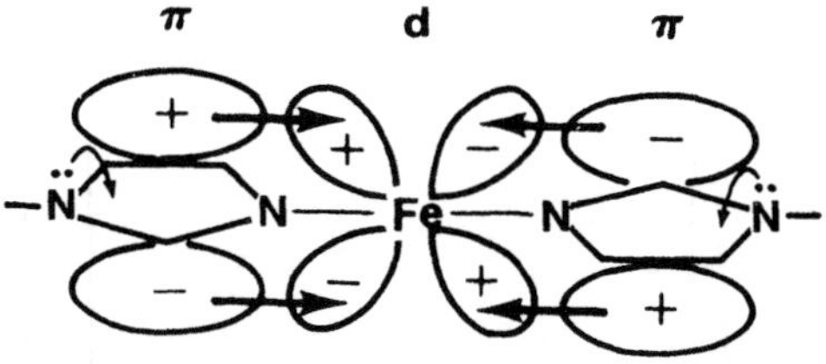

It should be pointed out that the charge density distributions among iron, axial ligands and ring do not necessarily correlate with the polarographic redox potential of the complex—which is basically a measure of the highest occupied orbital energy level. Even though the iron of iBC heme tends to transfer more electron density to the bound ligand, this is not always reflected by the redox potentials of the complex per se. More discussion on redox properties will be given later.

Nitrite - Binding One of the most important coordination reactions of heme d and siroheme would certainly be the nitrite binding to ferric iron. Nitrite complexes of Fe(III)porphyrins have not been studied previously. We have recently succeeded in preparing such compounds by reacting hemin chloride with either silver nitrite or tetra-n-butylammonium nitrite in suitable solvents (42). The high spin-hemin nitrite ligates easily with almost any donor ligands. Formation constants of the nitrite complex in THF are given in Table 3 and spectra are shown in Figure 7.

TABLE 3. Formation Constants of Hemin Nitrites

$$\mathrm{Fe(III)Cl} + \mathrm{Bu_4NNO_2} \xrightarrow[20^\circ C]{THF} \mathrm{(Solvent)Fe(III)(NO_2)}$$

Etio hemin	$K(M^{-1})$ = 790
FeMeOEC	990
FeDMOEiBC	2200

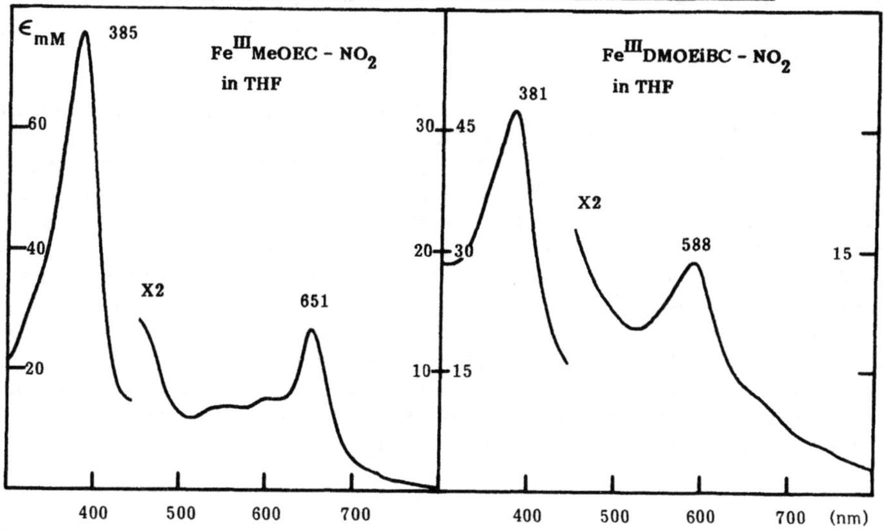

Figure 7. Absorption spectra of ferric nitrite complex of chlorin and iBC.

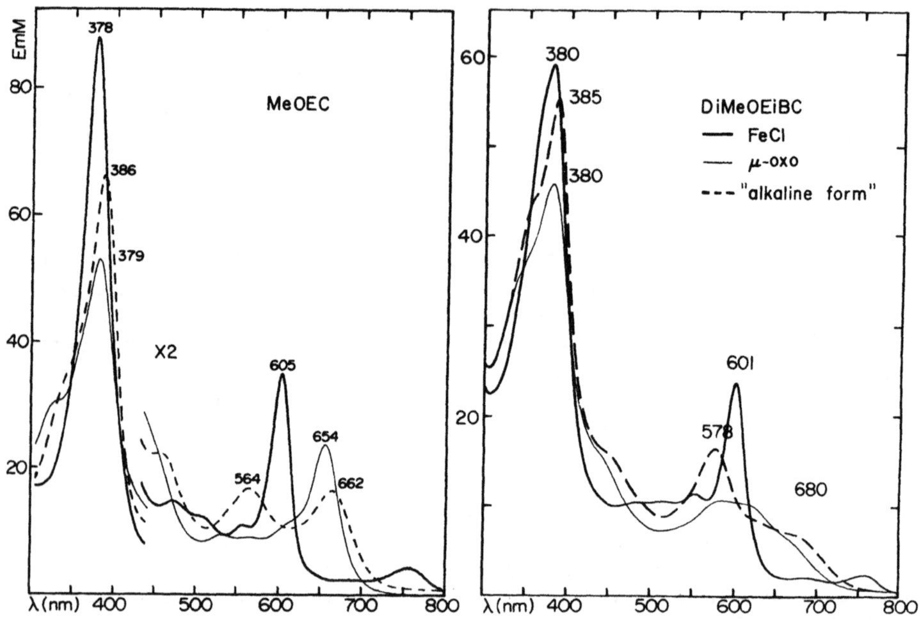

Figure 8. The three ferric species of chlorin and iBC. Absorption spectra were obtained in methylene chloride. The alkaline form was prepared by addition of one drop of dilute $(n\text{-}Bu)_4NOH$.

Hemin nitrites prepared from highly purified reagents are very stable in solution. EPR spectra showed that they have a principal g-value around 5 but there is always a small amount of Fe(II)NO present at the g 2 region (less than 10% of total heme). Demonstration of the nitrite reduction in organic solvents has not been very successful. At present the structure of the hemin nitrite complex has not been fully characterized, although preliminary IR studies suggest a Fe-ONO linckage.

The Alkaline Form During the preparation of iron chlorin and iBC, we have noticed that if a benzene solution of hemin chloride is washed with dilute (5%) NaOH solution the common μ-oxo dimer is formed. However, if concentrated base solution (20% or higher) is used, a different species is produced (Fig. 8). This "alkaline form" is also the dominant form of Fe(III)chlorins and siroheme in aqueous solution or in micelle suspension at high pH. Several curious properties are observed with this species. They exhibit a very broad EPR signal, spanned over 1000 G but centered at g 2. Both magnetic susceptibility and NMR spectra showed that they are not the s=5/2, 1/2, nor the strong exchange-coupled μ-oxo form. Attempts to isolate this species in solid form so far only yielded mainly the μ-oxo dimer. We speculate that this alkaline form is likely some kind of aggregated form of hydroxy hemin, possibly a bis-hydroxy dimer.

REDOX PROPERTIES

The polarographic redox potentials of a large number of free base chlorins, iBC, and their metal derivatives have been examined (50-52,28) by cyclic voltammetry. The prominent features of hydroporphyrins, as compared with porphyrins (Table 4) are that the redox potentials of hydroporphyrins are invariably more negative, i.e., they are easier to oxidize but harder to reduce. The redox span for formation of the cation and anion radicals nevertheless remains relatively constant throughout the porphyrin-chlorin-iBC series. Since such redox span, to a first approximation, is related to the energy differences between the HOMOs and LUMOs, this suggests that saturation of the pyrrole ring would have the same effect as substituting electron-rich metal ions to porphyrin, i.e., they do not alter the energy gap between HOMO and LUMO.

As a result of the extreme ease in oxidizing the iBC ring, it becomes possible to generate Fe(II)-iBC-π-cations. This has been demonstrated with the carbonyl complexes (50). Cyclic voltammetry shows that the pyridine-carbonyl complex undergoes a reversible one-electron oxidation at 0.13 v (vs SCE), whereas oxidations of the corresponding porphyrin or chlorin complexes are irreversible. The ring oxidized species exhibits a CO stretching frequency at 2010 cm^{-1} (to be compared with the ν_{CO} of 1970 cm^{-1} in the parent compond), indicating that the axial CO has not been lost on oxidation. The free radical character of the oxidized product is evident from its EPR spectrum, which consists of a three-line pattern centered at g=2.003 (Fig. 9). Such spectra can be readily assigned to a π-cation radical of the iBC ring on the basis of molecular calculations which show that there are more unpaired spin densities on the two protons at the 1,4 positions than there are at the 2,3 positions (52).

TABLE 4. Redox Potentials ($E_{\frac{1}{2}}$ volt) of Heme Derivatives (vs SCE)

		$Ring^{+\cdot}$/Ring	Fe^{III}/Fe^{II}	$Ring/Ring^{-\cdot}$
Porphyrin (OEP)		0.83		-1.45
FeCl	0.94*		-0.5	
$Fe(Py)_2$			-0.02	
Fe(Py)(CO)			0.52 (irreversible)	
Etiochlorin		0.63		-1.45
FeCl	0.68*		-0.4	
$Fe(Py)_2$			-0.04	
Fe(Py)(CO)			0.35 (irreversible)	
DMOEiBC		0.37		-1.72
FeCl	0.44*		-0.4	
$Fe(Py)_2$			-0.04	
Fe(Py)(CO)		0.13		
Fe(Im)(CO)		0.05		

*Presumably ring oxidation to generate Fe(III) π-cation radical.

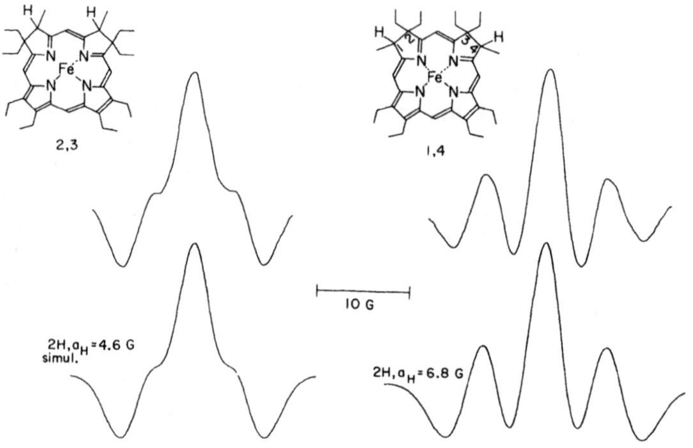

Figure 9. Experimental (upper) and simulated (lower) second derivative EPR spectra of two isomeric OC-Fe(II)-Py(DMOEiBC) π-cation radicals in CH_2Cl_2 at 23°C.

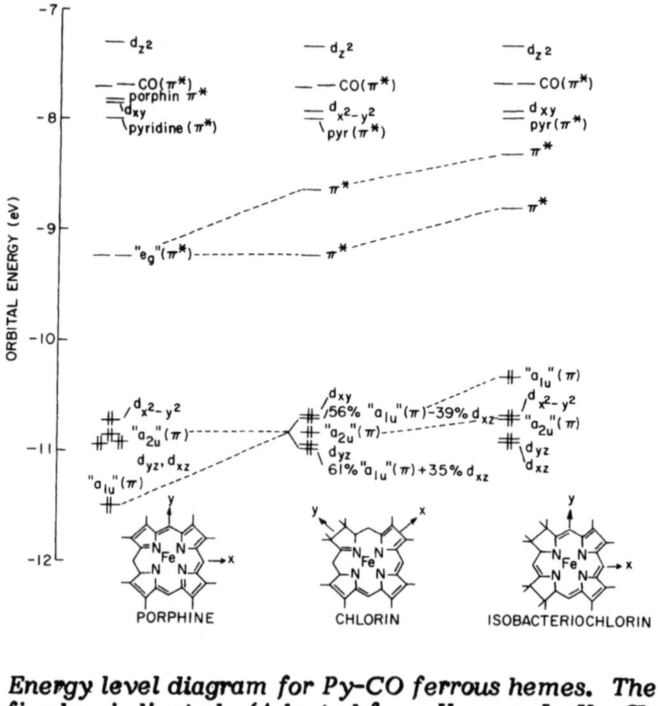

Figure 10. Energy level diagram for Py-CO ferrous hemes. The x and y axes are defined as indicated. (Adapted from Hanson, L. K., Chang, C. K., Davis, M. S., & Fajer, J. in INTERACTIONS BETWEEN IRON AND PROTEINS IN OXYGEN AND ELECTRON TRANSPORT, C. Ho, Editor, Elsevier, New York, in press.)

Extended Huckel MO calculations have also offered an explanation for both the redox properties of the complexes and the site of oxidation. Energy level diagram (Fig. 10) for the HOMOs and LUMOs of porphine, chlorin, iBC complexes shows that as the porphyrin skeleton is saturated, the a_{1u} orbitals are progressively raised (52). Whether an electron is abstracted from the metal or the ring depends on the relative energies of the d t_{2g} and the highest filled ring π orbitals. As seen in Figure 10, it is in the iBC, not the pophine or chlorin, that the π orbital surpasses the metal d t_{2g} orbital in their energy level and becomes the preferred site of oxidation. This energy diagram is further supported by the observed oxidation potentials of 0.52, 0.35 (both irreversible due to loss of CO during metal oxidation), and 0.13 v for porphin, chlorin, and iBC, respectively. This parallels the relative HOMO energy level of the 3 complexes.

The unique redox properties of iron iBC raised the possibility that the macrocyclic ring, in addition to the iron, is involved during the enzymatic cycles. A crucial point here is that the macrocycle system determines the π orbital energies whereas the axial ligands influence the d orbital energies. Whether an iBC heme complex would undergo ring oxidation or metal oxidation ultimately depends on the nature of the axial ligands. Although the presence of cation radical intermediates in the enzyme has not been confirmed and oxidation of the CO complxes of the siroheme-containing nitrite and sulfite reductases is reported to result in oxidation of the iron and loss of CO, one cannot rule out the possible involvement of the cation radical in the enzymes. For one reason, the CO heme π-cation radical is not very stable; once CO dissociates, the heme yields electron to the ring. More importantly, the ring may merely serve as an electron shuttle during the enzymatic cycle without ever accumulating substantial amounts of the radical intermediate. Indeed, the concept of electron transfer through the porphyrin ring in cytochromes is not new (53), but until now there has been no precedence for such electron shuffling between heme iron and ring at the same oxidation level. (Previously, a case of $Ni(II)OEP^{+\cdot} \longrightarrow Ni(III)OEP$ is known (54).) The alkaline form of ferric iBC provided an interesting example of this electron redistribution. Addition of pyridine-HCl salt to this alkaline form in methylene chloride under CO atomsphere resulted in the formation of the π-cation radical. This reaction must have taken place according to the following equation:

$$\left((iBC)Fe^{III}\text{-}OH\right) + Py\text{-}HCl \xrightarrow{CO} OC\text{-}Fe^{II}\text{-}Py(iBC)^{+\cdot} + H_2O + Cl^-$$

If the same experiment is carried out with chlorin, a very slow autoreduction can occur but no radical is observed. This again illustrates that such macrocycle-assisted redox chemistry is a very unique and characteristic property of the iBC hemes.

EPR CHARACTERISTICS

High-spin ferric forms of the three hemes have remarkably similar EPR spectra; the signals at g 6 and g 2 regions have very little variations from porphyrin to iBC. EPR spectra of low-spin ferric complexes, on the other

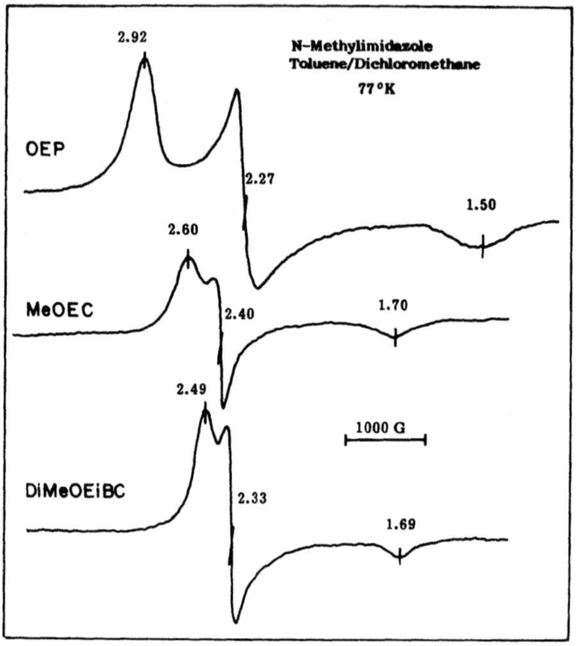

Figure 11. EPR spectra of Fe(III) (N-MeIm)$_2$ complexes at 77°K.

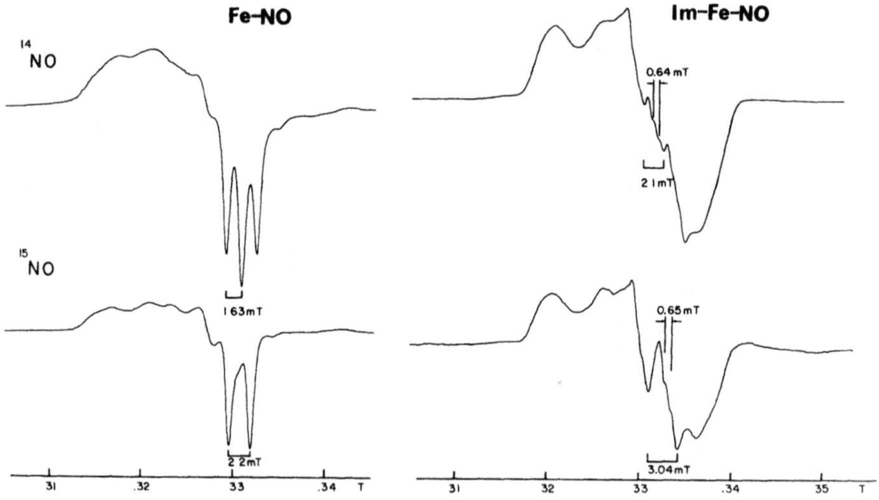

Figure 12. EPR spectra of 5-coordinate and 6-coordinate Fe(II)-NO iBC complexes. The hexacoordinate complexes were obtained by addition of N-MeIm to the nitrosyl heme. Spectra were recorded at 77°K in CH_2Cl_2-toluene, microwave power, 5 mW; frequency, 9.223 GHz.

hand, are quite characteristic for each heme. The spectra of the three 6-coordinate N-MeIm complexes, for example, are shown in Figure 11. The g-values can be related to the crystal field parameters: the tetragonal splitting V/λ and the rhombic splitting Δ/λ . These parameters have been used extensively by Blumberg and Peisach to describe the nature of the ligand field environment of heme iron and could be useful in identifying the native ligand in the enzyme (55-57). Table 5 summarizes the ligand field parameters of the model complexes and a few low-spin forms of the oxidized enzymes (58). From this comparison, it seems to suggest that the siroheme-containing sulfite reductase and nitrite reductase do not have the same axial ligand and that the presence of imidazole base in the spinach nitrite reductase is also unlikely. A cd-type dissimilatory nitrite reductase from T. denitrificans (8), however, do show the possibility of histidine ligation.

EPR spectra of paramagnetic nitrosyl-ferrous heme complexes may also have diagnostic value in identifying the native ligand. Two types of spectra are usually encountered depending whether there is another ligand at the 6th coordination site (59,60). Figure 12 illustrates such spectra: the hyperfine splitting caused by NO and the superhyperfine lines due to the imidazole nitrogen are clearly seen. For all practical purposes, the spectra obtained from ferrous porphyrins are superimposable with those derived from chlorins and iBC. EPR spectra of the NO complex of nitrite reductase have been reported —— but, discrepancies exist. Aparicio et al. (61) and Lancaster et al. (62) have reported NO-nitrite reductase spectra which are very similar to the 6-coordinate nitrosyl model except that the superhyperfine lines are not discernible. On the other hand, Fry et al. (63) have reported a very different spectrum with the hyderfine splittings significantly smaller than those shown in Figure 12. This may reflect a different NO bond in the enzyme (e.g. NO-iron sulfur cluster?) or the presence of an unusual sixth ligand whose influence on the nitrosyl heme spectra, to our knowledge, has not been mimicked by any heme models.

TABLE 5. g-Values and Ligand Field Parameters for Low-spin Ferric Heme

	g_x	g_y	g_z	V/λ	Δ/λ	V/Δ
FeOEP-Cl + MeIm	1.50	2.27	2.92	1.90	3.12	0.61
				(2.17	2.98	0.73)
FeMeOEC + MeIm	1.70	2.40	2.60	3.01	2.57	1.17
				(1.06	3.54	0.30)
FeDMOEiBC + MeIm	1.69	2.33	2.49	3.27	2.63	1.24
				(0.99	3.77	0.26)
Nitrite Red. + CN^-		2.35	2.71		(4.0	0.88)
Sulfite Red. + CN^-		2.31	2.39		(4.2	0.16)
T. denitrificans	1.73	2.45	2.52	3.45	2.12	1.62
				(0.40	3.65	0.11)

Spectra of models were measured in toluene glass, 77°K. Ligand field parameters were calculated according to Taylor (57). Values in parenthesis were obtained using the axis system of Blumberg and Peisach (56).

POSSIBLE MECHANISMS FOR MULTIELECTRON REDUCTIONS

A common feature of redox enzymes which catalyze multielectron reduction of substrate is that they contain more than one redox center per functional unit. The 4-electron reduction of oxygen in cytochrome c oxidase is made possible by the cooperative electron transfer among two hemes (a and a_3) and two copper ions. Various lines of evidence have suggested that heme a_3, where dioxygen binds, is electronically and magnetically coupled to one of the copper ion and thus the ligated oxygen can receive electrons from both heme iron and copper (64). A possible oxygen binding mode which involves the 2 metal centers is the bridged structure: Fe-OO-Cu, which now seems to be amply supported by experimental evidence (65). Pseudomonas cytochrome oxidase also contains more than one heme; it has one heme c and one heme d in each unit. The chlorin heme d is the binding site for oxygen. The siroheme-containing nitrite and sulfite reductases consist of a tetranuclear iron-sulfur cluster per siroheme (62). Recent EPR and Mossbauer studies even showed that the siroheme and the Fe_4S_4 cluster are strongly exchange-coupled (66). This is proven by the observations that the two chromophores share a single electronic spin and that the addition of 1 electron to oxidized enzyme changes the environment of 5 iron atoms. Spin sharing is apparently present in all three important redox states: oxidized, 1 e reduced (goes to siroheme), and 2 e reduced enzyme. Such a coupling certainly provides the means for 2-electron transfer to bound nitrite, which is probably reduced to ammonia via a succession of 2-electron processes (Scheme 3). The nature of this coupling, possibly through a common bridging ligand or through the ring, is yet to be established. Model systems incorporating both Fe_4S_4 cluster and heme molecule are needed to gain more insight.

Scheme 3

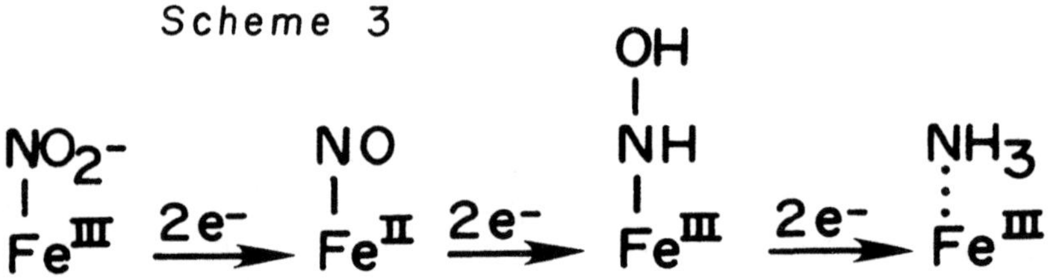

An interesting redox process involving a chemical reaction of the siroheme side chains has also been observed. During isolation of the metal free sirohydrochlorin, the acetic acid side chains on the reduced rings cyclized to form lactone (20,67). This is a reaction presumably initiated by a nucleophilic attack of the acetate on the π- cation (68) as similar reactions using external nucleophiles can be demonstrated with synthetic iBCs. During each ring cyclization, 2 protons and 2 electrons are released (Scheme 4). This may provide yet another possibility to abstract electrons and protons from siroheme. The absorption spectra and redox potentials of such iBC lactones do not deviate significantly from those of the parent compound (We have tested with synthetic acetate), and therefore may not exibit significant spectral changes should it happen in the enzyme. Similar processes could also occur in vitamin B_{12}.

Scheme 4

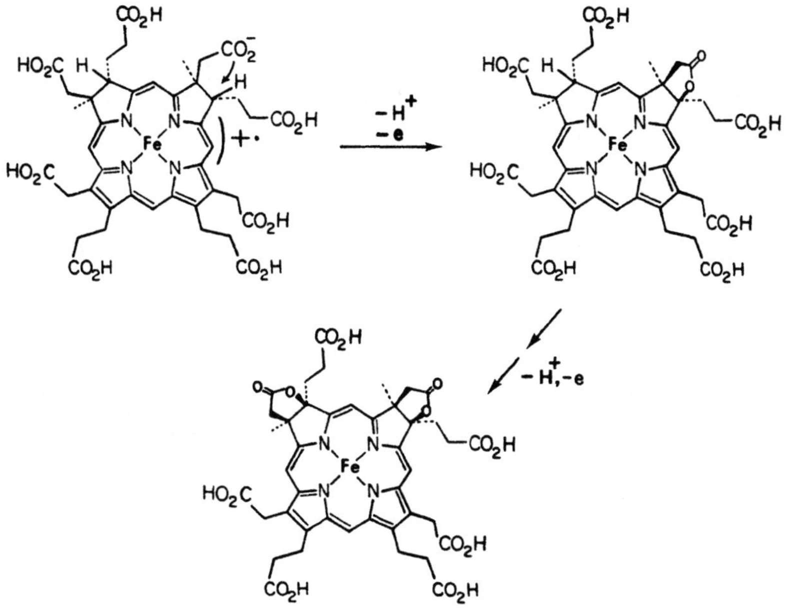

CONCLUSION

The comparative study on hydroporphyrin versus porphyrin hemes has yielded two significant findings:

(a) The affinities for ligands and the electron donating ability of the heme iron are greater in ring saturated hemes.

(b) The redox potential for the one electron oxidation process follows the order iBC<chlorin<porphyrin. The greater ease of oxidation of iBC heme creates the unusual existence of the ferrous heme cation radical.

Theses findings would inevitably lead to the assertion that iron hydroporphyrins, especially siroheme, are indeed better electron donor for the coordinated substrates and for that matter, they are eminently suitable to serve as multielectron redox catalysts. However, this must be considered only as intrinsic properties of these hemes. Whether these results can be extrapolated to the enzymatic reactions, of course, will depend crucially on the coordinatioin environment within the protein. As progress has been made in exploring the nature of the enzymes, it is certain that better and more accurate models can be developed to test and to amplify various aspects of the reaction. So long as one recognizes the advantages as well as the limitations of the model approach, these synthetic hemes are indispensable tools for understanding the natural systems.

ACKNOWLEDGMENTS

This work was generously supported by U. S. Department of Agriculture Grant 59-2261-0-1-437-0 and also in part by NSF Grant CHE 7815285 for instrumentations. I would like to thank my coworkers Brian Ward, Richard Young at MSU and Drs. Jack Fajer, Louise Hanson, and Kathy Barkigia at Brookhaven National Laboratory for their contributions. This manuscript was typed by the author with his new toy of the 80's.

REFERENCES

1. *Yaoi, H. & Tamiya, H. (1928) Proc. Imp. Acad., Japan* ***4****, 436.*
2. *Keilin, D. (1933) Nature (London)* ***132****, 783.*
3. *Barrett, J. (1956) Biochem. J.* ***64****, 626.*
4. *Yamanaka, T., Ota, A. & Okunuki, K. (1960) Biochim. Biophys. Acta* ***44****, 397.*
5. *Yamanaka, T., Kihimoto, S. & Okunuki, K. (1963) J. Biochem.* ***53****, 416.*
6. *Kuronen, T. & Ellfolk, N. (1972) Biochim. Biophys. Acta* ***275****, 308.*
7. *Newton, N. (1969) Biochim. Biophys. Acta* ***185****, 316.*
8. *LeGall, J., Payne, W. J., Morgan, T. V. & DerVartanian, D. V. (1979) Biochem. Biophys. Res. Commun.* ***87****, 355.*
9. *Iwasaki, H. & Matsubara, T. (1971) J. Biochem.* ***69****, 847.*
10. *Cox, C. D., Payne, W. J. & DerVartanian, D. V. (1971) Biochim. Biophys. Acta* ***253****, 290.*
11. *Murphy, M. J., Siegel, L. M., Tove, S. R. & Kamin, H. (1974) Proc. Natl. Acad. Sci. USA* ***71****, 612.*
12. *Vega, J. M., Garrett, R. H. & Siegel, L. M. (1975) J. Biol. Chem.* ***250****, 7980.*
13. *Vega, J. M. & Kamin, H. (1977) J. Biol. Chem.* ***252****, 896.*
14. *Siegel, L. M., Murphy, M. J. & Kamin, H. (1973) J. Biol. Chem.* ***248****, 251.*
15. *Murphy, M. J., Siegel, L. M. & Kamin, H. (1973) J. Biol. Chem.* ***248****, 2801.*
16. *Murphy, M. J. & Siegel, L. M. (1973) J. Biol. Chem.* ***248****, 6911.*
17. *Siegel, L. M., Davis, P. S. & Kamin, H. (1974) J. Biol. Chem.* ***249****, 1572.*
18. *Murphy, M. J., Siegel, L. M. & Kamin, H. (1974) J. Biol. Chem.* ***249****, 1610.*
19. *Scott, A. I., Irwin, A. J., Siegel, L. M. & Shoolery, J. N. (1978) J. Am. Chem. Soc.* ***100****, 7987.*
20. *Battersby, A. B. & McDonald, E. (1978) Bioorganic Chem.* ***7****, 161.*
21. *Deeg, R., Kriemler, H. P., Bergmann, K. H. & Muller, G. (1977) Z. Physiol. Chem.* ***358****, 339.*
22. *Scheer, H. & Inhoffen, H. H. (1978) in "The Porphyrins" (Dolphin, D. Editor) Vol. II, Academic Press, New York, p. 45.*
23. *Jacob, G. S. & Orme-Johnson, W. H. (1979) Biochemistry* ***18****, 2975.*
24. *Ballantine, J. A., Psaila, A. F., Pelter, A., Murray-Rust, P., Ferrito, V., Schembri, P. & Jaccarini, V. (1980) J. Chem. Soc., Perkin Trans. 1, 1080.*
25. *Agius, L. Ballantine, J. A., Ferrito, V., Jaccarini, V., Murray-Rust, P. Pelter, A., Psaila, A. F. & Schembri, P. J. (1979) Pure & Appl. Chem.* ***51****, 1847.*

26. Eisner, U. (1957) J. Chem. Soc., 3461.
27. Fischer, H. & Helberger, H. (1929) Justus Liebigs Ann. Chem. **471**, 285.
28. Stolzenberg, A. M., Spreer, L. D. & Holm, R. H. (1980) J. Am. Chem. Soc. **102**, 364.
29. Whitlock, H. W., Hanauer, R., Oester, M. Y. & Bowry, B. K. (1969) J. Am. Chem. Soc. **91**, 7458.
30. Barkigia, K. M., Fajer, J., Spaulding, L. D. & Williams, G. J. B. (1981) J. Am. Chem. Soc. **103**, 176.
31. Inhoffen, H. H., Buchler, J. W. & Thomas, R. (1969) Tetrahedron Lett. **14**, 1141.
32. Chang, C. K. (1980) Biochemistry **19**, 1971.
33. Chang, C. K., Ward, B. & Young, R., J. Am. Chem. Soc., in press.
34. Montforts, F.-P., Ofner, S., Rasetti, V., Eschenmoser, A., Woggon, W. D., Jones, K. & Battersby, A. R. (1979) Angew. Chem., Int. Ed. Engl. **18**, 675.
35. Harrison, P. J., Fookes, C. J. R. & Battersby, A. R. (1981) J. Chem. Soc., Chem. Comm., 797.
36. Snow, R. J., Fookes, C. J. R. & Battersby, A. R. (1981) J. Chem. Soc., Chem. Comm., 524.
37. Barkigia, K. M., Chang, C. K., Fajer, J. & Williams, G. J. B., J. Am. Chem. Soc., in press.
38. Rougee, M. & Brault, D. (1973) Biochem. Biophys. Res. Commun. **55**, 1364.
39. White, D. K., Cannon, J. B. & Traylor, T. G. (1979) J. Am. Chem. Soc. **101**, 2443.
40. Ward, B., Wang, C.-B. & Chang, C. K. (1981) J. Am. Chem. Soc. **103**, 5236.
41. Traylor, T. G., White, D. K., Campbell, D. H. & Berzinis, A. P. (1981) J. Am. Chem. Soc. **103**, 4932.
42. Chang, C. K. & Ward, B., manuscrpit in preparation.
43. Chang, C. K. & Traylor, T. G. (1975) Proc. Natl. Acad. Sci. USA **72**, 1166.
44. Seybert, D. W., Moffat, K., Gibson, Q. H. & Chang, C. K. (1977) J. Biol. Chem. **252**, 4225.
45. Wallace, W. J., Maxwell, J. C. & Caughey, W. S. (1974) Biochem. Biophys. Res. Commun. **57**, 1104; Wallace, W. J. & Caughey, W. S. (1975) ibid. **62**, 561.
46. Chang, C. K. (1979) in "Oxygen, Biochemical and Clinical Aspects" (Caughey, W. S., Editor) Academic Press, New York, p. 437.
47. Phillips, J. N. (1960) Rev. Pure & Appl. Chem. **10**, 35.
48. Chang, C. K. & Dolphin, D. (1976) Proc. Natl. Acad. Sci. USA **73**, 3338.
49. Parmely, R. C. & Goff, H. M. (1980) J. Inorg. Biochem. **12**, 269.
50. Chang, C. K. & Fajer, J. (1980) J. Am. Chem. Soc. **102**, 841.
51. Richardson, P. F., Chang, C. K., Spaulding, L. D. & Fajer, J. (1979) J. Am. Chem. Soc. **101**, 7736.
52. Chang, C. K., Hanson, L. K., Richardson, P. F., Young, R. & Fajer, J. (1981) Proc. Natl. Acad. Sci. USA **78**, 2652.
53. Castro, C. E. (1978) in "The Porphyrins" (Dolphin, D., Editor) Vol. V, Academic Press, New York, p. 1.
54. Johnson, E. C., Niem, T. & Dolphin, D. (1978) Can. J. Chem. **56**, 1381.

55. *Peisach, J. & Blumberg, W. E. (1971) in " Probes of Structure and Function of Macromolecules and Enzymes" (Chance, B., Yonetani, T. & Mildvan, S., Editors) Vol. II, Academic Press, New York, p. 231*
56. *Peisach, J., Blumberg, W. E. & Adler, A. (1973) Ann. N. Y. Acad. Sci.* **206**, *310.*
57. *Taylor, C. P. S. (1977) Biochim. Biophys. Acta* **491**, *137.*
58. *Lancaster, J. R., Salerno, J. C., Siegel, L. M. & Kamin, H., Biochemistry, in press.*
59. *Wayland, B. B. & Olson, L. W. (1974) J. Am. Chem. Soc.* **96**, *6037.*
60. *Kon, H. & Kataoka, N. (1969) Biochemistry* **8**. *4757.*
61. *Aparicio, P. J., Knaff, D. B. & Malkin, R. (1975) Arch. Biochem. Biophys.***169**, *102.*
62. *Lancaster, J. R., Vega, J. M., Kamin, H., Orme-Johnson, N. R., Orme-Johnson, W. H., Krueger, R. J. & Siegel, L. M. (1979) J. Biol. Chem.* **254**, *1268.*
63. *Fry, I. V., Cammack, R., Hucklesby, D. P. & Hewitt, E. J. (1980) FEBS Lett.* **111**, *377.*
64. *Palmer, G., Babcock, G. T. & Vickery, L. E. (1976) Proc. Natl. Acad. Sci. USA* **73**, *2206.*
65. *Chance, B. & Powers, L. Biophys. J. (1981)* **33**, *95a; Reed, C. A. & Landrum, J. T. (1979) FEBS Lett.* **106**, *265.*
66. *Christner, J. A., Munck, E., Janick, P. A. & Siegel, L. M. (1981) J. Biol. Chem.* **256**, *2098.*
67. *Battersby, A. R., Jones, K., McDonald, E., Robinson, J. A. & Morris, H. R. (1977) Tetrahedron Lett., 2213.*
68. *Smith, K. M., Barnett, G. H., Evans, B. & Martynenko, Z. (1979) J. Am. Chem. Soc.* **101**, *5953.*

SECTION G

HEME ENZYMES

PEROXIDASES

H. B. Dunford, T. Araiso,[1] D. Job, J. Ricard,[2]
R. Rutter, L. P. Hager,[3] R. Wever, W. M. Kast,
R. Boelens,[4] N. Ellfolk and M. Rönnberg[5]

[1]University of Alberta, Edmonton, Canada
[2]CNRS, CBM, Marseille, France
[3]University of Illinois, Urbana, U.S.A.
[4]University of Amsterdam, Amsterdam, The Netherlands
[5]University of Helsinki, Helsinki, Finland

A brief description is made of properties of a wide variety of peroxidases which illustrates differences as well as similarities. The article begins with monoheme enzymes and ends with myeloperoxidase and Pseudomonas cytochrome c peroxidase which have two heme groups per molecule. It encompasses two types of plant peroxidases (horseradish and turnip), two from microorganisms (chloro- from Caldariomyces fumago and Pseudomonas cytochrome c peroxidase) and two mammalian peroxidases (lacto- and myelo-). Reactions are discussed of the mammalian peroxidases which are important in the oral cavity.

1. INTRODUCTION

Peroxidases catalyze the oxidation of a variety of substrates, both organic and inorganic, by hydrogen peroxide (or by organic substrates of the type ROOH) (1). The enzymatic cycle is usually

$$\text{Peroxidase} + H_2O_2 \longrightarrow \text{Compound I} + H_2O$$

$$\text{Compound I} + AH_2 \longrightarrow \text{Compound II} + \cdot AH$$

$$\text{Compound II} + AH_2 \longrightarrow \text{Peroxidase} + \cdot AH + H_2O$$

$$2\ \cdot AH \longrightarrow A_2H_2\ (\text{or } A_2 + A_2H_4)$$

H. B. Dunford et al. (eds.), The Biological Chemistry of Iron, 337–355.

The overall reaction is

$$H_2O_2 + 2AH_2 \longrightarrow A_2H_2 + 2H_2O$$

Compound I is formed by oxygen atom addition and so possesses two oxidizing equivalents more than the native enzyme. Compound II has one more oxidizing equivalent than the native enzyme. The reducing substrate AH_2 is converted into a free radical species $\cdot AH$; therefore AH_2 is a hydrogen atom donor. The peroxidase cycle is thus a modified form of ping pong kinetics in which one molecule of oxidizing substrate and two molecules of reducing substrate are involved

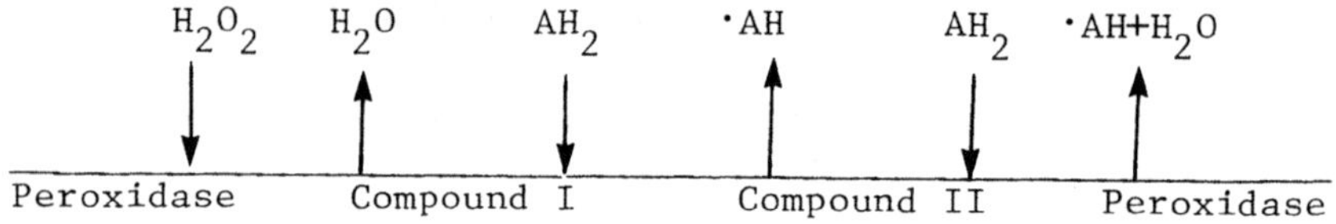

The active site of the native enzyme contains a ferric heme, usually ferriprotoporphyrin IX, the trivial name for [1,3,5,8-tetramethyl-2,4-divinyl-6,7-dipropionate-porphinato] Fe(III) (Fig. 1).

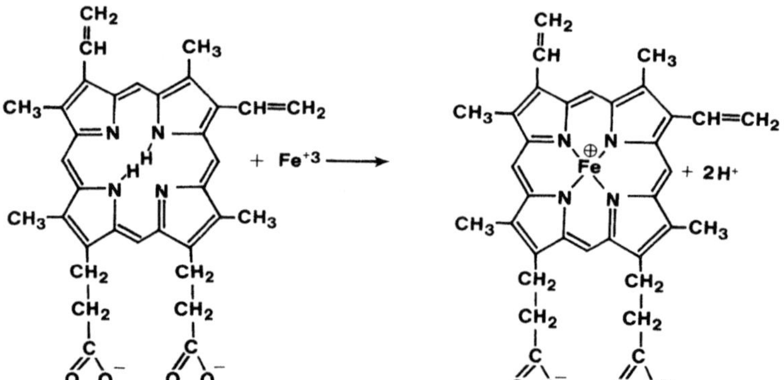

Fig. 1. Protoporphyrin IX + Fe^{3+} → ferriprotoporphyrin IX + $2H^+$

Addition of a ferric ion to the protoporphyrin IX displaces two protons from pyrrole ring nitrogen atoms and so the charge on the ferric ion is effectively reduced from +3 to +1. The four pyrrole nitrogen atoms provide ligands in coordination positions 1 to 4 for the ferric ion. Coordination position 5 is known to be occupied by the neutral imidazole group of a histidine residue (2). (See also the detailed discussion in the following two articles by LaMar and Kitagawa). This side of the heme plane is commonly called the proximal side. Coordination position 6 is either occupied by water (in metmyoglobin) or is probably vacant (in the peroxidases) (3) and this side of the heme plane is referred to as distal. In either case the total charge in the immediate environment of the Fe(III) is +1.

Since only four ligands are equivalent, the ferric ion does not have an octahedral environment. If the 5th and 6th ligands were equivalent then the ferric ion would be in a tetragonal environment. If one of the 5th or 6th ligands is missing then a square pyramid is obtained.

The porphyrin ring system is highly aromatic; the magnitude of its resonance energy is discussed in detail in this volume by George. Because the porphyrin is so electron deficient, it has been difficult for many inorganic chemists to accept the possibility that Fe(IV) is produced in the oxidation reaction between native peroxidase and hydrogen peroxide. Why should a system, already so electron deficient, give up another electron? However, the experimental evidence for Fe(IV) is now overwhelming (4) and our wonderment at nature's chemistry is changed to open admiration. Given a system that is already electron deficient, imagine the potency of an oxidation catalyst which is made even more electron deficient! Since Fe(IV) is present in compound I it therefore follows that the second oxidizing equivalent obtained from hydrogen peroxide is located elsewhere in the molecule. There are three possibilities (i) the porphyrin ring, (ii) the protein moiety and (iii) a ligand in the 5th or 6th coordination position of the iron. In all cases a free radical species must be involved. These possibilities are subject to detailed scrutiny elsewhere in this volume. It will suffice to say here that the available evidence favors the porphyrin ring (a π-cation radical) except for the cytochrome *c* peroxidases. Compound II possesses Fe(IV) but has lost the free radical.

With this brief introduction we turn our attention in this article to a more detailed discussion of properties of a variety of peroxidases: (i) horseradish, (ii) turnip (iii) chloro, (iv) myelo and lacto and (v) *Pseudomonas* cytochrome *c*. Yeast cytochrome *c* peroxidase is discussed in detail elsewhere in this volume by Hoffman.

2. HORSERADISH PEROXIDASE[1]

2.1 The Native Enzyme

The primary structure of horseradish peroxidase, HRP, (EC. 1.11.1.7, donor H_2O_2, oxidoreductase) is known. Two calcium ions are present in each enzyme molecule (6) and there are eight carbohydrate side chains. These facts refer to isoenzyme C which is the dominant isoenzyme (7).

The Soret spectra for native HRP and compounds I and II are shown in Figure 3. The molar absorptivity of native isoenzyme C is 1.021 x 10^5 $M^{-1}cm^{-1}$ at 402.5 nm and 25° (8). The purity number (reinheitzahl or RZ), the ratio of absorbance at 403 to 280 nm (corresponding to the ratio of hemin to protein absorbance), is an excellent criterion of enzyme purity. The number should approach the value 3.55 as an upper limit.

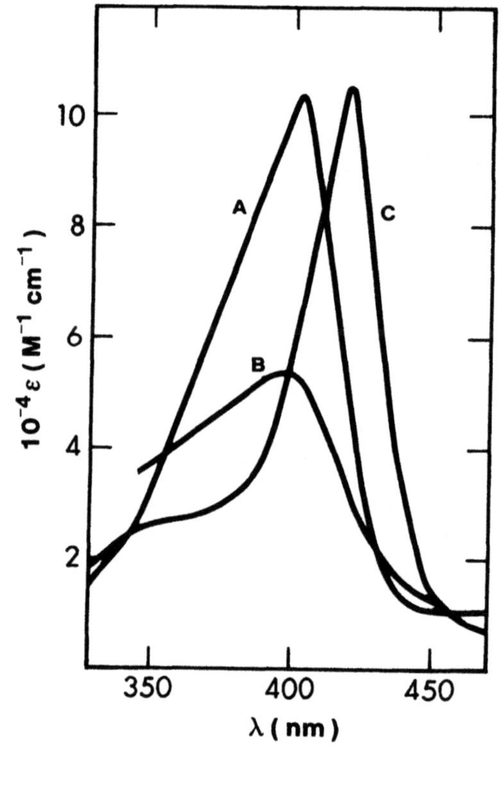

Fig. 2. Molar absorptivity in the Soret region of (A) native HRP, (B) Compound I and (C) Compound II.

2.2. Compound I Formation

A group with a pK_a of $\leq$ 3 is necessary to form HRP compound I and it has been assigned to Asp 43 (9). Such a pK_a value appears too small for a distal His. We extend the mechanism of compound formation of horseradish peroxidase to include Arg 38 for two reasons (i) Arg has been demonstrated to be present in the active site of yeast cytochrome *c* peroxidase and sequence homologies indicate its presence in the active site of HRP (10, 11) and (ii) a nitrogenous ligand has been demonstrated to occupy the 6th coordination position of horseradish peroxidase in its alkaline transition (12). Since the pK_a of this transition is 11 the ionization of Arg 38 at high pH followed by coordination to the ferric ion appears plausible. Between pH values of 3 and 11 compound I formation occurs at a constant rate of 2 x 10^7 $M^{-1}s^{-1}$ (13). We propose the following mechanism for HRP compound I formation: (i) nucleophilic attack of H_2O_2 on Fe(III) accompanied by loss of the α proton to Asp 43 (ii) transfer of the α proton by carboxylic acid group rotation to the β oxygen (iii) departure of β OH_2 (iv) sometime during processes (i) - (iii) electron transfer from the porphyrin ring to Fe(V). Arg 38 is believed to stabilize charge on β O as an emerging OH^- leaving group is converted to H_2O (10). It is also possible that a push-pull mechanism occurs, αH^+ to Asp 43, Arg 38 H^+ to β O.

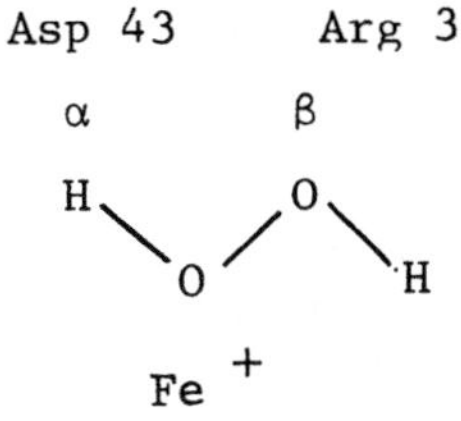

3. TURNIP PEROXIDASES[2]

Various peroxidases have been isolated from plant tissues and extensively studied over the last years. The same type of plant cell may contain several isoperoxidases. Their synthesis appears to be dependent upon different systems of regulation which establishes that they are truly different enzyme species and not degradation products of the same hemoprotein (14). In turnip roots seven isoperoxidases can be isolated and five of them were purified to homogeneity (15). Three are anodic at pH 6.8 (TP_1, TP_2, TP_3) and two are cathodic at this pH (TP_6, TP_7). They were compared in terms of their structure and chemical reactivity. All are glycoproteins and possess ferriprotoporphyrin IX as their prosthetic group. Five oxidation-reduction states are known, as it is for other peroxidases. These states are ferrous, ferric (native state), Compound II (ferryl form, secondary product in the peroxidation process), Compound I (primary product of peroxidase and H_2O_2) and Compound III (or oxy-ferrous). Ferric iron is readily accessible to ligands such as cyanide and fluoride (16). The peroxidases undergo ionizations in alkaline solutions, resulting in the conversion of the high spin form (neutral pH) to a low spin form (17). TP_1, TP_2 and TP_3 have similar molecular weights (51000 ± 2500); they are also similar in their amino acid and sugar compositions. TP_7 is much lighter (38700 ± 2500) and possesses a distinct amino acid composition and a lower sugar content (8). Peptide mapping studies showed that TP_1, TP_2, TP_3 and horseradish peroxidase isoenzyme C (HRP C) form a group which appears to be closely related in the amino acid sequences around four disulfide bridges. This suggested a similar overall spatial structure of TP_1, TP_2, TP_3 and HRP C. TP_7 differs from this group, at least around two of its disulfide bridges, and therefore from the other isoperoxidases in parts of its three dimensional structure (19, 20). CD studies indicate that turnip peroxidases have low α helical contents, as compared to other hemoproteins such as hemoglobin and myoglobin, which might be due to the presence of the carbohydrate moiety in the plant peroxidases (21).

It is well known that hemoproteins can participate in various types of reactions: binding of molecular oxygen, peroxide reduction and electron transfer between dehydrogenases and various acceptors. The same hemoprotein sometimes participates in more than one of these reactions. Thus besides the typical peroxidation reaction catalyzed by peroxidases, there is ample circumstantial evidence that at least some of the isoperoxidases in the plant cells control the level of the hormone indole-3-acetate. Thus is one of the reasons for our interest in these hemoproteins. Typical peroxidases such as TP_1, TP_2, TP_3 and HRP C have a very low oxidation-reduction potential (22) and are able to catalyze at a very high rate peroxidation of various electron donors. Rate constant values for reaction with H_2O_2, yielding Compound I, are very similar of the order of 10^7 $M^{-1}s^{-1}$ (23). Rate constant values for reactions of oxidized enzyme complexes (Compound I and Compound II) with organic substrates can be very high, larger than 10^8 $M^{-1}s^{-1}$. These reactions appear to be mainly dependent of the relative ease of oxidation of the substrate (24). These typical peroxidases can also participate at a much lower rate, in the oxidation of indole-3-acetate, in the absence of H_2O_2. It has been proposed that "indole-acetic acid oxidase" activity of a peroxidase is partly due to the reduction process of the ferric enzyme by indole-3-acetate followed by a reversible oxygenation of the ferrous form of the enzyme (Compound III formation)(25-27). The high value of the oxidation-reduction potential of TP_7 and the unexpected stability of the oxygenated form of its ferrous form are compatible with an oxygenase function of this hemoprotein (22). Moreover TP_7 has comparatively lower peroxidase activity. This is due to a particularly small rate constant of reaction of this hemoprotein with H_2O_2 (23), of the order of 10^6 $M^{-1}s^{-1}$. Compound I and Compound II of TP_7 also react more slowly than that of TP_1 with organic substrates (27).

Compound I formation studies with turnip peroxidases and H_2O_2 (23) or peroxybenzoic acids (28) have been helpful for understanding this redox process of considerable complexity. Comparison of the heme distal sequences of myoglobin, plant peroxidases and leghemoglobins has suggested the importance of protein residues for the reaction (28-30): whereas free ferriheme is constrained to utilizing HO_2^- as substrate (31,32), the specific enzymes such as hydroperoxidases have acquired a mechanism for the equally effective utilization of molecular H_2O_2, the only form present at physiological pH (pK ~ 11.5). Thus in the active site of hydroperoxidases ionization of H_2O_2 is decoupled from H_2O_2 ionization in the external solution (33). This process is an essential feature of hydroperoxidase reactions. TP_7 presents an interesting transitional case since, although reaction with HO_2^- cannot be detected (23), it does react with both molecular peroxy acids and their peroxy anions (28). The remaining peroxidases TP_1 or HRP C react exclusively with molecular hydroperoxidases (28,33). The sequences of 34 residues about histidine-42, which is probably distally located, are highly homologous for TP_1, TP_2, TP_3, TP_7 and HRP C. Acid base catalysis mediated by the invariant residue aspartate-43 of plant peroxidases (33) has been invoked for the conversion of H_2O_2 into the much better nucleophile HO_2^- (29). Histidine-40 is conserved except in TP_7 where it is replaced by phenylalanine and it has been suggested that this substitution may be crucial in determining the unusual properties of TP_7 (34).

4. CHLOROPEROXIDASE[3]

4.1 Introduction

Chloroperoxidase (CPO) shows a great deal of catalytic similarity to catalase and horseradish peroxidase (HRP). The enzyme also shows similarities with cytochrome P-450 (P-450) in its physical properties such as EPR, Mössbauer and visible absorption. Because of its similarities and differences with catalase, HRP, and P-450, CPO may be important in understanding the relationship between these three proteins and in understanding the factors which cause them to catalyze different reactions.

4.2 A Brief History of Chloroperoxidase

Chloroperoxidase is a glycohemoprotein of 45,000 daltons that is isolated from the mold *Caldariomyces fumago*. The enzyme has been crystalized, its amino acid composition has been determined, and it has been the subject of a variety of physical studies (35,36). The enzyme contains one mole of ferriprotophyrin IX and contains about 25 to 30 percent carbohydrate by weight.

Chloroperoxidase catalyzes a large number of oxidative reactions. The enzyme can chlorinate, brominate, and iodinate organic substrates as well as catalyze classical peroxidase and catalase reactions (37-40). It can also perform catalatic type reactions using organic peroxide substrates (40). The initial reaction of chloroperoxidase with peroxide substrates can be formulated as follows:

$$\text{CPO} + \text{ROOH} \longrightarrow \text{Compound I} + \text{ROH} \quad (1)$$

where R = H, methyl, ethyl, m-chlorobenzoic acid, or several other groups. Compound I in turn may react with several different electron donors to complete a catalytic sequence and return to the native resting state. One of the most important reactions of chloroperoxidase is its halogenating activity

$$\text{Compound I} + X^{-} \longrightarrow [\text{Compound X}] \quad (2)$$

$$[\text{Compound X}] + \text{HA} \longrightarrow \text{CPO} - Fe^{+3} + X - A + H^{+} \quad (3)$$

In equations 2 and 3, X^{-} represents iodide, bromide or chloride and A represents a halogen acceptor molecule. A large number of halogen acceptor molecules have been categorized (39-41).

4.3 Similarities to Cytochrome P-450

CPO shows a great deal of similarity to cytochrome P-450. P-450 cytochromes obtain their name from the location of the light absorption maximum near 450 nm when reduced and liganded with carbon monoxide (42). The Fe^{2+} CPO carbon monoxide complex also shows this 450 nm band (43). CPO and P-450 have been compared by a number of spectroscopic techniques. These include visible, Mössbauer, Resonance Raman, EXAFS, EPR, and magnetic circular dicroism (44-48). These various techniques have added further proof to the similarity of the heme environments of P-450 and CPO. By analogy, it has been assumed that CPO must contain the thiolate ligand which has been established for the P-450 cytochromes (49). However, an extensive search for a free sulfhydryl group using specific reagents and amperometric titrations has failed to detect any such group in CPO (50). Until further proof on the nature of CPO ligands arises, this area of study will remain controversial.

4.4 Comparison of HRP and CPO Compound I's

The chemical nature of peroxidase Compound I's has been the centre of much recent study. It has been shown by visible, Mössbauer, EPR, and ENDOR spectroscopy that horseradish peroxidase Compound I stores two oxidation equivalents by converting the iron in its heme prosthetic from Fe^{+3} to low spin Fe^{+4} and by removing an electron from the porphyrin A_{2u} orbital (4,51). In contrast to horseradish peroxidase, CPO forms an unstable Compound I which is difficult to study by conventional techniques. The instability of CPO Compound I probably arises from the fact that it stores enough oxidation potential to oxidize the carbohydrates which are part of its chemical composition. The visible absorption spectra of CPO Compound I has been obtained by stopped flow rapid scan methods (52). By analogy to Dolphin's model studies with cobalt, magnesium, and zinc porphyrins, the general character of the CPO Compound I visible spectra would put it into the class of compounds which have a porphyrin electron removed from the A_{1u} orbital (53-56). This would be in contrast to horseradish peroxidase Compound I which has a porphyrin electron removed from the A_{2u} orbital. Since the A_{1u} orbital has its major electron density farther away from the iron atom the prediction was made that in CPO there should be less of the coupling between the radical and the iron than with horseradish peroxidase (4).

CPO Compound I Mössbauer and EPR samples were prepared by freeze quench techniques (57). The Mössbauer spectra (Fig. 1), in opposition to the above prediction, show a much stronger magnetic effect than does horseradish peroxidase. Figure 1 shows the Mössbauer spectra obtained at 4.2° and 80°K respectively and a 400 gauss perpendicular magnetic field. The spectra are consistent with low spin Fe^{+4} species as in HRP but the electronic interaction is radically different. It has been reported previously that the EPR of CPO Compound I shows a sharp free radical signal consistant with the predicted result (58). Although this signal can be associated with changes that occur in the visible spectra which are characteristic of Compound I formation, it only represents about 4% of the theoretical spins calculated for a Compound I containing an Fe^{+4} and a free radical species. Another EPR signal has been found which is also associated with visible spectral changes associated with Compound I. This signal accounts for 95% of the theoretical spins. This broad signal, which occurs at $g = 1.72$, was not detected in the initial EPR studies because it was overshadowed by the presence of a large signal associated with a Mn^{+2} impurity that has consistently interfered with previous EPR studies (47). Fig. 2 shows the EPR spectrum of the CPO Compound I broad signal at 9.45 gigaherz and 7.0°K with 29 dB of microwave power. This broad signal is more consistent with the Mössbauer results although no explanation has been formulated for the deviation from predicted theory. The other important component of Compound I is the presence of a peroxide derived oxygen which remains associated with CPO Compound I. This was first shown in studies using O^{18}-O^{16} labeled peroxides (59). ENDOR spectra of horseradish peroxidase Compound I formed with 50% O^{17} labeled hydrogen peroxide also have recently proven the existence of a peroxide derived oxygen in this Compound I (60).

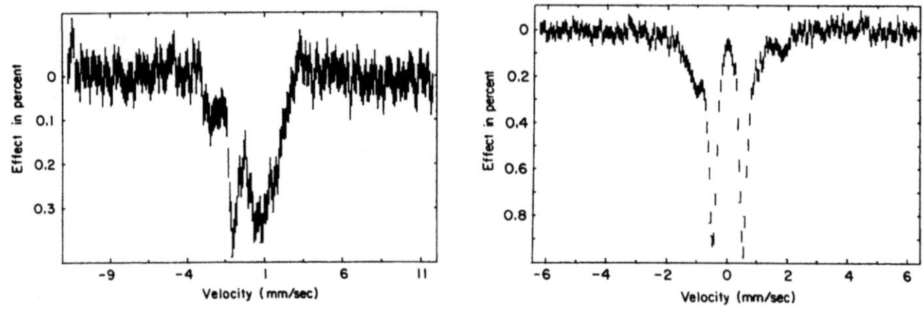

Figure 3. Mössbauer spectra of chloroperoxidase Compound I (~4.0 x 10^{-4} m/e). Left spectrum: 4.3°K, 400 gauss perpendicular field. Velocity in mm/sec relative to iron foil. Right spectrum: 800°K, 400 gauss perpendicular field. Isomer shift 0.12 mm/sec, quadrapole splitting 1.02 mm/sec.

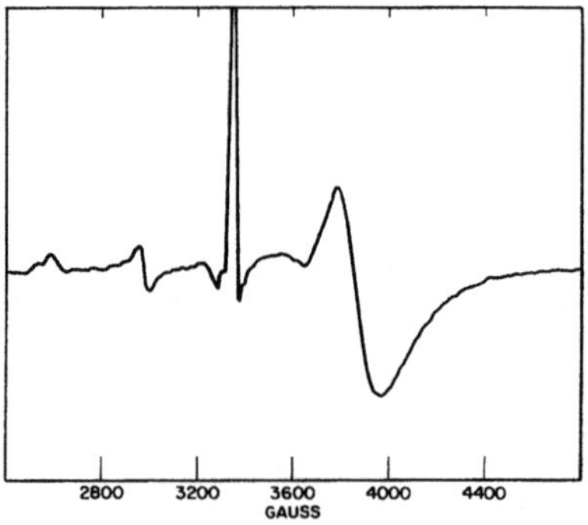

Figure 4. EPR spectrum of chloroperoxidase Compound I prepared by reacting 1.2 x 10^{-3} M CPO with 4.8 x 10^{-3} M peracetic acid and freeze quenching after 10 milliseconds.

5. THE PEROXIDATION OF THIOCYANATE CATALYSED BY MYELOPEROXIDASE AND LACTOPEROXIDASE[4]

5.1 Introduction

Neutrophils contain large amounts of myeloperoxidase, a hemoprotein which is involved in the killing of ingested micro-organisms and fungi (61). The antimicrobial and fungicidal activity of myeloperoxidase is due to the ability of the enzyme to oxidize the chloride ion to hypochlorous acid in the presence of H_2O_2 (62). Myeloperoxidase can also be found in the oral cavity and in particular in the gingival fluid, probably as a result of degradation of neutrophils (63). The enzyme, like lactoperoxidase (64), may therefore be involved in the production of hypothiocyanite ($OSCN^-$) from thiocyanate, which is present in saliva, and H_2O_2 generated by the bacterial oral flora. Hypothiocyanite inhibits the glycolysis of oral bacteria and is at low pH bactericidal (65). However, the factors which influence the enzymic activity of myeloperoxidase and lactoperoxidase, like pH and substrate concentrations, have not been studied, although such a study may give a clue to the question how the bacterial growth in the oral cavity is regulated. It is shown here that both myeloperoxidase and lactoperoxidase catalyse the peroxidation of SCN^- over a broad pH range. The position of the pH optima, which may vary from pH 5.0 to pH 7.5 is governed by the concentration of thiocyanate and H_2O_2. A kinetic model is proposed with inhibition by both substrates and hydrogen ions, which explains the observed enzymic behaviour.

5.2 Experimental Procedure

Myeloperoxidase was purified from human blood leucocytes as previously described (66). Lactoperoxidase was from Sigma. The concentrations were determined spectrophotometrically using absorbance coefficients of 89 $mM^{-1} \cdot cm^{-1}$ at 430 nm (67) for myeloperoxidase and 110 $mM^{-1} \cdot cm^{-1}$ at 413 nm for lactoperoxidase (68), respectively. The rate of formation of $OSCN^-$ was determined by following the rate of oxidation of 5-thio-2-nitrobenzoic acid (TNB) at 412 nm and using an absorbance coefficient for this compound of 13.6 $mM^{-1} \cdot cm^{-1}$ at 412 nm (64).

5.3 Results

As expected the rate of formation of $OSCN^-$ from H_2O_2 and SCN^- catalyzed by myeloperoxidase and lactoperoxidase is proportional to the enzyme concentration (0 - 2.5 nM) and independent of the concentration of TNB (10 - 100 μM). However, the rate of the reaction is strongly pH dependent with sharp optima. These optima shift about 1 pH unit to higher pH when the concentration of SCN^- is increased 10-fold, but decreases 1 pH unit when the concentration of H_2O_2 is increased 10-fold. In fact, the position of the pH optima for both enzymes is a linear function (slope = 1) of the logarithm of the ratio of $[SCN]/[H_2O_2]$. At the same substrate concentrations the pH optima of myeloperoxidase and lactoperoxidase differ about 0.4 pH unit, lactoperoxidase being more active at acid pH values. Similar relationships between the position of pH optima and substrate concentration were already observed in the peroxidation of chloride, bromide and iodide catalyzed by both myeloperoxidase (69,70) and eosinophil peroxidase (71,72).

At pH values lower than the optima, SCN^- not only acts as a substrate, but also as a competitive inhibitor with respect to H_2O_2. For myeloperoxidase the inhibition constant for SCN^- is 2 mM whereas that for lactoperoxidase is 12 mM. Both inhibition constants are pH independent (pH 4 - 5.7). However, the K_m's for H_2O_2 are strongly pH dependent. The pK_m's for H_2O_2 increase linearly with pH from pH 4.5 to pH 5.5. At higher pH a deviation of the linearity is observed. According to Dixon (73) this may indicate ionisation of a group on the enzyme or on the substrate. Since the pK of H_2O_2 is 11.8 an ionisable group on these enzymes must be present with a pK_a of about 6 which controls the enzymic activity. At pH values lower than the pK_a of this group the hydrogen ion should behave as a competitive inhibitor with respect to H_2O_2 and the K_i for H^+ should correspond to the K_a. This is indeed observed. The maximal velocity (V) is not affected when the hydrogen ion concentration is varied but only the K_m for H_2O_2. For both enzymes a K_i for H^+ of about 1 μM is found which corresponds to an ionisable group on these enzymes with an pK_a of 6. The K_m for H_2O_2 which can be calculated from the data is the same at low $[H^+]$ for lactoperoxidase and myeloperoxidase (30 μM). The turnover values are, however, different. For myeloperoxidase a value of 850 s^{-1} is found, whereas that for lactoperoxidase is about 3 times higher. Since myeloperoxidase (69) and eosinophil peroxidase (72) also show competitive inhibition in the peroxidation of chloride by H^+ at low pH, it is likely that ionisable groups on these peroxidases regulate the enzymic activity.

At pH values higher than the optima, the kinetics are more difficult to analyse. It is, however, observed that in this pH region, H_2O_2 inhibits the reaction competitively with respect to SCN^-. To explain the pH optimum in the formation of $OSCN^-$, we have to assume that the pK_m for SCN^- decreases, when the pH is increased. This seems to be justified since in the peroxidation of Cl^- catalysed by myeloperoxidase, it is observed, that the pK_m for chloride decreases linearly with pH (69). The following equation is proposed which explains most of the kinetic data.

$$\frac{V}{v} = 1 + \frac{K_m}{H_2O_2}\left(1 + \frac{SCN^-}{K_i}\right)\left(1 + \frac{H^+}{K_a}\right) + \frac{K_m}{SCN^-}\left(1 + \frac{H_2O_2}{K_i}\right)\left(1 + \frac{K_b}{H^+}\right)$$

It is obvious that such a function will pass through a minimum as H^+ is varied. The position of this minimum can be determined by differentiating the function with respect to H^+. It is found that when the substrate concentrations are high compared to the inhibition constants for H_2O_2 and SCN^-, that the optimal pH is a linear function of log $[SCN^-]/[H_2O_2]$ as is indeed found experimentally.

5.4 Discussion

What are the physiological consequences of this kinetic mechanism? When myeloperoxidase or lactoperoxidase are present in dental plaque and no carbohydrates are available, the bacterial metabolism will be low and as a consequence the production of H_2O_2 and lactic acid is low. Thus the pH in plaque will be near neutral and exactly at low concentration of H_2O_2, both peroxidases are active in the neutral region. If, however, carbohydrates are present, these will be rapidly metabolised by the bacteria, H_2O_2 will be produced and as consequence of lactic acid production by the bacteria, the pH in dental plaque will drop. Under these conditions i.e. high concentration of H_2O_2, the peroxidases are optimally active in the production of $OSCN^-$ in the more acid region. Since $OSCN^-$ inhibits bacterial glycolysis (65), this suggests that the peroxidase systems act as a natural defence mechanism in the mouth which regulates the oral flora.

6. *PSEUDOMONAS* CYTOCHROME *c* PEROXIDASE[5]

6.1 Introduction

Cytochrome *c* peroxidase (cytochrome *c* : H_2O_2 oxidoreductase, EC 1.11.1.5) catalyzes the oxidation of ferrocytochrome *c* to ferricytochrome *c* in the presence of H_2O_2 according to the equation

$$H_2O_2 + 2\ \text{ferrocyt.}\ c \longrightarrow 2\ \text{ferricyt.}\ c + 2\ OH^-$$

The enzyme was first discovered in baker's yeast by Altschul *et al.* in 1940 (74). In 1953, Lenhoff and Kaplan showed cytochrome *c* peroxidase activity in cells of *Pseudomonas aeruginosa* cultivated at low oxygen pressure (75,76). Purification of a homogeneous preparation of *Pseudomonas* cytochrome *c* peroxidase (PaCcP) was accomplished by Ellfolk and Soininen in 1970 (77).

6.2 Properties of PaCcP

Pseudomonas cytochrome *c* peroxidase, a membrane-bound protein, is a monomer of molecular weight of 43,200 containing two molecules of covalently bound heme *c* as prosthetic groups (78,79). Heme *c* or mesoheme is usually encountered in *c*-type cytochromes, whereas the prosthetic group of "normal" peroxidases, e.g. yeast chrochrome *c* peroxidase and horseradish peroxidase, is a protoheme.

The absorption spectrum of oxidation PaCcP exhibits a Soret band at 407 nm and a maximum at 525 nm. The enzyme can be reduced with dithionite to a ferrous form with a Soret band at 420 nm, an α-band at 524 nm and a split β-band at 557 nm with a shoulder at 551 nm (80). The spectrum, which is characteristic of a low-spin hexacoordinated structure, differs from that of other peroxidases, and appears to be in contrast with the peroxidatic function of PaCcP. However, in addition to the typically low-spin bands of the spectrum, a hematin band at 640 nm, suggestive of a weak-field ligand, can be observed in the spectrum of the oxidized enzyme (81). Circular dichroism (82) as well as resonance Raman (83) spectra further indicate the presence of one low-spin and one high-spin heme *c* in the enzyme. Titration of PaCcP with CO provides evidence that only the high-spin heme is accessible to ligands (81). The low-spin heme seems to form a cytochrome unit which is capable of electron transfer (84) whereas the high-spin heme has the peroxidative function (81).

Rapid-scan spectrometry and stopped-flow measurements have shown that the totally ferric enzyme reacts with H_2O_2 to form a complex, but the reaction is slow with a half-life of about 100 ms. This primary complex is slowly converted to a secondary one (half-life 16 s). Both of these species are catalytically inactive toward reduced substrates like cytochrome *c*-551 (85,86). This is in agreement with earlier steady-state kinetic observations that the enzyme has to be partially reduced before the peroxidatic oxidation of reduced substrates can be initiated (84).

In contrast, reaction of partially reduced enzyme with H_2O_2 leads to highly reactive intermediates. Rapid-scan spectrometry revealed two different spectral forms, one assigneable to Compound I and the other to Compound II as found in the reaction cycle of other peroxidases. The primary compound between the peroxidase and H_2O_2 is formed a few milliseconds after mixing. The combination of the reactants is nearly diffusion-controlled, the second-order rate constant being $1.2 \times 10^8\ M^{-1}s^{-1}$. The formation of Compound I is maximal when the enzyme is half-reduced (85,86).

EPR measurements have finally confirmed that in the half-reduced enzyme the low-spin heme group only is reduced while the high-spin heme is in the ferric state (87). Ferric PaCcP is readily reduced to half-reduction by various electron donors, e.g. reduced azurin and NADH in the presence of suitable mediators. The reduction causes a conformational change in the geometry of the high-spin heme showing that an interaction between the two hemes does exist.

From optical, resonance Raman and EPR spectra, the low-spin heme is suggested to have an imidazole nitrogen: heme-iron: methionine-sulfur hemochrome structure, and the high-spin heme is suggested to be mainly pentacoordinated with a histidine residue at the fifth co-ordination position. A comparison with the g- values of EPR spectra of different cytochrome c derivatives suggests that the iron remaining oxidized in the partially reduced enzyme has as the proximal ligand a deprotonated histidine rather than a neutral one as in the fully ferric protein. This change in ligation may cause a decrease in the oxidation-reduction potential of the heme, making this better suited to react with H_2O_2 in the half-reduced than in the fully oxidized form of the enzyme.

7. FOOTNOTE

The superscripts 1 - 5 on the subject titles indicate the authors and their addresses as shown on the beginning page of this article.

REFERENCES

1. Recent reviews: Dunford, H.B. and Stillman, J.S.: 1976, *Coord. Chem. Rev.* 19, pp. 187-251; Williams, R.J.P., Moore, G.R. and Wright, P.E.: 1977, "Biological Aspects of Inorganic Chemistry", pp. 369-441, Wiley-Interscience, N.Y.: Hewson, W.D. and Hager, L.P.: 1979, "The Porphyrins", 7, pp. 295-332, Academic Press, N.Y.
2. LaMar, G.N. and deRopp, J.S.: 1979, *Biochem. Biophys. Res. Commun.* 90, pp. 36-41.
3. Kobayashi, K., Tamura, M., Hayashi, K., Hori, H. and Morimoto, H.: 1980, *J. Biol. Chem.* 255, pp. 2239-2242.
4. Schulz, C.E., Devaney, P.W., Winkler, H., Debrunner, P.G., Doan, N., Chiang, R., Rutter, R., and Hager, L.P.: 1979, *FEBS Lett.* 103, pp. 102-108.
5. Welinder, K.G.: 1976, *FEBS Letters,* 72, pp. 19-23; Welinder, K.G.: 1979, *Eur. J. Biochem.* 96, pp. 483-502.
6. Haschke, R.H. and Friedhoff, J.M., *Biochem. Biophys. Res. Commun.,* 80, pp. 1039-1042: Ogawa, S., Shiro, Y. and Morishima, I.: 1979, *Biochem. Biophys. Res. Comm.,* 90, pp. 674-678.

7. Clarke, J. and Shannon, L.M.: 1976, *Biochem. Biophys. Acta,* 427, pp. 428-442.
8. Ohlsson, P.-I. and Paul, K.-G.: 1976, *Acta Chem. Scand.* B30, pp. 373-375.
9. Araiso, T. and Dunford, H.B.: 1980, *Biochem. Biophys. Res. Commun.* 94, pp. 1177-1182.
10. Poulos, T.L. and Kraut, J.: 1980, *J. Biol. Chem.* 255, pp. 8199-8205.
11. Takio, K., Titani, K., Ericsson, L.H. and Yonetani, T.: 1980, *Arch. Biochem. Biophys.* 203, pp. 615-629.
12. Morishima, I., Ogawa, S., Inubushi, T., Yonezawa, T. and Izuka, T.: 1977, *Biochemistry,* 16, pp. 5109-5115.
13. Dunford, H.B., Hewson, W.D. and Steiner, H.: 1978, *Can. J. Chem.* 56, pp. 2844-2852.
14. Ricard, J., Teisserre, M., Azou, Y. and Penon, P.: 1976, *J. Microscopic Biol. Cell.,* 26, pp. 139-150.
15. Mazza, G., Charles, C., Bouchet, M., Ricard, J. and Reynaud, J.: 1968, *Biochem. Biophys. Acta,* 167, pp. 89-98.
16. Job, D. and Ricard, J.: 1975, *Arch. Biochem. Biophys.,* 170, pp. 427-437.
17. Job, D., Ricard, J. and Dunford, H.B.: 1977, *Arch. Biochem. Biophys.* 179, pp. 95-99.
18. Mazza, G., Job, C. and Bouchet, M.: 1973, *Biochem. Biophys. Acta* 322, pp. 218-223.
19. Welinder, K.B. and Mazza, G.: 1975, *Eur. J. Biochem.* 57, pp. 415-424.
20. Mazza, G. and Welinder, K.G.: 1980, *Eur. J. Biochem.* 108, pp. 481-489.
21. Job, D. and Dunford, H.B.: 1977, *Can. J. Biochem.* 55, pp. 804-811.
22. Ricard, J., Mazza, G. and Williams, R.J.P.: 1972, *Eur. J. Biochem.* 28, pp. 566-578.
23. Job. D., Ricard, J. and Dunford, H.B.: 1978, *Can. J. Biochem.* 56, pp. 702-707.
24. Job, D. and Dunford, H.B.: 1976, *Eur. J. Biochem.* 66, pp. 607-614.
25. Ricard, J., Job, D., Santimone, J. and Cambe, H.: 1969, *C.R. Acad. Sci. Paris* 269, pp. 1445-1448.
26. Ricard, J. and Job, D.: 1974, *Eur. J. Biochem.* 44, pp. 359-374.
27. Job, D.: 1975, *Doctoral Thesis, Aix-Marseille University, France.*
28. Job, D. and Jones, P.: 1978, *Eur. J. Biochem.* 86, pp. 565-572.
29. Dunford, H.B. and Araiso, T.: 1979, *Biochem. Biophys. Res. Commun.* 89, pp. 764-768.
30. Job, D., Zéba, B., Puppo, A. and Rigaud, J.: 1980, *Eur. J. Biochem.* 107, pp. 491-500.
31. Jones, P., Robson, T. and Brown, S.B.: 1973, *Biochem. J.* 135, pp. 353-359.

32. Kelly, H.C., Davies, D.M., King, M.J. and Jones, P.: 1977, *Biochemistry* 16, pp. 3543-3549.
33. Davies, D.M., Jones, P. and Mantle, D.: 1976, *Biochem. J.* 157, pp. 247-253.
34. Welinder, K.G. and Mazza, G.: 1977, *Eur. J. Biochem.* 73, pp. 353-358.
35. Hager, L.P.: 1970 in "Methods in Enzymology" (Colowick, S.P. and Kaplan, N.O., eds.) Vol. 17, Part A, pp. 648-652, Academic Press, New York.
36. Morris, D.R. and Hager, L.P.: 1966, *J. Biol. Chem.* 241, pp. 1763-1768.
37. Shaw, P.D. and Hager, L.P.: 1961, *J. Biol. Chem.* 236, pp. 1626-1630.
38. Hager, L.P., Morris, D.R., Brown, F.S. and Eberwein, H.: 1966, *J. Biol. Chem.* 241, pp. 1769-1777.
39. Thomas, J.A., Morris, D.R. and Hager, L.P.: 1970, *J. Biol. Chem.* 245, pp. 3129-3134.
40. Thomas, J.A., Morris, D.R., and Hager, L.P.: 1970, *J. Biol. Chem.* 245, pp. 3135-3142.
41. Morris, D.R. and Hager, L.P.: 1966, *J. Biol. Chem.* 241, pp. 3582-3589.
42. Stern, J.O. and Peisach, J.: 1974, *J. Biol. Chem.* 249, pp. 7495-7498.
43. Hollenberg, P.F. and Hager, L.P.: 1973, *J. Biol. Chem.*, 248, pp. 2630-2633.
44. Champion, P.M., Lipscomb, J.D., Münck, E., Debrunner, P., and Gunsalus, I.C.: 1975, *Biochemistry* 14, pp. 4151-4158.
45. Champion, P.M., Chiang, R., Münck, E., Debrunner, P. and Hager, L.P.: 1975, *Biochemistry* 14, pp. 4159-4166.
46. Cramer, S.P., Dawson, J.H., Hager, L.P. and Hodgson, K.O.: 1978, *J. Amer. Chem. Soc.* 100, pp. 7282-7290.
47. Hollenberg, P.F., Hager, L.P., Blumberg, W.E. and Peisach, J.: 1980, *J. Biol. Chem.* 255, pp. 4801-4807.
48. Dawson, J.H., Trudell, J.R., Barth, G., Linder, R.E., Bunneberg, R.E., Djerassi, C., Chiang, R. and Hager, L.P.: 1976, *J. Am. Chem. Soc.* 98, pp. 3709-3710.
49. See G. Wagner's paper in this volume.
50. Chiang, R., Makino, R., Spomer, W.E. and Hager, L.P.: 1975, *Biochemistry* 14, pp. 4166-4171.
51. Roberts, J.E., Hoffman, B.M., Rutter, R. and Hager, L.P.: 1981, *J. Biol. Chem.* 256, pp. 2118-2121.
52. Palcic, M.M., Rutter, R., Araiso, T., Hager, L.P. and Dunford, H.B.: 1980, *Biochem. Biophys. Res. Comm.* 94, pp. 1123-1127.
53. Dolphin, D. and Felton, R.H.: 1974, *Accts. Chem. Res.* 7, pp. 26-32.
54. Dolphin, D., Forman, A., Borg, D.C., Fajer, T. and Felton, R.H.: 1971, *Proc. Natl. Acad. Sci. U.S.A.* 68, pp. 614-618.
55. DiNello, R.K. and Dolphin, D.: 1979, *Biochem. Biophys. Res. Comm.* pp. 190-194.

56. DiNello, R.K. and Dolphin, D.H.: 1981, *J. Biol. Chem.* 256, pp. 6903-6912.
57. Ballou, D.P. and Palmer, G.A.: 1974, *Anal. Chem.* 46, pp. 1248-1253.
58. Rutter, R., Hager, L.P. and Debrunner, P.: 1981, in "Interaction Between Iron and Proteins in Oxygen and Electron Transport" (Gibson, Q.H. and Ho, C., eds.), Elsevier North Holland, in press.
59. Hager, L.P., Doubek, P., Silverstein, R., Hargis, J.H. and Martin, J.C.: 1972, *J. Am. Chem. Soc.* 94, pp. 4364-4366.
60. Roberts, J.E., Hoffman, B.M., Rutter, R. and Hager, L.P.: 1981, Submitted to *J. Amer. Chem. Soc.*
61. Klebanoff, S.J. and Clark, R.A.: 1978, "The Neutrophil: Function and Clinical Disorders", pp. 409-448, North-Holland Publishing Company, Amsterdam.
62. Agner, K.: 1970, in "Structure and Function of Oxidation-Reduction Enzymes" (Åkesson, Å. and Ehrenberg, A. eds.) pp. 329-335, Pergamon Press, Oxford.
63. Makinen, K.K. and Tenovou, J.: 1976, *Acta Odont. Scand.* 34, pp. 141-150.
64. Aune, T.M. and Thomas, E.L.: 1977, *Eur. J. Biochem.* 80, pp. 209-214.
65. Kersten, H.W., Moorer, W.R. and Wever, R.: 1981, *J. of Dental Res.* 60, pp. 831-837.
66. Bakkenist, A.R.J., Wever, R., Vulsma, T., Plat, H. and Van Gelder, B.F.: 1978, *Biochim. Biophys. Acta* 524, pp. 45-54.
67. Ehrenberg, A. and Agener, K.: 1958, *Acta Chem. Scand.* 12, pp. 95-100.
68. Morrison, M., Hamilton, M.B. and Stotz, E.: 1957, *J. Biol. Chem.* 228, pp. 767-776.
69. Bakkenist, A.R.J., De Boer, J.E.G., Plat, H. and Wever, R.: 1980, *Biochim. Biophys. Acta* 613, pp. 337-348.
70. Bos, A.J., Wever, R., Hamers, M.N. and Roos, D.: 1981, *Infect. Immun.* 32, pp. 427-431.
71. Wever, R., Plat, H. and Hamers, M.N.: 1981, *FEBS Lett.* 123, pp. 327-331.
72. Wever, R., Hamers, M.N., De Graaf, C.J., Weening, R.S. and Roos, D.: 1980, Proc. Scand. Eur. Conf. on "Phagocytic Leucocytes Aurisina", (Trieste, Italy) Sept. 15-18, Plenum Press (In press).
73. Dixon, M.: 1953, *Biochem. J.* 55, pp. 161-170.
74. Altschul, A.M., Abrams, R. and Hogness, T.R.: 1940, *J. Biol. Chem.* 136, pp. 777-794.
75. Lenhoff, H.F. and Kaplan, N.O.: 1953, *Nature* 172, pp. 730-731.
76. Lenhoff, H.F. and Kaplan, N.O.: 1956, *J. Biol. Chem.* 220, pp. 967-982.
77. Ellfolk, N. and Soininen, R.: 1970, *Acta Chem. Scand.* 24, pp. 2126-2136.

78. Soininen, R., Sojonen, H. and Ellfolk, N.: 1970, *Acta Chem. Scand.* 24, pp. 2314-2320.
79. Ellfolk, N. and Soininen, R.: 1971, *Acta Chem. Scand.* 25, pp. 1535-1540.
80. Soininen, R. and Ellfolk, N.: 1973, *Acta Chem. Scand.* 27, pp. 35-46.
81. Rönnberg, M. and Ellfolk, N.: 1979, *Biochim. Biophys. Acta* 581, pp. 325-333.
82. Rönnberg, M., Ellfolk, N. and Soininen, R.: 1979, *Biochim. Biophys. Acta* 578, pp. 392-400.
83. Rönnberg, M., Osterlund, K. and Ellfolk, N.: 1980, *Biochim. Biophys. Acta* 626, pp. 23-30.
84. Rönnberg, M. and Ellfolk, N.: 1978, *Biochim. Biophys. Acta* 504, pp. 60-66.
85. Araiso, T., Rönnberg, M., Dunford, H.B. and Ellfolk, N.: 1980, *FEBS Lett.* 118, pp. 99-102.
86. Rönnberg, M., Araiso, R., Ellfolk, N. and Dunford, H.B.: 1981, *Arch. Biochem. Biophys.* 207, pp. 197-204.
87. Aasa, R., Ellfolk, N., Rönnberg, M. and Vänngård, T.: 1981, *Biochim. Biophys. Acta* 670, in press.

THE STATE OF PROTONATION OF THE PROXIMAL HISTIDYL IMIDAZOLE IN HORSERADISH PEROXIDASE

Gerd N. La Mar, V.P. Chacko, and Jeffrey S. de Ropp

Department of Chemistry
University of California
Davis, California 95616

ABSTRACT

The ease of formation of iron(IV) complexes in horseradish peroxidase as well as in other hemoproteins has been cited to support the view that in these proteins the proximal histidyl imidazole is deprotonated. However, most of the supporting evidence for this proposal has been indirect. From a detailed comparison of the proton nmr spectra of horseradish peroxidase in several oxidation/spin/ligation states with other hemoproteins and model compounds, it is shown here, by direct detection of the N_1H signal, that the proximal histidyl imidazole in horseradish peroxidase is definitely not deprotonated, but most likely involved in hydrogen bonding interactions with the protein in at least the reduced and the cyanide-ligated states.

INTRODUCTION

One of the basic hypotheses of the action of hemoproteins is that the reactivity of the iron center is largely controlled by the state of the protein ligand *trans* to the vacant site on the iron. In the majority of hemoproteins whose function, at least in part, is to coordinate a ligand, this protein ligand is a histidyl imidazole, commonly referred to as the proximal histidyl imidazole, as shown in A of Figure 1. The two control mechanisms for heme reactivity most frequently discussed are some form of steric strain (1) on the Fe-N_3 bond, as originally proposed by Perutz for the tense or T state of hemoglobin, and strong hydrogen bonding (B in Figure 1) or complete deprotonation (2-6) of the proximal histidyl imidazole N_1H (C in Figure 1). Essentially

H. B. Dunford et al. (eds.), The Biological Chemistry of Iron, 357–373.

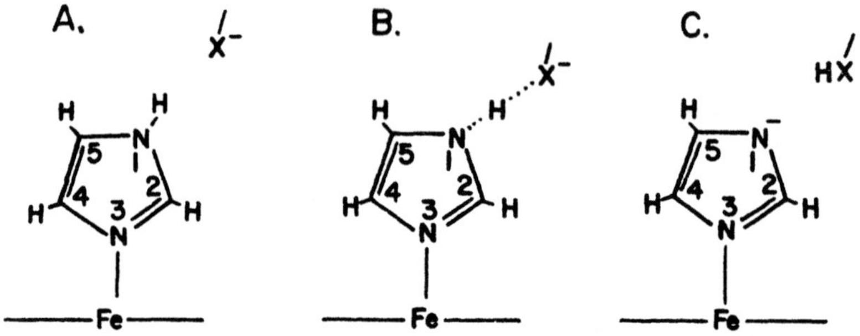

Figure 1: Structure of an imidazole coordinated to an iron porphyrin. A, no interaction between N_1H and the protein; B, hydrogen bonding between N_1H and X^- of protein; and C, complete deprotonation leading to coordinated imidazolate.

every x-ray crystallographic study of hemoglobin has revealed some form of interaction between the proximal histidyl imidazole N_1H and a backbone carbonyl (4). Studies of the CO recombination kinetics for neutral imidazole and imidazolate anion porphyrin complexes of iron(II) have demonstrated a very strong *trans* effect consistent with the imidazole acting as a stronger σ base (7).

For the protein of present interest, horseradish peroxidase, HRP, in particular, and heme peroxidases in general, the facility with which they form iron(IV) complexes has favored the view that the proximal histidyl imidazole is deprotonated (2). Evidence supporting this view is the requirement of proton uptake upon ligand binding (3), the closer similarity of the optical spectra of reduced HRP and its CO adduct to those of model compounds containing the imidazolate anion rather than neutral imidazole (8), and the much higher $Fe-N_3$ resonance Raman stretching frequency, ($\nu(Fe-N_3) \sim 244\ cm^{-1}$) in reduced HRP (5) as compared to deoxy myoglobin ($\nu(Fe-N_3) \sim 220\ cm^{-1}$). Recent resonance Raman studies of high-spin ferrous model compounds have shown that $\nu(Fe-N_3)$ indeed varies with the hydrogen bonding acceptor strength of the solvent (6). In the case of cytochrome b_5, evidence for a deprotonated imidazole has been based on the esr spectrum of the low-spin ferric form of the protein.

However, all of the evidence cited in support for deprotonated imidazole in hemoproteins in general, and for HRP in particular, has been indirect and hence necessarily ambiguous since it focuses on a portion of the heme crevice not directly involving the proximal histidine N_1H itself. While infrared spectroscopy could, in principle, yield direct information on the state of any $N_1H \cdots X^-$ interaction, the presence of a large number of such interactions in a protein makes resolution and assignment of the specific residue of interest a virtual impossibility. The other logical spectroscopic method whose parameters lend themselves to direct interpretation in terms of the state of a hydrogen bond is proton nuclear magnetic resonance, nmr. In a diamagnetic protein, hydrogen bonding can be followed by the well-characterized downfield shift, and deprotonation established by loss of the N_1H signal intensity. Unfortunately, the large number of such hydrogen bonding interactions present even in small proteins would make resolution of individual sites improbable due to the narrow spectral window and preclude unambiguous assignment of the N_1H signal of interest.

A special situation, however, exists for the case of paramagnetic hemoproteins which facilitates both the resolution and unambiguous assignment of the N_1H signal from a coordinated imidazole. Since all but one, (low-spin iron(II)), of the known oxidation/spin/ligation states of hemoproteins are paramagnetic (9), proton nmr is particularly suitable for investigating the state of the proximal histidyl imidazole N_1H.

INFORMATION CONTENT OF HYPERFINE SHIFTS

The hyperfine interaction between nuclei and the unpaired spin(s) on the iron are short-ranged and therefore influence the magnetic field primarily of protons close to the iron center (10). The net hyperfine shift, $(\Delta H/H)_{hf}$, defined as the shift from the normal diamagnetic resonance position of the same nucleus, is the sum of the dipolar and contact contribution. The former arises from a through-space interaction, falling off as the inverse cube of the distance from the iron, and is significant only for systems where the iron possesses appreciable magnetic anisotropy (*i.e.*, low-spin iron(III), intermediate-spin iron(II)); it can generally be neglected for both ferric and ferrous high-spin models and hemoproteins (9). The contact contribution to the hyperfine shift arises from direct delocalization of the iron unpaired spin(s) into molecular orbitals of the ligands coordinated to the iron; hence contact shift can arise only for the porphyrin, axial histidyl imidazole, and any sixth ligand.

The nature of the contact shift yields information as to the type of molecular orbital (σ or π) involved in the iron-ligand covalency, and the extent of the interaction (11). The characteristic probes for the type of transferred unpaired spin density are a proton and a methyl group attached to a given carbon or nitrogen. For dominant π spin density, the proton and methyl shifts have comparable magnitudes but opposite sign or direction. This is illustrated in the proton nmr spectrum of the low-spin di-cyanoferric deuterohemin complex (12) by the pyrrole-H and -CH_3 shifts, as shown in A of Figure 2.

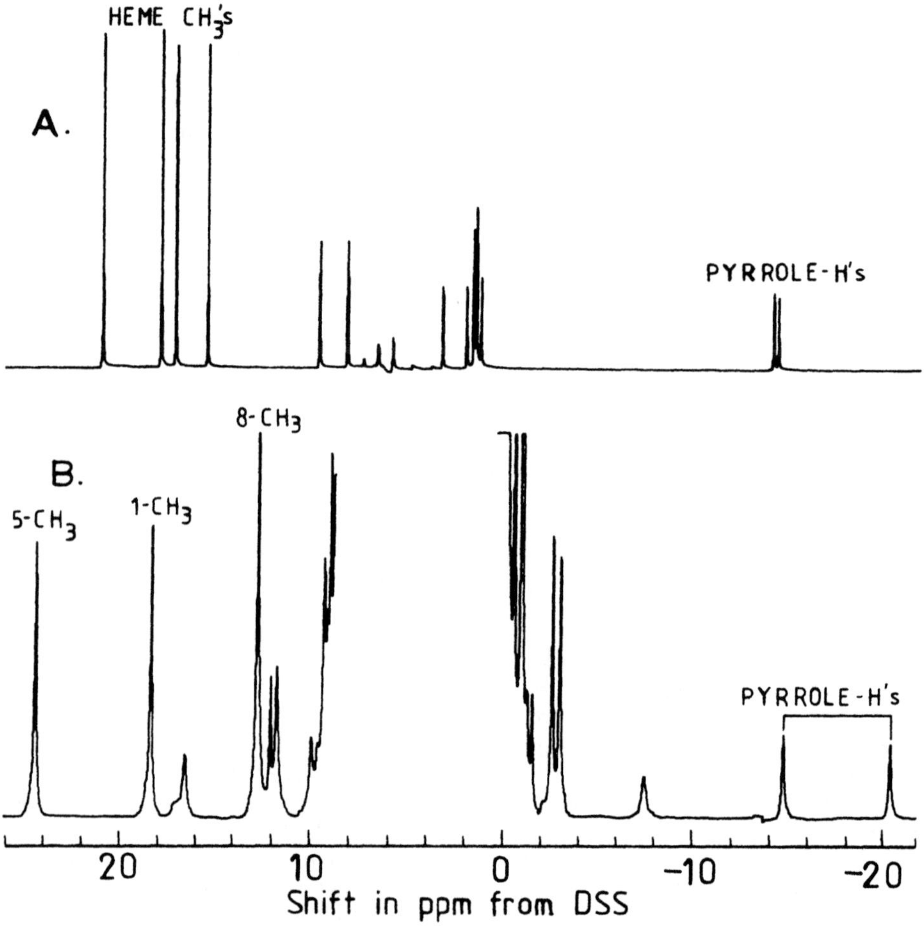

Figure 2: 360 MHz ^1H nmr spectrum of: A, biscyanoferric deuterohemin complex in D_2O at 25°C, pH 9; and B, deuterohemin-reconstituted metcyano-sperm whale myoglobin (metSWMbCN) in D_2O at 25°C, pH 8.5; the diamagnetic region between 8 and 0 ppm is omitted in B (sodium 2,2-dimethyl-2-silapentane-5-sulfonate, DSS, is used as internal reference).

In the case of σ spin density, the shifts are generally much larger than for π spin density and are all downfield, with a tendency for the shift magnitudes to attenuate with distance from the iron (11). Thus high-spin iron(III) porphyrin complexes exhibit substituent shifts consistent with dominant σ spin density (9). For both low-spin and high-spin ferric systems, similar hyperfine shift patterns are observed in models and proteins for the same functional group, as illustrated by comparison of the nmr spectrum of the model with met-cyano deuterohemin-reconstituted myoglobin (B in Figure 2). Comparable similarities have been observed for high-spin ferrous and ferric systems (13). It is apparent that the paramagnetic interactions, and hence the iron-porphyrin bonding in models and proteins, are very similar. The major influence of the protein is seen to be the presence of a much larger rhombic or in-plane asymmetry in the protein environment, probably due to the effect of the iron-imidazole bond (13,14).

Because the hyperfine shifts for identical functional groups for a given oxidation/spin state of iron are so similar in models and proteins (13), model compounds have served as important initial indicators of protein heme resonance assignments. This method is most valuable in high-spin *c*-type hemoproteins (15) where even the indirect Overhauser methods usable in low-spin cytochromes *c* (16) are of little value. In the *b*-type hemoproteins, where the heme can be reversibly removed, specific deuterium labeling has been found necessary for establishing the single-proton and individual methyl assignments (17-19).

The nature of the iron-imidazole bonding can be probed in a very similar fashion. For example, in five-coordinated, high-spin ferrous model complexes, the hyperfine shifts for the axial ligand have been shown to be very large and all in one direction, consistent only with dominant unpaired spin transfer due to iron-imidazole σ covalency (20). The proton nmr spectrum of the mono-2-methylimidazole complex of ferrous tetraphenylporphyrin is illustrated in A of Figure 3. While the 2-H and 4-H peaks tend to be broad due to their proximity to the iron (the 5-H is replaced by a $-CH_2$ in histidine) the exchangeable N_1H peak is readily and uniquely identified by comparison with the spectrum of the same complex with the N_1H exchanged with deuterium (21), as shown in B of Figure 3. In the case of low-spin ferric iron, the imidazole hyperfine shif-s exhibit some dipolar contribution (9,14), but the dominant effect is due to π contact shifts, as evidenced by shifts of opposite sign for -H and $-CH_3$ at various positions (22, 23). Again, the exchangeable N_1H of the coordinated imidazole is readily assigned by adding minute amounts of CD_3OD which replaces N_1H with N_1D, as shown in A and B of Figure 4.

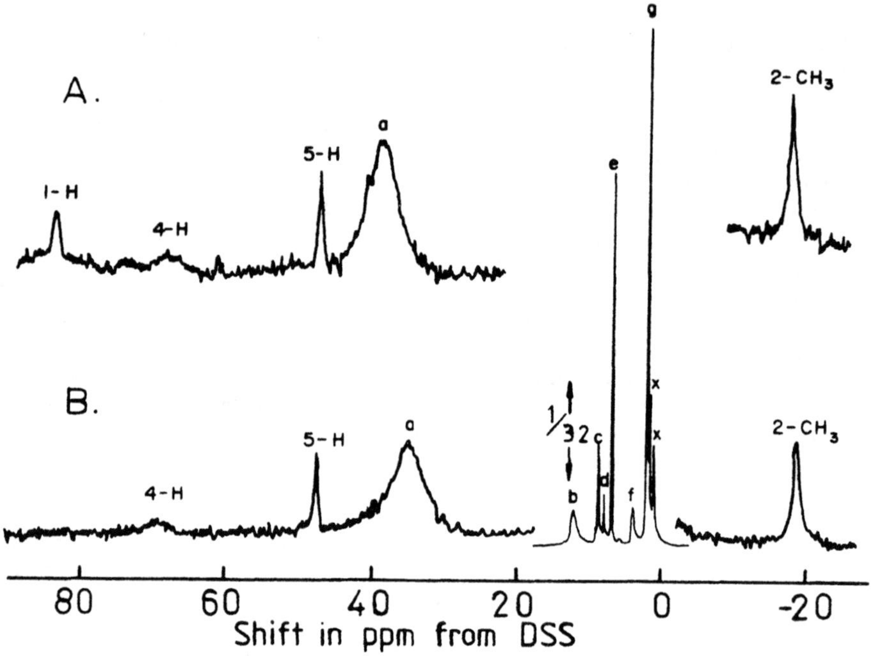

Figure 3: 100 MHz ^1H nmr spectrum of the mono-2-methylimidazole complex of ferrous tetraphenylporphyrin, PFe, in benzene-d_6 at 25°C; A, 0.010 M PFe and 0.003 M 1-D-2-CH_3-Im; peaks a-d and g are the average porphyrin peaks; e is the residual solvent peak.

Thus, the most likely axial imidazole resonance which can be expected to be resolved and assigned in a protein is that due to the exchangeable N_1H; the peak should be present in H_2O solution but disappear in D_2O solution. Although the spectral windows in which N_1H peaks are found in most oxidation/spin states of models and hemoproteins are relatively unique, their assignments can be further substantiated on the basis of measured spin-lattice relaxation times of the N_1H, which when compared to that for an assigned heme methyl, yields a distance from the iron which has been shown to be consistent with the exchangeable proton being attached to the imidazole ring (24). The assignments are further strengthened by observing solely base catalysis for the exchange of N_1H with bulk water protons (24,25).

The detection of a hyperfine-shifted N_1H in a hemoprotein is direct evidence against deprotonation of the proximal histidyl imidazole. Very similar hyperfine shifts for the N_1H in a protein and its model complex also argue against important contributions from hydrogen bonding. Thus, the large downfield bias for N_1H in sperm whale deoxy myoglobin (A and B in Figure 5) is also due to

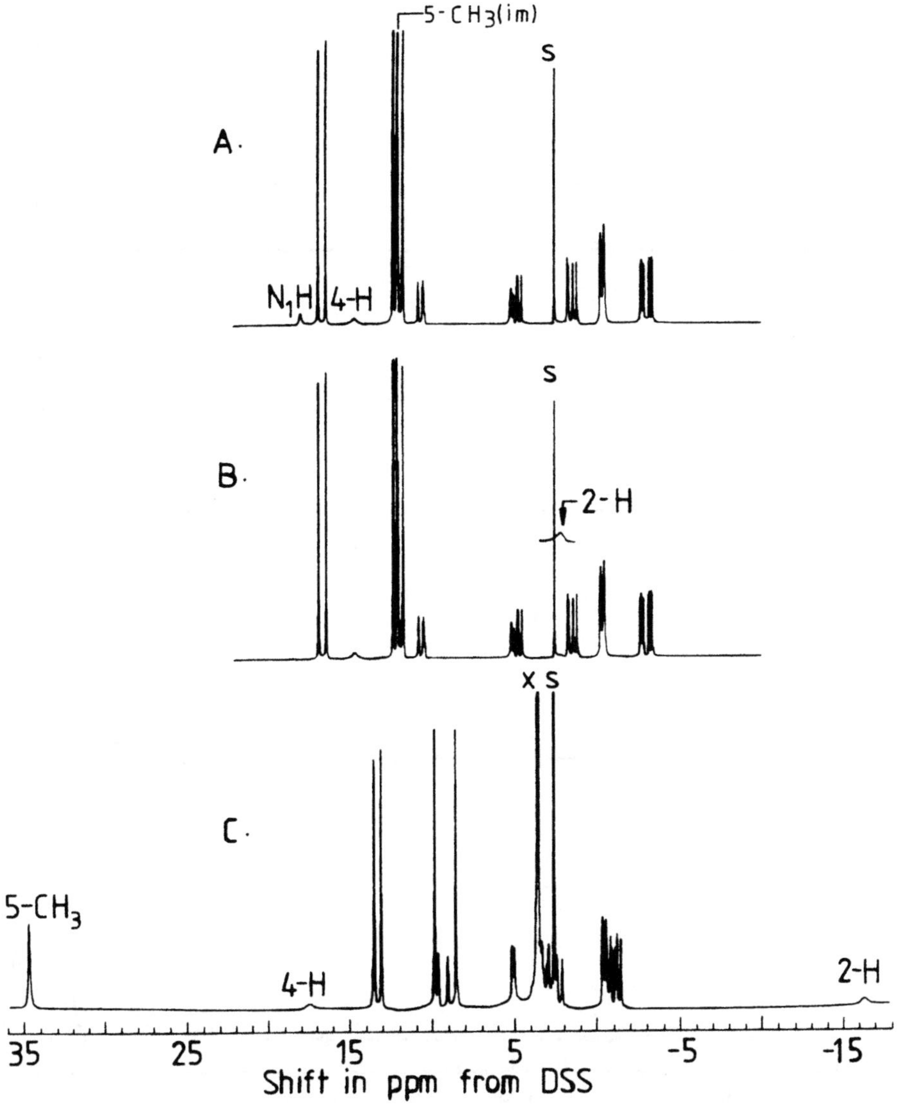

Figure 4: 360 MHz ^1H nmr spectrum of: A, ferric protoporphyrinIX mono-cyano-mono-5-methylimidazole complex in DMSO-d_6 at 25°C; B, same as A, except 1-D-5-methylimidazole is used; and C, ferric protoporphyrinIX monocyano-mono-5-methylimidazolate complex in DMSO-d_6 at 25°C, produced by addition of NaOD/D_2O to the sample in A; the position of 2-H, as determined by deuterium nmr, is indicated by a dotted line at ~2 ppm in B; for clarity, peaks from other complexes present in the solution are deleted from this figure; s is the residual solvent peak and x marks the peak due to HOD; unassigned peaks are the porphyrin resonances.

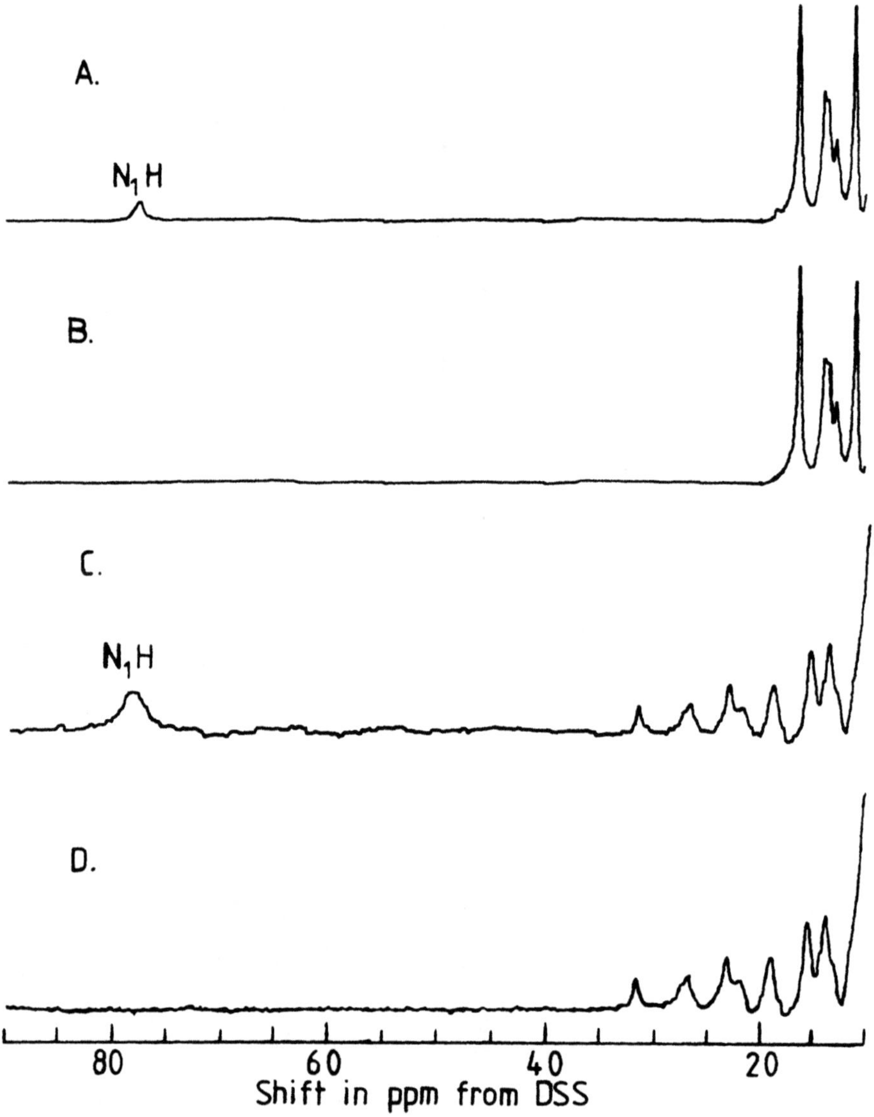

Figure 5: Low-field portions of the 200 MHz ^1H nmr spectra of: A, sperm whale deoxymyoglobin in H_2O at 25°C, pH 8.6; B, same as A, but in D_2O; C, ferrous HRP in H_2O at 25°C, pH 5.8; and D, same as C, but in D_2O.

σ spin transfer (21), and the similarity of the shift to that in the model (A in Figure 3) indicates negligible proton donation to a protein acceptor residue. Strong hydrogen bonding would necessarily lead to a sharp decrease in the N_1H hyperfine shift due to the interruption of the interaction between this proton and the unpaired spin density on the imidazole (26). In the limit of complete deprotonation, the hyperfine shift must go to zero. The downfield N_1H hyperfine shift found in high-spin ferric met-aquo-myoglobin (27) is similar to that found in deoxy myoglobin (21) and hence is also indicative of σ covalency. An appropriate model compound remains to be developed. Comparison of the spectra of met-aquo-myoglobin (27) in H_2O (A) and D_2O (B) in Figure 6 clearly reveals the N_1H signal. Again, extensive hydrogen bonding should strongly decrease the downfield shift.

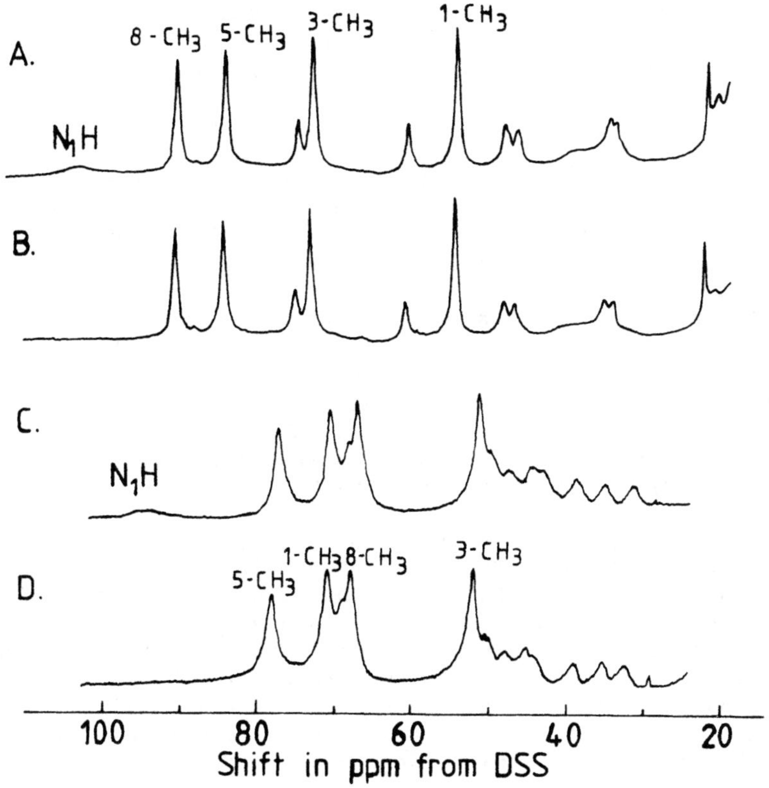

Figure 6: Low-field portions of 360 MHz ^1H nmr spectra of: A, sperm whale met-aquo-myoglobin at 35°C in H_2O, pH 6.2; B, same as A, but in D_2O; C, the resting state of HRP at 35°C in H_2O, pH 7.0; and D, same as C, but in D_2O.

In the case of low-spin iron (III) systems, the interpretation of the N_1H shift is less direct since analysis of the hyperfine shift origins has revealed that the dipolar interaction makes substantial contributions to the N_1H shift (14,22). In such a case the N_1H hyperfine shift is unlikely to yield direct information on any $N_1H \cdots X^-$ hydrogen bonding. However, as we show below, indirect methods utilizing the shift position of the non-exchangeable proximal histidyl imidazole protons can yield evidence in certain proteins for partial imidazolate character for the axial ligand, and hence infer the presence of relatively strong hydrogen bonding.

DIRECT EVIDENCE AGAINST N_1H DEPROTONATION

Comparison of the downfield portions of the proton nmr spectra of reduced HRP in H_2O (C) and D_2O (D) in Figure 5 identifies the N_1H peak at 78 ppm from DSS; the signal is observed with unit intensity up to pH 10, although it changes its position (26). Similarly, resting state HRP in H_2O (C) and D_2O (D) in Figure 6 differ in their nmr spectra only in the peak at 96 ppm, leading to its unique assignment as the N_1H of the proximal histidine (27). Again, the intensity is maintained to pH 10, above which the protein converts to a low-spin form due to a second internal ligand binding at the sixth site. However, even in this alkaline, low-spin form the N_1H signal is clearly detectable, indicating that the imidazole is not deprotonated. Not only do these nmr data clearly prove that resting state and reduced HRP are not deprotonated, contrary to earlier interpretations based on indirect evidence (5,6,8), but show that the N_1H shift in HRP has essentially the same magnitude as found in myoglobin (21,27) in the same oxidation/spin state, arguing against major imidazolate character for the axial ligand due to proton donation in a hydrogen bond to the presumed glutamate acceptor, as found in the cytochrome *c* peroxidase, CcP (28).

It has been proposed that the binding of substrates such as indole propionic acid to HRP involves a hydrogen bond to the proximal histidyl N_1H (29). While substantial hyperfine shift changes are observed for a number of heme resonances of resting state HRP upon addition of indole propionic acid, the N_1H peak is unique in its total insensitivity to the substrate (17,27). Hence the nmr data exclude direct interaction between substrate and the axial histidine but suggest a probable binding site near the carboxyl groups of the propionic acid side chains.

A feature of interest not apparent in the HRP nmr spectra presented in Figure 5 and 6 is that when the protein is first dissolved in D_2O, the N_1H peaks can actually be detected at full intensity. Only after a time is there exchange between the N_1H and the D_2O solvent which leads to loss of the peak intensity and its eventual disappearance (26,27,30), as shown in D of Figures 5 and 6.

The half-life for exchange for resting state HRP is over one hour at 50°C, while reduced HRP exhibits a N_1H exchange half-life of approximately 13 minutes at 35°C and pH 6.2. Thus, not only is the proximal histidyl imidazole not deprotonated in these two forms of HRP, but the proximal side of the heme crevice is strongly shielded against penetration by either water or hydroxyl ions. The exchange rates for myoglobins in the same oxidation/spin states are on the order of tens of milliseconds (24,25). Since exchange rates with bulk water of internal labile protons are direct indicators of the flexibility or dynamic stability of the heme pocket (31), it can be concluded that HRP is a much tighter and less flexible protein than myoglobin in its high-spin ferric and ferrous states. The necessarily tighter interactions in the heme cavity, also noted from the absence of internal motions of heme peripheral side chains (17), may play some role in stabilizing the highly oxidized compound I which cannot be formed in myoglobin.

All attempts to resolve the N_1H signal for the proximal histidyl imidazole in compound I of HRP failed. The assigned proton nmr trace in D_2O is reproduced in Figure 7; no additional peaks could be detected in H_2O solution (19). The inability to detect this peak could have several bases; the signal could be shifted and broadened beyond detection (considered unlikely since calculations put little spin density on the axial imidazole in compound I), the signal is unshifted and hence not resolvable in the crowded diamagnetic

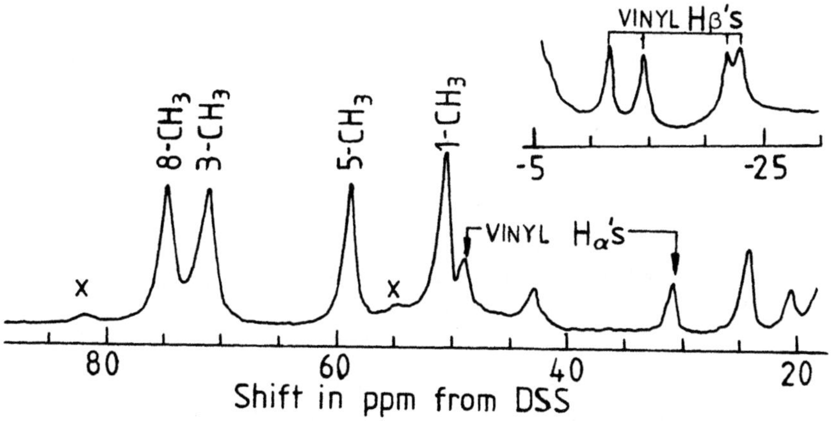

Figure 7: 360 MHz ^1H nmr spectrum of HRP compound I in D_2O at 25°C; x marks peaks from re-emerging HRP; all peak assignments were determined by deuterium labeling; diamagnetic portion is not shown.

envelope (the more likely reason), or the imidazole is deprotonated when compound I is formed. Evidence against this last possibility can be obtained by taking advantage of the very slow exchange with D_2O of the N_1H of freshly dissolved HRP in the resting state. When such freshly dissolved HRP in D_2O is treated with hydrogen peroxide to yield compound I for up to 20 minutes, and allowed to decay back to the resting state, no loss of N_1H intensity was observed when compared to a control sample which was dissolved in D_2O and kept in the resting state at the same pH and temperature for the same duration (30,32). Hence, N_1H exchange with solvent must be even slower in compound I than in resting state HRP. These results strongly support that the axial imidazole is not deprotonated even in the highly oxidized compound I. Since the decay of compound I goes by one electron steps through compound II, it is highly likely that these conclusions also pertain to that state. Thus deprotonation of the proximal histidine does not appear to occur in any of the physiologically relevant states of this enzyme.

Some evidence for hydrogen bonding of the N_1H is obtained from the nmr data of unligated reduced HRP. In the reduced form, HRP undergoes a conformational change modulated by a single proton with pK 7.4 (26). On converting from the acid to the alkaline forms, the N_1H contact shift increases by approximately 8 ppm. For this same transition, the resonance Raman $\nu(Fe-N_3)$ decreases by 3 cm^{-1} (5). Simple analysis of the influence of both steric strain and N_1H hydrogen bonding on the nmr and resonance Raman parameters has shown (26) that conformational transitions which modulate primarily the iron-imidazole bond tension necessarily change the magnitude of the N_1H contact shift and resonance Raman $\nu(Fe-N_3)$ in the same direction, while perturbations on the histidine involving primarily changes in hydrogen bonding require that the magnitude of the nmr and resonance Raman parameters vary in opposite directions. Since the latter situation is found to occur for reduced HRP, the 8 ppm change in N_1H shift must arise mainly from changes in hydrogen bonding. While we can conclude that N_1H hydrogen bonding must be stronger in the acid than the alkaline form of the protein, the absence at this time of any established theoretical framework for interpreting the nmr data precludes any more quantitative conclusions as to the extent of hydrogen bonding in either state.

Comparison of the proton nmr spectra of the non-physiologically relevant low-spin HRP-CN in H_2O and D_2O (A and B in Figure 8) reveals two exchangeable peaks of which the one at ~32 ppm from DSS has been shown to arise from the proximal histidyl N_1H based on relaxation data and the mechanism of exchange with bulk solvent (30,32). The shift is somewhat larger than that displayed by metmyoglobin cyanide (24,25,33) in H_2O (C in Figure 8). Also, the distal histidyl imidazole NH is much further from the iron center in HRP-CN than metMbCN where it interacts with the coordinated ligand (24). The proximal histidyl imidazole N_1H in HRP-CN is

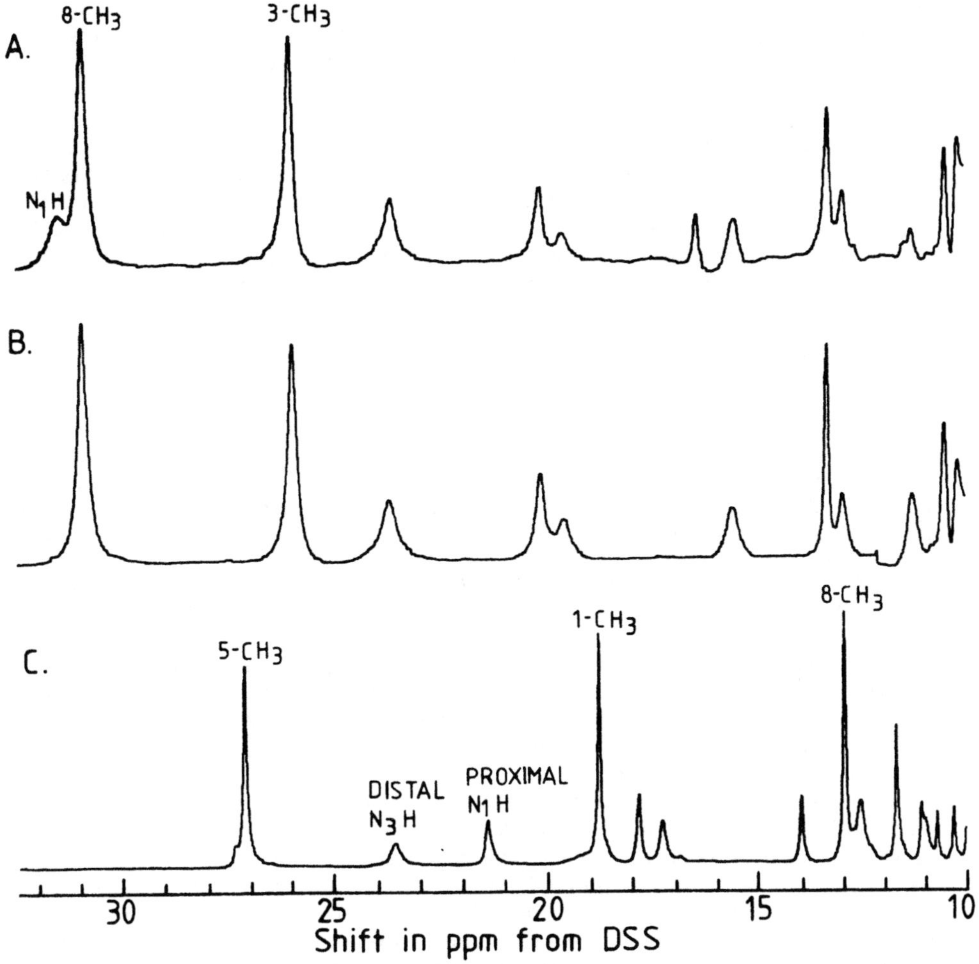

Figure 8: Low-field portions of the 360 MHz ^1H nmr spectra of: A, HRP-CN in H_2O at 25°C, pH 7.0; B, HRP-CN in D_2O at 25°C, pH 6.8; and C, met-SWMb-CN in H_2O at 25°C, pH 8.5.

detectable to pH 9 where the protein undergoes a structural transition to some alkaline form which is still low-spin and retains the cyanide ligand (30,32). The N_1H signal is observed in H_2O even in this alkaline form at pH 11.2 Thus we can conclude that deprotonation is absent for the cyanide complex over the complete accessible pH range. The major difference between HRP-CN and all of the other forms discussed above is that half-life for exchange with bulk solvent for the N_1H is only approximately 30 milliseconds at pH 7 and exhibits strong base catalysis (30,32), as previously found for metMbCN (24). Hence the cyanide complex of HRP must have a much more flexible structure which permits ready access of both water and hydroxyl ions to the proximal side of the heme cavity.

Although the N_1H hyperfine shift is not as amenable to interpretation in this protein form as in the high-spin species because of comparable contact and dipolar and pseudocontact contributions (9,14), closer analysis of the hyperfine shifts for the non-exchangeable protons on the proximal histidyl imidazole provides some indirect evidence for strong hydrogen bonding in this state.

INDIRECT EVIDENCE FOR N_1H HYDROGEN BONDING

Proton nmr spectra of low-spin model porphyrin complexes of iron(III) possessing a cyanide and neutral imidazole ligand exhibit coordinated imidazole peaks with the 4-H peak near 16 ppm and the 2-H resonance near 0 ppm (just downfield from DSS), independent of any other imidazole or porphyrin substituents (A and B in Figure 4). Upon deprotonating the imidazole with potassium hydroxide to yield the cyanide/imidazolate complex (C in Figure 4), the 4-H peak is affected little but the 2-H peak moves upfield dramatically (34). The extent of this upfield shift depends somewhat on the nature of any imidazole substituent and is illustrated in Figure 4 for 5-methyl-imidazole, which bears the methyl where histidine has a methylene group. This substantial upfield bias of the axial imidazole 2-H peak upon deprotonation can, in principle, serve as a sensitive indicator of the degree of proton donation of the axial imidazole in hemoproteins. Unfortunately, assignment of the non-exchangeable imidazole peaks in proteins is difficult and not unambiguous. However, the studies of model complexes (22,23,24), as well as the predictions of dipolar relaxation (35) indicate that the 2-H and 4-H resonances whould exhibit 6-8 times the broadening of a heme methyl. Such peaks would necessarily be much broader than any proton arising from the heme itself. In fact, a pair of such broad resonances completely consistent with arising from the imidazole 2-H and 4-H have recently been observed at high field strength (36). Spectra of metMbCN (25), CcP-CN (37), and HRP-CN (30), depicting at least the upfield resonances are illustrated in Figure 9; the peak presumed to arise from the 2-H is marked as such. In the case of the first two proteins, the presumed downfield 4-H's are also clearly seen. Consistent with expectations, metMbCN exhibits a 2-H peak near 0 ppm, as found in the model complex containing neutral imidazole (22,36). HRP-CN, on the other hand, displays a 2-H shift closer to that of the model with the imidazolate ligand. CcP-CN exhibits a shift close to but slightly smaller than that of HRP-CN. While the direct observation of the N_1H peak, (at least in HRP-CN; the spectrum in H_2O of CcP-CN has not yet been reported) precludes deprotonation, the 2-H shift is suggestive of

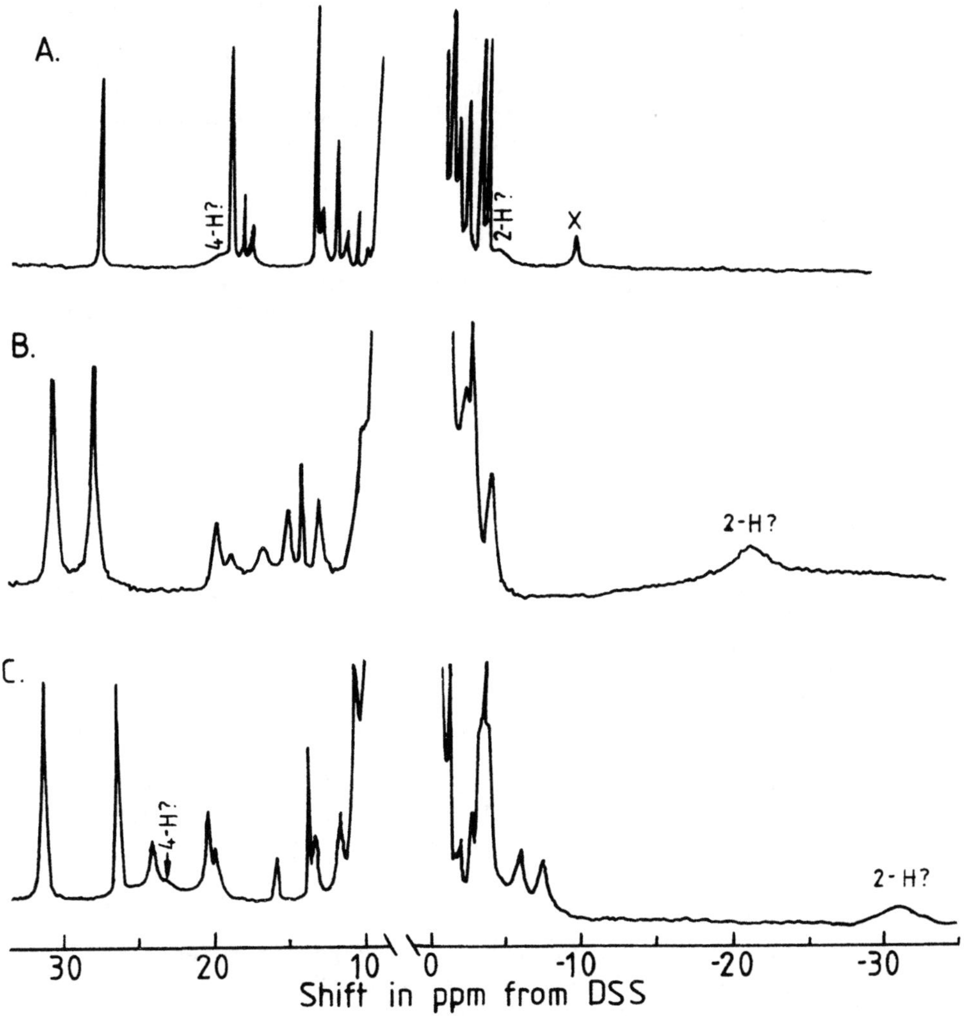

Figure 9: 360 MHz ^1H nmr spectra of: A, met-SWMb-CN in D_2O at 25°C, pH 7.5; B, HRP-CN in D_2O at 25°C, pH 7.0: and C, CcP-CN in D_2O at 25°C, pH 7.2; the diamagnetic region is omitted; x marks resonance due to a non-coordinated amino acid CH near the heme.

strong hydrogen bonding to some acceptor residue, presumably a glutamate, as found in the crystal structure of CcP (28). Again, quantitative interpretation of the 2-H shift is hampered by lack of knowledge of how other possible structural pertubations (*i.e.*, imidazole tilting) affect the 2-H hyperfine shifts.

However, in spite of current shortcomings in the quantitative interpretation of the proton nmr hyperfine shifts, the data unequivocally show that the proximal histidyl imidazole is not deprotonated in HRP, but that the N_1H is likely to be involved in some form of strong hydrogen bonding not present in myoglobins. This partial proton donation may aid in stabilizing the higher oxidation states in heme peroxidases.

ACKNOWLEDGEMENTS

This work was supported by grants from the National Institutes of Health (HL-16087, GM-26226) and National Science Foundation (CHE-81-08766).

REFERENCES

1. Perutz, M.F. and Ten Eyck, L.F.: 1971, *Cold Spring Harbor Symp. Quant. Biol.* 36, pp. 295-310.
2. Morrison, M. and Schonbaum, G.R.: 1976, *Ann. Rev. Biochem.* 45, pp. 86-127.
3. Nicholls, P.: 1962, *Biochim. Biophys. Acta,* 60, pp. 217-228.
4. Valentine, J.S., Sheridan, R.P., Allen, L.C. and Kahn, P.: 1979, *Proc. Nat. Acad. Sci. U.S.A.* 76, pp. 1009-1013.
5. Teraoka, J and Kitagawa, T.: 1980, *Biochem. Biophys. Res. Commun.* 93, pp. 694-700.
6. Stein, P., Mitchell, M. and Spiro, T.G.: 1980, *J. Am. Chem. Soc.* 102, pp. 7795-7797.
7. Swartz, J.C., Stanford, M.A., Moy, J.N., Hoffman, B.M. and Valentine, J.S.: 1979, *J. Am. Chem. Soc.* 101, pp. 3396-3398.
8. Mincey, T. and Traylor, T.G.: 1979, *J. Am. Chem. Soc.* 101, pp. 765-766.
9. La Mar, G.N. and Walker, F.A.: 1978, *The Porphyrins* 4, pp. 61-157.
10. Jesson, J.P.: 1973, in "NMR of Paramagnetic Molecules," La Mar, G.N., Horrocks, Jr., W.D. and Holm, R.H., eds., Academic Press, New York, pp. 1-52.
11. La Mar, G.N.: 1973, in "NMR of Paramagnetic Molecules," La Mar, G.N. Horrocks, Jr., W.D. and Holm, R.H., eds., Academic Press, New York, pp. 85-126.
12. La Mar, G.N., Viscio, D.B., Smith, K.M., Caughey, W.S. and Smith, M.L.: 1978, *J. Am. Chem. Soc.* 100, pp. 8085-8092.
13. La Mar, G.N.: 1979, in "Biological Applications of Magnetic Resonance," Shulman, R.G., ed., Academic Press, New York, pp. 305-343.
14. Shulman, R.G., Glarum, S.H. and Karplus, M.: 1971, *J. Mol. Biol.* 57, pp. 93-115.
15. La Mar, G.N., Jackson, J.T. and Bartsch, R.G.: 1981, *J. Am. Chem. Soc.* 103, pp. 4405-4410.

16. Keller, R.M. and Wüthrich, K.: 1978, *Biochim. Biophys. Acta* 533, pp. 195-208.
17. La Mar, G.N., de Ropp, J.S., Smith, K.M. and Langry, K.C.: 1980, *J. Biol. Chem.* 255, pp. 6646-6652.
18. La Mar, G.N., de Ropp, J.S., Smith, K.M. and Langry, K.C.: 1980, *J. Am. Chem. Soc.* 102, pp, 4833-4835.
19. La Mar, G.N., de Ropp, J.S., Smith, K.M. and Langry, K.C.: 1981, *J. Biol. Chem.* 256, pp. 237-243.
20. Goff, H. and La Mar, G.N.: 1977, *J. Am. Chem. Soc.* 99, pp. 6599-6606.
21. La Mar, G.N., Budd, D.L. and Goff, H.: 1977, *Biochem. Biophys. Res. Commun.* 77, pp. 104-110.
22. Satterlee, J.D. and La Mar, G.N.: 1976, *J. Am. Chem. Soc.* 98, pp. 2804-2808.
23. La Mar, G.N., Frye, J.S. and Satterlee, J.D.: 1976, *Biochim. Biophys. Acta* 428, pp. 78-90.
24. Cutnell, J.D., La Mar, G.N. and Kong, S.B.: 1981, *J. Am. Chem. Soc.* 103, pp. 3567-3572.
25. La Mar, G.N., Cutnell, J.D. and Kong, S.B.: 1981, *Biophys. J.* 34, pp. 217-226.
26. La Mar, G.N. and de Ropp, J.S.: *J. Am. Chem. Soc.* in press.
27. La Mar, G.N. and de Ropp, J.S.: 1979, *Biochem. Biophys. Res. Commun.* 90, pp. 36-41.
28. Poulos, T.L. and Kraut, J.: 1980, *J. Biol. Chem.* 255, pp. 8199-8205.
29. Schejter, A., Lanir, A. and Epstein, N.: 1976, *Arch. Biochem. Biophys.* 174, pp. 36-44.
30. de Ropp, J.S.: 1981, Ph.D. Thesis, University of California, Davis, pp. 191-224.
31. Woodward, C.K. and Hilton, B.D.: 1979, *Ann. Rev. Biophys. Bioeng.* 8, pp. 99-127.
32. La Mar, G.N. and de Ropp, J.S.: unpublished results.
33. Sheard, B., Yamane, T. and Shulman, R.G.: 1970, *J. Mol. Biol.* 53, pp. 35-48.
34. Chacko, V.P. and La Mar, G.N.: unpublished results.
35. Swift, T.J.: in "NMR of Paramagnetic Molecules," La Mar, G.N., Horrocks, Jr., W.D. and Holm, R.H., eds., Academic Press, New York, pp. 53-83.
36. La Mar, G.N. and de Ropp, J.S.: to be published.
37. Satterlee, J.D. and Erman, J.E.: 1981, *J. Am. Chem. Soc.* 103, pp. 199-201.

COORDINATION CHARACTERISTICS OF PROXIMAL HISTIDINE OF PLANT PEROXIDASES AND THEIR RELEVANCE TO THE HEME-LINKED IONIZATION

Teizo Kitagawa and Junji Teraoka

Department of Molecular Physiological Chemistry
Osaka University Medical School
Nakanoshima, Kitaku, Osaka, 530 Japan

The information obtainable from the resonance Raman spectra of hemeproteins is reviewed first, and then a correlation between the state of iron-bound histidine and the Fe-His stretching frequency is discussed. Finally a distinctive feature of the iron-coordinated histidine of plant peroxidases and its change upon distal heme-linked ionization is pointed out.

1. INTRODUCTION

Raman scattering arises from two simultaneous electronic transitions of a molecule namely, as shown in Fig. 1(A), excitation of a molecule from |m> to |e> with absorption of a photon ($h\nu_0$) and its relaxation from |e> to |n> with emission of a photon ($h\nu$). Generally |e> is a virtual state and $h\nu_0$ is not always equal to the true excitation energy ($= E_e - E_m$). Suppose that |m> and |n> represent the molecular states with v = 0 and v = 1, respectively, within a ground electronic state (v : a vibrational quantum number), ν would be shifted from ν_0 by the vibrational frequency. When the intensity of the scattered

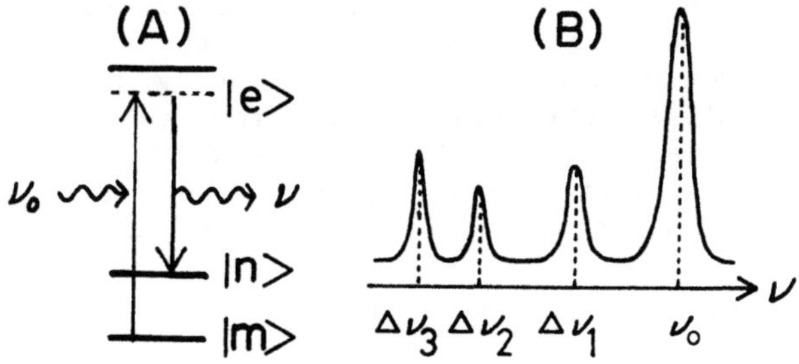

Fig. 1
Raman scattering mechanism (A) and a Raman spectrum (B)

H. B. Dunford et al. (eds.), The Biological Chemistry of Iron, 375–389.

radiation is plotted against the frequency shift ($\Delta\nu = \nu_0 - \nu$), a spectrum as shown in Fig. 1(B) is obtained, which has peaks at individual vibrational frequencies. Accordingly, from the measurements of Raman spectra, we learn the vibrational frequencies of a molecule in the ground electronic state. Note that the intensity of Raman lines depends upon the transition probability from $|m\rangle$ to $|n\rangle$ through $|e\rangle$ and thus upon the electronic excited state.

When ν_0 is very close to an absorption frequency of the scattering molecule, some of the Raman lines associated with the chromophore are markedly intensified. This is called resonance Raman scattering. In the resonance condition, the sample concentrations of 1 mM to 1 μM are practical, and there only the resonance-enhanced Raman lines are observed. Since the chromophore of a biological molecule often serves as an active site for its function, analysis of its resonance Raman spectra may be expected to yield important structural information about the active site. Already the technique has been extensively applied to iron proteins including nonheme iron dioxygenases, iron sulfur proteins, hemerythrin, transferrin, and various kinds of hemeproteins, and the biological applications have been reviewed recently (1).

2. RESONANCE RAMAN SPECTRA OF HEMOPROTEINS

2.1 Vibrations of Metalloporphyrins

Resonance Raman scattering from hemoproteins has revealed the vibrational spectra of the iron porphyrin interacting *in-situ* with the surrounding proteins (2-4). To make the best use of these data for elucidating structural details of the heme chromophore, assignments of Raman lines are required. For this purpose, isotope labeled octaethylporphyrinato-Ni [Ni(OEP)] was investigated and the fundamental frequencies were determined from *ca.* 90 combination bands of the individual isotope-labeled molecules (5). The force constants for the Urey-Bradley force field were adjusted to reproduce the observed frequencies and their isotopic frequency shifts, and finally the normal vibrations responsible for the resonance Raman lines of Ni(OEP) were determined (6). The vibrational modes of the Raman lines which are often used as indicators are displayed in Fig. 2.

2.2 A Marker Line for Electron Delocalization

In most Raman spectra of hemoproteins excited at 400-500 nm, there is an intense line which appears around 1355-1362 cm^{-1} or 1370-1375 cm^{-1} according to the ferrous or ferric heme present, respectively and therefore it has been called an oxidation state

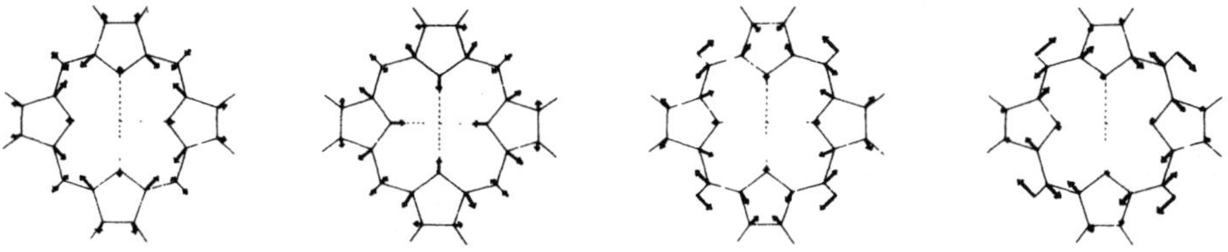

Fig. 2 The ν_3, ν_4, ν_{10}, and ν_{19} modes of metalloporphyrin, their symmetry species under the D_{4h} point group, and the observed frequencies for Ni(OEP) (6).

marker (7,8). The Raman line is assigned to the ν_4 mode, which involves, as shown in Fig. 2, the in-phase displacements of four pyrrole nitrogen atoms toward the metal ion by 0.01 A (9) and the C_α-N symmetric stretching vibrations.

The frequencies of this Raman line for various hemeproteins are diagrammatically represented in Fig. 3. So far, compound II of horseradish peroxidase (HRP) and reduced cytochrome P-450 gave the highest and the lowest frequencies, respectively. Some of the ferrous low spin complexes such as HbO_2, MbO_2, HbCO, $HbCNC_2H_5$ and MbNO exhibit the ν_4 band at the frequency region of the ferric low spin complexes such as ferricytochrome *c* and ferricytochrome *b*$_5$. This indicated the existence of two kinds of ferrous low-spin complexes, and they were interpreted in terms of two kinds of Fe-L (L: axial ligand) interactions as described below (7,10).

Since the $d_{x^2-y^2}$ and d_{xy} orbitals are in the porphyrin plane, they are considered to be less important for the interaction with axial ligands. Four possible cases of the Fe^{2+}-L interactions using d_π(Fe) (=d_{xz},d_{yz}), d_{z^2}(Fe), σ(L), and π(L) orbitals in the low spin complexes are schematically illustrated in Fig. 4 (a-d). The interaction of d_π(Fe)-σ(L) (Fig. 4a) and d_{z^2}(Fe)-π(L) (Fig. 4d) are of nonbonding type because of symmetry. Coordination of a lone pair orbital of an axial ligand to the heme iron (Fig. 4b) results in the formation of the σ bond in the sense that the interaction is symmetric around the bond axis. Delocalization of the iron d_π electrons to the empty π^* orbitals of the axial ligand (Fig. 4c) forms another type of chemical bond which may conveniently be called the π bond.

As the d_{z^2} orbital belongs to the a_{1g} species in the D_{4h} symmetry, the σ bond affects little a state of the porphyrin π

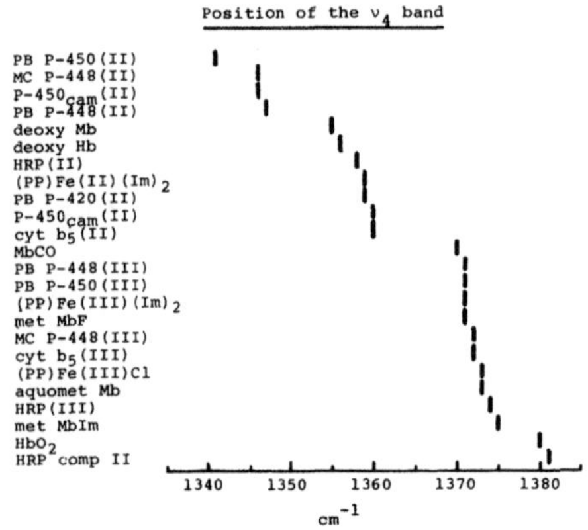

Fig. 3.
The ν_4 band position of various derivatives of hemoproteins and iron-porphyrins.

orbitals. In contrast, the formation of the π bond causes a change in delocalization of π electrons, because the d_π(Fe) orbitals belong to the same e_g species as the lowest empty π* orbital of porphyrin. In other words, the π*(porphyrin) and d_π(Fe) orbitals form a molecular orbital even in the absence of L, but when L is bound, π*(L) is also involved in the d_π(Fe)-π*(porphyrin) molecular orbital. As a result, the electrons which were originally delocalized to the π*(porphyrin) orbital are partially transferred to the empty π*(L) orbital. Since the π*(porphyrin) orbital is antibonding with regard to the $C_\alpha N$ bonds, reduction of electron occupation in the π*(porphyrin) orbital results in strengthening the $C_\alpha N$ bonds, and causes the upward shift of the $C_\alpha N$ stretching frequency. Accordingly, the stronger the π acidity of the axial ligand is, the higher the ν_4 frequency appears. We emphasize here that the two kinds of ferrous low spin complexes are classified into the σ and π type of complexes, and only the latter undergoes photodissociation (11).

Conversely, with the π basicity of an axial ligand is higher, ν_4 is expected to shift to a lower frequency. This is, in fact, observed for reduced cytochrome P-450 (12,13) for which coordination of the sulfur anion to the heme iron is postulated (14-16). As illustrated in Fig. 4f, one (p_z) of two lone pair orbitals of sulfur anion would constitute the σ bond with the iron d_z2 orbital but the other (p_y) would constitute the π bond with the iron d_π orbitals, donating electrons to the π*(porphyrin) orbital through the d_π(Fe)-π*(porphyrin) molecular orbital, and thus lowering the ν_4 frequency. This peculiarity would be removed if the sulfur anion were protonated (Fig. 4e) or replaced by another ligand. In accord, the ν_4 frequency returns to the normal frequency upon conversion to the catalytically in-inactive form (P-420).

Recently a device to measure very small frequency shifts

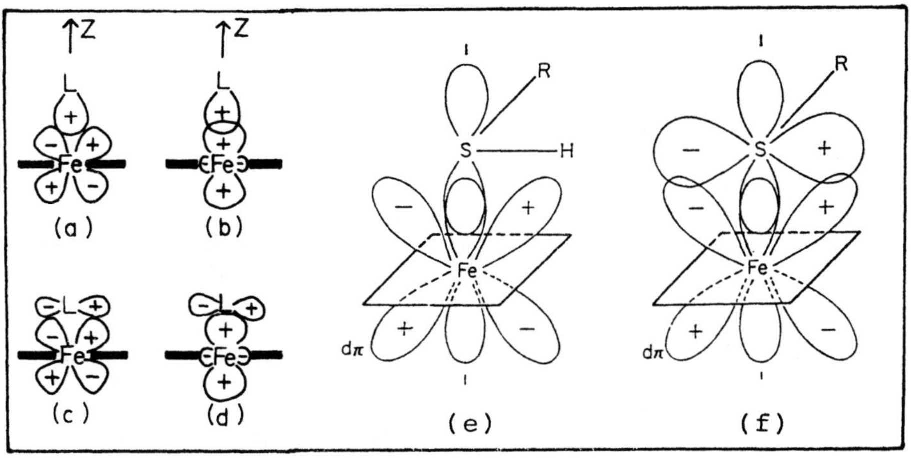

Fig. 4. Four types of orbital overlap between Fe and L (L: axial ligand) (a-d), and difference in the bonding between L = RSH and L = RS^- where R is an alkyl group (e,f).

(0.1 cm^{-1}) of Raman lines was completed (17), and with its use a 1.5 cm^{-1} difference was found for the ν_4 modes of deoxy Hb with the T and R structures (18). Furthermore, delicate differences in the heme environments among various C-type cytochromes are also discussed on the basis of the small but reliable data of frequency shifts (19).

2.3 Marker Lines for Axial Ligation and Core Expansion

A few Raman lines have been noted as spin state indicators (7,8), which classify hemeproteins as one of four groups, namely, the ferrous or ferric low- or high-spin states. Recently, however, the marker lines are being reinvestigated to interpret those which cannot be satisfactorily classified by the empirical rule. One of a curious phenomenon is demonstrated in Fig. 5, where Raman spectra of $Fe^{3+}(OEP)(ClO_4)$ in various solvents are displayed (20).

The Raman lines of $Fe^{3+}(OEP)(ClO_4)$ in CH_2Cl_2 at 1645 and 1377 cm^{-1} are assigned to ν_{10} and ν_4, respectively (see Fig. 2). The frequency of the ν_{10} line exhibits a dramatic change with solvent while the ν_4 frequency remains unaltered. The ν_{10} line is usually located 1629-1631 cm^{-1} for the 5-coordinate ferric high-spin complexes whereas it is shifted to 1615-1618 cm^{-1} for the 6-coordinate ferric high-spin complexes (20,21). Furthermore, it has been demonstrated that there is no simple relation between the magnetic moment and the ν_{10} frequency as shown in Table I.

On the other hand, the ν_{10} frequency as well as ν_3 and ν_{19}

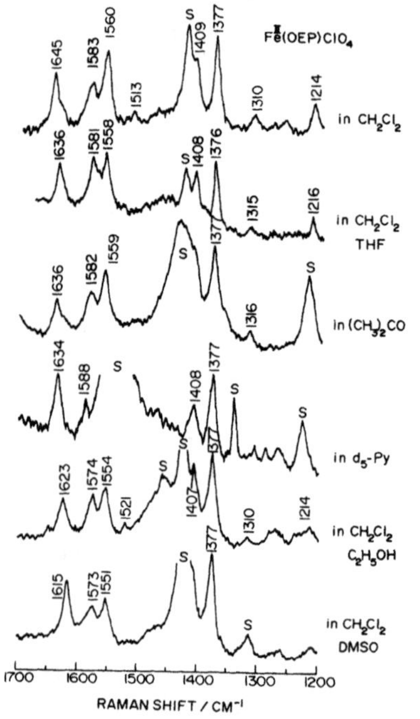

Fig. 5.
Solvent effect in the resonance Raman spectrum of $Fe^{3+}(OEP)(ClO_4)$. The Raman lines marked by S are due to solvent. For THF/H_2O, C_2H_5OH, and DMSO solvents, the maximum amount was mixed with CH_2Cl_2 to the solubility limit of $Fe^{3+}(OEP)(ClO_4)$. Excitation at 488.0 nm (Ref. 20).
(Reproduced with permission from the American Chemical Society)

serve as core-size indicators (21,22). As seen in Fig. 3, all these bands are associated mainly with the methine bridge stretching vibrations, and in the case of ν_{10}, reduction of frequency by 7 cm^{-1} is anticipated for a decrease of the $C_\alpha C_m$ stretching force constant by 0.1 mdyn/A (1.8%)(6). If the FeN_4 core were expanded, a distortion in the $C_\alpha C_m C_\alpha$ portion of the porphyrin would be induced unless the pyrrole ring is distorted. The invariance of the ν_4 frequency implies a negligible distortion of the $C_\alpha N$ bonds. Therefore, it seems quite reasonable that ν_3, ν_{10}, and ν_{19} are shifted to a lower frequency as the core-size becomes larger. Since the contributions of the $C_\alpha C_m$ stretching term to the ν_3, ν_{10}, and ν_{19} are calculated to be as large as 41%, 49%, and 67%, respectively (6), ν_{19} is expected to be the most sensitive to stretch of the $C_\alpha C_m$ bond, in good agreement with the experimental results (21).

The origin of the core expansion is considered to be repulsion between the axial ligand and pyrrole nitrogen atoms. It is therefore not unreasonable that the ν_{10} frequency may be categorized according to the axial ligands coordinated. A good guess for the ligated species at the axial position of the heme iron could be made from the observed ν_{10} frequency if more experimental data were available. This is the theoretical basis for the fact that the ν_{10} frequency seems to have a close correlation with the coordination number and the coordinated

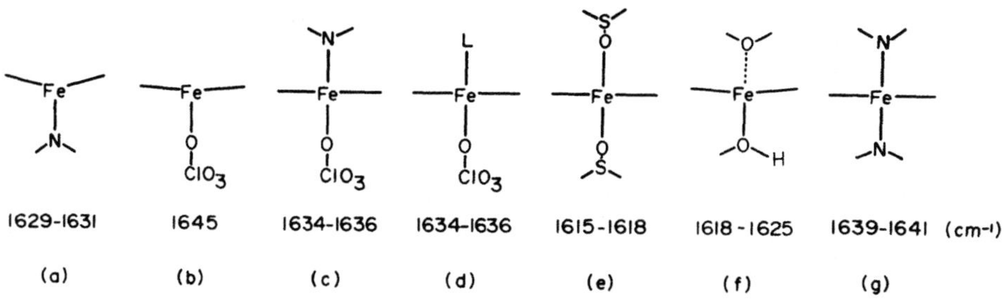

Fig. 6. Categorization of the ν_{10} frequency of the Fe^{3+}(OEP)-(ClO_4)LL' complexes in terms of the axial ligation (20).
(Reproduced with permission from the American Chemical Society)

Table I. The ν_{10} and ν_4 Frequencies and Magnetic Moments of Fe^{3+}(OEP)(L)(ClO_4) Complexes in the Solid State (20).
(Reproduced with permission from the American Chemical Society)

	275 K			77 K		
L	ν_{10}, cm^{-1}	ν_4, cm^{-1}	μ, μ_B	ν_{10}, cm^{-1}	ν_4, cm^{-1}	μ, μ_B
none	1629	1377	4.78	1635	1378	4.16
4-CHOPy	1629	1377	4.49	1635	1377	3.60
4-CNPy	1630	1377	3.55	1634	1377	2.31
Py	1630	1377	2.99	1634	1377	2.05

species. Currently, an empirical relationship illustrated in Fig. 6 is proposed (20).

2.4 Fe-Ligand Stretching Vibration

The greatest advantage of resonance Raman spectroscopy lies in measuring directly the bond strength of a chemical bond by its stretching frequency. In the protoheme containing proteins, the Fe-fifth ligand bond is extremely important as it is the sole chemical bond between the heme and protein, and enzymatic reaction takes place at its opposite side. Therefore, we (23) and others (24,25) have done experiments to assign the Fe-His (proximal) stretching Raman line of deoxyMb.

On the basis of the ^{54}Fe isotopic frequency shift (23) and also of studies on suitable model system (26), the Raman line of deoxyMb at 220 cm^{-1} was assigned to the Fe-His stretching mode. Note that this Raman line exhibits a significant frequency shift upon change of the quaternary structure of deoxyHb (27,28), and also that the lower frequency shift upon the transition from the R to T structure is appreciably larger for the α subunit than for the β subunit (29). Furthermore, for nitrosylHb Raman spectroscopy provided clear evidence for cleavage of the Fe-His-(proximal) bond of the α subunit in the T structure; but no cleavage for the β subunit in the T structure nor for the α and β subunits in the R structure (30).

3. RESONANCE RAMAN STUDY ON PLANT PEROXIDASE

3.1 Heme-Linked Ionization of Horseradish Peroxidase

HRP is a protoheme containing peroxidase which catalyzes the oxidation reaction of a wide variety of substrates by H_2O_2 as its specific oxidizer. The reaction mechanism of HRP has been studied in detail (31 and references therein). The ferric enzyme is oxidized by H_2O_2 to compound I which is considered to be a Fe^{4+} porphyrin π cation radical; and then compound I is reduced by a substrate to ferric enzyme through compound II.

The existence of a heme-linked ionizing group with pK_a = ~7 was suggested for the reduced form of HRP (32) and later confirmed from the pH dependence of E_0' (ferric/ferrous) of HRP (33). The pK_a of the ionizing group depends on the redox states of HRP as shown in Table 2. The close parallelism of the pK_a values between HRP A_2 and HRP C isozymes (31) suggests the identity of the ionizing group in the ferrous, ferric, and ferryl states. To elucidate the structural implication of the heme-linked ionization, we examined the resonance Raman spectra (34,35).

3.2 Coordination Characteristics of the Proximal Histidine Ligand

In Figure 7 a comparison is made between the resonance Raman spectra of native HRP C and the reconstituted HRP C with the ^{54}Fe-incorporated heme, both in the reduced state. The Raman line of native HRP C at 244 cm^{-1} was shifted to higher frequency by 2 cm^{-1} upon the ^{54}Fe substitution while all the other lines did not exhibit a detectable frequency shift. The amount of the observed frequency shift was close to the expected frequency shift (2.4 cm^{-1}) for an uncoupled harmonic vibration of the iron-imidazole stretching mode. Therefore, the 244 cm^{-1} line of ferroHRP was assigned to the Fe-His (proximal) stretching vibration.

The Raman spectrum of HRP A was quite close to that shown in Fig. 7, but the frequencies of individual Raman lines were appreciably different from those of HRP C. The pH dependence of the frequencies of a few selected Raman lines of the two isozymes is illustrated in Fig. 8. It is evident that only the Fe-His (proximal) stretching Raman line displays a pH dependent frequency shift while all the other lines do not show a systematic frequency shift. Furthermore, this frequency shift takes place at the pH

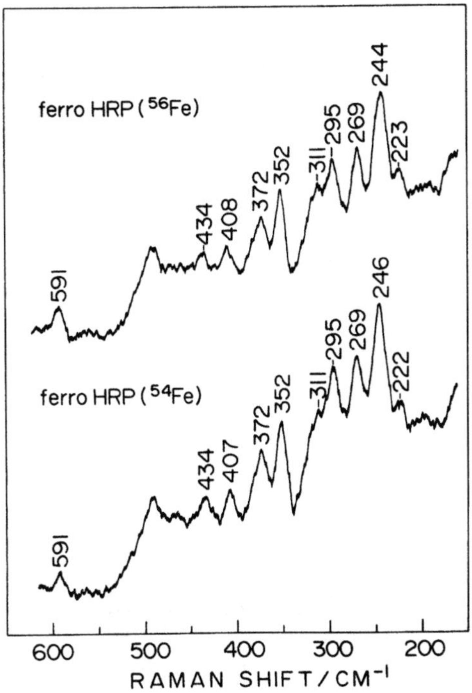

Fig. 7
The low frequency Raman spectra of native ferroHRP (upper) and the ^{54}Fe-incorporated reconstituted ferroHRP (lower). Excitation at 441.6 nm (35).

Table 2. Heme-Linked Ionization Constants (pK_a) of HRP (31)

	ferrous	ferrous·CO	ferric	compound II
HRP C	7.3	8.2	11.1	8.5
HRP A_2	5.8	6.7	9.3	6.9
$pK_a(C) - pK_a(A_2)$	1.5	1.5	1.8	1.6

values specific to each isozyme, and their midpoint pH values (pH 7 for HRP C and pH 5.5 for HRP A) are remarkably close to the reported pK_a values (see Table 2). Consequently, it is concluded that the heme-linked ionization of ferroHRP is definitely sensed by the Fe-His bond on the proximal side.

For deoxyMb, with a similar Fe-His linkage, the Fe-His stretching Raman line as well as other lines did not show any systematic frequency shift between pH 4.7 and 11.8. Also the Fe-2MeIm (2MeIm: 2-methylimidazole) stretching band of the aqueous solution of Fe^{2+}(PP)(2MeIm) (PP: protoporphyrin IX) observed at 219 cm^{-1} as well as other Raman lines showed no pH dependence between pH 9.1 and 13.1. On the other hand, when a mole ratio of 2MeIm/porphyrin was changed from 54 to 1088, the 219 cm^{-1} line was shifted to a lower frequency by 4 cm^{-1} and then saturated. This line was shifted to 208 cm^{-1} in the presence of 0.01% SDS, 0.02% TX-100 or 0.01% CTAB (35). These observations suggest the possibility that hydrogen bonding of the bound 2MeIm at the N_3 position significantly affects the Fe-2MeIm stretching frequency.

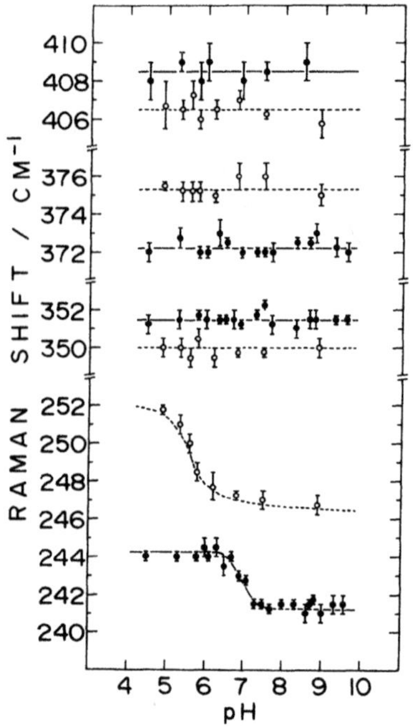

Fig. 8
The pH dependence of frequencies of a few selected Raman lines of ferroHRP: closed circles, HRP C; open circles, HRP A (35).

An effect of ionization of the coordinated 2MeIm on the Fe-2MeIm stretching frequency is demonstrated in Fig. 9, where Fe^{2+}(OEP) in DMF was examined. For the normal 2MeIm complex of Fe^{2+}(OEP), three Raman lines were observed at 208, 257 and 336 cm^{-1}, and only the 208 cm^{-1} line showed a downward frequency shift by 3 cm^{-1} upon perdeuteration of 2MeIm. As the mass of the axial ligand increased by 5 units upon perdeuteration, a 2.5 cm^{-1} shift is theoretically expected for the isolated Fe-2MeIm stretching vibration, in close agreement with the observed frequency shift. Therefore, the 208 cm^{-1} line of Fe^{2+}(OEP)(2MeIm) was assigned to the Fe-2MeIm stretching mode. As shown in the lower two traces of Fig. 9, three lines were observed for Fe^{2+}(OEP)($2MeIm^-$) in DMF at 233, 257, and 336 cm^{-1} and only the 233 cm^{-1} line exhibited a downward frequency shift for the perdeuterated $2MeIm^-$ complex. Consequently, the 233 cm^{-1} line was assigned to the Fe-$2MeIm^-$ stretching mode of Fe^{2+}(OEP)($2MeIm^-$). The higher frequency shift of the Fe-2MeIm stretching mode upon ionization of the coordinated imidazole was also recently reported by Stein *et al* (36).

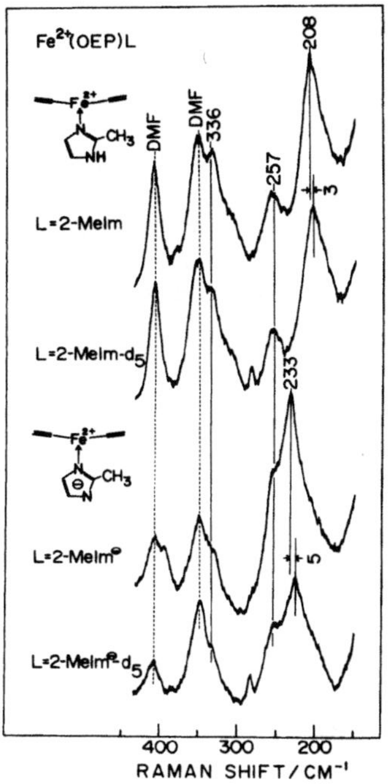

Fig. 9.
A change of the Fe-2MeIm stretching frequency upon the deprotonation of 2MeIm at the N_3 position. The spectrum for the perdeuterated 2MeIm (2MeIm-d_5) complex is also included. The Raman lines marked by DMF are due to solvent (dimethyl formamide). Excitation at 441.6 nm (35).

It is stressed that the difference between the Fe-2MeIm and Fe-2MeIm$^-$ stretching frequencies ($\Delta\nu$ = 25 cm^{-1}) is strikingly close to the difference in the Fe-His stretching frequencies of deoxyMb and ferroHRP ($\Delta\nu$ = 24 cm^{-1}). This may suggest that the histidyl imidazole of ferroHRP is ionized as previously pointed out from the similarity in some spectroscopic properties between HRP and the imidazolate-heme complexes (37). However, La Mar and de Ropp (38) found an NMR signal of a paramagnetically shifted imidazole proton and identified it as the N_δ-proton of proximal histidine. Therefore, the presence of a proton near the N_δ atom is apparent. A plausible model compatible with these facts would be that the N_δ-proton is very strongly hydrogen bonded to some residues in HRP, and this would be the origin of some of the large differences between deoxyMb and ferroHRP. Recent results of x-ray crystallographic analysis of cytochrome *c* peroxidase (CcP) revealed the presence of strong hydrogen bonds among proximal histidine, Gln (239), and Glu (187) (39). Since there are considerable similarities in the primary structures of CcP and HRP, the presence of a strong hydrogen bond of the proximal histidine in HRP appears most likely.

3.3 Structural Implication of Heme-Linked Ionization

Yamazaki *et al.* (31) have postulated that the heme-linked ionization arises from deprotonation of the distal histidine.

The Raman spectrum of ferroHRP in the 1200-1700 cm^{-1} region indicates the absence of the sixth ligand of the heme iron (34). Hence the ionization of the distal histidine must be communicated to the Fe-His (proximal) bond without passing through the heme iron. As a model which interprets those facts consistently, we propose a change of a peptide conformation upon the heme-linked ionization as schematically illustrated in Fig. 10, although the molecular mechanism by which the ionization of the distal histidine influences the proximal histidine is not yet known.

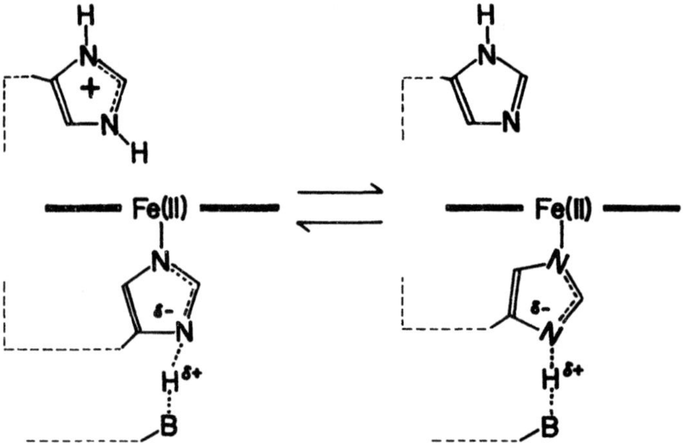

Fig. 10. Structural implication of the heme-linked ionization of HRP. B denotes an arbitrary amino acid residue (35).

If a change in the hydrogen bonding of the N_δ proton of the proximal histidine is directly connected with the pH dependent frequency shift, we have to refer to the results shown in Fig. 9 to predict a direction of the frequency shift. The Fe-2MeIm stretching frequency becomes lower as the N_δ-proton is more tightly bound to the imidazole ring. As the Fe-His stretching frequency of ferroHRP becomes lower on the alkaline side of the transition, we are forced to assume that the hydrogen bond is weaker at the alkaline side. This seems to be unnatural.

The Fe-His (proximal) stretching Raman line of deoxyHb shows an appreciable frequency shift upon change of the quaternary structure (27); and this is attributed to a strain from the globin. If ionization of the distal histidine of HRP caused a conformational change of the polypeptide chain resulting in a strain of the Fe-His bond similarly to the case of deoxyHb, the 3 to 5 cm^{-1} shift of the Fe-His (proximal) stretching frequency seems reasonable. However, we cannot rule out a possibility that the observed frequency shift is a result of a superposition of two opposite effects, namely distortion of the proximal histidine and change of hydrogen bonding.

Recently we found the same kind of frequency shift of the Fe-His (proximal) stretching mode for Japanese radish peroxidase. The frequency of the Fe-His (proximal) stretching frequency was higher similar to that of HRP. Consequently, we are tempted to conclude that the strong hydrogen bonding of the proximal histidine and its change upon heme-linked ionization is a common characteristic feature of plant peroxidases. Presumably the ionization of the distal histidine plays an essential role in peroxidase catalysis, because it results in a significant change of the whole polypeptide chain so as to be observed as a change of the Fe-His (proximal) bond placed on the opposite side of the heme.

ACKNOWLEDGMENT

The authors thank Dr. Denis Rousseau for critical reading of this manuscript. T.K. is indebted to Bell Laboratories for the stay there, where this manuscript was prepared.

REFERENCES

1. Carey, P.R. and Salares, V.R., 1980, in "Advances in Infrared and Raman Spectroscopy" (Clark, R.J.H. and Hester, R.E. eds.) Vol. 7, pp. 1-58, Heydon, London.
2. Spiro, T.G. and Loehr, T.M., 1975, in "Advances in Infrared and Raman Spectroscopy" (Clark, R.H.H. and Hester, R.E., eds.) Vol. 1, pp. 98-142, Heydon, London.
3. Yu, N.T.: 1977, *Crit. Rev. Biochem.* 4, p. 229.
4. Kitagawa, T., Ozaki, Y., and Kyogoku, Y.: 1978, *J. Chem. Phys.* 11, pp. 153-192.
5. Kitagawa, T., Abe, M. and Ogoshi, H.: 1978, *J. Chem. Phys.* 69, p. 4516.
6. Abe, M., Kitagawa, T. and Kyogoku, Y.: 1978, *J. Chem. Phys.* 69, p. 4526.
7. Spiro, T.G. and Strekas, T.C.: 1974, *J. Am. Chem. Soc.* 96, p. 338.
8. Yamamoto, T., Palmer, G., Gill, D., Salmeen, I.T. and Rimai, L.: 1973, *J. Biol. Chem.* 248, p. 5211.
9. Kitagawa, T., Abe, M., Kyogoku, Y., Ogoshi, H., Sugimoto, H. and Yoshida, Z.: 1977, *Chem. Phys. Lett.* 48, p. 55.
10. Kitagawa, T., Iizuka, T., Saito, M. and Kyogoku, Y.: 1975, *Chem. Lett.* p. 849.
11. Kitagawa, T., Kyogoku, Y., Iizuka, T. and Ikeda-Saito, M,: 1976, *J. Am. Chem. Soc.* 98, p. 5169.

12. Ozaki, Y., Kitagawa, T., Kyogoku, Y., Shimada, H., Iizuka, T. and Ishimura, Y.: 1976, *J. Biochem. (Tokyo)* 80, p. 1447; Ozaki, Y., Kitagawa, T., Kyogoku, Y., Imai, Y., Hashimoto-Yutsudo, C. and Sato, R.: 1978, *Biochemistry* 17, 5826.
13. Champion, P.M. and Gunsalus, I.C.: 1977, *J. Am. Chem. Soc.* 99, p. 2001.
14. Stern, J.O. and Peisach, J.: 1974, *J. Biol. Chem.* 249, p. 7495.
15. Chang, C.K. and Dolphin, D.: 1975, *J. Am. Chem. Soc.* 97, p. 5948.
16. Collman, J.P. and Sorrell, T.N.: 1975, *J. Am. Chem. Soc.* 97, p. 4133.
17. Rousseau, D.L.: 1981, *J. Raman Spectrosc.* 10, p. 94.
18. Rousseau, D.L., Shelnutt, J.A., Henry, E.R. and Simon, S.R.: 1980, *Nature* 285, p. 49.
19. Shelnutt, J.A., Rousseau, D.L., Dethmers, J.K. and Margoliash, E.: 1979, *Proc. Natl. Acad. Sci. U.S.A.* 76, p. 3865.
20. Teraoka, J. and Kitagawa, T.: 1980, *J. Phys. Chem.* 84, p. 1928.
21. Spiro, T.G., Stong, J.D. and Stein, P.: 1979, *J. Am. Chem. Soc.* 101, p. 2648.
22. Felton, R.H., Yu, N.T., O'Shea, D.C. and Shelnutt, J.A.: 1974, *J. Am. Chem. Soc.* 96, p. 3675.
23. Kitagawa, T., Nagai, K. and Tsubaki, M.: 1979, *FEBS Lett.* 104, p. 376.
24. Desbois, A., Lutz, M. and Banerjee, R.: 1979, *Biochemistry* 18, p. 1510.
25. Kincaid, J., Stein, P. and Spiro, T.G.: 1979, *Proc. Natl. Acad. Sci. U.S.A.* 76, p. 549; p. 4156.
26. Hori, H. and Kitagawa, T.: 1980, *J. Am. Chem. Soc.* 102, p. 3608.
27. Nagai, K., Kitagawa, T. and Morimoto, H.: 1980, *J. Mol. Biol.* 136, p. 271.
28. Ondrias, M.R., Rousseau, D.L. and Simon, S.R.: 1981, *Science* 213, p. 657.
29. Nagai, K. and Kitagawa, T.: 1980, *Proc. Natl. Acad. Sci. U.S.A.* 77, p. 2033.
30. Nagai, K., Welborn, C., Dolphin, D. and Kitagawa, T.: 1980, *Biochemistry* 19, p. 4755.
31. Yamazaki, I., Araiso, T., Hayashi, Y., Yamada, H. and Makino, R.: 1978, *Adv. Biophys.* 11, p. 249.
32. Theorell, H.: 1947, *Adv. Enzymol.* 7, p. 265.
33. Harbury, H.A.: 1957, *J. Biol. Chem.* 225, p. 1009.
34. Teraoka, J. and Kitagawa, T.: 1980, *Biochem. Biophys. Res. Commun.* 93, p. 694.
35. Teraoka, J. and Kitagawa, T.: 1981, *J. Biol. Chem.* 256, p. 3969.
36. Stein, P., Mitchell, M. and Spiro, T.G.: 1980, *J. Am. Chem. Soc.* 102, p. 7795.
37. Mincey, T. and Traylor, T.G.: 1980, *J. Am. Chem. Soc.* 101, p. 765.

38. La Mar, G.N. and de Ropp, J.S.: 1979, *Biochem. Biophys. Res. Commun.* 90, p. 36.
39. Poulos, T.L. and Kraut, J.: 1980, *J. Biol. Chem.* 255, p. 8199.

COMPOUNDS I OF HORSERADISH AND YEAST CYTOCHROME C PEROXIDASES

Brian M. Hoffman

Department of Chemistry
Northwestern University
Evanston, Illinois 60201 USA

Hemoprotein Compounds I form a class of enzymic intermediates with the formal oxidation level of Fe(V), yet exhibiting diverse properties. Two examples which illustrate this diversity are the Compounds I of horseradish peroxidase and cytochrome c *peroxidase. This report first presents the general characteristics of the latter protein. Next it describes and contrasts the properties of the Compounds I of these enzymes and discusses the recent determinations of their structure by electron-nuclear double resonance (endor), after first briefly describing this technique. These, and likely all, Compounds I are comprised of an oxy-ferryl (Fe^{IV}=O) center interacting to some degree with a free radical. Differences between the proteins reflect, in significant part, differences in radical, in coupling strength, and possibly in the properties of the oxy-ferryl center.*

Peroxidases are ferri-hemoproteins which react with hydrogen peroxide to form an enzymic intermediate, called Compound I, which contains both oxidizing equivalents of the H_2O_2, and is thus at the formal oxidation level of Fe(V). Compound I can react with diverse organic, inorganic, or macromolecular substrates in two consecutive one-electron steps. The product of the first step, having the oxidation level of Fe(IV), is called Compound II (1-3). The characterization of these two Compounds has been a fascinating challenge of long standing and is of even greater importance because of the related problems presented by other enzyme classes. Catalases decompose hydrogen peroxide to water in a two step process, the product of the first step also being a Compound I (2-4). Cytochromes P450, which play a significant role in the metabolic activation and detoxification of

H. B. Dunford et al. (eds.), The Biological Chemistry of Iron, 391–403.

environmental toxins, employ dioxygen as an oxidant and follow a very different enzymatic cycle than do either peroxidases or catalase. Nevertheless, this cycle involves an enzymic intermediate at the same oxidation level of Compounds I (3).

Little as yet is known about the detailed properties of the "Fe(V)" state of cytochrome P450. However, it has become clear that the peroxidases and catalases can be discussed in common and fall into two major classes. The larger class contains and is typified by horseradish peroxidase and catalase. The second is represented by cytochrome c peroxidase. In this report we first present the general properties of this latter protein (Section A). Then we discuss the properties of the Compounds I formed by HRP and CCP (Section B), and describe the recent determinations of their structure as afforded by electron-nuclear double resonance (endor) (Sections D,E). Before so doing, we digress briefly to describe this technique (Section C).

A. Cytochrome c peroxidase (CCP).

CCP is isolated from aerobically grown yeast, with the commercially available bakers and brewers yeasts being the most convenient sources for large-scale preparation. The protein is present in the mitochondria at concentrations comparable to that of cytochrome oxidase, but its physiological role has not been established, although the suggestion has been made that it may serve to prevent accumulation of harmful peroxides. CCP can catalyze the reduction of H_2O_2 by a number of reducing agents such as ascorbate and ferrocyanide, but the reaction rates with mammalian and yeast ferrocytochromes c are far larger than for other reductants (5,2). This inter-protein electron-transfer process is unique in that it proceeds by way of an observable one-to-one complex which is so stable that the ferri CCP-ferri cytochrome c complex has been isolated chromatographically (6).

The enzyme is monomeric, being comprised of a single noncovalently bound heme and a polypeptide chain of 293 amino acids having a calculated apoprotein molecular weight of 33,419 and known sequence (7). In contrast to the other peroxidases described in this volume, CCP is not a glycoprotein and probably for this reason it is readily crystallized. Its crystal structure has been determined (8-10) and, as the only structure of a peroxidase to date, provides the best available starting point for investigating the means by which a protein environment confers peroxidase activity upon the heme prosthetic group (9).

The heme environment in resting state (ferri) CCP has been considered in detail and found to show significant similarities

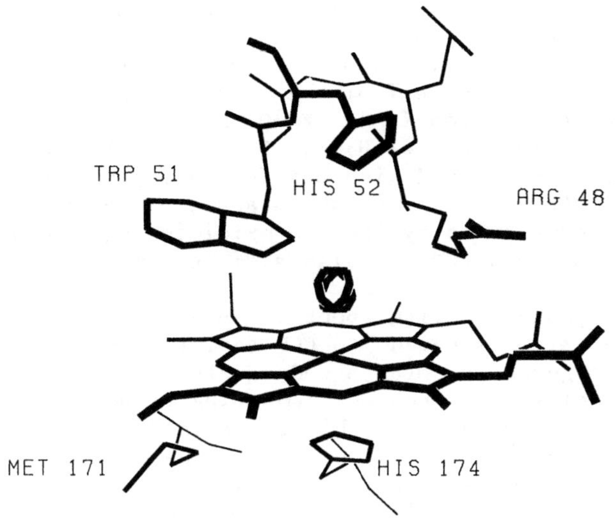

Fig. 1. View of the heme crevice showing interactions between the heme, its axial ligands, a histidine and a water, and other nearby residues of significance (Ref. 10).

with that of met-myoglobin (8,9). The heme is located in a crevice and oriented so that the two proprionate sidechains extend toward the surface; only an edge of pyrrole ring IV is accessible to solvent. The fifth, or proximal, ligand is the NE_2 atom of histidine-174 and the iron atom appears to be displaced in this direction. As in myoglobin, the sixth coordination position of the heme is occupied not by an endogenous ligand, but by a water molecule which is hydrogen-bonded to distal histidine-52. Thus, the immediate coordination environment is essentially the same in myoglobin, a protein which reversibly binds O_2, and in CCP, an enzyme which catalyzes the reductive cleavage of oxygen-oxygen bonds.

Clearly, the differences in the properties of CCP and myoglobin must arise from the several key features which distinguish one structure from the other (8,9). Distal arginine-48 in CCP has no analogue in myoglobin, and has been proposed to act in consort with HIS-52 to catalyze heterolytic cleavage of the peroxide O-O bond. It is suggested that H_2O_2 first binds to the ferric iron, with histidine functioning as a base and removing a proton. Heterolytic O-O bond cleavage is then facilitated as arginine-48 stabilizes a developing negative charge on the oxygen atom remote from iron and histidine donates its newly acquired proton to that oxygen. In this proposal, which gives molecular detail to many earlier conjectures, the bond-breaking step produces a water molecule from the remote oxygen; the near oxygen remains bonded to iron in an (Fe-O) moiety which retains the two oxidizing equivalents of H_2O_2. This or analogous mechanisms are plausibly associated with any ferrihemoprotein which is oxidized to an "Fe(V)" intermediate state upon reaction with H_2O_2. The final forms taken by the two oxidizing equivalents in Compounds I of HRP and CCP are discussed in section B.

Distal tryptophan-51, lying about 3.6 Å above and parallel to the heme plane, also is unique to CCP and, as discussed below, has been suggested to play a redox role (9). The structure of the proximal side of the CCP heme ring is not yet fully established, but also appears to exhibit a number of distinguishing features (8-10). The sulfur atom of methionine-171 lies 5-6 Å below the proximal surface of the heme, and appears to be effectively in contact with the buried charge of a nearby aspartate. In addition, an unusual adjacent pair of methionyl residues, at positions 229 and 230, appears to loop back toward the heme proximal side. A suggested redox role for proximal residues (11,12) is also discussed below.

B. Nature of Peroxidase Compounds I (and II)

Compounds I and II of HRP have been described previously (1,2), as well as elsewhere in the volume. We recall key features for convenient reference, 1) Their optical spectra are significantly different, HRPI being green and HRPII, red, and differ from that of the resting state. 2) Magnetic susceptibility measurements indicate three and two unpaired electrons respectively. Compound I shows an unusual, broad epr signal which is stoichiometric with the heme and observable only below 30K, but HRPII is epr silent (13). 3) Their ^{57}Fe quadrupole splittings, isomer shifts, and their Mossbauer spectra in high applied magnetic fields are very similar, but their spectra in low applied field differ significantly (13).

The two-equivalent oxidized form of CCP offers distinct contrasts with HRPI, as follows (5). 1) It is red, not green, with an optical spectrum like that of HRPII. 2) Three unpaired electrons are present, as in HRPI, and a stoichiometric epr signal is also observed, but the signal from H_2O_2-oxidized CCP is relatively narrow and free-radical like and is observable at 77K and above. 3) The ^{57}Fe Mossbauer results are like those of HRPII (14). The CCP intermediate has historically been called

ES. Because it is at the same oxidation level as HRPI, but differs so significantly, we propose calling it CCPI^R, for reasons which will become clear.

What is to be concluded about the nature of Compounds I and II? Analysis of the Mossbauer and susceptibility data indicates that in all instances the heme-iron is in the ferryl, Fe(IV) state, with two unpaired electrons. Thus, Compound II in fact contains a S = 1 and formally quadrivalent iron, oxidized by one equivalent above the ferric iron resting state. The Mossbauer data also shows Compound I, though at an oxidation level corresponding to "Fe(V)", nevertheless to contain an Fe(IV). Therefore, peroxidase Compounds I must store a second oxidation equivalent in the form of a free radical, R·, somewhere other than at the iron. We can thus describe the active site by a scheme

$$\left.\begin{matrix} \mathrm{Fe(IV)} \\ (S = 1) \end{matrix}\right\} \xleftrightarrow{\;J\;} \left\{\begin{matrix} \mathrm{R}\cdot \\ (S = 1/2) \end{matrix}\right.$$

<u>Scheme I</u>

involving a ferryl iron, a free radical, and a possible coupling between them. This result is reconciled with the mechanism of section A by assuming that the "Fe(V)" moiety produced by O-O bond cleavage reversibly oxidizes a nearby redox center. The task of characterizing the active site then can be formulated with three questions: i) What is the identity of the radical center, ii) What is the detailed character of the iron center, and iii) What is the description and origin of any coupling between centers?

C. ENDOR

Electron-nuclear double resonance (endor) is performed by using a frequency-swept radiofrequency field to induce nuclear transitions in a paramagnetic center whose epr signal is simultaneously being monitored; the occurrence of a nuclear transition is indicated by a change in the epr signal intensity (15). Thus, an endor sweep in fact amounts to a solid-state nmr spectrum of an epr-observable paramagnetic center, but one in which resonances from <u>all</u> nuclei may be obtained simultaneously. As such, an assignment of nuclear resonances to a particular molecular entity provides a direct identification of the radical center, and an analysis of the measured hyperfine couplings provides detailed information about its electronic structure. The main advantage to the endor technique is that its spectral resolution can be more than three orders of magnitude greater than that of conventional epr, thus permitting the measurement of hyperfine couplings under conditions where hyperfine structure is not observable in the epr spectrum. In addition, the quadrupole interaction term, which reflects the deviations from spherical symmetry of the charge density around the nucleus, contributes to the spectrum on an equal footing with the hyperfine terms, unlike ordinary epr, and is as readily observed as are hfs.

The normal endor pattern for a set of magnetically equivalent protons is a single pair of lines separated in frequency by the orientation-dependent hyperfine coupling constant, A^H, and mirrored about the free-proton Larmor frequency, ν_H,[15] which varies in proportion to the epr spectrometer microwave frequency. In contrast, the $\Delta m = \pm 1$ transitions for nucleus i, i = ^{14}N (I=1) or ^{17}O (I=5/2), are expected to obey the expression[15]

$$\nu_{\pm}(m) = \left| A^i/2 \pm \nu_i + P^i\,(2m-1) \right| \qquad (1)$$

where $-I + 1 \leq m \leq I$, P^i is the orientation dependent quadrupole interaction constant, and all parameters are to be taken as positive (absolute values). When the parameters obey the inequality, $A^i/2 > \nu_i > P^i\,(2I-1)$, as is typically true for ^{14}N and ^{17}O, the equation describes a pattern of 4I lines centered at $A^i/2$, which is a molecular parameter independent of spectrometer microwave frequency, and with splittings determined by ν_i and P^i.

D. HRP I

The Compounds I of HRP, and of catalase and many other peroxidase as well, are widely believed to contain the second oxidizing equivalent as the π-cation free radical of the porphyrin ring. This identification was proposed by Dolphin, Fajer, Felton, and coworkers because the optical spectra of Compounds I resemble those of a series of synthetic porphyrin π-cation radicals. In addition, they further characterized the optical spectra into two classes and assign them to porphyrin radicals with different ground state symmetries (3,16).

However, Morishima and Ogawa concluded that their proton nmr spectra of the heme peripheral substituents were inconsistent with the localization of an odd-electron on the ring and suggested the data might better be interpreted in terms of a free radical located on a nearby amino acid residue (17). LaMar and deRopp reexamined the matter and concluded that the observed hyperfine shifts are consistent with but not proof of the porphyrin cation radical identification (18). It is thus clear that high resolution nmr studies do not provide a direct identification of the radical.

The stoichiometric epr signal associated with HRPI was only recently seen by Schulz et al (13). The resonance was interpreted in terms of a Fe(IV) ferryl iron in which the two unpaired electrons experienced a strong interaction with the environment (large zero field splitting) and a very weak interaction with a free radical, the strength of this interaction only being equivalent to several degrees Kelvin. This result has several features which are of particular note. Firstly, a successful simulation of the epr spectrum required the assumption of a Gaussian distribution for the values of the exchange tensor components, and the physical basis of this assumption was unclear; indeed it was not definitively proven that the entire signal is associated with a single coupled center. Secondly, and of more fundamental nature, even though the analysis be correct in its essentials, the model of heme-radical coupling does not afford an identification of the radical involved. Finally, so weak a coupling between iron and porphyrin spin systems would be a major surprise.

We therefore undertook an endor study of HRPI in order to see whether the technique could resolve the issue (19). The epr absorption spectrum of HRPI is broad and totally feature less (Fig 2 inset). In contrast HRPI exhibits a rich endor pattern, with three types of signals discernible. The intense resonances located in the frequency range of 11.5-18 MHz occur in pairs whose center frequency falls at the free proton Larmor frequency. These signals are proton resonances, as is proven by the shift with magnetic field dramatically evident in a comparison of Figs. 2A and 2B. The strength and shape of the resonances, as well as other characteristics, suggest that they arise from β-protons, those attached to carbon atoms that are in turn bonded to a spin-bearing carbon.

Two weak features at frequencies above 18.5 MHz (Fig. 2A) also shift appropriately with field (Fig. 2B) and thus are proton resonances as well. Their low intensity and other characteristics suggest that they arise from α-protons, those directly attached to a spin-bearing carbon atom. Moderately strong resonances are also observed in the frequency region 0-9.5 MHz. Ignoring the possibly small contributions from the partners to the high-frequency α-proton peaks, a comparison of Figs. 2A and 2B shows that these signals do not appreciably shift with field.

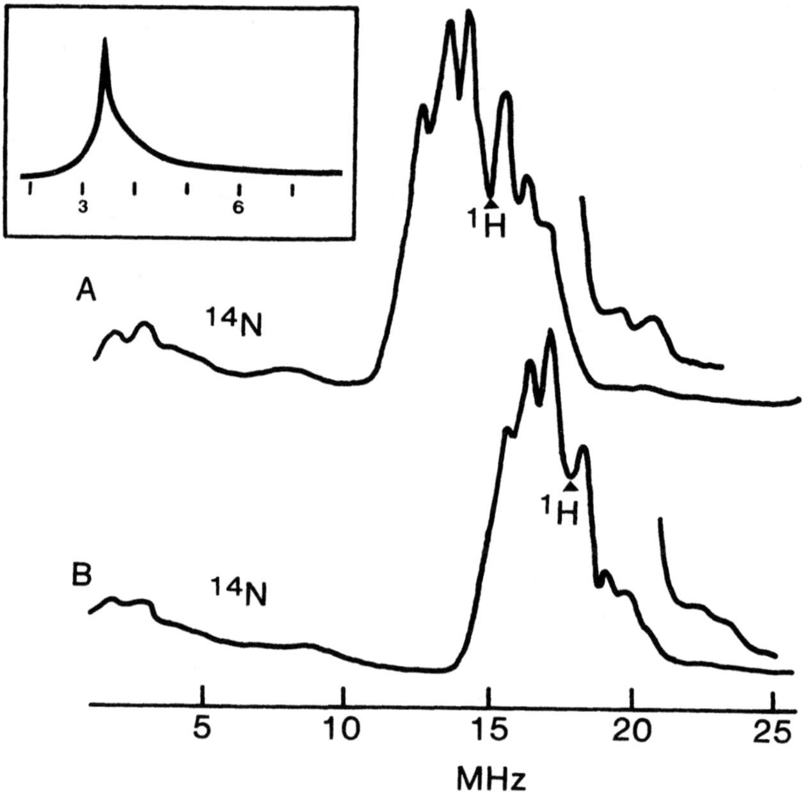

Fig 2. ^{1}H and ^{14}N endor spectra of horseradish peroxidase Compound I taken at g ≅ 2.0, as indicated by an arrow in the inset, with microwave frequencies 9.60 GHz (A) and 11.66 GHz(B). Arrows on the spectra indicate ν_H, the free proton Larmor frequency. Inset, epr signal of horseradish peroxidase Compound I taken at 2 K under rapid passage conditions, 0.2 mW microwave power, and 100 KHz filed modulation of 3 G amplitude (Ref 19).

Thus, they obey eq 1 and are not proton resonances, and are assigned to ^{14}N.

The spectra presented here clearly show the radical site of HRP Compound I to contain ^{14}N and β-protons, and probably α-protons as well, which immediately suggests that the S = ½ portion of the HRP Compound I intermediate is indeed the porphyrin π-cation radical. Endor measurements following selective deuterium substitution on the porphyrin periphery prove this conclusion to be correct, since several of the resonances disappear upon partial deuteration of the heme ring (20).

With the identity of the HRPI radical site resolved, we next turned to the task of characterizing the ferryl-iron center (21). Proposals such as that given in section A, have frequently led to the assumption that the Fe(IV) state is stabilized through bonding to oxygen, in the form of an oxy-ferryl (Fe^{IV}=O) or a hydroxy-ferryl (Fe^{IV}-O-H) center, and more recently it has been noted that an oxygen might bridge the Fe a porphyrin nitrogen (22). However, there was no direct evidence of the center's properties.

We therefore prepared HRPI(^{17}O) by oxidation of HRP with $H_2{}^{17}O_2$. The observation of endor from ^{17}O, proves that one oxygen atom from the oxidant remains bound to iron (Figure 3).

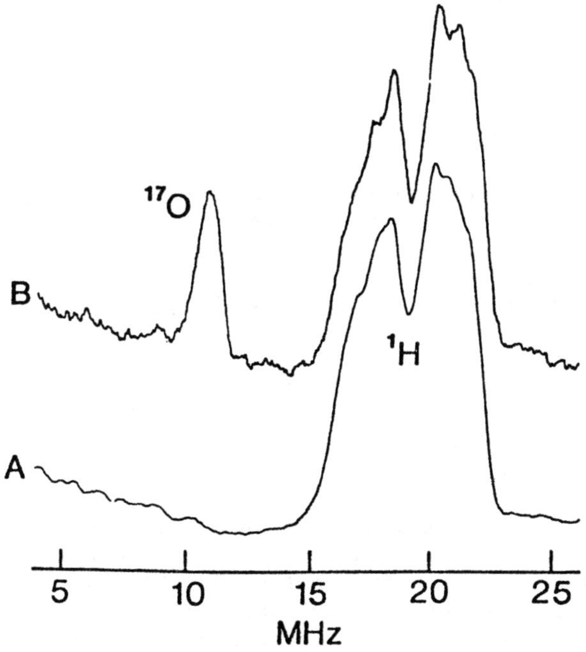

FIGURE 3 - ENDOR spectra of HRPI. A) HPR(^{16}O) and B) HRP(^{17}O) at g ≅ 1.5. The emergence of strong ^{17}O signals is apparent. Conditions: 2μW microwave power, T = 2K, field modulation ≃ 3 G, scan rate ~ 2 MHz/sec. Each spectrum represents the average of approximately 150 scans (Ref. 21).

The magnetic characteristics and optical properties of HRPI led us to interpret the data in terms of a triplet (S=1) oxy-ferryl (Fe^{IV}=O) center whose axis lies normal to the porphyrin cation plane, and to discard to alternative models noted above. The ^{17}O hyperfine couplings obtained from endor further allowed us to analyze the bonding in (Fe^{IV}=O). The two odd electrons are in antibonding π-molecular orbitals

$$\psi_x = (1-c^2)^{1/2}\, d_{xz}^{Fe} - c\, p_x^{o}$$

$$\psi_y = (1-c^2)^{1/2}\, d_{yz}^{Fe} - c\, p_y^{o} \qquad (2)$$

and the analysis implies that $c^2 \sim 0.25$, corresponding to a quadrivalent iron with substantial covalency to oxygen through $d_\pi - p_\pi$ bonding.

E. CCPI^R

The identification of the radical site in CCPI^R would appear to be easier than that of HRPI. The free radical-like character of the CCPI^R epr signal led to the early suggestion that the second oxidizing equivalent is not associated with the porphyrin, but resides in a spatially removed paramagnetic species, R·, presumably the stable free radical formed from an amino acid side chain (5). This suggestion is supported by the similarity of the CCPI^R optical and Mossbauer spectra to those of HRPII (5,2,3,14). Our own measurements of the g-values and spin lattice relaxation properties of the center confirm that the radical interacts at most very weakly with the oxy-ferryl center, with a coupling strength of no more than a few hundredths of a degree (12).

On the basis of the finding that TRP-51 lies near the heme, it was recently suggested that R· is the tryptophanyl radical formed by H-atom abstraction from NE1 of TRP-51 (9). This suggestion was supported by an examination of the amino acid sequence homologies between yeast cytochrome c and horseradish peroxidase (7). It was concluded that tryptophanyl residue 51 of cytochrome c peroxidase is represented by a less reactive phenylalanine in horseradish peroxidase, a substitution which may account for the differences between the enzymic intermediates formed by the two proteins.

However, the g-values of the R· epr signal are large and temperatures dependent ($g_{\parallel}$ = 2.09 at 77K; $g_{\parallel}$ = 2.05 at 4K) with

$g_{\parallel} > g_{\perp}$, features which contradict an identificiation with a π-cation radical. Furthermore, proton endor results from $CCPI^R$ are not consistent with the expectations for a tryptophanyl radical, and neither epr nor endor gives evidence for the large interaction with nitrogen which such a radical should exhibit. From this and other evidence we concluded that R· is not the radical of any aromatic amino acid, in particular not of a tryptophan (11,12).

The properties of the R˙ epr signal are in fact reminiscent of those seen for sulfur-based radicals. We were therefore led to propose that R˙ is the thioether cation radical based on the sulfur atom of methionine 171 (11,12). We believe that O—O bond cleavage is followed by electron transfer from porphyrin to metal to form a ferryl-iron/heme-radical center identical with that of horseradish peroxidase Compound I, and that this subsequently oxidizes the nearby methionyl side chain. However, the proton hyperfine couplings observed by endor appear to be too large to admit the assignment to an isolated methionyl radical, and require nucleophilic stabilization of the charge and delocalization of the spin over more than the single amino acid residue (11,12). Available models for such a delocalized center suggest that the nucleophile might be a second methionyl sulfur, an oxygen, or a nitrogenous base. At the present stage of the CCP structure refinement (10), it is not possible to unambiguously establish the residues comprising this sulfur-based redox cluster. Electrochemical studies with model compounds (23) make the methionine-171, aspartate-234 grouping an attractive candidate. However, the best-characterized model compounds involve two sulfur atoms (24), and it will be fascinating to see whether the sulfur atoms of methionyl residues 171, 229, and/or 230 are appropriately positioned to form a redox cluster.

F. Discussion

The two Compounds I discussed here can be described by the same structural scheme, scheme I, yet their physical and chemical properties are quite different. It seems likely that most Compounds I can be described by this scheme, and that differences in chemistry in large part reflect differences in the identity and properties of the radical, in the strength of the radical (Fe^{V}=O) coupling, and, perhaps most intriguingly, in the bonding within the (Fe^{IV}=O) moiety. It is tempting to speculate that protein-induced changes in iron-oxygen covalency (c^2, eq 2) control, say, the degree of oxygen electrophilicity and thus the type of reactivity exhibited by the enzymic intermediate.

Acknowledgments: Profound thanks are due to my collaborators, in particular Dr. James E. Roberts. This work was supported by National Science Foundation Grants PCM 76-81304 and PCM 79-10656 and National Institutes of Health Grant HL-13531. The measurements were performed in the magnetic resonance facility of Northwestern University's Material Research Center, supported in part under the National Science Foundation NSF-MRL Program (Grant DMR876-80847).

References

1) Dunford, H.B. and Stillman, J.S.: 1976, Coord. Chem. Revs. 19, pp 187-251.
2) Hewson, W.D. and Hager, L.P.: 1979, in "The Porphyrins" (Dolphin, D., ed.) Vol. 7, pp 295-332, Academic Press, New York.
3) This volume.
4) Schonbaum, G.R. and Chance, B.: 1976 in "The Enzymes" (Boyer, P.D., ed.) Vol. 13, pp 303-408, Academic Press, New York.
5) Yonetani, T.: 1976 in "The Enzymes" (Boyer, P.D., ed.) Vol. 13, pp 345-361, Academic Press, New York.
6) Kang, C.H., Ferguson-Miller, S., Margoliash, E.: 1977, J. Biol. Chem. 252, pp 919-926.
7) Takio, K., Titani, K., Ericsson, L.H., and Yonetani, T.: 1980, Arch. Biochem. Biophys. 203, pp 615-629.
8) Poulos, T.L., Freer, S.T., Alden, R.A., Edwards, S.L., Skogland, U., Takio, K., Eriksson, B., Xuong, N.-h., Yonetani, T., and Kraut, J.: 1980, J. Biol. Chem. 255, pp 575-580.
9) Poulos, T.L. and Kraut, J.: 1980 J. Biol. Chem. 255, pp. 8199-8205.
10) Poulos, T.L.: Personal Communication.
11) Hoffman, B.M., Roberts, J.E., Kang, C.H., and Margoliash, E.: 1979, Proc. Nat. Acad. Sci. 76, 6132-6136.
12) Hoffman, B.M., Roberts, J.E., Kang, C.H., and Margoliash, E.: 1981, J. Biol. Chem. 256, pp 6556-6594.
13) Schulz, C.W., Devaney, P.W., Winkler, H., DeBrunner, P.G., Doan, N., Chiang, R., Rutter, R., and Hager, L.P.: 1979, FEBS Lett. 103, pp 102-105.
14) Lang, G., Spartalian, K., and Yonetani, T.: 1976, Biochim. Biophys. Acta 451, pp 250-258.
15) Atherton, N.: 1973, "Electron Spin Resonance", John Wiley, New York.
16) Dolphin, D. and Felton, R.H.: 1974, Accts. Chem. Res. 7, pp 26-32.
17) Moroshima, I. and Ogawa, S.: 1978, Biochemistry 17, pp 4384-4388.

18) La Mar, G.N. and De Ropp, J.S.: 1980, J. Am. Chem. Soc. 102, pp 395-397.
19) Roberts, J.E., Hoffman, B.M., Rutter, R., and Hager, L.P.: 1981, J. Biol. Chem. 256, pp 2118-2121.
20) To be published.
21) Roberts, J.E., Hoffman, B.M., Rutter, R. and Hager, L.P.: J. Am. Chem. Soc., in press.
22) Chevrier, B., Weiss, R., Lang, M., Chottard, J.-C., and Mansuy, D.: 1981, J. Am. Chem. Soc. 103, pp 2899-2901.
23) Glass, R.S., Duchek, J.R., Klug, J.T., and Wilson, G.S.: 1977, J. Am. Chem. Soc. 99, pp 7349-7350.
24) Asmus, K.D.: 1979, Accts Chem. Res. 13, pp 200-206.

CYTOCHROME P450: STRUCTURE AND STATES

G.C. Wagner and I.C. Gunsalus

Department of Biochemistry
University of Illinois
Urbana, Illinois 61801

The cytochrome P450 monoxygenases assume a significant role in the elucidation of heme-iron mechanisms of oxygenation and O_2 reduction. The P450 monoxygenases mediate a two-electron activation of O_2 resulting in a molecule of H_2O and the insertion of a single O atom into a hydrocarbon substrate (1). The P450 reaction cycle illustrates a redox controlled pathway of Fe-O reactivity which supplements our understanding of 1) heme-O_2 binding in Mb and Hb, with 2) electron transfer in respiratory cytochromes and 3) oxygen activation of the peroxidases. A broad base of interdisciplinary study has contributed to the characterization of this physiologically important process. The P450 cytochromes refer to a general class of mixed function oxidases, who are widely distributed in nature and display diverse biological activity towards hydrocarbon substrates (2,3). In eukaryotes, and mammals in particular, the range of P450 functions is extensive and includes detoxification of xenobiotics, the activation of carcinogenes and the biosynthesis of regulatory steroid hormones. The prokaryotes are best represented by a P450 monoxygenase from *P. putida* that initiates the degradation of its growth substrate camphor (4).

The bacterial camphor monoxygenase mediates a stereospecific oxygenation of the bicyclic terpene to the corresponding 5-*exo*-hydroxycamphor. The complete system summarized in Figure 1 is comprised of three interacting protein components: 1) an FAD containing flavoprotein, fp; 2) an $Fe_2S^*_2Cys_4$ redox effector putidaredoxin, Pd; and 3) a b-type hemeprotein, cytochrome $P450_{CAM}$, abbreviated m̲ for monoxygenase. The flavo- and iron-sulfur proteins transfer two reducing equivalents (e_1 and e_2) derived from NADH to the camphor monoxygenase. The reaction cycle has been resolved into various redox and heme-ligand intermediates, which are

H. B. Dunford et al. (eds.), The Biological Chemistry of Iron, 405–412.

characterized by the wavelength maximum of the Soret absorption band. Also included is the catalytically inactive reduced CO complex, m_{CO}^{rs}, whose unique absorption near 450 nm provides the criterion of classification. The heme active site properites of these intermediates are common to all P450 cytochromes, despite the various functions, substrate specificities and modes of electron transport in these systems (5). Based on this homology, the camphor monoxygenase has served as a useful biological model system for structural and mechanistic study. This account presents a general overview into the structures of cytochrome $P450_{CAM}$, and the development into a molecular mechanism for the activation of oxygen. Since the molecular parameters of the $P450_{CAM}$ system have been reviewed periodically (4,6-8) over the past decade, a comprehensive survey is not intended.

Advances in the camphor monoxygenase rely in part on intrinsic properties of the bacterial system. The protein components are isolated in high yield as soluble and monomeric forms (9), which contrasts to the integral membrane association of mammalian cytochrome P450 (2,5). Bacterial culture on minimal media further allows protein labelling with specific isotopes of Fe, S, and N for spectroscopic study. The total enrichment technique is complemented by heme reconstitution which selectively labels the active site. The prosthetic group can be removed from $P450_{CAM}$ and the resultant apoprotein used for reconstitution with native (Fe) and artificial (Co) hemes (10,11). The Fe reconstituted P450 is

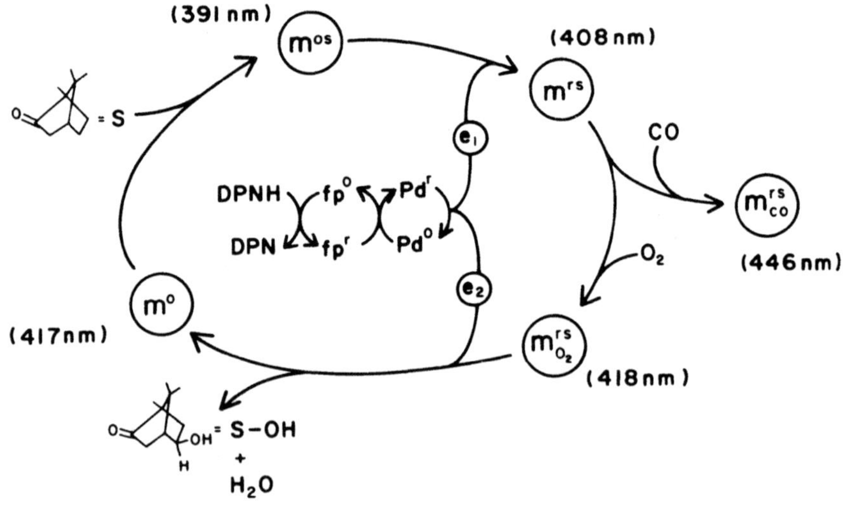

Figure 1. Camphor 5-monoxygenase reaction cycle.

characterized by enzymatic and spectroscopic forms which are indistinguishable from native enzyme. Finally the camphor monoxygenase is the only known P450 to be isolated in crystalline form. To date, three different crystalline forms of $P450_{CAM}$ have been obtained from modification of the chemical and physical parameters controlling crystal growth (8). Crystallization provides a source of homogeneous protein for composition and sequence analysis (12). Recently the primary structure of $P450_{CAM}$ has been deduced from the sequence of proteolytic and CNBr peptide fragments (K. Yasunobu, personal communication). The protein is a single polypeptide chain with 412 amino acid residues and a molecular weight near 47000. Approximately half of the residues are nonpolar, with the remaining polar groups divided equally between ionized and neutral functions. Predictions of secondary structure, made by a Chou and Fasman analysis of the sequence (S.G. Sligar, personal communication), indicate slightly less than half of the protein forms α-helical domains, including half of the 8 Cys residues. A high-resolution X-ray diffraction study of $P450_{CAM}$ (8) is in progress which utilizes an orthorhombic crystal measured in native form and in Pt and Hg heavy-atom derivatives. Supplementing these measurements, spectroscopic probes of oriented single crystals have established that the heme active site retains a native conformation in crystalline protein. The results from polarized absorption and EPR spectra (13) reveal that the redox and heme-ligand states of the monoxygenase cycle can be prepared in isomorphic form suitable for X-ray diffraction (T.L. Poulos, personal communication). Thus the $P450_{CAM}$ system offers the opportunity to develop a molecular 3D model of catalysis.

In advance of the crystallographic analysis, a summary of the structural properties of the P450 heme active site has been determined from spectroscopic probes of solution states (4,6-8). The axial ligand structures assigned to the intermediates of the native monoxygenase cycle are indicated in Figure 2. An artificial peroxygenase shunt (14) is also included, which utilizes peroxides or peracids as an oxygen source, without an external electron donor, to form the stereospecific 5-*exo* product (15). The magnetic and redox assignments of the native cycle are well-documented (4, 6-8), and based on various physical probes including EPR, Möss-bauer, NMR and susceptibility measurements. The assignments of low-spin configurations to six-coordinate structures and high-spin states with five-coordinate complexes are consistent with the central dogma relating magnetic state to heme structure. There are three main aspects of this scheme which are unique among the hemeproteins so far investigated. First, a sulfur ligand is retained throughout the reaction cycle. Second, coordination changes are associated with substrate binding, although there is no evidence the hydrocarbon is a heme ligand. And last, an ionized mercaptide state is assigned to the axial ligand.

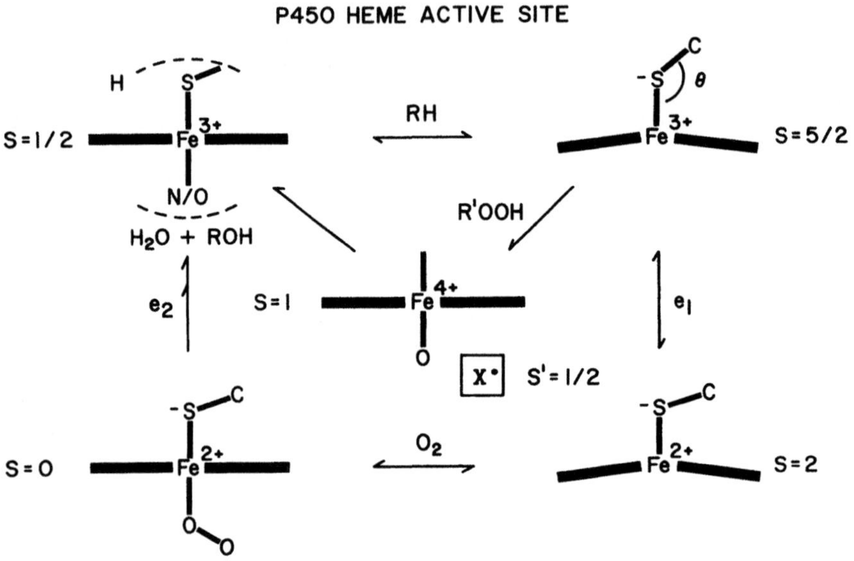

Figure 2. Cytochrome $P450_{CAM}$ heme ligand coordination.

For over 15 years, dating to the original efforts of Mason and colleagues, a cysteinyl sulfur ligand has been implicated in P450 hemeproteins. A direct observation of sulfur coordination has been given only recently by a resonance Raman study, in Figure 3, of the ferric enzyme substrate complex ($\underline{m}^{os}$) labelled with heavy and light isotopes of Fe and S (16). A close examination of the intense low-frequency mode at 351 cm^{-1} by a Raman difference technique reveals frequency shifts expected by the mass effect. The sensitivity of the 351 cm^{-1} mode to both Fe and S unambiguously confirms an Fe-S axial ligand bond in a P450 hemeprotein. Evidence for an ionized mercaptide ligand was previously presented from resonance Raman spectra of the oxidized $\underline{m}^{os}$ and reduced $\underline{m}^{rs}$ camphor states (17). There a weakening of high frequency porphyrin ring modes was interpreted as due to coordination of a strongly electron donating ligand ($R\text{-}S^-$). The absence of a sixth ligand in the $\underline{m}^{os}$ high-spin structure was inferred from the ENDOR of the meso-protons on the periphery of the porphyrin (18). Comparison of the ENDOR proton coupling with those of known strucures revealed the P450 coupling was consistent with a five-coordinate complex. Supporting this conclusion, no exchangeable protons in the ENDOR spectrum of $\underline{m}^{os}$ were found in D_2O solvent. This contrasts to the analogous Mb state, with a weak distal H_2O ligand in the sixth position, whose large H_2O proton couplings are readily lost with D_2O exchange.

Substrate induced structural changes at the heme active site have been proposed since the early magnetic characterizations of

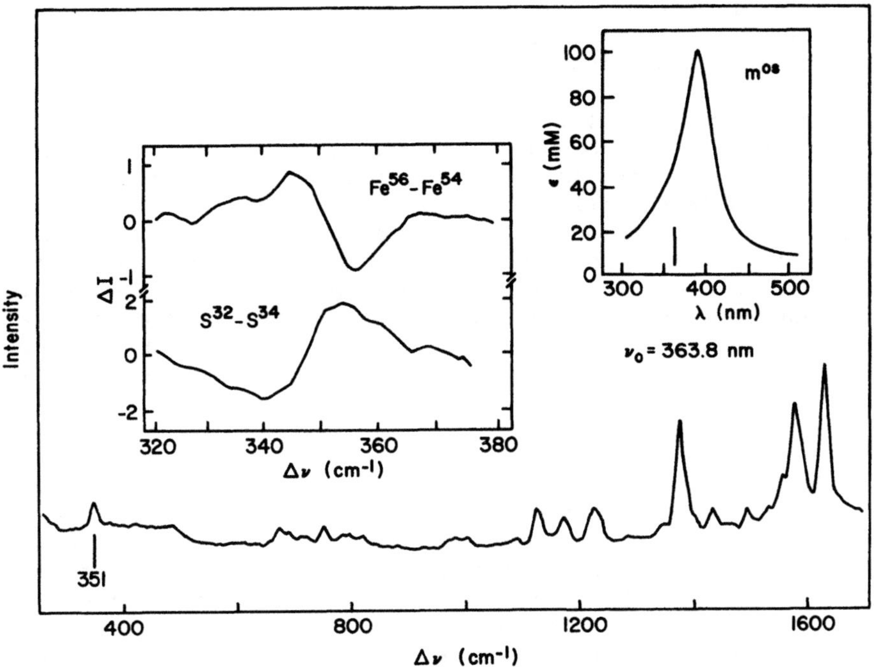

Figure 3. Resonance Raman spectra of $P450_{CAM}$.

the S=5/2 $\underline{m}^{os}$ and S=1/2 native $\underline{m}^{o}$ states (4,6-8). The low-spin configuration in $\underline{m}^{o}$ is attributed to the coordination of two axial ligands, one sulfur and the second from either an oxygen or nitrogen based residue. Further insight into the active site of $\underline{m}^{o}$ developed from the effect of $\underline{m}^{o}$ on the NMR relaxation rate of H_2O protons (19,20). A proton exchange mechanism mediated through a heme iron ligand was proposed and subsequently confirmed by an ENDOR investigation (18). The ENDOR spectrum was characterized by large H-couplings which were exchangeable with deuterated solvent. Analysis of the coupling was consistent with an Fe to H distance of 2.5-3 Å and permitted assignment of the exchangeable proton to a heme axial ligand. The absence of a heme-linked proton exchange mechanism (vida supra) in the high-spin $\underline{m}^{os}$ state is compatible with a specific change in heme coordination with camphor binding. Since hydrocarbon binding is associated with a transition from low- to high-spin states, considerable emphasis has been directed at correlating the magnetic state with the position of the Soret absorption (21,22). Camphor binding will induce a 26 nm blue-shift of the Soret maximum which can be related to the fraction of high-spin state. A comparable 24 nm shift was also observed with the Co-P450 analog (11). Since Co^{3+} porphyrins are without exception in a low spin d^6 configuration, the substrate induced optical shift can only be correlated directly with a change in heme coordination.

The Co-P450 analog has also provided evidence of the ligand coordination in the reduced and oxygenated states of the monoxygenase cycle. EPR spectra of Co-$\underline{m}^{rs}$, like those of Co-Mb and Co-HRP, are indicative of a five-coordinate complex with a strong Lewis base for an axial ligand (11). In Mb and HRP, where the axial ligand is a nitrogenous His, each of the eight Co hfs lines in the EPR spectra are split into triplets due to a shfs interaction with N in the axial position. In contrast, EPR spectra of Co-P450 reveal an absence of triplet structure and thus are not consistent with a nitrogenous ligand. Although sulfur and oxygen based ligands are compatible, the very small hfs coupling constants in Co-P450 reflect coordination by a Lewis base stronger than a thiol or alcohol. An ionized thiolate ligand, in both the reduced and O_2 adduct of P450, is the most consistent interpretation of the available data.

The addition of the second reducing equivalent (e_2) to the O_2 adduct results in regeneration of native enzyme without evidence for a stable intermediate (23). Insight into possible mechanisms of the critical oxygen insertion step(s) has developed from studies of the peroxygenase pathway (14,15,23,24). The reaction of Fe^{3+} $\underline{m}^{o}$ with peracid produces a transient intermediate which can be trapped by freeze-quenching. The EPR spectrum of this intermediate in Figure 4 reveals a narrow free-radical signal at g=2.0, whose integrated intensity after 30 ms of reaction time approaches 60% of the initial Fe^{3+} signal. In these early

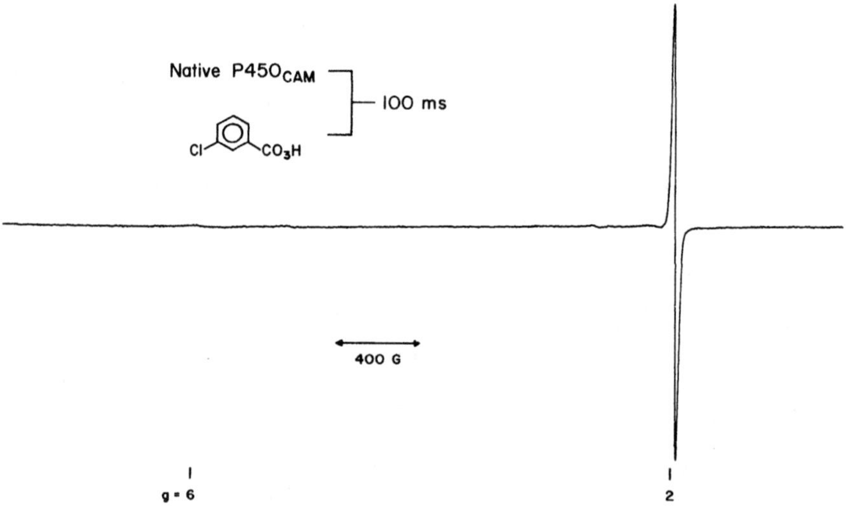

Figure 4. EPR spectrum of peroxygenase intermediate.

phases of reaction the radical species is formed stoichiometrically with the decrease in Fe^{3+} $\underline{m}^{o}$ spin intensity. This observation is consistent with the EPR silent Fe^{4+} species assigned to various peroxidase intermediates of higher oxidation state (25). The g=2 species is characterized by a peak-to-peak line width of 24 G and apparent hyperfine structure. Resolution of the hfs depends slightly on the presence of the camphor substrate, but is not affected by deuteration (2,4,6-d_3-$\underline{m}$-chloroperbenzoic acid) of the peracid reagent. These preliminary findings provide no evidence for an exogenous acid radical intermediate in the peroxygenase pathway. The relatively slow spin-lattice relaxation time inferred from power saturation studies indicate the g=2 intermediate differs from a compound I state of HRP (25). The limited characterizations of the P450 free radical species are compatible with an EPR active intermediate localized on a non-heme protein function.

REFERENCES

1. Hayaishi, O.: 1974, "Molecular Mechanisms of Oxygen Activation", Academic Press, New York.
2. Sato, R. and Omura, T.: 1978, "Cytochrome P450", Academic Press, New York.
3. Coon, M.J., Conney, A., Estabrook, R., Gelboin, H., Gillette, J., and O'Brien, R.: 1980, "Microsomes, Drug Oxidations, and Chemical Carcinogenesis", Academic Press, New York.
4. Gunsalus, I.C., Meeks, J.R., Lipscomb, J.D., Debrunner, P.G., and Münck, E.: 1974 in "Molecular Mechanisms of Oxygen Activation", (Hayaishi, O., ed) Chap. 14, Academic Press, New York.
5. Coon, M.J. and White, R.E.: 1980 in "Dioxygen Binding and Activation by Metal Centers", (Spiro, T.G., ed), pp. 73-123, John Wiley & Sons, New York.
6. Debrunner, P.G., Gunsalus, I.C., Sligar, S.G., and Wagner, G.C.: 1978 in "Metal Ions in Biological Systems", (Sigel, H., ed), Vol. 7, pp. 241-275, Marcel Dekker, New York.
7. Gunsalus, I.C. and Sligar, S.G.: 1978, Adv. Enzymol. 47, pp. 1-44.
8. Gunsalus, I.C., Wagner, G.C., and Debrunner, P.D.: 1980 in "Microsomes, Drug Oxidations and Chemical Carcinogenesis" (Coon, M.J., Conney, A., Estabrook, R., Gelboin, H., Gillette, J. and O'Brien, R., eds) pp. 233-242, Academic Press, New York.
9. Gunsalus, I.C. and Wagner, G.C.: 1978, Methods Enzymol. 52, pp. 166-188.
10. Wagner, G.C., Toscano, W.A., Perez, M., and Gunsalus, I.C.: 1981, J. Biol. Chem. 256, pp. 6262-6265.
11. Wagner, G.C., Gunsalus, I.C., Wang, M.-Y., and Hoffman, B.M.: 1981, J. Biol. Chem. 256, pp. 6266-6273.

12. Yasunobu, K., Armes, L.G., Crabb, J.W., Hanui, M., Gunsalus, I.C., and Coon, M.J.: 1980 in "Proceedings of the Biochemistry, Biophysics and Regulation of Cytochrome P450", Elsevier/North Holland, Amsterdam, in press.
13. Devaney, P., Wagner, G.C., Gunsalus, I.C., and Debrunner, P.G.: 1981, J. Biol. Chem., in preparation.
14. White, R.E., Sligar, S.G., and Coon, M.J.: 1980, J. Biol. Chem. 255, pp. 1108-1111.
15. Sligar, S.G., Shastry, B.S., and Gunsalus, I.C.: 1977 in "Microsomes and Drug Oxidations", (Ullrich, V., Roots, I., Hildebrandt, A., Estabrook, R.W., and Conney, A., eds), pp. 202-209, Pergamon Press, Oxford.
16. Champion, P.M., Stallard, B.R., Wagner, G.C., and Gunsalus, I.C.: 1981, Nature, in press.
17. Champion, P.M., Gunsalus, I.C., and Wagner, G.C.: 1978, J. Am. Chem. Soc. 100, pp. 3743-3751.
18. LoBrutto, R., Scholes, C.P., Wagner, G.C., Gunsalus, I.C., and Debrunner, P.G.: 1980, J. Am. Chem. Soc. 102, pp. 1167-1170.
19. Griffen, B.W. and Peterson, J.A.: 1975, J. Biol. Chem. 250, pp. 6445-6451.
20. Philson, S.B., Debrunner, P.G., Schmidt, P.G., and Gunsalus, I.C.: 1979, J. Biol. Chem. 245, pp. 10173-10179.
21. Sligar, S.G.: 1976, Biochemistry 15, pp. 5399-5406.
22. Lange, R., Pierre, J., and Debey, P.: 1980, Eur. J. Biochem. 107, pp. 441-445.
23. Pederson, T.C., Austin, R.H., and Gunsalus, I.C.: 1977 in "Microsomes and Drug Oxidations", (Ullrich, V., Roots, I., Hildebrandt, A., Estabrook, R.W., and Conney, A., eds), pp. 275-283.
24. Blake, R.C., and Coon, M.J.: 1980, J. Biol. Chem. 255, pp. 4100-4111.
25. Schulz, C.E., Devaney, P.W., Winkler, H., Debrunner, P.G., Doan, N., Chiang, R., Rutter, R., and Hager, L.P.: 1979, FEBS Let. 103, pp. 102-105.

CYTOCHROME P450 AS A REDUCTASE AND OXENE TRANSFERASE: WHICH IS ITS CHARACTERISTIC FUNCTION ?

V.Ullrich, H.J.Ahr, L.Castle, H.Kuthan, W.Nastainczyk, H.H.Ruf

Department of Physiological Chemistry, University of Saarland, 6650 Homburg-Saar, Fed. Republic of Germany

The reaction mechanism of cytochrome P450 comprises two phases: the reduction of dioxygen with two electrons to an active oxygen species and its subsequent transfer to organic molecules. Other compounds with appropriate redox potentials can substitute for dioxygen and accept electrons from the reduced cytochrome. Geminal polyhalogenated hydrocarbons are thus reduced by one electron to free radicals which can be liberated from the active site. Further reduction at the iron leads to carbanions or carbenes which may form complexes with cytochrome P450. From a comparison with other hemoproteins and from models it is concluded that the electron transfer to dioxygen or halogenated hydrocarbons is not peculiar to cytochrome P450. The oxene transfer, however, is dependent on the presence of a thiolate ligand which is a feature common to all P450 enzymes. The biosynthesis of prostacyclin is proposed as a physiological oxene transfer for which a strong indication for P450 catalysis could be obtained.

The hemoprotein with the unusual carbon monoxide complex absorbing at 450 nm has been called "cytochrome P450", since its two subsequent electron transfers to dioxygen have been considered as a main and intrinsic function of this hemoprotein. Unlike other hemoproteins, the many cytochrome P450 enzymes known so far are all capable of hydroxylating organic molecules like amino acids, steroids or drugs and hence they were characterized as monooxygenases (1). The reaction cycle of all P450 isozymes consists of five major steps: i) the binding of a substrate under conversion of the low spin cytochrome P450 to a high spin enzyme-substrate complex ii) its reduction to a ferrous complex iii) dioxygen addition to form an oxycomplex iv) transfer of a second electron

H. B. Dunford et al. (eds.), The Biological Chemistry of Iron, 413–425.

to give the active oxygen species and v) oxene transfer to the organic molecule (Fig.1) (2).

Mechanism of Cytochrome P450-dependent Monooxygenases

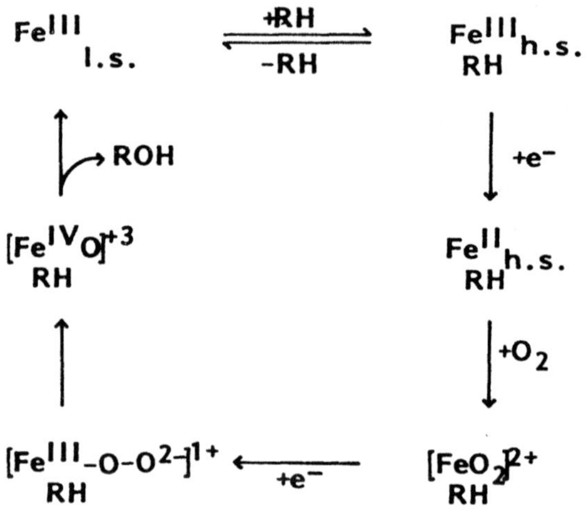

The active oxygen species is considered to be an oxenoid complex rather than a peroxo-complex, since hemoproteins like peroxidases and catalases generally split the peroxide bond under release of water (3). Also, certain organic peroxides (4,5) and iodosoarenes (6) have been discovered as artificial oxene donors that can support P450-dependent monooxygenations in the absence of dioxygen and electrons.

These reactions made it evident that electron transport and oxene transfer are two distinct and separable processes in the reaction cycle of cytochrome P450. Therefore the question became interesting, whether the unusual coordination sphere of cytochrome P450, namely thiolate sulfur as an axial ligand of iron, was required for either the process of the two electron reduction of dioxygen or for the oxene transfer process, or as a third possibility, for both reactions. This problem shall be discussed here in greater detail and an answer from recent experiments will be given.

The unusual spectral properties of cytochrome P450 are, without doubt, linked to the presence of a special ligand at the heme which has now been identified by indirect evidence as a thiolate group, obviously from a cysteinyl residue (7, 8, 9, 10). The unusually red-shifted Soret bands of cytochrome P450 and its various complexes can be explained by a charge transfer to the porphyrin ring and were classified as hyper-porphyrin spectra (11). Removal of the thiolate ligand, which can be achieved by protonation or detergent treatment will prevent this charge transfer and results in normal hemoprotein spectra ("P420") (12). Concomitant with the transition to P420, catalytic activity is lost.

Several possible rôles for the thiolate ligand may be suggested. It could mediate electron transport to the iron or could maintain the low redox potential necessary for the one electron reduction of dioxygen. As a third possibility, it could be involved in the transfer of the active oxygen to the substrate. Since both functions of cytochrome P450, electron transfer and oxene transfer, can be investigated separately, we have attempted to evaluate the participation of the thiolate ligand in both processes.

1. THE REDUCTION OF DIOXYGEN AND HALOGENATED HYDROCARBONS

Although dioxygen is the physiological substrate for P450 dependent monooxygenases, other acceptors of suitable redox potential can also compete for the electrons. Table 1 gives examples of polyhalogenated hydrocarbons which cause an oxidation of NADPH in anaerobic liver microsomes. The inhibition by carbon monoxide seen in Table 1 proves the participation of cytochrome P450. It can be seen that hexa- and pentachloroethane are reduced almost as rapidly as dioxygen (13).

NADPH oxidation in anaerobic NADPH reduced liver microsomal fractions from phenobarbitone pretreated rats in the presence of polyhalogenated hydrocarbons.

Incubations in 2,5 ml Tris HCl buffer pH 7.6 contained 2,5 mg microsomal protein (2.3 nmol cytochrome P450/mg), NADPH 0.2 mM and the haloalkane (1 mM) at 30°C in a stoppered cuvette.Dioxygen was removed by bubbling with N_2 for 15 min before addition of the haloalkane and NADPH.
NADPH oxidation was measured using the wavelength pair 366/510 nm and an extinction coefficient of 3.3 $mM^{-1}cm^{-1}$.

Compound	NADPH Oxidation nmol/mg Protein/min	CO-Inhibition %
$Cl_3C - CCl_3$	34.9	99
$Cl_3C - CHCl_2$	30.6	99
$Cl_3C - CF_2Cl$	29.7	98
$Cl_3C - C_6H_5$	13.1	98
$Cl_3C - CF_3$	9.2	81
DDT	5.1	94
CCl_4	2.6	81
$F_3C - CHClBr$	1.2	98

Other compounds like carbon tetrachloride or halothane cause an initial fast phase in the oxidation of NADPH and then the reactions slow down and finally stop. In these cases, one can observe spectral intermediates, which in the case of carbon tetrachloride has been identified as a carbene complex of ferrous cytochrome P450 (14). The evidence for this assignment is the absence of an epr signal (15) and the stoichiometric conversion of the complex into carbon monoxide in which the carbon atom was derived from CCl_4 (14). At temperatures below 4°C, the complex was stable in microsomes from phenobarbital pretreated animals and no hydrolysis to CO was observed. Final confirmation for the existence of such carbene complexes comes from the work of Mansuy et al. (16) who have crystallized a corresponding five-coordinated CCl_2-Fe (TPP)-porphyrin complex and established its molecular structure. Unlike the cytochrome P450 CCl_2-complex, with its Soret absorption at 459 nm, the complex obtained under the same conditions but with halothane absorbed at 470 nm (17). Since diazotrifluoroethane produced an identical spectrum a carbene complex of the structure S ·· Fe=CH-CF_3 was proposed. Doubt was cast upon this after we were able to detect a rather broad low spin epr signal in microsomes to which halothane and dithionite were added (Fig. 2).

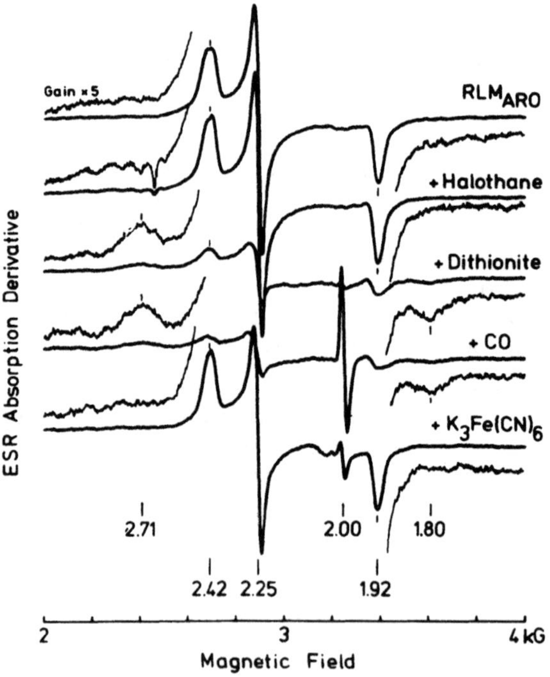

EPR spectra of the interaction of halothane with rat liver microsomes.

To 300 µl of microsomes (51 mg protein/ml, 200 µM P450) from rats pretreated with a single dose of Arochlor 1254 (500 mg/kg body weight i.p., 4 days prior to use) were subsequently added 10 mM halothane, 3 mM dithionite (incubated for 2.5 min at 20°C), CO (bubbled for 20 s and frozen within another 20 s) and 10 mM $K_3Fe(CN)_6$.
EPR was measured at 100 K using 9.17 GHz microwave frequency, 40 mW power and 25 G modulation amplitude. The gain was 250 and 1250 for the inserted wings of the spectra, respectively. The P450-halothane complex at g = 2.71 and 1.80 was not observed when the sample was bubbled with CO before the addition of halothane and reductant.

This signal was clearly dependent on halothane and was not decreased after addition of carbon monoxide. It reflected the behaviour of the species absorbing at 470 nm in every respect (15). Although this epr-signal was clearly derived from cytochrome P450 it did not show the characteristic narrow splitting of the g-tensor. It exhibits the largest anisotropy of the g-tensor observed so far for P450 and its ligand complexes. Its nature is not immediately apparent but certainly its low spin ferric characteristics are not consistent with the ferrous carbene complex originally proposed.

More information on the structure of the 470 nm complex came from studies of the products of halothane reduction (15). The main product was 2-chloro-1,1,1-trifluoroethane (CTE) which was derived from halothane by a one-electron reduction followed by hydrogen abstraction. In a completely analogous reaction, chloroform is formed from carbon tetrachloride (14). A second metabolite was identified as 2-chloro-1,1-difluoroethylene (CDE). This metabolite was a direct degradation product of the 470 nm complex. Therefore the structure of the complex was tentatively assigned as a ferric cytochrome P450 carbanion complex arising from a two-electron reduction of halothane according to the following scheme (15).

This pathway seems to have a general applicability for geminal polyhalogenated hydrocarbons, since the metabolism of hexa- and pentachloroethane to yield the corresponding ethylene derivatives should proceed in the same way but without the intermediate formation of a stable complex. It seems that only the CF_3 group can stabilize such a carbanion complex sufficiently, whereas in all other cases ß-elimination of a halide anion is preferred (13).

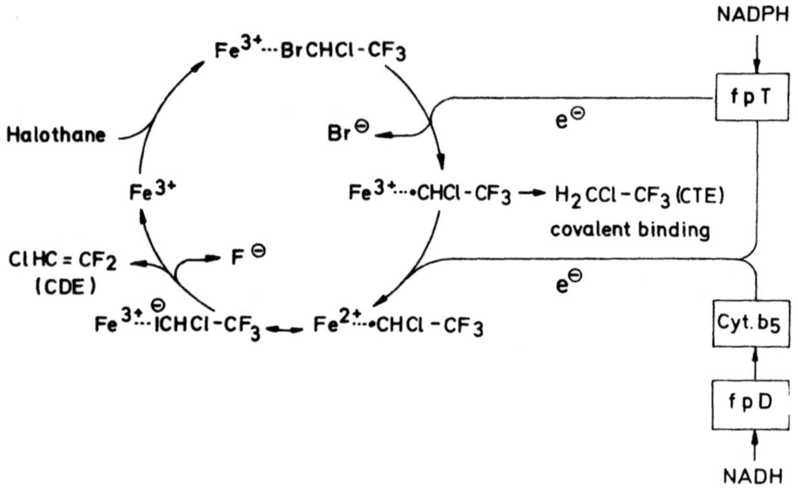

In order to synthesize a model complex, Mansuy et al. succeeded in preparing a complex upon reductive treatment of halothane and iron-tetraphenylporphyrin (FeTPP) (personal communication). The epr spectrum of this species is shown in Fig. 3.

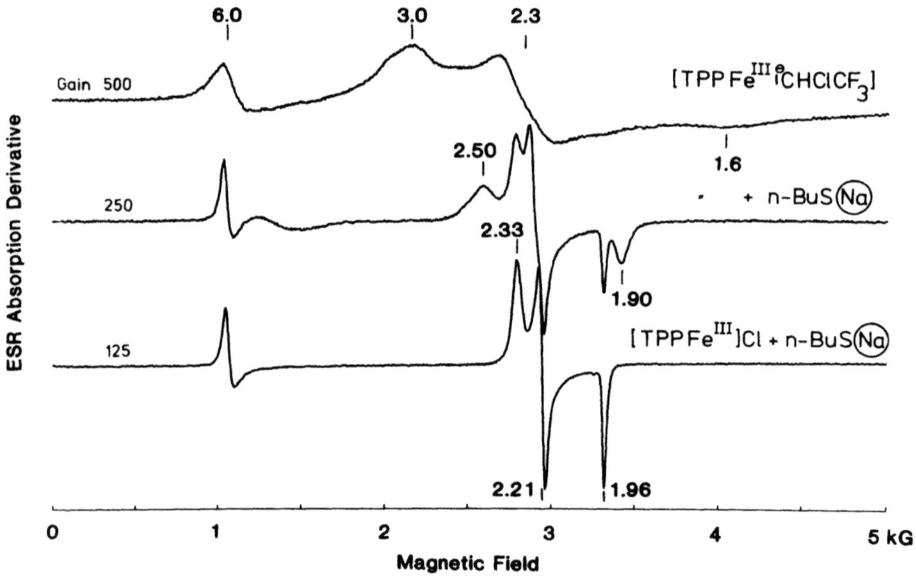

EPR spectra of Fe^{III}TPP complexes.

The samples contained about 1 mM Fe^{III}TPP. The complex with halothane was prepared according to Mansuy et al. and dissolved in CH_2Cl_2 under strictly anaerobic conditions (upper trace). n-BuSNa (about 0.3 M in DMF with dibenzo-18-crown-6) was added to give a final concentration of 30 mM at -78°C under anaerobiosis (middle trace). The Fe^{III}TPP dithiolate complex was prepared according to Ruf et al. (18) by mixing the iron porphyrin with an about 5-fold excess of n-BuSNa at -78°C (lower trace). The EPR parameters are described in Fig. 2 except that a modulation amplitude of 10 G was used.

It clearly indicates a low spin ferric complex. If a solution of this complex is treated at low temperature with n-butane thiolate a new species is formed with g-values at 1.902, 2.251 and 2.501. At best 70 % of the original complex could be formed. The remaining complex formed a dithiolate complex of Fe^{III}TPP (lower spectrum) under these conditions (18). If plotted according to its ligand field parameters (Fig. 4) the new species is analogous to the species found in microsomes.

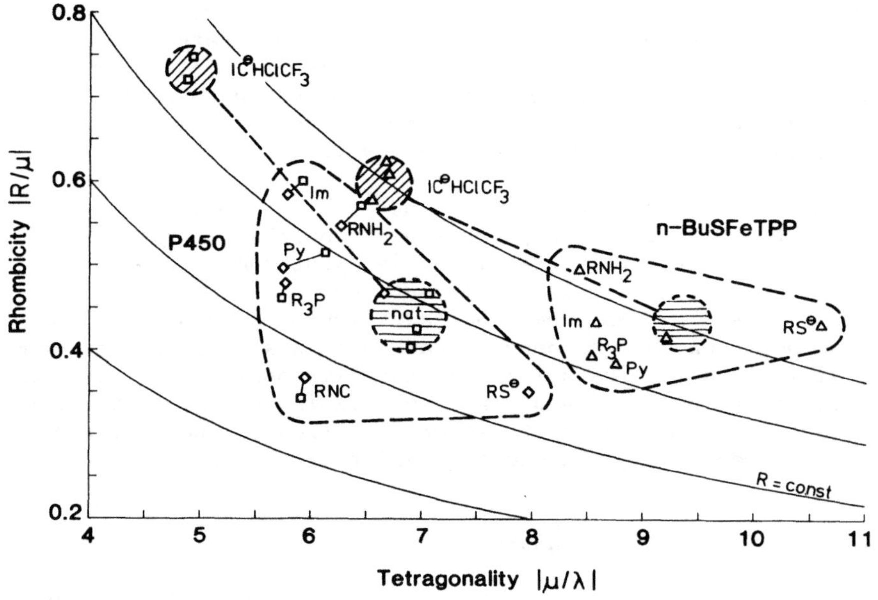

Comparison of the ligand field parameters of P450 complexes with Fe^{III}TPP thiolate model complexes with various sixth ligands.

We used $P450_{CAM}$ and $P450_{RLM}$ with the ligands RNH_2, n-octylamine; Im, imidazole or N-methylimidazole; Py, pyridine or metyrapone;

R_3P, diethylphenylphosphine; RNC, benzylisonitrile; RS^-, n-butylthiolate and nat, native.
The ligand field parameters were calculated according to Blumberg et al. (19) using Bohan's convention III for the ordering of g-values (20). The orbital reduction factor k was found to vary between 1.0 and 1.3.

The optical spectra of this complex are difficult to record, but only a 383 and 466 nm absorption were observed. Its assignment to the new epr-species is not clear-cut, however, since the ferric TPP-dithiolate complex also absorbs at these wavelengths.

Although the structure of the starting complex has still to be established, we tentatively suggest that the coordination sphere of the cytochrome P450-halothane complex and the model complex are the same and that a ferric-porphyrin-thiolate-2-chloro-1,1,1-trifluoro carbanion structure is most likely.

Such a novel chemistry at cytochrome P450 looks exciting, but is in fact not very different from the processes occuring during dioxygen reduction. The oxy-complex corresponds to the carbon radical formed after the first electron transfer to the halogenated hydrocarbons. The second electron generates the carbanion which corresponds to the peroxide anion and both can undergo elimination reactions to form the olefin (or a carbene complex) or the oxenoid complex. Thus the two subsequent electron transfers to the halogenated hydrocarbons can be considered as a model for the first four steps in the P450 reaction cycle (Fig. 1).

The reductive processes as well as the complex formation of the intermediates with the iron-porphyrin are unlikely to be a special function of the thiolate ligand in cytochrome P450 since model iron porphyrin complexes without the thiolate perform the same reactions. It even seems that addition of a thiolate ligand in trans-position to a carbene or carbanion destabilizes the complex, similar to the instability of the oxy-complex of cytochrome P450. If the thiolate ligand plays no rôle in the reductive processes one can expect that its importance lies in the last step of the P450 mechanism, the oxene transfer process.

2. OXENE TRANSFER BY CYTOCHROME P450

The first compound found to mediate hydroxylations and epoxidations by ferric cytochrome P450 was cumene hydroperoxide (4,5). Other hydroperoxides and even hydrogen peroxide can serve also (21). We then reported that iodosobenzene and its derivates can also act as oxene donors (6,22). The mechanism of these systems is not clear yet, but the activity is dependent on intact cytochrome P450. Cytochrome P420, which presumably has lost the thiolate bond, is inactive, so that the presence of the sulfur ligand seems required for the oxene transferase activity. Experimental support for this assumption was obtained from the finding of an intramolecular oxene transfer within nitromethane, which gives rise to formaldehyde and nitrous oxide (23, 24). This reaction could be mimicked by a five-coordinated heme-thiolate model. The ferric porphyrin alone was much less active, even in the presence of thiol (25). We have interpreted this as a trans-effect of the thiolate which, by its negative charge and/or its d-orbital interaction with the FeO^{3+} species, allows the oxygen atom to attack organic compounds. Formally, Compound I of horse radish peroxidase and oxenoid complex of cytochrome P450 are equivalent in oxidation state (Fe^{IV} plus one oxidizing equivalent), but the thiolate ligand causes a difference in electron distribution which is indicated by the following structures:

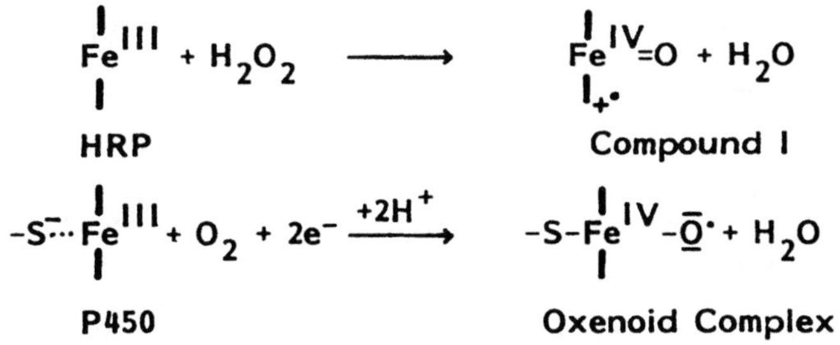

One can therefore consider Compound I as a rather stable π-cation radical with an electronically saturated oxo group whereas the oxenoid complex of P450 may have a partial electron deficiency with a partially unpaired spin giving it electrophilic properties. The mechanism of attack of the oxenoid complex of cytochrome P450 on saturated hydrocarbon moieties seems to be a radical abstraction process followed by a rapid rearrangement of the crypto-radical complex (26):

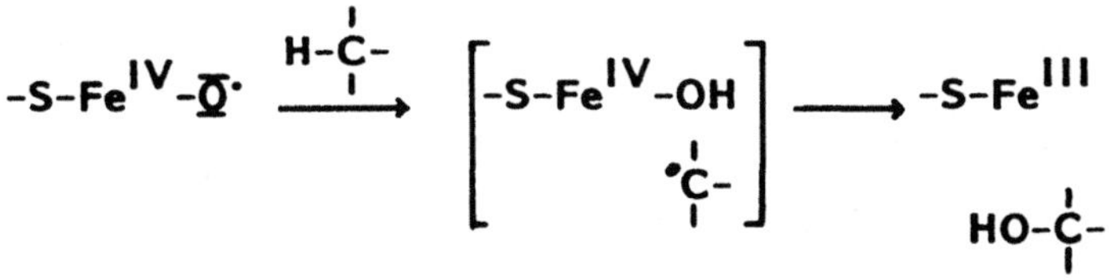

For unsaturated substrates, an addition process generating first a carbon-centered radical can be envisaged:

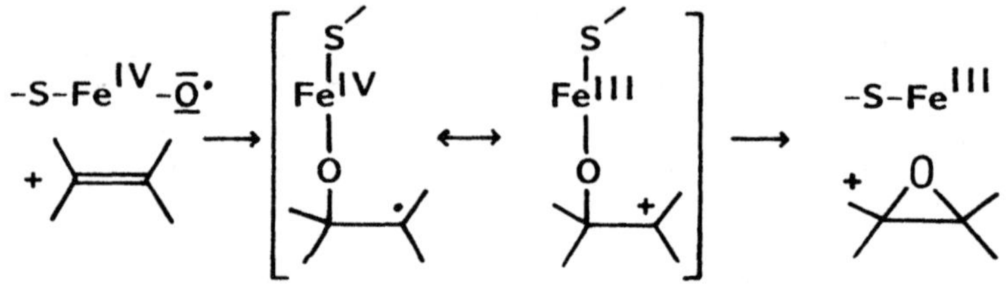

An extremely short life time for the radical intermediates is indicated by the following observations: 1) only partial epimerisation occurs during the hydroxation of norbornane (26) 2) alkene epoxidation proceeds with absolute syn-stereospecificity. Thus, we find greater than 99,9 % retention of cis-configuration when cis-but-2-ene is epoxidised by rat liver microsomes supported either by $NADPH/O_2$ or by the oxene donors PhIO, $PhC(CH_3)_2OOH$, H_2O_2 or 3-Cl-PhCOOOH.

We conclude that the characteristic function of cytochrome P450 would be that of an oxene transferase and the properties of an electron acceptor/donor would only be required to generate the oxene from dioxygen.

Following these lines of thought, we were looking for physiological functions of cytochrome P450 as an oxene transferase without electron transfer to the iron. Biologically important peroxides that undergo intramolecular oxene transfer reactions are occuring in the biosynthesis of prostacyclin and thromboxane A_2 from the 9, 11-endoperoxide of 15-hydroxy arachidonic acid (27).

Prostacyclin is synthesized mainly in arterial walls and the corresponding synthase is localized in the microsomal fraction (28). Indeed cytochrome P450 was found in this fraction but its rôle was described as that of a monooxygenase (29). We were unable to verify this, since neither 7-ethoxycoumarin was O-dealkylated nor was cytochrome P450 reduced enzymatically by NADPH in the presence of CO (30).

The cytochrome was found to react spectrally with the prostaglandin endoperoxide (PGH_2) (30). In the presence of tranylcypromin (trans-2-phenylcyclopropylamine, an inhibitor of prostacyclin synthesis), this spectral interaction with PGH_2 was abolished by the formation of a ligand complex with the inhibitor. Also. the cytochrome was found to interact spectrally with 15-hydroperoxy arachidonic acid at about the same micromolar concentrations that are known to inhibit prostacyclin synthesis (31).

Unlike liver microsomal cytochrome P450, the P450 protein in aorta microsomes does not form an enzyme-substrate complex with drugs, but forms a ligand complex with cyanide already at low concentrations. All these data allow the conclusion that the aorta P450 is different from a drug monooxygenase but may be identical to prostacyclin synthase. Similar work is now in progress with the thromboxane A_2 synthase of platelets. Platelets contain cytochrome P450 that has properties similar to the aortic P450.

In conclusion, we would like to propose that the thiolate ligand of cytochrome P450 is required for its oxene transferase activity. At least for the P450 isozyme in aorta, which is not reduced, the term "cytochrome" would be misleading since no electron transfer is involved.

ACKNOWLEDGMENT

Supported by the Deutsche Forschungsgemeinschaft, Sonderforschungsbereich 38, project L1.

REFERENCES

1. Ullrich, V.: 1979, Topics in Current Chemistry 83, pp. 67-104.
2. White, R.E., and Coon, M.J.: 1980, Ann. Rev. Biochem. 49, pp. 315-356.
3. Schonbaum G.R., and Lo, S.: 1972, J.Biol.Chem. 247, pp. 3353-3360
4. Kadlubar, F.F., Morton, K.C., and Ziegler, D.M.: 1973, Biochem. Biophys. Res. Commun. 54, pp. 1255-1261.
5. Rahimtula, A.D., and O'Brien, P.J.: 1974, Biochem. Biophys. Res. Commun. 60, pp. 440-447.
6. Lichtenberger, F., Nastainczyk, W., and Ullrich, V.: 1976, Biochem. Biophys. Res. Commun. 70, pp. 939-946.
7. Tang, S.C., Koch S., Papaefthymiou, G.C., Foner, S., Frankel, R.B., Ibers, J.A., and Holm, R.H.: 1976, J. Am. Chem. Soc.98, pp. 2414-2434.
8. Collman, J.P., and Sorrell, T.N.: 1975, J. Am. Chem. Soc. 97, pp. 4133-4134.
9. Cramer, S.P., Dawson, J.H., Hodgson, K.O., and Hager, L.P.: 1978, J. Am. Chem. Soc. 100, pp. 7282-7290.

10. Ullrich, V., Ruf, H.H., and Wende, P.: 1977, Croat. Chem. Acta 49, pp. 213-222.

11. Hanson, L.K., Eaton, W.A., Sligar, S.G., Gunsalus, I.C., Gouterman, M., and Connel, C.R.: 1976, J. Am. Chem. Soc. 98, pp. 2672-2674.

12. Jung, C., und Ristau, O.: 1977, Chem. Phys. Lett. 49, pp. 1o3-1o8.

13. Nastainczyk, W., Ahr, H.J., and Ullrich, V.: Biochem. Pharmacol., in press.

14. Ahr, H.J., King, L.J., Nastainczyk, W., and Ullrich, V.: 1980, Biochem. Pharmacol. 29, pp. 2855-2861.

15. Ahr, H.J., King, L.J., Nastainczyk, W., and Ullrich, V.: Biochem. Pharmacol., in press.

16. Mansuy, D., Lange, M., Chottard, J.C., Bartoli. J.F., Chevrier, B., and Weiss, R.: 1978, Angew. Chemie 90, pp. 828-829.

17. Mansuy, D., Nastainczyk, W., and Ullrich, V.: 1974, Naunyn Schmiedeberg's Arch. Pharmacol. 285, pp. 315-324.

18. Ruf, H.H., Wende, P., and Ullrich, V.: 1979, J. Inorg. Biochem. 11, pp. 189-204.

19. Blumberg, W.E., and Plisach, J.: 1971, Bioanorganic Chemistry, R.F. Gould, Ed., Adv. Chem. Ser. 100, pp. 271-291, Am. Chem. Soc., Washington.

20. Bohan, T.L.: 1977, J.Magn. Res. 26, pp. 1o9-118.

21. Estabrook, R.W., and Werringloer, J.: 1977, Microsomes and Drug Oxidations, pp. 748-757, (V.Ullrich, I. Roots, A. Hildebrandt, R.W. Estabrook, and A.H. Conney, eds.) Pergamon Press, Oxford, New York.

22. Lichtenberger, F., and Ullrich, V.: 1977, Microsomes and Drug Oxidations, pp. 218-223, (V.Ullrich, I. Roots, A. Hildebrandt, R.W. Estabrook and A.H. Conney,eds.) Pergamon Press, Oxford, New York.

23. Ullrich, V., Herrmann, G., and Weber, P.: 1978, Biochem. Pharmacol. 27, pp. 2301-2304.

24. Sakurai, H., Herrmann, G., Ruf, H.H. and Ullrich, V.: 1980, Biochem. Pharmacol. 29, pp. 341-345.

25. Ullrich, V.: 1980, J. Molec. Catalysis 7, pp. 159-167.

26. Groves, J.T., Mc Clusky, G.A., White, R.E. and Coan, J.: 1978, Biochem. Biophys. Res. Commun. 81, pp. 154-16o.

27. Samuelsson, B., Goldyne, M., Granström, E., Hamberg, M., Hammarström, S., and Malmsten, C.: 1978, Ann. Rev. Biochem. 47, pp. 997-1029.

28. Salmon, J.A., Smith, D.R., Flower, R.J., Moncada, S., and Vane, J.R.: 1978, Biochim. Biophys. Acta 523, pp. 250-262.

29. Juchau, M.R., Bond, J.A., and Benditt, E.P.: 1976, Proc. Natl. Acad. Sci. USA 73, pp. 3723-3725.

30. Ullrich, V., Castle, L., and Weber, P.: 1981, Biochem.Pharmacol. 3o, pp. 2033-2036.

31. Gryglewski, R.J., Bunting, S., Moncada, S., Flower, R.J., and Vane, J.R.: 1976, Prostaglandins 12, 685-708.

CATALASES AND IRON-PORPHYRIN MODEL SYSTEMS: ROLES OF THE COORDINATION ENVIRONMENT OF IRON IN CATALYTIC MECHANISMS

Peter Jones

University of Newcastle upon Tyne, U.K.

Catalases and protein-free iron (III)- porphyrins (ferrihaems) catalyze the decomposition of H_2O_2 by mechanisms in which the initial Fe(III) catalyst is oxidised by substrate to higher oxidation state intermediates. Comparison of detailed mechanistic pathways and reactivities of the enzymic and model species affords an approach to assessment of the roles played by catalase protein in the expression of enzyme function.

1. INTRODUCTION

The catalases and most of the peroxidases are haemoproteins in which the essential metal co-factor is present in the native enzymes as an iron (III)-porphyrin (ferrihaem). In these species two-thirds of the octahedral (or near-octahedral) first coordination sphere of iron is occupied by the porphyrin ligand (most commonly protoporphyrin-IX). The remaining coordination positions are (probably) occupied by ligands derived from the protein, solvent or solutes, although the exact nature of these aspects of the coordination environment of iron is known only in a few cases. In addition, the different proteins provide variable, more extended environments for the ferrihaem prosthetic group which are also probably involved in the expression of enzymic function. The resultant of these influences is a range of enzymes which display both interesting similarities in mechanism but also marked differences in specificity.

The phenomenon of 'catalatic' action, i.e. catalysis of the reaction: $2H_2O_2 = 2H_2O + O_2$, was observed by Thenard using animal tissues, shortly after his discovery of hydrogen peroxide

H. B. Dunford et al. (eds.), The Biological Chemistry of Iron, 427–438.

in 1818, and this specific function formed the basis of the classification of the enzymes as H_2O_2:H_2O_2 oxidoreductases (E.C.1.11.1.6). Interpretation of the many studies aimed at elucidating the mechanism of action of the enzymes has been limited by lack of knowledge of the intimate details of their molecular structures. The recent X-ray crystallographic studies are an immensely valuable new beginning in the studies of the catalases, although optimism must be tempered with caution since it has to be recognized that there are at least four structures with which the search for understanding of mechanism must be concerned, two of which are susceptible to direct experimental study and two which are not. These are the structures of native catalase, the major catalytic intermediate (Catalase Compound I) and the two transition states which lie on the reaction co-ordinates for their interconversion.

Although the catalases are the most highly selective of the peroxide metabolizing enzymes, 'catalatic' function is not their only observable catalytic property, certainly in vitro and possibly also in vivo. Thus catalases also exhibit significant 'peroxidatic' activities (peroxidase-like properties), most notably in catalysis of the oxidation of ethanol by hydrogen peroxide. Among the peroxidases, chloroperoxidase shows significant 'catalatic' activity. These overlaps of function between catalases and peroxidases are reflected in similarities of basic mechanism and the mechanism of 'catalatic' action is a special case of a general peroxidase mechanism in which hydrogen peroxide acts as both acceptor and donor substrate.

The central concept of these mechanisms is that hydrogen peroxide oxidises the native Fe(III) enzymes to form an intermediate (Compound I) which retains both oxidising equivalents of its parent peroxide molecule. Thus Compound I may also be described formally as an Fe(V) derivative of the enzyme - both nomenclatures revealing the limitations of present knowledge of the detailed structures of these intermediates. It has, nevertheless, become clear that the Compound I intermediates of different enzymes are significantly different in detailed electronic structure and probable that these differences are important determinants of functional diversity.

A limitation on developing understanding of the nature of the catalytic intermediates has been the almost total absence of similarly hyperoxidised derivates of simple Fe(III) compounds formed by peroxide oxidation in catalytically active systems. Although phenomenological, iron containing models of catalases abound it has by no means certain that their catalytic reactions bear any mechanistic similarity to the enzymic process. In the catalytic action of ferric salts in decomposing H_2O_2 it is widely (but not unanimously) held that the initial action of H_2O_2

upon Fe^{3+} is reductive:

$$Fe^{3+} + HO_2^- \longrightarrow Fe^{2+} + HO_2$$

and that this process initiates a complex chain reaction in which $HO_2^-/O_2^{\cdot -}$, $OH\cdot$ and Fe^{2+} are chain carriers. On this basis there is a dramatic change in catalytic mechanism associated with replacement of aqua-ligands in the coordination environment of Fe(III) by the environment present in the catalase enzymes. (Fig. 1).

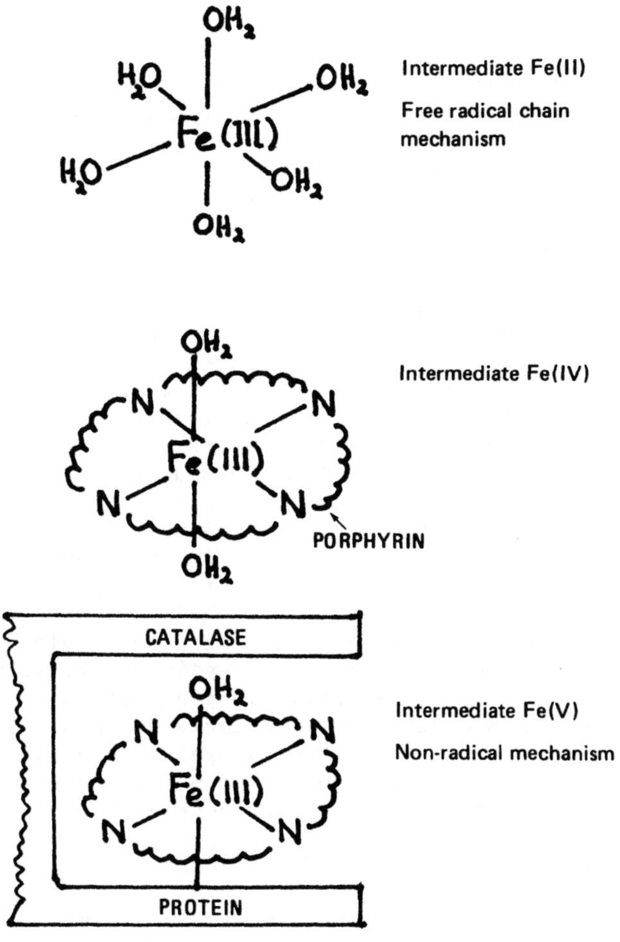

Figure 1. Influence of the coordination environment of iron on 'Catalatic' mechanism.

Protein-free, water soluble ferrihaems exhibit both 'catalatic' and 'peroxidatic' properties. Mechanistic studies imply that, although these reactions are kinetically more complex than the enzyme reactions, an important feature is the initial oxidation of the Fe(III) - porphyrin by H_2O_2 to form a higher oxidation state haem intermediate. The reactions show marked similarities to, but important differences from the enzymic processes. Detailed comparisons between the enzyme and model systems make possible an assessment of the roles played by protein in the evolution of the catalytic mechanisms and in the development of reactivity and specificity.

2. CATALASES

The extensive literature on catalases has been frequently reviewed in its various aspects (kinetic/mechanistic, biochemical and biological) (1). This account summarizes the salient features of the many studies which have yielded the present limited understanding of catalase mechanism.

2.1. General Properties.

Catalases are widely distributed in aerobic organisms. The molecular weights of the enzymes from various sources are very similar ($2.4 - 2.5 \times 10^5$) with four ferrihaem (iron (III) - protoporphyrin-IX) prosthetic groups per molecule. (In some cases, notably beef liver catalase, a proportion of the prosthetic groups may be damaged by porphyrin oxidation either before or during extraction). Genetic studies imply that the enzyme is a tetramer of identical subunits but the quaternary structure does not involve covalent bonding. The quaternary structure is essential for 'catalatic' activity and the prosthetic groups are equivalent and non-interacting during catalysis. Dissociation occurs at the extrema of pH (with loss of activity) and this denaturation is never completely reversible. Enzymes from different sources show variable resilience to such exposure. In a Swiss-type actalasaemia enzyme is synthesised which is active but which possesses a very unstable quaternary structure. Mouse liver catalase may be protected against denaturation by sialic acid binding. Reconstitution of native, active catalase from ferrihaem and apoprotein has not been achieved.

2.2. Coordination Environment of Ferrihaem.

Indirect arguments based on electronic and esr spectra have suggested that one axial ligand of the prosthetic group may be histidine. It has been widely assumed that the other axial position is occupied, in the native enzyme, by a water molecule (other suggestions have been hydroxide and carboxylate). The crystallographic data imply that these suggestions were very wide of the mark. Water proton nmr relaxation studies imply a fast-exchanging (>2 x $10^5 s^{-1}$) water molecule in the first coordination sphere of catalase ferrihaem. Measurements on water and formate protons in both the absence and presence of cyanide indicate that, not only are water and formate simultaneously bound to catalase, but also cyanide is bound in the presence of both water and formate. The calculated iron-formate proton distance (>7A) implies that formate (a good peroxidatic substrate) is not bound in the first coordination sphere of iron, at least in the native enzyme. A number of probes (magnetic susceptibility, esr, Mössbauer) agree that iron in native catalase is in the high-spin Fe(III) state. Changes in the esr spectrum can be induced, e.g. by small ions such as phosphate, which probably arise from transitions of the protein between at least two conformations that are very close in free energy. Catalases exhibit rapid ligand-binding reactions e.g. with cyanide and fluoride, which are typical reactions of high-spin Fe(III) centres except for the complication that the equilibria appear to involve either binding of the neutral conjugate acids of the ligands (e.g. HCN and HF) or binding of ligand anion by a conjugate acid form of the enzyme. This behaviour has close parallels in peroxidase chemistry where in some cases it is clear (and may indeed be true generally) that the reactant partners in the initial ligand-enzyme reaction are enzyme and ligand conjugate acid. It is difficult to conceive that in the cyanide complex, it is other than CN^- which is bound to iron, but the fate of the retained proton is not understood.

Selective chemical modification of amino-acid residues in the catalase protein has led to suggestions that arginine residues appear to be essentially related to catalatic activity, while tyrosine residues are associated with the maintenance of protein conformation and the ferrihaem group environment. The most striking result of this type is that 3-amino-1,2,4- triazole irreversibly inhibits catalase (by combination with a histidine residue) but the reaction occurs not with the native enzyme but with the catalytic intermediate, Compound I.

2.3. Kinetics of Catalase Action.

Assay of 'catalatic' activity of the enzymes is complicated by: (i) the very high specific activity of the enzyme; (ii) inactivation of the enzyme in the absence of substrate at convenient [catalase] (a phenomenon related to dissociation of tetramer followed by denaturation); (iii) a time dependant inactivation of the enzyme by H_2O_2. Factors (i) and (ii) restrict the range of [catalase] for accurate measurement with conventional techniques to 0.1 - 1 nM and factor (iii) requires that either fairly rapid techniques or extrapolation procedures be employed. In the convenient range of $[H_2O_2]$ ($\lesssim$0.1M) the reaction is first order in H_2O_2 and the second order catalytic constant (calculated on the basis of the concentration of tetrameric enzyme) has values near $10^7 M^{-1} s^{-1}$ for enzymes from a variety of sources and is pH-independent over a wide range (about 5-10). The temperature coefficient of the rate constant is remarkably low (activation energies in the range 0.6 - 1.7 kcal mol^{-1}) and the rate constant is independent of viscosity of the medium in glycerol-water mixtures (up to 70% glycerol). Using the double-mixer quenched-flow technique the range of $[H_2O_2]$ studied has been extended up to 5M. Under these conditions Michaelis-Menten kinetics are observed with K_M = 1.1M at 20°C.

Formation of Compound I upon addition of H_2O_2 to the enzyme is accompanied by a decrease in absorbance at the Soretband. The absorption decrement becomes independent of initial $[H_2O_2]$ and comparison with the absorption changes observed using organic hydoperoxides (where the reaction is near stoichiometric) indicate that the maximal formation of Compound I with H_2O_2 is 25-40%. Compound I is reduced and the native catalase spectrum recovered by adding a peroxidatic donor (ethanol). The data support the proposal advanced by Britton Chance that the catalatic reaction occurs by a 'peroxidatic mechanism':

$$\text{Catalase} + H_2O_2 \xrightarrow{k_1} \text{Compound I} + H_2O$$

$$\text{Compound I} + H_2O_2 \xrightarrow{k_4} \text{Catalase} + O_2 + H_2O.$$

Satisfactory calculation of k_1 and k_4 from the data is not straightforward, in particular because the steady-state approximation cannot realistically be applied to the maximal formation of Compound I under conditions where experiments are feasible (although this approximation was used in very early analyses). A recent analysis, avoiding use of the steady-state approximation, is promising but some puzzling internal inconsistencies appear in the analysis of published data.

Representation of Compound I formation as an elementary bimolecular reaction (diffusion or chemically controlled) is not an entirely adequate description of the process. The minimal acceptable mechanism involves formation of a reversible enzyme-substrate pre-cursor complex (Michaelis complex) preceding enzyme oxidation: $\text{Catalase} + H_2O_2 \rightleftharpoons (\text{Catalase}\cdot H_2O_2) \longrightarrow \text{Compound I} + H_2O$.

Studies of the formation of Compound I with organic hydroperoxides reveal significant steric constraints upon access of substrate to the active site. Only aliphatic hydroperoxides with small end groups are acceptable and there is a parallel between rate attenuation and increasing van der Waal's volume of the hydroperoxide. With the more readily ionizable peroxyacetic acid the results present a close parallel with the ligand-binding reactions, since the neutral peracid is an acceptable substrate but its anion is not. In contrast to the peroxidases, present knowledge of the electronic structure of Catalase Compound I is very limited. The intermediate shows no esr spectrum and the Mossbauer spectrum has not been obtained.

2.4. Reduction of Compound I.

In the basic 'peroxidatic' mechanism the reduction of Compound I by H_2O_2 is written as an elementary bimolecular reaction. The general theory of electron-transfer reactions implies that a reversible pre-cursor complex will be a intermediate in the process, although not necessarily kinetically significant. The view that the reaction is a one-step, 2e-equivalent redox process is supported by the observation that formation of the le-reduction product of Compound I, the Fe(IV) derivative Compound II inhibits the 'catalatic' reaction (Compound II reacts reversibly with H_2O_2 to form a further derivative Compound III). ^{18}O-isotopic tracer studies of oxygen produced in the 'catalatic' reaction imply that the O-O bond remains intact. Nevertheless, it is clear that both H_2O_2 and organic hydroperoxides can act as 1-e-donors to Compound I, forming Compound II.

In classifying mechanisms of Compound I reduction by other donors the appearance or non-appearance of Compound II during the reaction is used as a criterion (although not an entirely satisfactory one). On this basis the major donors (methanol, ethanol, formic acid and nitrous acid) react by one-step, 2e-equivalent mechanism. This behaviour is in striking contrast to the peroxidases where le-equivalent reactions are the usual (but not exclusive) pathways for Compound I reduction.

3. IRON (III) - PORPHYRIN MODELS.

3.1. Kinetics of 'Catalatic' Action.

In catalysis of H_2O_2 decomposition by water-soluble iron (III)-porphyrins the initial rate of decomposition shows a first order saturation kinetic dependence on H_2O_2. (Initial rates studies are necessary because the porphyrin ligands are oxidised, and the catalysts inactivated during the reaction). The apparent second order catalytic constants are not independent of [Ferrihaem] but decrease with increasing [Ferrihaem].

This complexity is associated with the dimerization of

ferrihaems according to:

$$2M \underset{}{\overset{K_D}{\rightleftharpoons}} D + H^+$$

(where M and D represent monomeric and dimeric ferrihaem respectively). It is probable (although not certain) that the dimers are of the μ-oxo-bridged type known to exist in the solid state. The dimerization constant, K_D, is markedly influenced by changes in the β-pyrrolic substituents in the porphyrin ring. Analysis of the [Ferrihaem] dependence of the catalytic kinetics implies: (i) monomeric ferrihaems are the predominant catalytic species (dimers are not entirely unreactive); (ii) the specific catalytic activity of monomeric ferrihaem is insensitive to change in β-pyrrolic substituents and is inversely proportional to $[H^+]$ in unbuffered solutions. Thus the rather low stoichiometric catalytic activity of the catalase prosthetic group derives from the high dimerization constant ($K_D = 4.5$) of this ferrihaem.

3.2. Formation of Intermediates.

The similarity in the 'catalatic' properties of a range of ferrihaem monomers suggests a common catalytic mechanism. The ferrihaem derived from deuteroporphyrin-IX (deuteroferrihaem, DFH) has a much lower dimerization constant ($K_D = 3.4 \times 10^{-2}$) than the native prosthetic group and has provided a useful system for the study of catalytic intermediates.

Reaction of H_2O_2 with DFH results in the formation of an intermediate accompanied by a decrease in absorbance at the Soret band. With equimolar DFH and H_2O_2 the extent of intermediate formation corresponds to about 0.3 oxidising equivalents per mol haem. This behaviour is consistent with the formation of an intermediate which is involved in the 'catalatic' turnover of H_2O_2 by DFH, in a similar manner to Catalase Compound I.

Stoichiometric studies of intermediate formation, using organic hydroperoxides and other oxidants (e.g. chlorite), under conditions where catalytic turnover does not limit intermediate formation, yield the surprising result that intermediate formation corresponds to a le-oxidation of DFH, i.e. to an Fe(IV) state,

$$2Fe(III) + ROOH = 2Fe(IV) + ROH$$

The kinetics of intermediate formation and stoichiometric data are consistent with a basic two-stage mechanism in which initial 2e-oxidation of the ferrihaem is followed by a rapid comproportionation reaction:

$$Fe(III) + ROOH \longrightarrow Fe(V) + ROH$$

$$Fe(V) + Fe(III) \longrightarrow 2Fe(IV)$$

3.3. Reduction of Intermediate.

The oxidised intermediate derived from DFH does not react with ethanol but is readily reduced by a range of typical peroxidase donors (iodide, phenols, anilines). Although the reduction of intermediate by H_2O_2 cannot be studied independently of the catalytic cycle, a mechanism in which it is supposed that H_2O_2 reacts in the form of its conjugate base with Fe(IV) is consistent with the data:

$$Fe(IV) + HO_2^- \longrightarrow Fe(III) + HO_2^{\cdot}$$

$$Fe(IV) + HO_2^{\cdot} \longrightarrow Fe(III) + O_2$$

The overall mechanism is a varient of the 'peroxidatic' mechanism in which, apart from the complications associated with catalyst dimerization, there are two important changes: (i) H_2O_2 reacts in the form of its conjugate base, HO_2^-; (ii) the oxidized catalyst intermediate is in an Fe(IV) state and therefore oxidises donor only by 1e reactions.

4. COMPARISON OF ENZYMES AND MODELS

4.1. Reactivity Correlations

Although the rate constant for decomposition of H_2O_2 by ferrihaem monomer is much lower than that of catalase at physiological pH, this is no longer the case at high pH. A very similar relationship holds between the rates for Catalase Compound I formation and formation of the oxidised iron-porphyrin intermediate. The reaction of aniline with the haem intermediate procedes at an almost identical rate to that with Horseradish peroxidase Compound I (HRP-I) and the rates of reaction of substituted anilines with the two intermediates are closely similar. Reaction of phenol with the haem intermediate is inversely proportional to $[H^+]$ and suggests that phenolate anion is the reactive species. Rate constants calculated on this basis correlate well with those for reaction of HRP-I with a series of phenols. Reaction of iodide with the haem intermediate is pH-independent in contrast to $[H^+]$-proportional reaction of HRP-I but the two species have comparable reactivity at pH 8.5.

4.2. Role of the Protein in Enzyme Catalysis

Two roles of the catalase protein are immediately apparent:

(i) protection of the prosthetic group ligand against oxidation (the extensive studies of haemoprotein catabolism suggest that this protection is at least partly, and perhaps largely, steric in origin); (ii) a dimer-breaking action, the prosthetic group being bound to enzyme as the intrinsically more reactive monomeric ferrihaem.

A catalyst which is capable of decomposing hydrogen peroxide efficiently at physiological pH must be capable of utilizing molecular H_2O_2 (since pK (H_2O_2) ~ 11.6). The inability of ferrihaem monomer to meet this requirement is a clear limitation on its catalytic performance at pH ~ 7. Conversely, the ability of catalases to use molecular H_2O_2 is a major factor in their catalytic effectiveness, since the maximal activities of enzyme and ferrihaem are closely similar. The latter observation suggests that the micro-environment encountered by a H_2O_2 molecule, entering the active site of catalase, is one of very high basicity, so that hydrogen peroxide is delivered to the iron centre as HO_2^-. There have been a number of speculations concerning the nature of possible acid-base functions, derived from the protein, which might participate in this process and this is one area where enzyme structural information should provide an immediate critical test of functional concept.

The inability of catalases and peroxidases to accept anionic ligands or anionic forms of the oxidising substrate (peroxyanions) has led to the suggestion that electrostatic factors may play an important role in Compound I formation. One hypothesis was that negatively charged protein residues near the entrance to the enzyme active site could act as an 'electrostatic gate', permitting neutral molecules to diffuse freely in and out of the active site, but reflecting anions (both present in the external environment and formed within the active site). Alternatively discrimination against charged ligands could arise from a hydrophobic environment around the ferrihaem and the crystal structure data suggest that this is the case in the catalases.

Comparison of enzymes (both catalases and peroxidases) and models implies that an important facet of the enzymic environment of iron is to provide a storage facility for the second oxidising equivalent of the peroxidic oxidant within the Compound I intermediate. Different peroxidases appear to have evolved different solutions to this problem; the particular solution(s) in the catalases remain(s) unknown. The comproportionation reaction proposed to occur rapidly in iron-porphyrin oxidation is known to occur also, but much more slowly, between HRP-I and native HRP.

It would be surprising if an enzyme, evolved to catalyze

the decomposition of hydrogen peroxide in organisms, should do so by a mechanism which involved formation of the free radical products of the le-oxidation or reduction of H_2O_2. The hydroxyl radical in particular is well-known, from radiobiological studies to be capable of inducing serious malfunction of organisms, and this problem is neatly avoided by the formation of Compound I as the initial step in the catalytic cycle. Much less is known about the second step, the oxidation of H_2O_2 by Compound I. The substrate for this reaction is again molecular H_2O_2 (a requirement for overall catalytic efficiency) and the unreactive nature of Compound II argues that the reaction is a 2e-equivalent process e.g. involving H^- and H^+ transfer. It is also possible, however, that the reaction involves consecutive le-transfer reactions and that the protein operates to trap within the active site the superoxide radical anion produced by the first electron transfer (HO_2 radical is a much stronger acid than H_2O_2; pK($HO_2^{\cdot}$) ~ 4.8). On this basis the formation of Compound II by the reaction of Compound I with H_2O_2 is to be regarded as a leakage of superoxide ion from the trap:

Catalase Compound I + H_2O_2 → [Catalase Compound II. $O_2^{\cdot -}$]

A → Catalase + O_2

B → Catalase Compound II + $O_2^{\cdot -}$

Calculation of the relative rates of the two pathways indicates that this trap (if it exists) must be almost 100% efficient (the rate constant ratio Path A : Path B exceeds 10^7). The problem of avoiding release of $O_2^{\cdot -}$ from the [Catalase Compound II.$O_2^{\cdot -}$] complex is very similar to that of avoiding $O_2^{\cdot -}$ release from oxy-haemoglobin (regarded as essentially a [$Fe^{3+} \cdot O_2^{\cdot -}$] species and the highly hydrophobic haem environment in both haemproteins may fulfil a similar function.

The close correlation of the oxidant properties of the 'Fe(IV)' intermediate in the model studies with those of HRP Compound I strongly suggests that the 'Fe(IV)' species is an [Fe(III)-porphyrin cation radical] entity. Further the closely comparable maximal 'catalatic' reactivities of monomeric ferrihaems and catalases suggests that Catalase Compound I may be an [Fe(IV)-porphyrin cation radical] species and that further esr studies at liquid helium temperatures are highly desirable.

REFERENCES

1. The following reviews provide detailed references to the primary journal literature: Schonbaum, G.R., and Chance, B.: 1976, in The Enzymes, 3rd Ed, Vol. 13, (P. Boyer, ed.). Academic Press, p. 363.

Jones, P., and Wilson, I.,: 1978, in Metal Ions in Biological Systems, Vol. 7, (H. Sigel, ed.), Dekker, p. 185, Deisseroth, A., and Dounce, A.L.,: 1970, *Physiol. Rev.*, 50, p. 319. Sies, H.,: 1974, *Agnew. Chem. Int. Ed.*, 13, p. 706. Aebi, H., and Suter, H.,: 1971, *Advan. Hum. Genet.*, 2, p. 143.

2. The literature to 1978 is reviewed in: Jones, P., and Wilson, I.: 1978, Metal Ions in Biological Systems, 7, p. 185. More recent work in Ph.D. Theses by I. Wilson and J.E. Frew (University of Newcastle upon Tyne). See also: Kelly, H.C., Parigi, K.J., Wilson, I., Davies, D.M., Jones, P., and Roettger, L.J.: 1981, *Inorg. Chem.*, 20, p. 1086.

THE STRUCTURE OF BEEF LIVER CATALASE

Mathur R. N. Murthy, Thomas J. Reid III[1], Andrew Sicignano[2], Nobuo Tanaka[3] and Michael G. Rossmann

Department of Biological Sciences, Purdue University, W. Lafayette, Indiana 47907, U.S.A.

The three-dimensional structure of beef liver catalase has been determined to 2.5 Å resolution by a combination of isomorphous and molecular replacement techniques. The tetrameric catalase molecule has 222 symmetry with one of its diads coincident with a crystallographic twofold axis. The known polypeptide sequence has been unambiguously fitted to the electron density map. The heme is well buried in a hydrophobic pocket, 20 Å below the surface of the molecule, and accessible through a hydrophobic channel. Residues that line the heme pocket belong to two different subunits. Tyr 357 is the proximal heme ligand and the catalytically important residues on the distal side are residues His 74 and Asn 147. The tertiary structure consists of four domains: an extended non-globular amino terminal arm which stabilizes the quaternary structure; an anti-parallel, eight-stranded β-barrel providing the residues on the distal side of the heme; a rather random "wrapping domain" around the subunit exterior including the proximal heme ligand; and a final α-helical structure.

INTRODUCTION

Catalase (EC 1.11.1.6 hydrogen peroxide:hydrogen peroxide oxidoreductase) occurs in almost all aerobically-respiring organisms and serves to protect cells from the toxic effects of H_2O_2 by catalyzing the reaction

$$2H_2O_2 \rightarrow 2H_2O + O_2 \quad . \qquad (1)$$

Obligate anaerobes cannot tolerate O_2 in part because they lack

H. B. Dunford et al. (eds.), The Biological Chemistry of Iron, 439–458.

this enzyme. The "catalatic" activity (reaction 1) distinguishes catalase from other peroxidases (1) except for chloroperoxidase which catalyzes a wider variety of substrates (2). Catalase can also oxidize a variety of compounds in the presence of H_2O_2 (3). This "peroxidatic" activity (reaction 2)

$$H_2O_2 + AH_2 \rightarrow A + 2H_2O \qquad (2)$$

is responsible for ethanol oxidation in liver, provided the H_2O_2 concentration is sufficiently high (4).

Catalase contains four identical subunits (5,6) each equipped with a high-spin Fe(III)-protoporphyrin IX (7-9). However, approximately one of the four subunits contains an inactive high-spin Fe(III)-biliverdin complex (10-12). Low angle X-ray scattering data showed beef liver catalase to be an ellipsoidal molecule with a 40 Å radius of gyration (13). One subunit of catalase has 506 amino acid residues with an additional ten to fifteen residues at the C-terminal of a small fraction of the subunits (W. A. Schroeder, private communication). The N-terminus is blocked by an as yet unidentified group (14). The 16 cysteine residues in the tetrameric mammalian catalase molecule do not form disulfide bridges (15,16).

Crystals of catalase were observed as early as 1937 by Sumner and Dounce (10). Reports of crystalline catalase from lamb (17), horse (18) and human (19) liver, from beef (20), horse (18) and human (18,21) erythrocytes, from spinach leaf (22) and from bacteria (23) followed. Electron microscopy (24-26) and electron diffraction (27-29) investigations of complexes with a habit similar to that obtained by Sumner and Dounce (10) revealed an orthorhombic lattice with space group $P2_12_12_1$. Human erythrocyte and beef liver catalase can also form cylindrical tubules (26).

The first catalase crystals to be studied by single crystal X-ray diffraction were a disordered orthorhombic form of horse erythrocyte catalase (30). Subsequently, two crystal forms of beef liver catalase suitable for high resolution structure determination were reported; the first was a trigonal form of space group $P3_121$ or $P3_221$ with six molecules in the unit cell (31-34) and the other was an orthorhombic form of space group $P2_12_12_1$, containing one molecule in the asymmetric unit (34,35). Burkey and McPherson (36) showed that beef and deer liver catalase crystals are isomorphous. Eventoff and Gurskaya (37) were unable to distinguish between the point groups D_2 and C_4 for the molecular symmetry in a rotation function study with 10 Å resolution data of beef liver catalase. The possibility of a fourfold symmetry was, however, discounted as catalase dissociates into dimers and monomers (5). This was confirmed in other crystal forms by the incorporation of a crystallographic twofold axis in the molecule of beef liver catalase (38), *Penicillium vitale* catalase (39) and *Micrococcus luteus* catalase (40). The structure of the beef liver catalase determined from the crystal form first reported by Eventoff *et al.* (38) is described here. A report of the crystallographic techniques used in this work is in press (41). An analysis of the structure in terms of catalase mechanism will be given elsewhere.

DESCRIPTION AND OVERALL ORGANIZATION OF THE MOLECULE

The molecule is roughly dumbbell shaped. The tetramer is 105 Å along the *P* and *R* axes, exhibits a waist of 50 Å in the *R* = 0 plane (Fig. 1) and is in reasonable agreement with the electron imaging results of Barynin *et al.* (42). Identity of the space groups in which beef liver and *Penicillium vitale* catalase crystallize together with their nearly equal *a* axial lengths of 142 and 145 Å, respectively, suggest that the same molecular two-fold axis is incorporated into the crystal symmetry in both cases. As far as can be ascertained from published information (43),

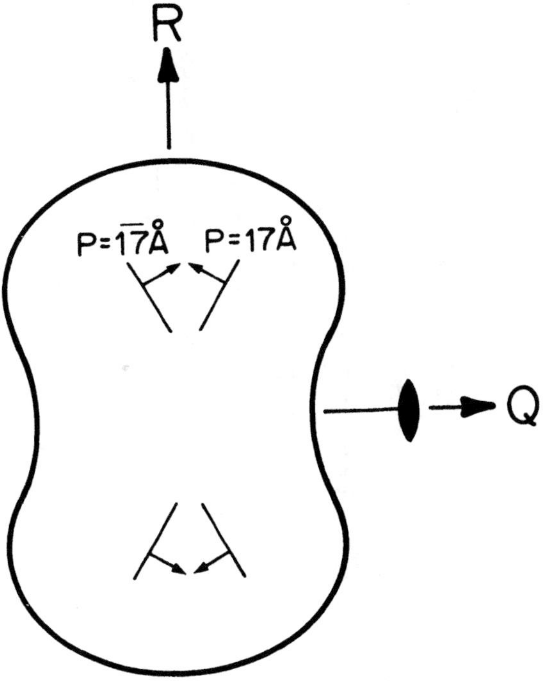

Figure 1. Overall shape of the catalase molecule showing the approximate location and orientation of the buried heme groups.

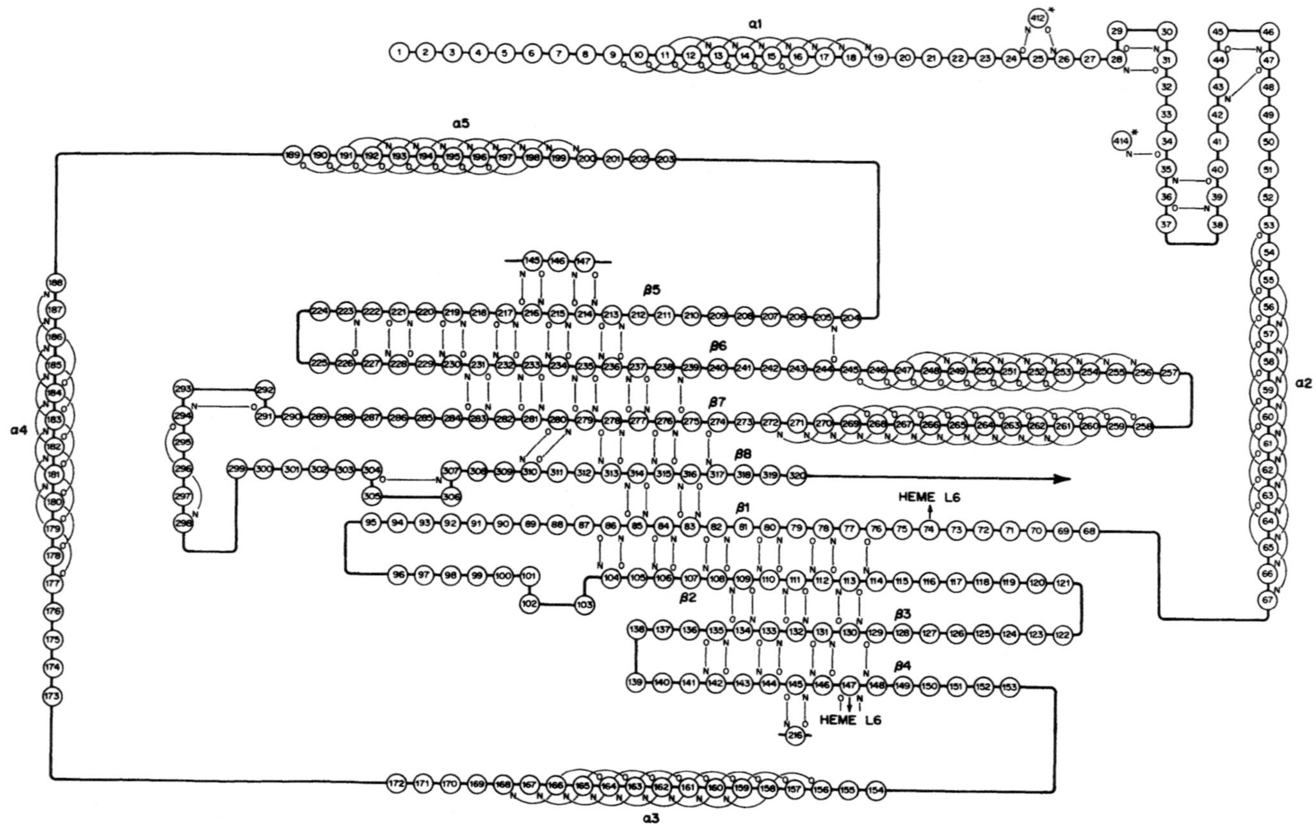

Figure 2. The main chain hydrogen bonding pattern of catalase. The secondary structure consists of two helices in the amino terminal arm, an eight-stranded β-barrel, two helices in the "wrapping" domain and four helices at the carboxy end of the molecule. A two-stranded β-structure is also

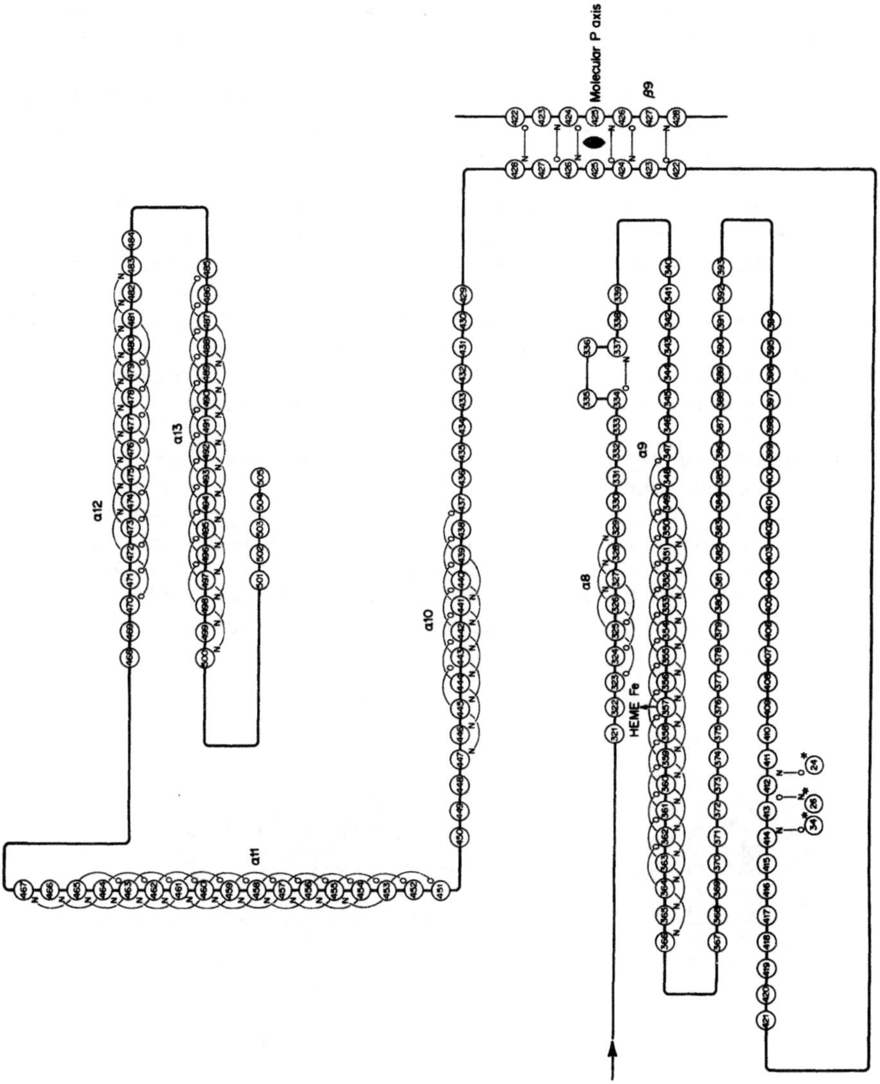

Figure 2 (continued): observed across the molecular P-axis. Helices have been designated α1 - α13 and extended strands in sheets as β1 - β9.

this is indeed the situation. The four heme groups are well buried in the tetramer, at a distance of 20 Å below the molecular surface and 23 Å away from the molecular center. It could be easily recognized in the multiple isomorphous replacement map as a flattened density, 38% higher than the largest protein density. The relative height of the heme iron peak with respect to the average protein value remained constant during the molecular replacement calculations. The heme iron atoms related by molecular *P*, *Q* and *R* axes are at a distance of 30.6, 45.3 and 34.5 Å, respectively. The heme normal makes a small angle of 19° to the molecular *P*-axis.

A scaled model of the molecule was constructed in a Richards optical comparator (44). The structure was in complete agreement with the amino acid sequence (14), with most side chains being clearly visible. However, the four amino terminal residues, six carboxy terminal residues and residues 18 - 21 had no appreciable density. The density corresponding to the external parts of the polypeptide chain that do not belong to the internal eight-stranded β-barrel domain was of somewhat lower quality.

The secondary structure of catalase is represented as a backbone hydrogen bonding network in Figure 2, as a stereodiagram of the C_α atoms (Fig. 3) and as a diagrammatic representation in Figure 4. According to usual conventions, α-helices have been numbered 1 through 13 and strands within β-sheets 1 through 9. Helices account for 26% and sheet residues for 12% of the molecule. Optical rotatory dispersion measurements had indicated

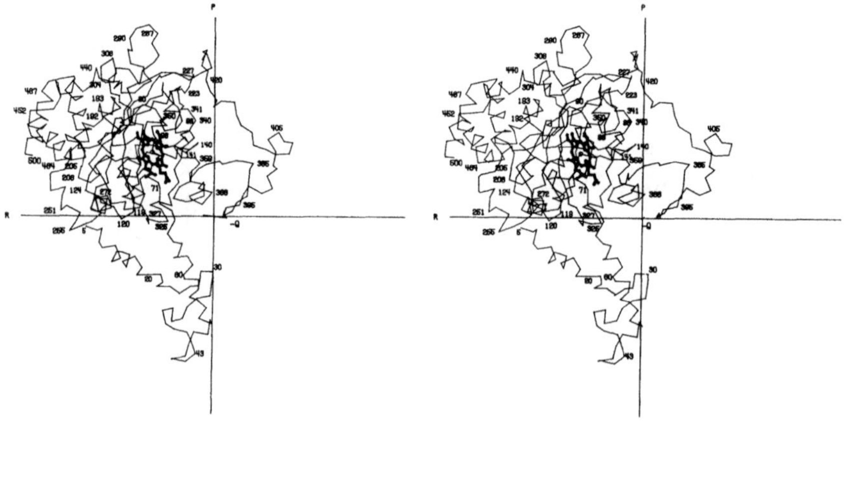

Figure 3. Alpha carbon backbone of one catalase subunit. The heme group is shown with thicker bonds.

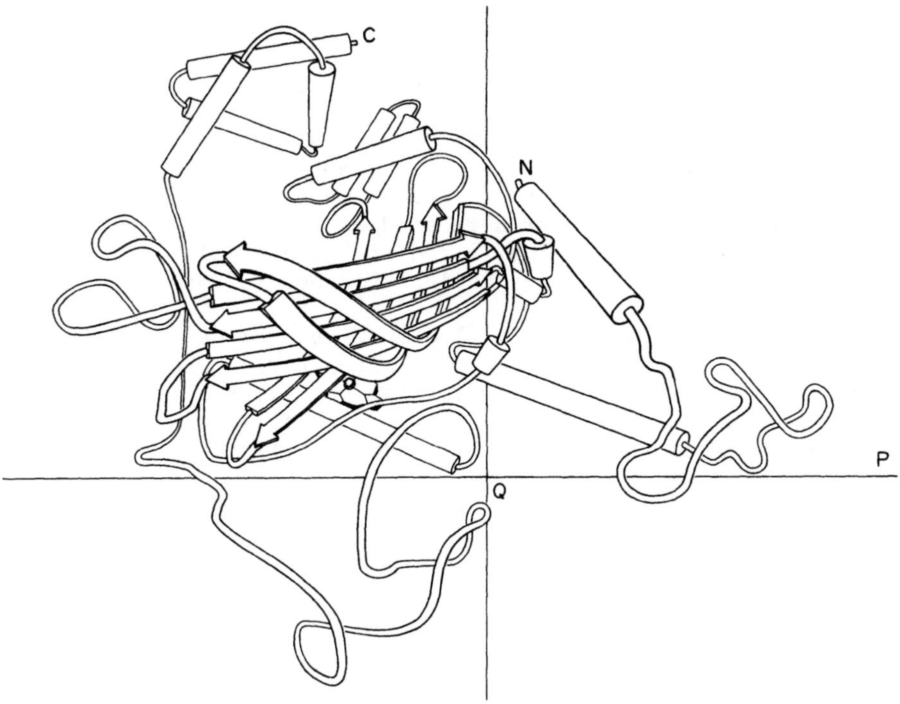

Figure 4. Diagrammatic representation of secondary structural elements.

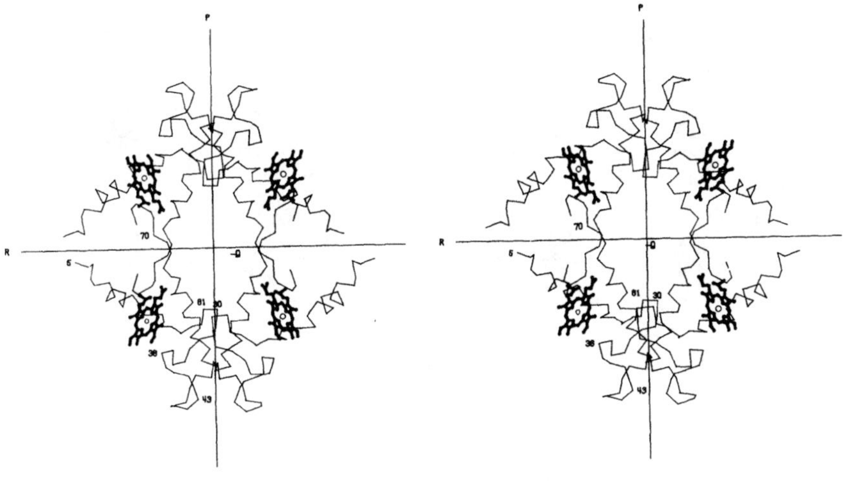

Figure 5. Organization of the amino terminal arm within the tetramer.

around 50% helical content (45). The structure of each subunit can be considered as being composed of four structural domains.

The first domain (Fig. 5) comprises the amino terminal 75 residues which form an arm extending from the globular region of each subunit. The arm consists of two helices (α1 and α2) both of which are involved in inter-subunits contacts, the first with the *Q*-axis-related subunit and the second with the essential helix and heme of the *R*-axis-related subunit. Other amino terminal arms have been implicated in the assembly of lactate dehydrogenase (46) and of spherical viruses (47,48).

The second domain (Fig. 6) comprises residues 76 - 320 which form a large eight-stranded anti-parallel sheet β-barrel. The heme's distal side interacts with the exterior of this barrel. The β-barrel constitutes two, topologically similar (β1 to β4 and β5 to β8), four-stranded, anti-parallel sheets such as occur in rubredoxin (49). Most residues on the distal side of the heme (Val 73, His 74, Val 115, Phe 160, Phe 152 and Asn 147) originate from this region of the molecule. The stretch of polypeptide chain between the topologically similar halves of the β-barrel contains three helices (α3 - α5) and contributes to a hydrophobic environment on the distal side of the heme.

The third domain (Fig. 7), residues 321 - 436 to be referred

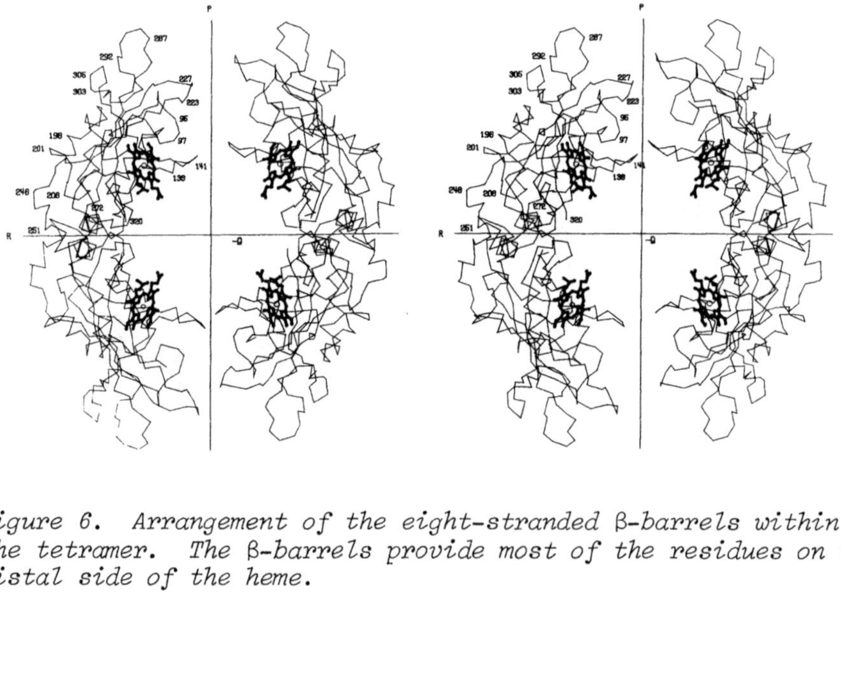

Figure 6. Arrangement of the eight-stranded β-barrels within the tetramer. The β-barrels provide most of the residues on the distal side of the heme.

to as the "wrapping domain", forms an outer layer to each subunit. This does not have obvious similarities to other previously observed structures. It lacks discernible secondary structure in a long stretch of polypeptide chain between residues 366 and 420. However, this domain contains the essential helix ($\alpha 9$) with the proximal ligand Tyr 357. The wrapping domain also forms a short β-structure ($\beta 9$) with an identical region of the *P*-axis-related subunit.

The carboxy terminal portion of the molecule (residues 437 - 506) is folded into a four-helical domain ($\alpha 10$, $\alpha 11$, $\alpha 12$, $\alpha 13$) with a topology and hand similar to the helices E, F, G and H of the globin fold (Fig. 8). This domain is removed from the active site and is not involved in inter-subunit interactions. However, it limits the accessibility of the active center by contributing to the formation of the hydrophobic channel leading to the heme.

The structure of beef liver catalase is similar to *Penicillium vitale* catalase (43), although it lacks the "flavodoxin"-like domain, found at the carboxy terminal region of the latter, indicating the conservation of a basic catalase structure during evolution. The additional domain in *Penicillium vitale* catalase might indicate an additional nucleotide binding function of this molecule (50). The obvious conservation of the basic catalase fold argues strongly in favor of divergent

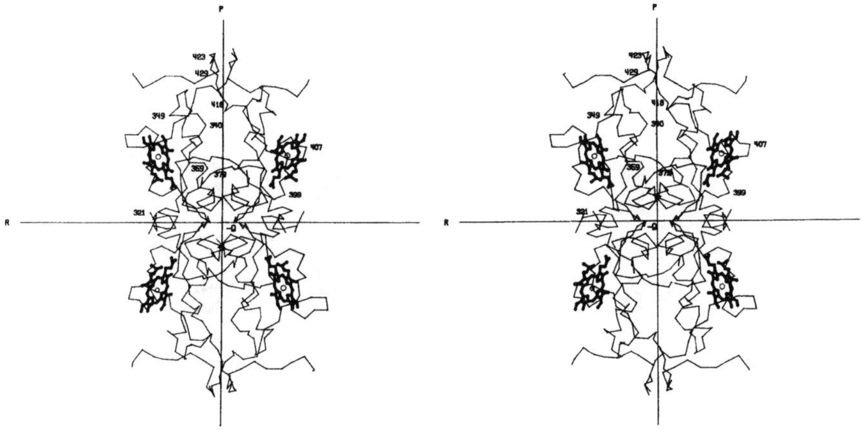

Figure 7. Structure of the four "wrapping domains" in one catalase molecule. These domains form an outer layer in each subunit. The essential residue Tyr 357 is contained in helix $\alpha 9$ of this domain.

evolution of this molecule from at least early eukaryotic organisms. The extra domain in fungal catalase may be the result of gene fusion with a nucleotide binding gene and would then be another example of the separate evolutionary development of functional protein domains (51), possibly supported by gene organization into coding and non-coding sequences (cf. 52).

The heavy atom compounds used for the isomorphous replacement method and the amino acids involved in their binding give some information on the relative activity of the cysteine residues (Table 1). The "A" and "B" sites are located at the exposed sulfhydryl residues, Cys 392 and Cys 459, respectively. The "C" site, which is also at a cysteine residue (376), is not fully substituted presumably because it is partially buried. However, only Cys 376 is reactive with respect to $PtCl_4$. Cys 231, which is totally buried, does not react with any of the heavy atom compounds used in this study. The minor site "E", found only in the $HgCl_2$ derivative, is associated with His 234. The "D" site is very close to the molecular *R*-axis and is associated with His 254, probably from both subunits. The "F" site is 6 Å away from the *R*-axis and relates to Glu 255 in only two of the four subunits.

Most of the charged residues are found on the outside of the molecule. All the internal charged residues, except for Lys 236, are either charge balanced or occur in hydrophilic environments. Lys 236 is a most unusual residue occurring in a hydrophobic environment in the interior of the β-barrel domain.

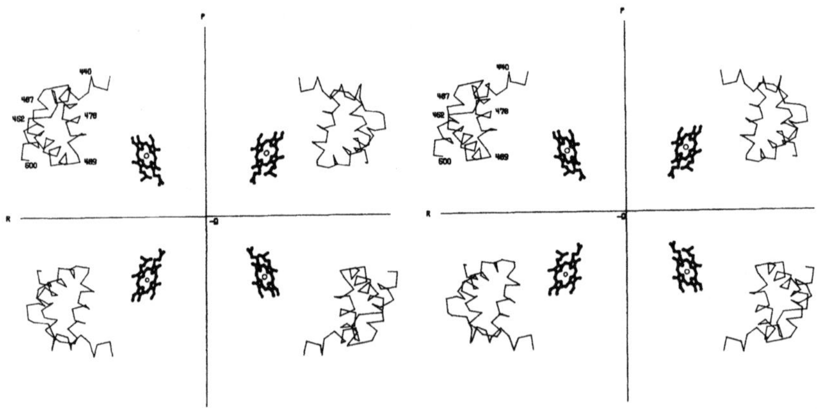

Figure 8. Organization of the carboxy terminal helical domain in the tetrameric molecule. This domain restricts the accessibility of the heme groups.

Table 1. Amino acids involved in binding heavy atom compounds.

Site	Average Occupancy	Residue	Subunit(s)	Compound(s)
A	47	Cys 392	1, 2	Mers, DMA, $HgCl_2$
B	46	Cys 459	1, 2	Mers, DMA, $HgCl_2$
C	22	Cys 376	1, 2	Mers, $PtCl_4$
E	25	His 234	1, 2	$HgCl_2$
D	34	His 254	1-2	$HgCl_2$ (on *R*-axis)
F	16	Glu 255	1	$HgCl_2$

QUATERNARY STRUCTURE

The subunit contacts generated by the molecular *P*, *Q* and *R* axes of catalase are illustrated by an interaction tetrahedron (53) in Figure 9. Most of the intersubunit contacts are confined to the amino terminal arm and the wrapping domain. The first section of the arm (residues 15 - 35) is in contact with the *Q*-axis-related subunit while the second section (residues 36 - 73) interacts with the *R*-axis-related subunit. Due to a change in direction between helices α1 and α2 the actual amino terminus is situated not far from the β-barrel domain of the same subunit (Fig. 3). The first 25 residues of the arm are inserted into a hole generated by excursion of residues 390 and 420 and by residues 138 - 141, 337 - 340 and 378 - 420 of the *Q*-axis-related subunit (Fig. 10). Arg 71 on the arm interacts with a propionic acid side chain of a heme belonging to the same subunit.

Each of the four hydrophobic channels, leading to the heme, is lined with residues from all subunits. Thus, removal of even one subunit leaves the heme exposed to water. Acatalasemia is known to occur in patients with a mutant catalase in which the tetramer has the same specific activity but there is a tendency to dissociate into dimers of low activity (54). Apparently, therefore, the burial of the heme within the molecule, removing it from water, is an essential component of catalase activity.

HEME ENVIRONMENT

The heme site is accessible by a channel 30 Å long and 15 Å wide that occurs between the β-barrel and the carboxy terminal domains. The channel opens toward the molecular *R*-axis and is lined with the polar residues Arg 126, Asp 127, Asn 167, Lys 168 and Lys 176 at the entrance and with the hydrophobic residues Val 115, Ala 116, Pro 128, Phe 152, Phe 153, Phe 163, Ile 164, Val 173 and Leu 198 as the channel descends toward the heme, narrowing as the heme is approached (Fig. 11). The limited

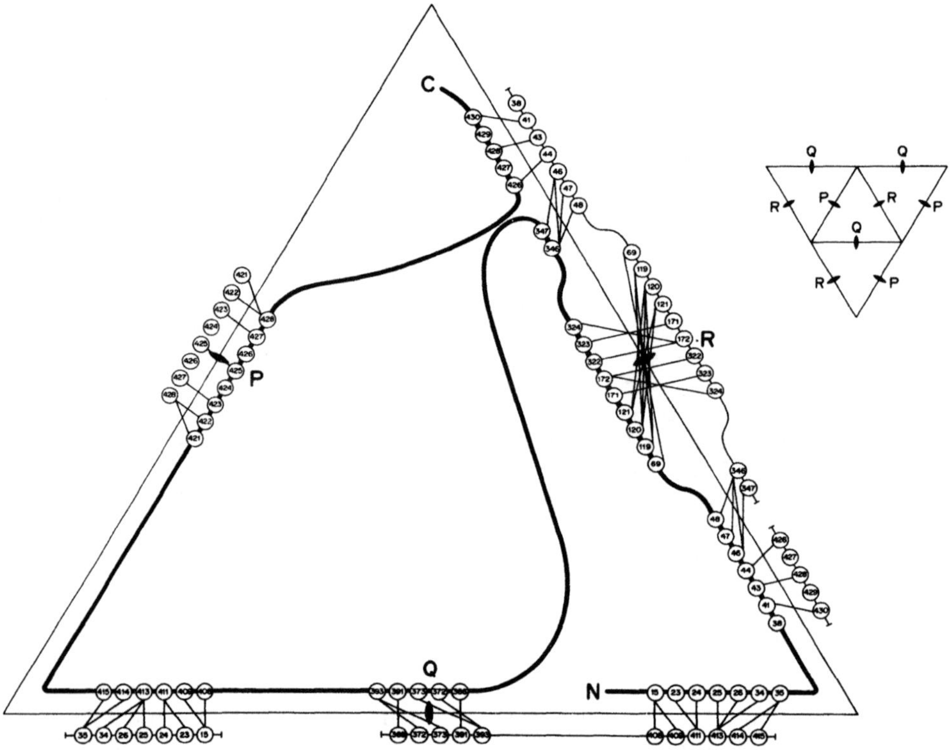

Figure 9. Diagrammatic representation of contacts between residues in neighboring subunits. The reference subunit is represented by a triangle with its edges representing the three different interactions generated by the molecular P, Q, R axes. Shown are all Cα-Cα distances less than 5 Å, although contacts between 51 - 51, 162 - 403, 171 - 398, 171 - 399 generated by the P-axis and between 31 - 140, 32 - 140, 323 - 398, 326 - 396 generated by the Q-axis have been omitted for clarity.

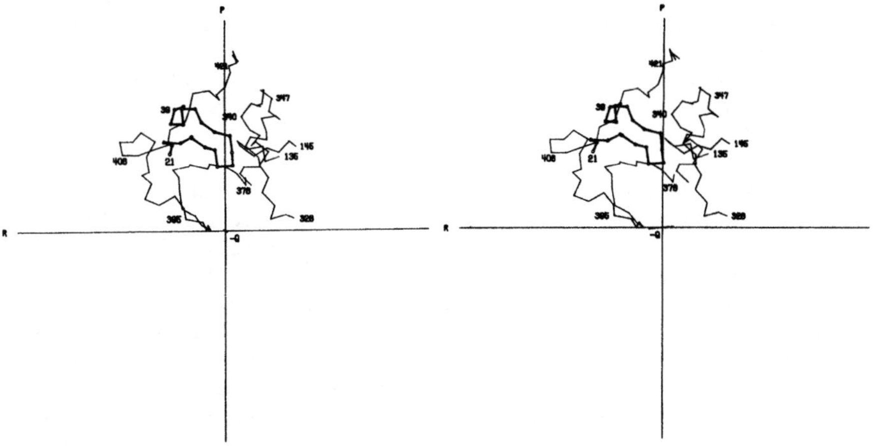

Figure 10. The first 25 residues of the arm in one subunit are inserted into an opening of another subunit related to the first by the molecular Q-axis.

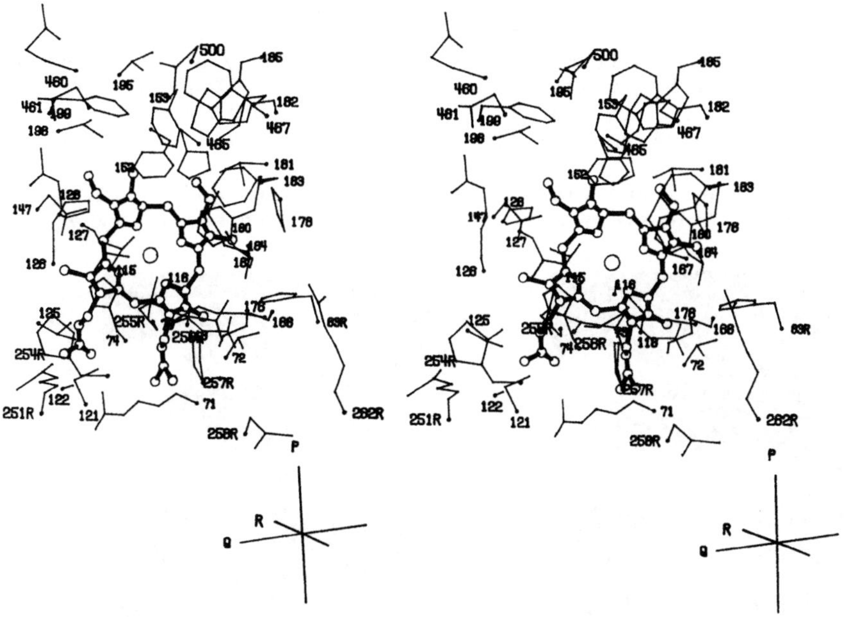

Figure 11. View of the hydrophobic channel through which the substrates must diffuse to the active center.

diameter of this channel can account, in part, for the reduction of the rate of oxidation by about an order of magnitude for each additional carbon in the R group of straight chain alcohols (55). The hydrophobicity of the channel interior can readily explain the requirement for the neutral species in ligand exchange reactions (56).

The heme Fe coordinates in the reference subunit are 16.34, 3.38, and 16.03 Å and the heme normal has direction cosines of -0.321, -0.718 and 0.617 with respect to the molecular *P*, *Q*, *R* axes. The heme handedness, which is determined by the asymmetric distribution of propionic and vinyl side chains, was easily identified (Fig. 12). The essential residues within the heme pocket differ substantially from that of other known hemoprotein structures (Fig. 13). The heme cavity is essentially hydrophobic with interspersed hydrophilic residues. On separation of the tetramer into monomers, pyrrole rings I and IV become partially exposed (Fig. 3). The heme edge formed by pyrrole rings II and III abuts on strands β2, β3 and β4 of the β-barrel which supports the heme. The faces of the heme are surrounded by helices. Both propionyl groups are directed toward the molecular center and, unlike most other hemoproteins, are buried. The propionic acid of pyrrole ring III is neutralized by the guanidinium group of Arg 71 while that of pyrrole IV is neutralized by Arg 364. Of all the other known hemoprotein structures, only the globins form protein-propionyl salt bridges (57). The heme vinyl and methyl groups interact by van der Waals contacts with nonpolar side chains of the protein. A list of heme side chain-protein interactions is given in Table 2.

The proximal and distal heme sides can be readily identified. Tyr 357 provides a protein ligand (Fe-phenolic oxygen

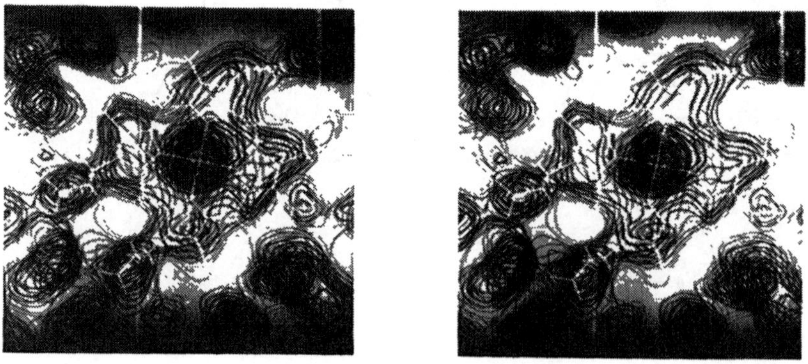

Figure 12. Superposition of heme structure on electron density, clearly identifying hand of the heme and its orientation. The view is from the distal side.

distance is 1.9 Å) on the proximal side, while there is no protein ligand on the opposite side. Furthermore, His 74 is located near the latter side. Modification of this residue by 3-amino-1,2,4-triazole (58,59) inhibits the enzyme by hindering the substrate binding. The proximal side of the heme is crowded and contains residues Val 145, His 211, Pro 235, Arg 353, Ala 356 (Fig. 14). The phenolic hydroxyl group of the proximal Tyr 357 is presumably deprotonated due to the electron withdrawing power of Fe which, although formally in the ferric oxidation state, has an effective charge of +1. However, the guanidinium group of Arg 353 is only 3.5 Å from the phenolic oxygen of Tyr 357 (60) and, therefore, probably will favor ionization of the tyrosine OH and lower its pK. This juxtaposition may also be responsible for the modified spectral behavior of the heme Fe in catalase as compared to model porphyrin complexes with anionic axial ligands (61). Interestingly, His 217, which had been suggested as a possible proximal ligand on the basis of sequence similarities between catalase and other hemoproteins (62), is only 12.5 Å from the heme Fe on its proximal side. However, it is improbable that this residue participates in catalysis. The residue following Tyr 357 is Pro 358 which does not terminate but re-directs helix α9. In the cytochrome b_5 family (63), one of the two Fe-coordinating histidines and an immediately following proline residue are in a conserved tetrapeptide. The occurrence of a proline adjacent to such a critical residue as a heme ligand raises the possibility that the former might play a role in the orientation of the ligand in both cytochrome b_5 and catalase.

The distal side of the heme is much less confined, includes residues His 74, Val 115, Asn 147, Phe 152 and Phe 160, but the L_6 site is unoccupied. In contrast, this position is occupied by a water in metmyoglobin (64) and in cytochrome *c* peroxidase (65). These observations can be correlated with the lack of pH

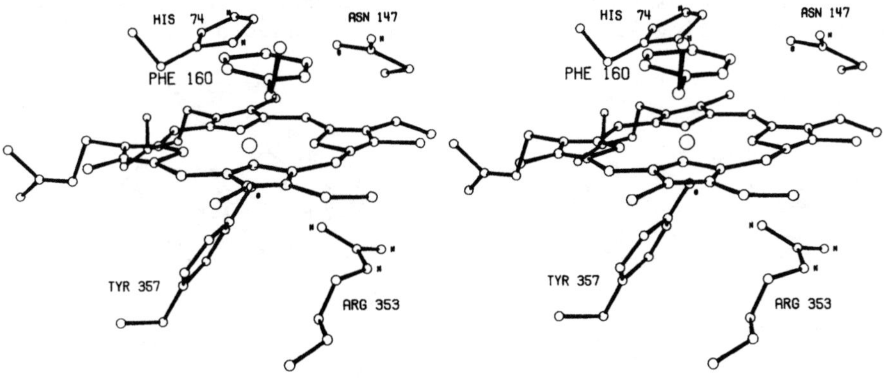

Figure 13. Essential residues in the heme environment.

Table 2. Heme-protein interactions.

Pyrrole Ring[a]	Side Group	Contacts
I	Methyl	Met 60[a], Phe 160
	Vinyl	Ala 157, Phe 160, Met 349, Gly 352
II	Methyl	Asn 147, Phe 152
	Vinyl	Gly 130, Gly 146
III	Methyl	Ser 113, Gly 130, Ala 132
	Propionic acid	Arg 71, His 74, Arg 111, Ser 113
IV	Methyl	Phe 63[a], Asp 64[a], Thr 360
	Propionic acid	Thr 360, Val 72, Arg 364, His 361

[a]Residue from the *R*-axis-related subunit.

dependence of the UV and visible spectrum in catalase in contrast to the other heme proteins. A small electron density peak between His 74 and Asn 147 suggests a water molecule hydrogen bonded to these residues. The oxygen of this putative water molecule is 3.8 Å from the Fe and, therefore, is too far away to coordinate to the Fe atom. The N_δ of His 74 is 4.3 Å from the heme iron. The catalytic importance of this residue has been indicated by the loss of enzyme activity and ligand exchange capacity after its irreversible modification with 3-amino-1,2,4-triazole (58,59). The Fe, His 74 and Asn 147, then, are the essential active site components participating in the catalytic mechanism (Fig. 13). In addition, Val 115 and Phe 152 are positioned to direct the stereospecific abstraction of the pro-R hydrogen of ethanol by compound I (66,67). The phenyl ring of Phe 160 is 3.5 Å above and parallel to pyrrole ring I. In cytochrome *c* peroxidase there is a tryptophan situated similarly above pyrrole ring II. These aromatic residues are oriented to

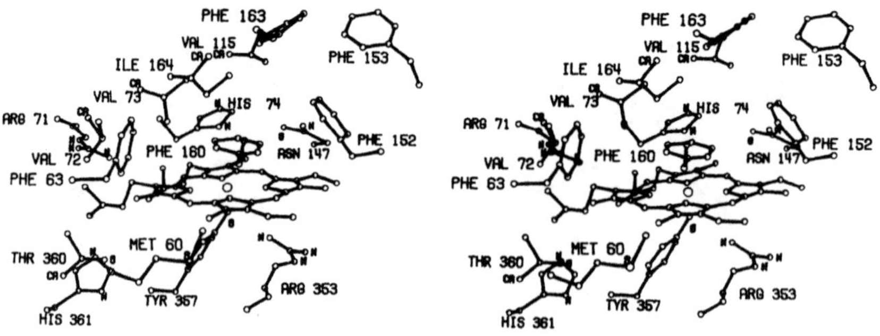

Figure 14. The heme pocket.

form π-π interactions with the heme. This situation does not occur in the globins.

A proposed catalytic mechanism in light of the structure and the wealth of chemical information on catalase will be presented in a subsequent paper (68).

ACKNOWLEDGMENTS

We are grateful for assistance over a short period of time in the early stages of this project to Dr. W. Donald L. Musick. We greatly appreciated a diagram of the chain tracing of *Penicillium vitale* catalase from Prof. B. K. Vainshtein at a time when we had not yet been able to trace the complete beef liver catalase polypeptide chain. We wish to thank Ms. Sharon Wilder, Kathy Shuster, Peggy Baker and Mr. William Boyle for help in the preparation of this manuscript. The work was supported by the National Institutes of Health (grant no. GM 10704) and the National Science Foundation (grant no. PCM78-16584). T.J.R. was supported by a National Institutes of Health Cellular and Molecular Biology grant and A.S. by a National Institutes of Health postdoctoral fellowship (no. 1 F32 GM 06364) during part of this work.

NOTES TO AUTHOR LIST

[1]Present address: Uniformed Services, University of the Health Sciences, Bethesda, Maryland 20014.
[2]Present address: Charter International Oil Co., P.O. Box 5008, Houston, Texas 77012.
[3]Present address: Institute for Protein Research, Osaka University, 5311 Yamada-kami, Suita, Osaka, Japan.

REFERENCES

1. Chance, B.: 1948, *Nature (London)*, 161, pp. 914-917.
2. Hewson, W.D., and Hager, L.P.: 1979, *The Porphyrins*, D. Dolphin, ed., Vol. VII, Academic Press, New York, pp. 295-332.
3. Keilin, D., and Hartree, E.F.: 1955, *Biochem. J.* 60, pp. 310-325.
4. Chance, B., Oshino, N., Sugano, T., and Jamieson, D.: 1974, *Alcohol and Aldehyde Metabolizing Systems*, R.G. Thurman, T. Yonetani, J.R. Williamson, and B. Chance, eds., Academic Press, New York, pp. 169-182.
5. Tanford, C., and Lovrien, R.: 1962, *J. Amer. Chem. Soc.* 84, pp. 1892-1896.

6. Sund, H., Weber, K., and Mölbert, E.: 1967, *Eur. J. Biochem.* 1, pp. 400-410.
7. Theorell, H., and Agner, K.: 1943, *Ark, Kemi Mineral. Geol.* 16A(7), pp. 1-14.
8. Torii, K., Iizuka, T., and Ogura, Y.: 1970, *J. Biochem. (Tokyo)*, 68, pp. 837-841.
9. Stern, K. G.: 1936, *J. Biol. Chem.* 112, pp. 661-669.
10. Sumner, J.B., and Dounce, A.L.: 1937, *J. Biol. Chem.* 121, pp. 417-424.
11. Lemberg, R., and Legge, J.W.: 1943, *Biochem. J.* 37, pp. 117-127.
12. Chang, J.Y., and Schroeder, W.A.: 1972,*Arch. Biochem. Biophys.* 148, pp. 505-508.
13. Malmon, A.G.: 1957, *Biochim. Biophys. Acta,* 26, pp. 233-240.
14. Schroeder, W.A., Shelton, J.R., Shelton, J.B., Robberson, B., and Apell, G.: 1969, *Arch. Biochem. Biophys.* 131, pp. 653-655.
15. Mörikofer-Zwez, S., Cantz, M., Kaufmann, H., von Wartburg, J.P., and Aebi, H.: 1969, *Eur. J. Biochem.* 11, pp. 49-57.
16. Pihl, A., Lange, R., and Evang, A.: 1961, *Acta Chem. Scand.* 15, pp. 1271-1279.
17. Dounce, A.L.: 1942, *J. Biol. Chem.* 143, pp. 497-501.
18. Bonnichsen, R.K.: 1947, *Arch. Biochem.* 12, pp. 83-94.
19. Bonnichsen, R.K.: 1947, *Acta Chem. Scand.* 1, pp. 114-117.
20. Laskowski, M., and Sumner, J.B.: 1941, *Science,* 94, p. 615.
21. Herbert, D., and Pinsent, J.: 1948, *Biochem. J.* 43, pp. 203-205.
22. Galston, A.W., Bonnichsen, R.K., and Arnon, D.I.: 1951, *Acta Chem. Scand.* 5, pp. 781-790.
23. Herbert, D., and Pinsent, J.: 1948, *Biochem. J.* 43, pp. 193-202.
24. Hall, C.E.: 1950, *J. Biol. Chem.* 185, pp. 749-754.
25. Valentine, R.C.: 1964, *Nature (London),* 204, pp. 1262-1264.
26. Kiselev, N.A., Shpitzberg, C.L., and Vainshtein, B.K.: 1967, *J. Mol. Biol.* 25, pp. 433-441.
27. Matricardi, V.R., Moretz, R.C., and Parsons, D.F.: 1972, *Science,* 177, 268-270.
28. Taylor, K.A., and Glaeser, R.M.: 1974, *Science,* 186, pp. 1036-1037.
29. Unwin, P.N.T.: 1975, *J. Mol. Biol.* 98, pp. 235-242.
30. Glauser, S., and Rossmann, M.G.: 1966, *Acta Crystallogr.* 21, pp. 175-177.
31. Labaw, L.W.: 1967, *J. Ultrastruct. Res.* 17, pp. 327-341.
32. Rossmann, M.G., and Labaw, L.W.: 1967, *J. Mol. Biol.* 29, pp. 315-316.
33. Longley, W.: 1967, *J. Mol. Biol.* 30, pp. 323-327.
34. Gurskaya, G.V., Karpukhina, S.Y., and Lobanova, G.M.:

1971, *Biofizika,* 16, pp. 553-555.

35. McPherson, A., Jr., and Rich, A.: 1973, *Arch. Biochem. Biophys.* 157, pp. 23-27.
36. Burkey, D.J., and McPherson, A.: 1977, *Experientia,* 33. pp. 880-881.
37. Eventoff, W., and Gurskaya, G.V.: 1975, *J. Mol. Biol.* 93, pp. 55-62.
38. Eventoff, W., Tanaka, N., and Rossmann, M.G.: 1975, *J. Mol. Biol.* 103, pp. 799-801.
39. Vainshtein, B.K., Melik-Adamyan, V.R., Barynin, V.V., Vagin, A.A., Nekrosov, Y.V., Malinina, L.V., Gulyi, M.F., Gudkova, L.V., and Degtyar, R.G.: 1979, *Doklady Akad. Nauk USSR,* 246, pp. 220-223.
40. Marie, A.L., Parak, F., and Hoppe, W.: 1979, *J. Mol. Biol.* 129, pp. 675-676.
41. Murthy, M.R.N., Reid, T.J., III, Sicignano, A., Tanaka, N., and Rossmann, M.G.: 1981, *J. Mol. Biol.,* in press.
42. Barynin, V.V., Vainshtein, B.K., Zograf, O.N., and Karpukhina, S.Y.: 1979, *Molek, Biol.* 13, pp. 1189-1196.
43. Vainshtein, B.K., Melik-Adamyan, V.R., Barynin, V.V., and Vagin, A.A.: 1980, *Doklady Akad. Nauk USSR,* 250, pp. 242-246.
44. Richards, F.M.: 1968, *J. Mol. Biol.* 37, pp. 225-230.
45. Yang, J.T., and Samejima, T.: 1963, *J. Biol. Chem.* 238, pp. 3262-3267.
46. Adams, M.J., Ford, G.C., Koekoek, R., Lentz, P.J., Jr., McPherson, A., Jr., Rossmann, M.G., Smiley, I.E., Schevitz, R.W., and Wonacott, A.J.: 1970, *Nature (London),* 227, pp. 1098-1103.
47. Harrison, S.C., Olson, A.J., Schutt, C.E., Winkler, F.K., and Bricogne, G.: 1978, *Nature (London),* 276, pp. 368-373.
48. Abad-Zapatero, C., Abdel-Meguid, S.S., Johnson, J.E., Leslie, A.G.W., Rayment, I., Rossmann, M.G., Suck, D., and Tsukihara, T.: 1980, *Nature (London),* 286, 33-39.
49. Watenpaugh, K.D., Sieker, L.C., and Jensen, L.H.: 1979, *J. Mol. Biol.* 131, pp. 509-522.
50. Rossmann, M.G., Moras, D., and Olsen, K.W.: 1974, *Nature (London),* 250, pp. 194-199.
51. Rossmann, M.G., and Liljas, A.: 1974, *J. Mol. Biol.* 85, pp. 177-181.
52. Artymiuk, P.J., Blake, C.C.F., and Sippel, A.E.: 1981, *Nature (London),* 290, pp. 287-288.
53. Rossmann, M.G., Adams, M.J., Buehner, M., Ford, G.C., Hackert, J.L., Liljas, A., Rao, S.T., Banaszak, L.J., Hill, E., Tsernoglou, D., and Webb, L.: 1973, *J. Mol. Biol.* 76, pp, 533-537.
54. Aebi, H., Wyss, S.R., Scherz, B., and Skvaril, F.: 1974, *Eur. J. Biochem.* 48, pp. 137-145.
55. Chance, B.: 1949, *J. Biol. Chem.* 179, pp. 1341-1369.

56. Chance, B.: 1952, *J. Biol. Chem.* 194, pp. 483-496.
57. Watson, H.C.: 1969, *Progress in Stereochemistry*, B.J. Aylett, and M.M. Harris, eds., Vol. 4, Butterworth & Co., London, pp. 299-333.
58. Margoliash, E., and Novogrodsky, A.: 1958, *Biochem. J.* 68, pp. 468-475.
59. Agrawal, B.B.L., Margoliash, E., Levenberg, M.I., Egan, R.S., and Studier, M.H.: 1970, *Fed. Proc.* 29, p. 22.
60. Reid, T.J.R., III, Murthy, M.R.N., Sicignano, A., Tanaka, N., Musick, W.D.L., and Rossmann, M.G.: 1981, *Proc. Natl. Acad. Sci. U.S.*, in press.
61. Mincey, T., and Traylor, T.G.: 1979, *J. Amer. Chem. Soc.* 101, pp. 765-766.
62. Schonbaum, G.R., and Chance, B.: 1976, *The Enzymes*, P.D. Boyer, ed., 3rd edn., Vol. XIII, Academic Press, New York, pp. 363-408.
63. Guiard, B., and Lederer, F.: 1979, *J. Mol. Biol.* 135, pp. 639-650.
64. Takano, T.: 1977, *J. Mol. Biol.* 110, pp. 537-568.
65. Poulos, T.L., Freer, S.T., Alden, R.A., Edwards, S.L., Skogland, U., Takio, K., Eriksson, B., Xuong, N., Yonetani, T., and Kraut, J.: 1980, *J. Biol. Chem.* 255, pp. 575-580.
66. Gang, H., Cederbaum, A.I., and Rubin, E.: 1973, *Biochem. Biophys. Res. Common.* 54, pp. 264-269.
67. Corrall, R.J.M., Rodman, H.M., Margolis, J., and Landau, B.R.: 1974, *J. Biol. Chem.* 249, pp. 3181-3182.
68. Murthy, M.R.N., Tanaka, N., and Rossmann, M.G.: 1981, manuscript in preparation.

THE SUBUNITS OF CYTOCHROME C OXIDASE

Howard S. Mason

Department of Biochemistry, School of Medicine
Oregon Health Science University, Portland Oregon
U.S.A. 97201

Cytochrome c oxidase, the key to aerobic life, has been the subject of a number of excellent recent reviews written from general points of view (e.g., 1-14) and there is no need for another. However, within the last five years it has become apparent that the structures and functions of the subunits that comprise the enzyme hold the secrets to ultimate organization of the enzyme, its biosynthesis and assembly, and its overall functions in electron transfer, proton pumping, and reduction of O_2 to water in the context of mitochondrial structure. The subunits have proved to be difficult experimental materials with which to obtain results consistent from laboratory to laboratory. What is really established about them? and what remains to be accomplished?

1. BACKGROUND

The question, are the identifiable prosthetic groups of cytochrome c oxidase, cytochrome a, cytochrome a_3, and the two forms of copper, attached to one or to separate proteins? was raised by Okunuki (11) and by Tzagaloff and MacLennon (15,16). Acetone-extracted oxidase, succinylated, becomes soluble and from such a solution four proteins can be extracted, of which one ("Fraction II") contains 40-50 mumoles Cu/mg protein (equivalent to a minimum molecular weight of 20,000-25,000). These early studies initiated an intense effort to purify and characterize the oxidase components. However, although the enzyme has a molecular weight according to light scattering criteria of 230,000 and each unit carries two molecules of heme a and two copper atoms (17,18), six to fifteen different proteins have been

H. B. Dunford et al. (eds.), The Biological Chemistry of Iron, 459–473.

observed when purified oxidase preparations are subjected to SDS-polyacrylamide gel electrophoresis in the presence of 6M urea (15,16) or recently, 3M urea, 10.6% glycerol (19).

2. DEFINITION OF THE OXIDASE

Cytochrome c oxidase in higher organisms has at least four functions: (1) to accept electrons from the respiratory chain, (2) to reduce O_2 to water, (3) to pump protons from the mitochondrial matrix to the intermembranal space, and (4) to neutralize reduced oxygen with protons within the matrix. The equation which describes these functions is:

$$4Fe^{2+}_{c\ cytosol} + O_{2\ matrix} + 8H^{+}_{matrix} = 2H_2O_{matrix} + 4Fe^{3+}_{c\ cytosol} + 4H^{+}_{cytosol}$$

The definition of "cytochrome c oxidase" from this point of view is an operational one (6). The oxidase may be that substance which catalyzes reduction of O_2 to $2H_2O$ by Fe^{2+}_c, as in assay of purified enzyme, or that substance which, oriented in the mitochondrial inner membrane, catalyzes the reduction of matrix O_2 to $2H_2O$ by C-side Fe^{2+}_c and transfers protons from one side of the membrane to the other. The number of subunits which comprises cytochrome c oxidase depends upon the number of functions regarded as characteristic of the oxidase cluster and their degree of mutual interdependence.

3. COMPOSITION OF THE OXIDASE

Values for molecular weights of purified oxidase have been reported from 72,000 to 520,000 M_r. The *minimum functional molecular weight* is an unsolved problem of major mechanistic significance (14). "It is not unreasonable to think that the functional unit of oxidase is likely to be composed of four heme and four copper chromophores in a protein molecule of approximately 34 x 10^4 (18) or 34.5 x 10^4 (20). A hydrodynamically monodisperse preparation of cytochrome oxidase was recently found to have a molecular weight of 21 x 10^4; it contained four heme a groups and four copper atoms per molecule of protein (21). Other investigators (22) have shown that monomeric cytochrome c oxidase containing two heme a and two copper atoms is fully functional. Dimeric aggregation of this oxidase is clearly not essential to its oxidative function *in vitro*.

4. PROSTHETIC GROUPS OF CYTOCHROME C OXIDASE

The four different metal centers of holo-cytochrome c oxidase, two containing Cu and two, heme a, are characterized briefly in Table 1. These are properties of the assembled oxidase, not necessarily those of native disassembled subunits.

TABLE 1. METAL CENTERS OF BEEF HEART CYTOCHROME C OXIDASE*

1. Cytochrome a, a heme a protein.

   SPECTRUM: oxidized, 426 nm (120) 550 nm 595 nm (19.5)

   reduced, 444 nm (113) 604 nm (39.5)

   SPIN STATE: low spin, oxidized and reduced.

   REDOX POTENTIAL; 220 nV (low potential for the oxidase).

   EPR SPECTRUM; g = 3.0, 2.2, 1.5

2. Cu_A, a copper protein loosely associated with cytochrome a.

   SPECTRUM: oxidized, 600 nm (15% Cu_A) 830 nm 85% Cu_A)

   reduced, none

   REDOX POTENTIAL: 215 nV (low potential for the oxidase)

   EPR SIGNAL: g = 2 signal, narrow $g_{//}$ region, similar to but but not identical to "blue" Cu^{2+} protein or Cu^+ - ·S

3. Cu_B, a copper protein tightly coupled to cytochrome a_3.

   SPECTRUM: oxidized, 655 nm, due to Cu_B-cytochrome a^{3+} interaction.

   EPR SIGNAL: none.

   REDOX POTENTIAL: 350 mV

   SPIN STATE: Cu^{2+} 1/2 coupled to high spin Fe^{3+}_{a3} , 5/2. Net spin, 2/2

4. Cytochrome a_3, a heme protein.

   SPECTRUM: oxidized, 414 nm (81) 600 nm (4)

   reduced, 442.5 (125) 602 nm (9)

   SPIN STATE: high spin 5/2 coupled to Cu^{2+}_B 1/2

   EPR SIGNAL: None, in fully oxidized enzyme.

   REDOX POTENTIAL: 380 nM (high potential for oxidase).

*Spectra were measured by W.H. Vanneste (22). Numbers in parentheses have the dimensions $mM^{-1}cm^{-1}$: holo-oxidase has spectra, oxidized, 421 nm (83), and 598 nm (11); reduced, 443 nm (107), and 603 nm (23.2).

A major effort has been made by several groups of investigators to explain the catalytic and structural properties of the holo-oxidase in terms of the structures of these prosthetic groups (reviewed in 5, 6, 24, 25). Two recent studies of prosthetic group structure employing iron and copper X-ray absorption edge and absorption fine structure spectroscopy (24,25) led to specific proposals for the ligands of the Cu and heme a groups and their arrangement in space (Figure 1).

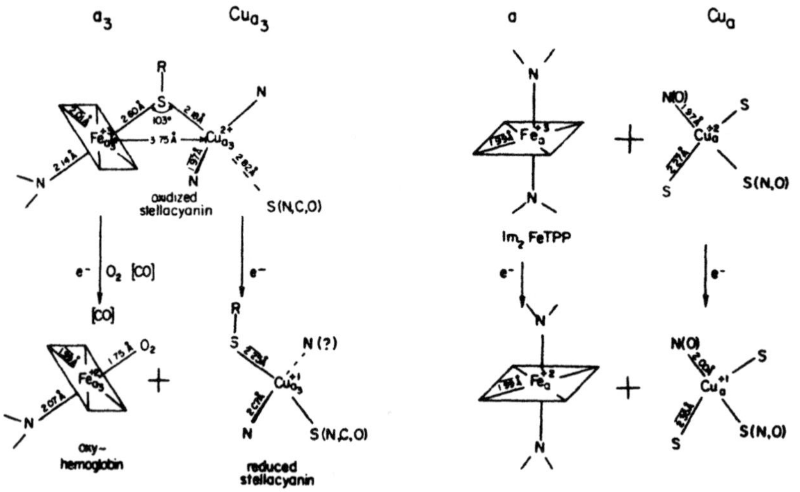

Figure 1. X-ray absorption results for redox centers in fully oxidized and reduced CO states of cytochrome c oxidase. Alternative assignments are in parentheses, and (?) indicates ligands not observed but postulated from the properties of model compounds. Bars of distances indicate average distances for the same type of ligands in that center (25).

As described (25) "The structure of the active site, a cytochrome a_3-copper pair .... contains a sulphur bridge with three ligands 2.18 Å from Fe_{a3} and 2.18 Å from Cu_{a3}. The distance between Fe_{a3} and Cu_{a3} is 3.75 Å, making the sulphur bond angle 103°, reasonable for sp sulphur bonding. The Fe_{a3} first shell has four typical heme nitrogens (2.01 A) with a proximal nitrogen at 2.14 A. The sixth ligand is the bridging sulphur. The Cu_{a3} first shell is identical to oxidized stellacyanin (a type I or blue copper protein) containing two nitrogens and a bridging sulphur. ...The Fe_{a3} first shell is identical to oxyhemoglobin. The other redox centers, Fe_a and the other "EPR detectable" Cu are not observed in higher shells of Fe_{a3}. Fe_a has six equidistant nitrogens and Cu_a has one (or two) nitrogens and three (or two) sulphurs with typical distances. These structures afford the basis for an oxygen reduction mechanism involving oxy- and peroxy-intermediates."

5. THE MULTIPLICITY OF CYTOCHROME OXIDASE SUBUNITS

Good reviews of this topic through 1981 are available (14, 19, 26, 28). As discussed in Section B, the oxidase molecule is defined in terms of its measurable functions. The subunits necessary to these functions can be determined by classical resolution, purification, and reconstitution but so far this has not been accomplished with certainty. Current estimates of the number of subunits in cytochrome c oxidase (mammalian) range from six (21,26) to fifteen (28) but these are based primarily upon subunit composition revealed by various analytical procedures (14, 19) rather than functional reconstitutions.

The three largest polypeptides, subunits, I, II, and III, are coded for by the mitochondrial genome. Their complete structures have been determined by sequencing the human and yeast mitochondrial genes (30). Subunits I and II appear to bear the Cu and heme a prosthetic groups (27,34) and subunit III contains sequence homologies to the hydrophobic channel protein of ATP synthetase (29). The functions of the remaining subunits are not known; not all may be subunits of the oxidase aggregate. However, most of the polypeptides which are found in purified cytochrome c oxidase have been sequenced. Results from three laboratories with beef heart oxidase (37,38) are compared with results from one laboratory (30) with human placenta oxidase subunits I, II, and III in Table 2. Human subunit I, M_r = 57,000, is apparently substantially larger than beef heart oxidase subunit I, but subunit II from beef heart and human placenta are of similar molecular weights. The total molecular weights of these oxidases, assuming one subunit of each size in the functional aggregate, and six to fifteen subunits, is from $11.7\text{-}15.4 \times 10^4$ for beef heart enzyme, and $13.4\text{-}17.9 \times 10^4$ for human enzyme. The content of Cu and heme a, assuming two heme a molecules and two atoms of copper per single aggregate unit, is in the range of 13-17 nmoles per mg protein for beef heart enzyme and 11-15 nmoles per mg for human enzyme. Purified enzymes prepared by standard procedures (e.g., 18, 36) contain amounts of prosthetic groups in this range, but the minimum concentration required to reduce O_2 to $2H_2O$, a mechanistically important value, waits upon determination of the minimum number of subunits required to catalyze the reaction. The large variation in reported M_r for subunit I is puzzling and remains to be resolved; since human and beef heart oxidase share formylmethionine and lysine as N- and C-terminal amino acids it is probable that the value determined by sequencing, 57,000 M_r, is the correct one for mammalian oxidases, rather than 40,000 (37) or 36,000 (38).

TABLE 2. SUBUNIT STRUCTURES OF BEEF HEART AND OTHER CYTOCHROME OXIDASES

#	Source	M_r	N-term-ius	C-term-ius	Reference
I	Beef Heart	40,000	FMet	Lys	Wei & King (37)
		36,000	FMet	Lys	Buse et al. (38)
		30,000	FMet	Lys	Verheul et al. (38)
	Human placenta	57,000	FMet	Ser	Anderson et al. (30)
II	Beef Heart	26,000	FMet	Leu	Verheul et al. (38)
		26,000	FMet	Leu	Buse et al. (38)
		21,000	FMet	Leu	Wei & King (37)
	Human placenta	26,021	FMet	Ser	Anderson et al. (30)
III	Beef Heart	21,000	Met/Thr		Verheul et al. (38)
		22,000	Thr/Met		Buse et al. (38)
	Human placenta	30,000	FMet	Ser	Anderson et al. (30)
IV	Beef Heart	17,000	Ala	Lys	Verheul et al. (38)
	"III"	14,800	Ala	Lys	Wei & King (37)
		17,153	Ala	Lys	Buse et al. (38)
V	Beef Heart	12,000	Ser	Val	Verheul et al. (38)
		12,000	Ser	Val	Buse et al. (38)
		11,600	Ser	Val	Wei & King (37)
VI	Beef Heart	12,000	Ala		Verheul et al. (38)
		11,000-14,000	Ala		Buse et al. (38)
	"IV"	13,500	Ala	His	Wei & King (37)
VII	Beef Heart	10,000	blocked Ala	Ile	Verheul et al. (38)
		10,000	AcAla	Ile	Buse et al. (38)
	"VI"	9,500	Ala	Ile	Wei & King (37)
VIIIa	Beef Heart	5,000-6,000	Ser		Verheul et al. (38)
			Ser	Lys	Buse et al. (38)
VIIIb	Beef Heart	5,000-6,000	Ile		Verhuel et al. (38)
			Ile		Buse et al. (38)
	"VII"	7,600	Acetyl blocked	Val	Wei & King (37)
VIIIc	Beef Heart	5,000-6,000	Phe		Verheul et al. (38)
			Phe		Buse et al. (38)

6. THE DILEMMA IN SUBUNIT NOMENCLATURE

Capaldi et al. (32) made the interesting observation that the order of migration of oxidase subunits changes under different conditions of SDS polyacrylamide gel electrophoresis. They point out that membrane proteins occasionally migrate at anomolous rates because they are not completely unfolded by the denaturing detergent, or because they bind more SDS (or less) than usual, particularly when they contain attached polysacchardies. They observed that subunits II and III change positions, depending upon whether the Weber-Osborn, Swank-Munkres, or Fairbanks SDS gel electrophoresis methods are used. These methods differ in the intensity of denaturation. The Swank-Munkres procedure gave the following molecular weights: subunit I, 35,300; II, 25,200; and III, 21,000 (32), in order of migration distance, and Capaldi et al. adopted the nomenclature indicated in the absence of complete knowledge of the molecular weights of the subunits. However, the sequence work of Anderson et al. (30) has shown that "subunit II" is actually smaller than "subunit III", at least in human cytochrome c oxidase. We continue to use the Capaldi nomenclature for the sake of consistent communication, but with membrane proteins, migration distance on polyacrylamide gels in the presence of SDS cannot be used reliably as a measure of molecular weight or identity. We and others have begun to check N- and C-terminal amino acids as a primary assurance of identity. This is particularly important because several systems of numbering subunits are in use (37).

7. HEME A AND Cu LOCATIONS OF OXIDASE SUBUNITS

A number of investigators have attempted to determine which subunits bind the prosthetic groups in cytochrome c oxidase, but there is no consensus. Functional heme a binding has been ascribed to subunits I, II, III, IV, and V. Functional Cu binding has been ascribed to subunits I, II, III, IV, V, VI, and VII. The subject is reviewed in references 19 and 34. The Cu and heme a are readily mobilized by denaturing detergents, high ionic strengths, sulfhydryl compounds, and gel bound peroxides and free radicals generated by peroxidic initiation of gel polymerization (34). In addition, all subunits have some affinity for copper and heme a so that they are readily picked up by other subunits when released from their native binding sites. In our studies, the use of mixed ionic and non-ionic detergents, low temperature electrophoresis, pre-electrophoresis of the gels with thioglycolate to remove peroxides and free radicals from the polyacrylamide, low ionic strengths, and high protein-to-detergent ratios, made it evident that the four prosthetic groups of cytochrome oxidase are functionally bound to subunits I and II. Figure 2 illustrates by time lapse methods, the migration of heme a during

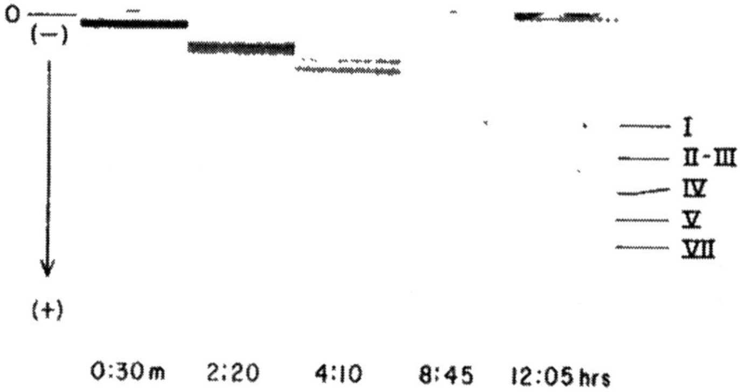

Figure 2. Migration of heme a during electrophoresis of cytochrome c oxidase with 25% glycerol as an antifreeze. The gel (12% polyacrylamide, 0.05% Triton X-100, 0.1% LiDS, 25% glycerol in 0.5 M Tris phosphate buffer, pH 6.5) was pre-electrophoresed and cooled to -4°C. Cytochrome c oxidase, beef heart, 1.50 mg protein was incubated for 1 min in 1% LiDS, 0.05% Triton X-100, in 50 ul of 0.05 M Tris phosphate, pH 6.5, and applied evenly to two gel slots each 3.8 cm long by 0.32 cm wide. Electrophoresis was interrupted at the intervals shown and the gel photographed. Positions of subunit polypeptides determined by protein stain at the end of the experiment are labelled on the right. From (34) with permission of the J. Biol. Chem.

electrophoresis of cytochrome c oxidase at -4°C with 25% glycerol as an antifreeze, in the presence of dodecylsulfate-Triton X-100. The yellow green band at zero time broadens and splits into two bands of approximately equal intensity but after 12 hours of electrophoresis substantial proportion of the heme a is associated with subunit V. The visible spectra of subunits I and II-III in polyacrylamide gel from this experiment are shown in Figure 3. Both subunits I and II-III show normal absorption maxima for the oxidized, reduced, and reduced-CO states of cytochrome oxidase, while subunit V showed extra absorption bands indicative of free heme a bound non-specifically to protein (Figure 4). We concluded that heme a is probably bound to subunits

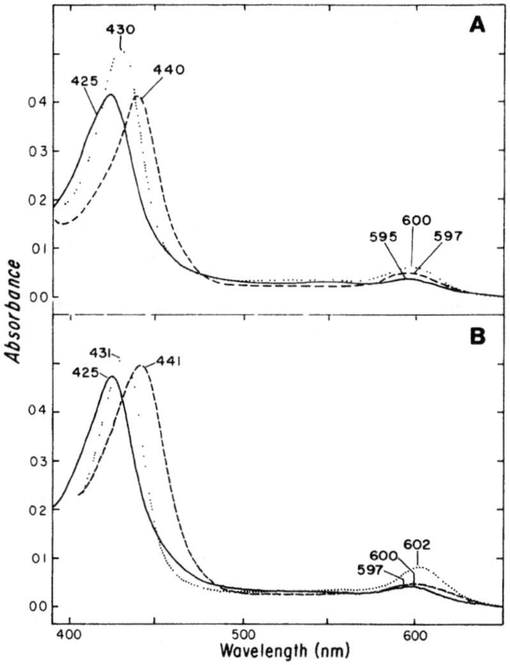

Figure 3. Visible spectra of subunits I and II-III in polyacrylamide gels. Gels prepared as described for Figure 2 were prepared. Bands I and II-III were sliced out and immersed in 0.05 M Tris phosphate buffer pH 6.5 containing 0.025% Triton X-100 deoxygenated by equilibrium with prepurified N. Absorption spectra were measured at 0°C. _____, oxidized subunit, as isolated; -------, dithionite-reduced subunit; ······, reduced-CO complexes of the subunits. A, spectra of Band I; B, spectra of Band II-III. From (34) with permission of the J. Biol. Chem.

I and II and Cu to subunit II, and proposed an organization of the oxidase shown in Figure 5.

8. PREPARATION OF NATIVE SUBUNITS AND THEIR PROPERTIES

My coworkers and I have adapted analytical gel electrophoresis such as that shown in Figure 2 to preparative conditions, taking into account the short time (about 3 hours) available for separation and elution before Cu and heme a are substantially mobilized from their native locations. We used the Hoeffer preparative gel electrophoresis apparatus, with a combination

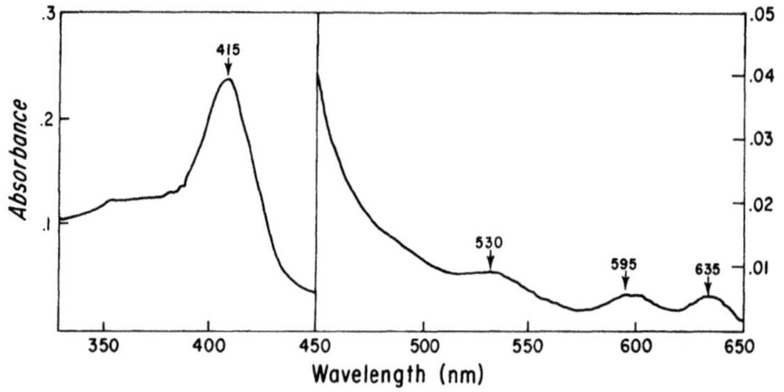

Figure 4. Visible spectrum of subunits V (M_r = 12,436). Cytochrome c oxidase, 1.2 mg, was incubated 30 min at 0° with 1% LiDS, 0.025% Triton X-100, in 50 ul of 0.05 M Tris phosphate buffer, pH 6.5, and then applied to a single gel slot. The pre-electrophoresed gel contained 10% acrylamide with 0.1% SDS, 0.05 M Tris phosphate, pH 6.5, and 0.025% Triton X-100. The incubated cytochrome c oxidase was placed on the gel, capped and electrophoresed 4 h at -0.2°C. Slices of Band V were placed in 0.05 M Tris phosphate, 0.025% Triton X-100, and scanned at 1°C. From reference (34) with permission of the J. Biol. Chem.

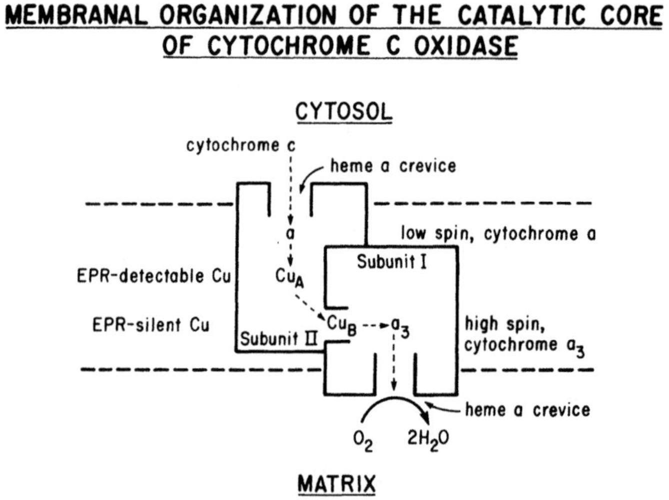

Figure 5. Membranal organization of the catalytic core of cytochrome c oxidase. There are two heme a crevices, at cytochrome a and at cytochrome a_3. Cytochrome a is low spin. Cytochrome a_3 is high spin, interacting with EPR silent copper. Cu_A is EPR detectable. With permission, from the J. Biol. Chem. (34).

of low temperatures, mixed detergents, and an elution channel through which appropriate elution buffers can be passed. Some separation was achieved. Figure 6 shows an elution pattern at

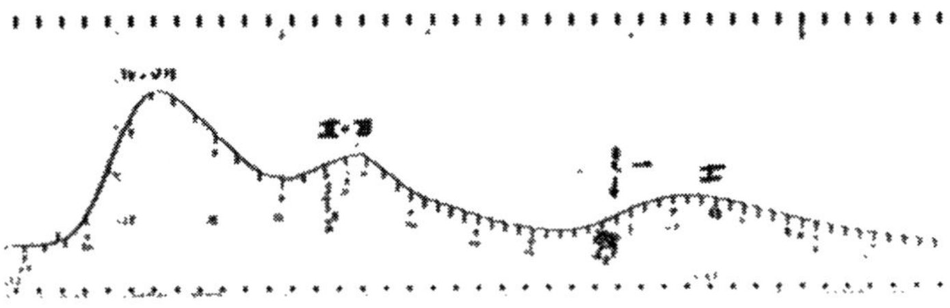

Figure 6. Elution pattern of cytochrome c oxidase subunits under preparative conditions adapted from (34). The first fractions to be eluted are at the left of the diagram, monitored at 280 nm.

280 nm in which the first fraction to be eluted consists largely of the smaller subunits IV-VII. The second fraction consists of subunit II with some smaller subunits still present. The last fraction to be eluted consists largely of subunit I. This is illustrated by the analytical gel electrophoresis performed according to the Swank-Munkres procedure (Figure 7). The absorption spectra of freshly eluted subunits I and II are shown

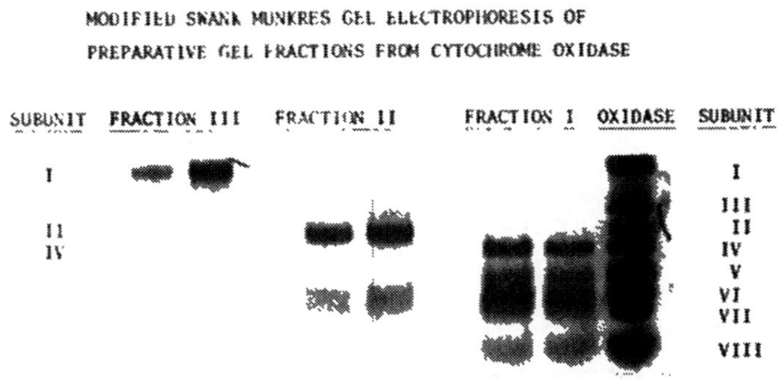

Figure 7. Swank-Munkres analytical gels of the fractions eluted in Figure 6. Fraction I contains the subunits of lower molecular weight. Fraction II contains subunit II predominantly, with some contamination from subunits VI and VII. Fraction III contains almost homogeneous subunit I. The two columns for each fraction represent two concentrations of eluate, that to the right being 5-fold higher than that to the left in each case.

in Figure 8). There is no clear 830 nm band characteristic of Cu_A (Table 1). The spectra are both very similar to that of native

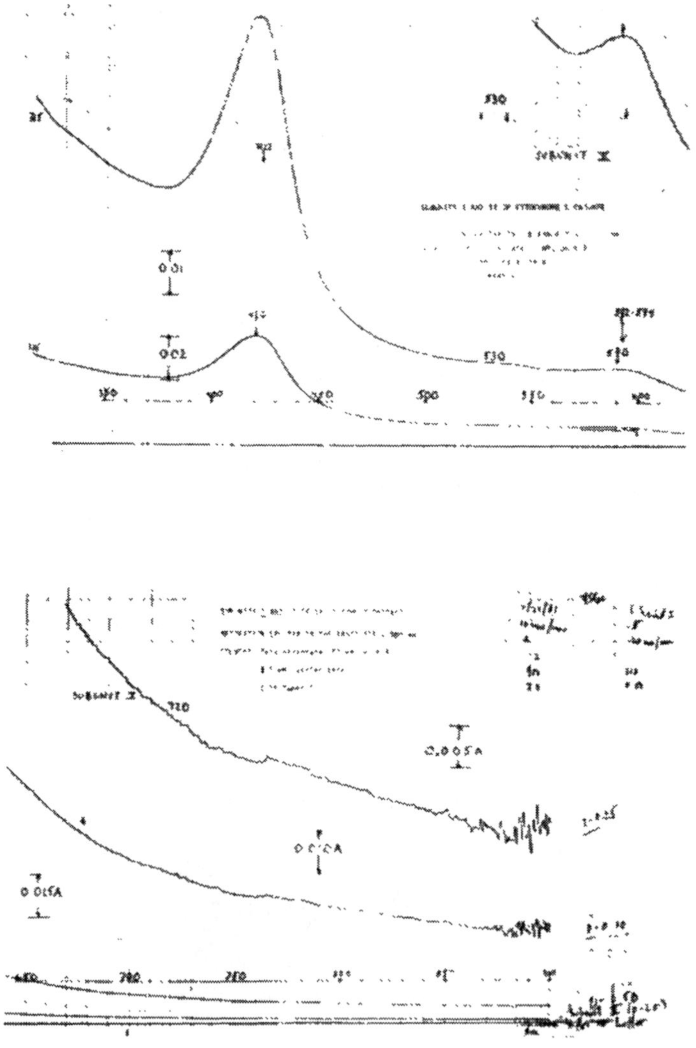

Figure 8. Absorption spectra of freshly eluted subunits I and II, oxidized, as prepared. The absorbancy range was systematically varied from 0-0.05 full scale to 0-2.0 full scale. The spectra are almost identical. Heavy line, I; light line, II.

oxidized cytochrome c oxidase, as was observed earlier (Figure 2). In addition, the EPR spectra of subunit I at -160°C at a rather low signal to noise level shows a lack of hyperfine structure in the $g_{//}$ region but subunit II shows classical axial Cu^{2+} spectra. The structural meaning of this observation is unclear. At present we are continuing to characterize the subunits isolated by this procedure, using various physical and chemical methods, and have carried out the first attempts at reconstitution of the holo-oxidase from isolated subunits cytochrome a_3. There is evidence that the stability of isolated subunits is low, and this problem is being studied now.

9. FOOTNOTE

Recent results described in this preliminary paper were carried out with Caroline Woody, Freddie F.P. Novak and Timothy Vincent. Research on preparative purification of oxidase subunits was carried out with Nicholas Grinich and Dr. Daryl Winter. This study was supported by a grant, AM0718, from the National Institute of Arthritis and Metabolic Disease to HSM, for which the author is grateful.

REFERENCES

1. Lemberg, M.R. and Barrett, J.: 1973, "Cytochromes", Academic Press: New York.
2. Nichols, P. and Chance, B.: 1974, in "Molecular Mechanisms of Oxygen Activation", Hayaishi, O. ed., Academic Press: New York, pp. 479-543.
3. Caughey, W.S., Wallace, W.J., Volpe, J.A. and Yoshikawa, S.: 1976, in "The Enzymes", 3rd ed., Boyer, P.D., ed., XIII Part C, pp 299-344.
4. Malmstrom, B.G.: 1977, in "Bioinorganic Chemistry", Raymond, K.H., ed., Adv. Chem., Series 162, Volume II, pp. 173-206.
5. Malmstrom, B.G.: 1979, *Biochim. Biophys. Acta,* 549, pp. 281-303.
6. Malmstrom, B.G.: 1980, in "Metal Ion Activation of Dioxygen", Spiro, T.G., ed., pp. 183-207.
7. "Cytochrome Oxidase": 1979, Vol. 5 of "Developments in Biochemistry", King, T.E., Orii, Y., Chance, B. and Okunuki, K., eds., Elsevier-North Holland, Amsterdam.
8. Vanderkooi, G.: 1974, *Biochim. Biophys. Acta,* 344, pp. 307-345.

9. Poyton, R.O.: 1980, "Current Topics in Cell Regulation", 17, pp. 231-295.
10. Wikstrom, M.F., Harmon, H.J., Ingedew, W.J. and Chance, B.: 1976, *FEBS Lett.* 65, pp. 259-277.
11. Okunuki, K.: 1972, in "Aspects Cell Mol. Physiol.", pp. 57-73.
12. Erecinska, M. and Wilson, D.: 1978, *Arch. Biochem. Biophys.* 188, pp. 1-14.
13. Beinert, H.: 1980, *Coord. Chem. Rev.* 33, pp. 55-85.
14. Azzi, A.: 1980, *Biochim. Biophys. Acta,* 594, pp. 231-252.
15. Tzagaloff, A. and MacLennon, D.H.: 1965, in "The Biochemistry of Copper", Peisach, J., Aisen, P. and Blumberg, W.M., eds. Academic Press, pp. 253-265.
16. MacLennon, D.H. and Tzagaloff, A.: 1965, *Biochim. Biophys. Acta,* 240, pp. 166-168.
17. Tzagoloff, A. and Wharton, D.C.: 1965, *J. Biol. Chem.* 240 pp. 2628-2633.
18. Kuboyama, M., Yong, F.C. and King, T.E.: 1972, *J. Biol. Chem.* 247, pp. 6375-6383.
19. Kadenbach, B. and Merle, P.: 1981, *FEBS Lett.* 135, pp. 1-9.
20. Robinson, N.C. and Capaldi, R.A.: 1977, *Biochemistry* 16, pp. 375-381.
21. Saraste, M., Pentilla, T. and Wikstrom, M.K.F.: 1981, *Eur. J. Biochem.* 115, pp. 261-268.
22. Wilson, M.T., Lalla-Maharajh, W., Darley-Usmar, V., Bonaventura, J., Bonaventura, C. and Brunori, M.: 1980, *J. Biol. Chem.* 255, pp. 2722-2728.
23. Vanneste, W.H.: 1966, *Biochemistry* 5, pp. 838-848.
24. Powers, L., Blumberg, W.E., Chance, B., Barlow, C.H., Leigh, J.S., Smith, J., Yonetani, T., Vik, S. and Peisach, J.: 1979, *Biochim. Biophys. Acta* 546, pp. 520-538.
25. Powers, L., Chance, B., Ching, Y. and Angiolillo, P.: 1981, *Biophys. J.* 34, pp. 465-498.
26. King, T.E., Yu, L., Yu, C.A. and Wei, Y.H.: 1979, in "Cytochrome Oxidase", King, T.E., et al., eds., Elsevier-North Holland, p. 53; Steffens, G.J. and Buse, G., ibid, p. 79; Yasunobu, K.T., Tanaka, M., Hanui, M. and others, ibid, p. 91.
27. Steffens, G.J. and Buse, G.: 1979, *Hoppe-Seyler's Z. Physiol. Chem.* 360, pp. 613-619.
28. Griffin, D.C. and Landon, M.: 1981, *Biochem. J.* 197, pp. 333-344.
29. Prochaska, L.J., Bisson, R., Capaldi, R.A., Steffens, G.C.M. and Buse, G.: 1981, *Biochim. Biophys. Acta,* 637, pp. 360-373.
30. Anderson, S., Bankier, A.T., Barrell, B.G., deBruin, M.H.L., Coulson, A.R., Drouin, J., Eperon, I.C., Nierlich, D.P., Roe, B.A., Sanger, F., Schrier, P.H., Smith, A.J.H., Staden, R. and Young, I.G.: 1981, *Nature* 290, pp. 457-465.
31. Steffens, C.M., Steffens, G.J. and Buse, G.: 1979, *Hoppe-Seyler's Z. Physiol. Chem.* 360, pp. 1641-1650.

32. Capaldi, R.A., Bell, R.L. and Branchek, T.: 1977, *Biochim. Biophys. Res. Commun.* 74, pp. 425-433.
33. Downer, N.W., Robinson, N.C. and Capaldi, R.A.: 1976, *Biochemistry* 15, pp. 2930-2936.
34. Winter, D.B., Bruyninckx, W.J., Foulke, F.G., Grinich, N.P. and Mason, H.S.: 1980, *J. Biol. Chem.* 255, pp. 11408-11414.
35. Steffens, G.J. and Buse, G.: 1976, *Hoppe-Seyler's Z. Physiol. Chem.* 357, pp. 1125-1137.
36. Hartzell, C.R., Beinert, H. Van Gelder, B.F. and King, T.: 1978, *Methods Enzymol.* 53, pp. 54-66.
37. Wei, Y.H. and King, T.E.: 1961, *J. Biol. Chem.* 265, pp. 10999-11003.
38. Verheul, F.E.A.M., Draijer, J.W., Dentener, I.K. and Muijsers, A.O.: 1981, *Eur. J. Biochem.* 119, pp. 401-408.

EXTENDED X-RAY ABSORPTION FINE STRUCTURE OF THE COPPER SITES IN CYTOCHROME C OXIDASE.

Robert A. Scott

School of Chemical Sciences, University of Illinois, Urbana, Illinois 61801

ABSTRACT

The extended x-ray absorption fine structure (EXAFS) technique is discussed with relevance to the determination of details of the local structural environment of the copper active sites of beef heart mitochondrial cytochrome c oxidase. By studying the copper x-ray absorption edges and EXAFS of the resting state, fully reduced, and mixed valence-formate derivatives of the enzyme, most if not all of the oxidation state-sensitive sulfur ligands can be assigned to Cu_A*, the epr-detectable copper. No evidence can be found for radiation induced oxidation state changes in these samples at -60 C under the flux conditions of these experiments.*

INTRODUCTION

Cytochrome c oxidase (ferrocytochrome c:O_2 oxidoreductase, EC 1.9.3.1), as the terminal enzyme of the mitochondrial respiratory chain, plays an extremely important role in electron transport (catalyzing the reduction of O_2 to H_2O) and in bioenergetics (as a proton pump)(1). It is a rather large enzyme (ca. 150,000 daltons) containing four metal active sites. Two of these sites (cytochrome a and cytochrome a_3) contain heme a but are spectroscopically and chemically distinct, indicating that their protein environments and, in particular, their axial ligands differ. The other two active sites contain copper atoms which are also spectroscopically (and presumably structurally) distinct. These sites are denoted as Cu_A and Cu_B. (Cu_B is the copper atom believed to be magnetically coupled to the heme of

H. B. Dunford et al. (eds.), The Biological Chemistry of Iron, 475–484.

cytochrome $\underline{a}_3$(2).) It has become increasingly clear that the unique set of catalytic functions carried out by cytochrome c oxidase are intimately related to unique structural features of the various active sites. Thus, elucidation of the detailed structural characteristics of the active sites will undoubtedly contribute to our understanding of the functioning of this enzyme. Toward this end, we have initiated a study of the structure of the copper active sites as revealed by x-ray absorption spectroscopy (XAS).

Interest in XAS (specifically, extended x-ray absorption fine structure, EXAFS) as a tool for structural determination has increased dramatically in the last decade due to the availability of synchrotron radiation as a high-intensity, tunable source of x-rays(3). The power of the XAS technique for biochemical applications resides in its sensitivity to the local structural environment of metal atoms in samples of any physical state. This allows the selective examination of a particular type of metal atom in an enzyme or protein and extraction of information regarding the types and numbers of ligands as well as accurate metal-ligand bond distances. An examination of the copper EXAFS of various oxidation states of cytochrome c oxidase is reported here. This has provided structural information on the copper sites which may have important implications regarding the catalytic functions of these sites.

MATERIALS AND METHODS

Beef heart mitochondrial cytochrome c oxidase was prepared as previously described(4). Three different derivatives have been examined: the resting state (fully oxidized), the fully reduced state, and the mixed valence-formate derivative. These were prepared by standard methods (reduction of the enzyme was carried out using ascorbate with a catalytic amount of cytochrome c). The integrity of each sample was checked before and after x-irradiation using electron paramagnetic resonance (epr) spectroscopy and low-temperature optical reflectance spectroscopy. All samples were kept in thin lucite sample cells (suitable for epr)(5) at ca. -60 C under a cold helium atmosphere during x-irradiation and cooled to liquid nitrogen temperature within five minutes after irradiation ceased. They were maintained at liquid nitrogen temperature until epr and optical reflectance spectra could be recorded (in the same cell used for x-irradiation).

Fluorescence excitation spectra were recorded at the Stanford Synchrotron Radiation Laboratory (SSRL) under dedicated conditions (3.0 GeV, ca. 70 mA) on one of the wiggler beam lines (IV-2, wiggler field of 10-13 kG)(6). Data were recorded with

Table 1. Summary of EXAFS Samples and Data Collection Conditions

Sample	Derivative[a]	Average Flux (photons/s) ($\times 10^{-10}$)	Irradiation Time (min)	Total Irradiation (photons) ($\times 10^{-14}$)
OX-B	resting	2.8	82	1.4
OX-C	"	--[b]	--[b]	--[b]
OX-D	"	2.3	59	0.8
OX-E	"	2.1	129	1.6
OX-F	"	2.6	27	0.4
OX-G	"	2.8	81	1.4
OX-H	"	2.7	183	3.0
OX-I	"	2.8	39	0.7
RD-A	reduced	5.7	660	22.5
RD-B	"	1.7	950	9.7
MVF-A	MV-form	11.4	750	51.1
MVF-B	"	2.7	910	15.0

[a] resting - fully oxidized resting state; reduced - fully reduced state; MV-form - mixed valence-formate state.
[b] sample OX-C was used as an unirradiated control.

both Si[111] and Si[220] monochromator crystals to assure that no specific crystal characteristics (e.g., "glitches") affected the data. The fluorescence data were collected using a 20-element filtered scintillation detector array(7) and weighted and averaged as described previously(5).

In this study we have employed internal calibrations of the energy scale of the x-ray absorption spectra, recording a high-resolution edge of a known standard (e.g., copper foil) simultaneously with each scan. This was accomplished by the use of three ionization chamber detectors, the sample residing between the first and second and the copper foil standard residing between the second and third. Thus, at the same time transmission (and fluorescence) data were being collected on the sample, the transmission of the standard was being collected in the third ionization chamber. This internal calibration procedure allowed the calibration of the energy scale to an accuracy of ± 0.1 eV.

Due to the possibility of photoreduction of the active sites of this enzyme during x-irradiation(8), care was taken to monitor irradiation times and photon fluxes and to check for radiation-induced oxidation state changes after data collection. For the resting state, which may be particularly sensitive to photoreduction(8), several samples were used to collect an overall average data set. Two samples each of the other

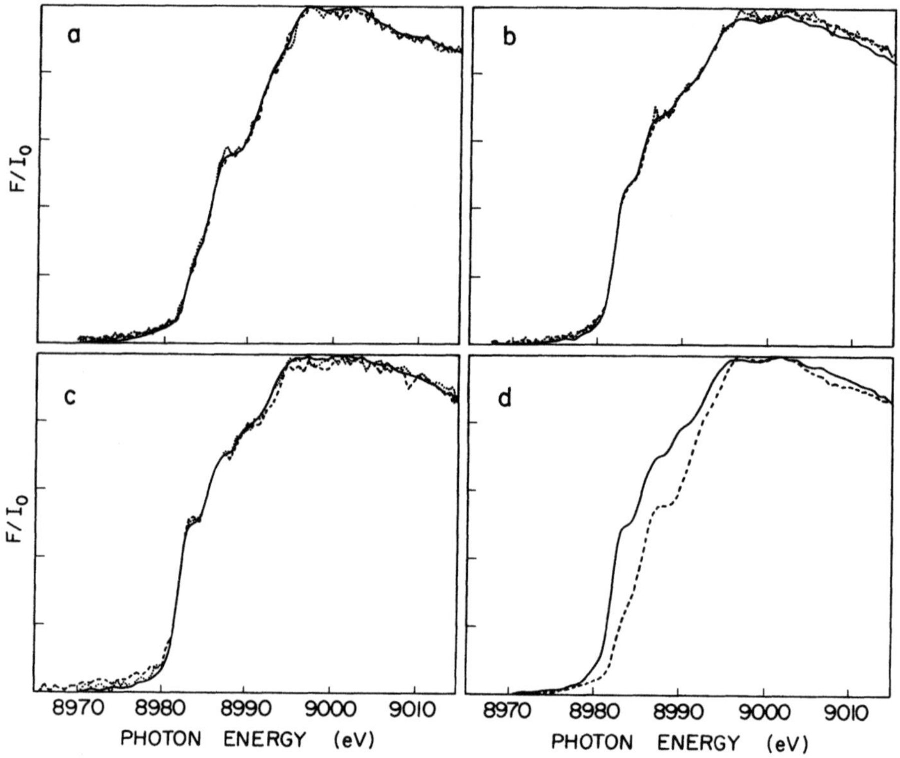

Figure 1. Comparison of Cu K-absorption edges recorded in: the first scan (- - -); the last scan (•••••); and the average of all scans (———) of: (a) sample OX-H; (b) sample MVF-B; (c) sample RD-B. (d) comparison of the edge from the average of all resting state samples (- - -); fully reduced sample RD-B (———).

derivatives (fully reduced and mixed valence-formate) were examined. The details of x-irradiation for each sample discussed are given in Table 1.

RESULTS

For data collected with Si[220] monochromator crystals, high resolution K-edges were recorded. The energy position of the copper edge is dependent on the oxidation state of the copper atoms in the enzyme(9,10). Thus, the position of the edge was used to monitor whether photoreduction was occurring during data collection. For three of the samples listed in Table 1 (OX-H, RD-B, MVF-B), Figure 1a-c shows a comparison of the edge recorded in the first scan (less than 30 minutes irradiation), the edge recorded in the last scan, and the edge from the average of all

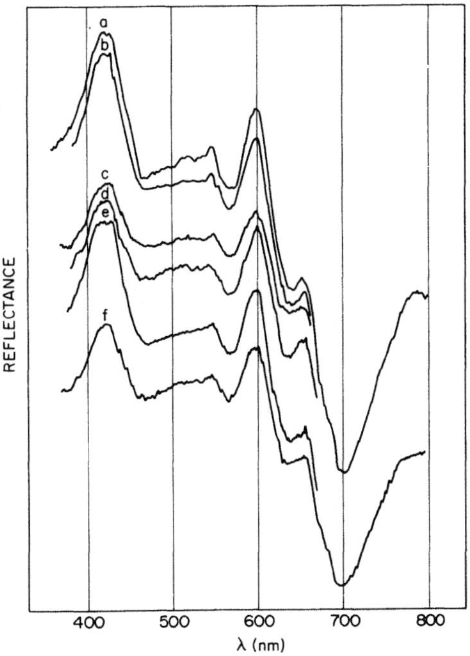

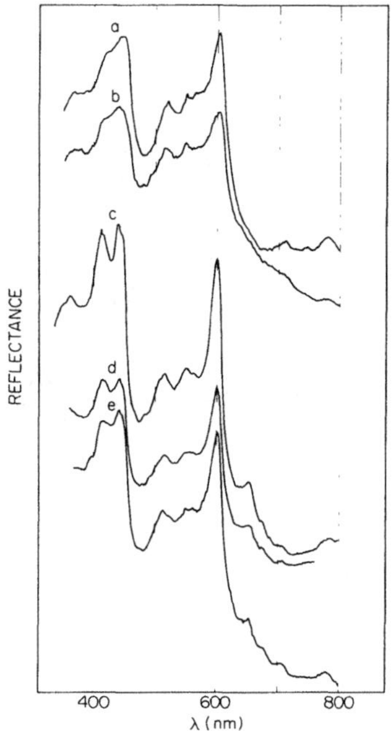

Figure 2. Low-temperature optical reflectance spectra of resting state samples after x-irradiation: (a) sample OX-C; (b) sample OX-F; (c) sample OX-D; (d) sample OX-G; (e) sample OX-E; (f) sample OX-H.

Figure 3. Low-temperature optical reflectance spectra of fully reduced and mixed valence-formate samples: (a) RD-A after irradiation; (b) RD-B after irradiation; (c) MVF-A before irradiation; (d) MVF-A after irradiation; (e) MVF-B after irradiation.

scans. Figure 1d compares the averaged edge from all of the resting state samples with the averaged edge from sample RD-B (fully reduced). This gives an indication of the energy shifts to be expected upon reduction. These plots (Figure 1a-c) are typical of the behavior observed for all of the samples listed in Table 1. In addition, low-temperature optical reflectance (Figures 2,3) and epr spectra (not shown) of all of the samples after data collection have been recorded.

Radiation-dependent generation of radicals is observed in the samples used in this study. The amount of radical epr signal (at ca. g=2) generated seems to be directly related to the number of photons which have been incident upon the sample (at least in the particular series of resting state samples listed in Table 1). However, it is also clear from the epr spectra that the

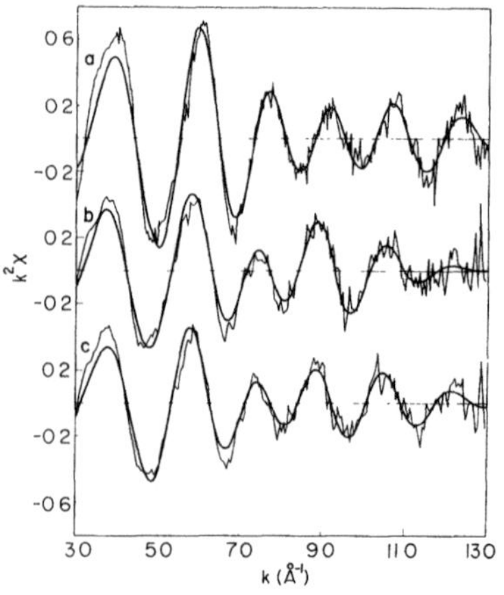

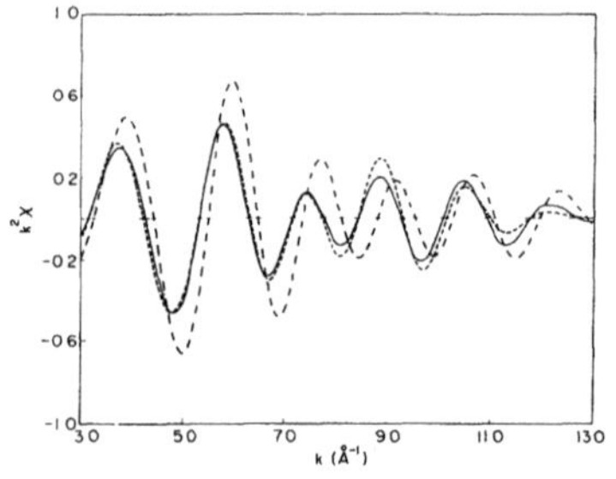

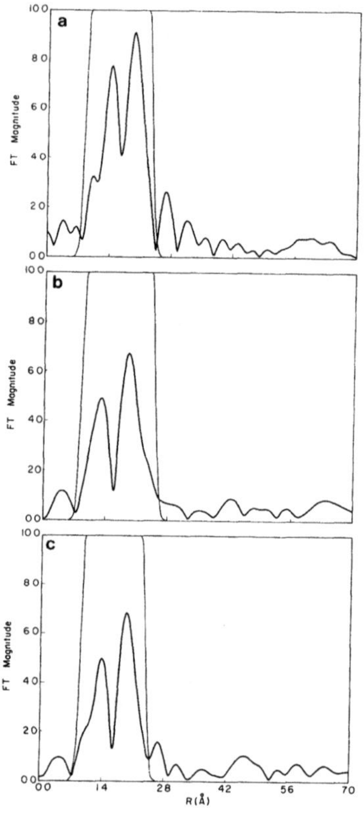

Figure 4. (upper left) EXAFS data of three derivatives of cytochrome c oxidase: raw data (light spectra); filtered EXAFS (heavy curve) of: (a) average of all resting state samples; (b) average of samples RD-A, RD-B; (c) average of samples MVF-A, MVF-B. The filtered data were obtained by k^3-weighted transformation of the raw EXAFS data over a k=3.0-13.0 Å^{-1} range followed by back-transformation of the data within the filter windows shown in Figure 6.

Figure 5. (lower left) Comparison of filtered EXAFS of cytochrome c oxidase derivatives: resting state (-•-•-); fully reduced (- - -); mixed valence-formate (———).

Figure 6. Fourier transforms of raw EXAFS data in Figure 4: (a) resting state; (b) fully reduced; (c) mixed valence-formate. The transforms were all done by k^3-weighting of the raw data over a k=3.0-13.0 Å^{-1} range. The filter windows for back-transformation to the filtered EXAFS of Figures 4,5 are also shown.

Table 2. Two Shell (Nitrogen and Sulfur) Fits to Filtered EXAFS

Sample	Cu-N (or -O)		Cu-S	
	R(Å)	#	R(Å)	#
a	1.97	1.8	2.27	1.4
b	1.95	1.8	2.26	1.7
RD-A	1.94	1.2	2.32	1.4
RD-B	1.96	1.1	2.31	1.4
MVF-A	1.97	1.3	2.32	1.2
MVF-B	1.96	1.3	2.32	1.5

[a]data of previous resting state sample (reference 5).
[b]averaged data from all resting state samples in Table 1.

observation of these radiation-produced radicals does not necessarily imply photoreduction of the cytochrome c oxidase active sites (since no significant increase in the g6 signal was observed in any of the resting state samples). Very similar radical epr signals are also generated in a sample of buffer irradiated under identical conditions. Thus, no evidence has been found for photoreduction of cytochrome c oxidase samples at -60 C under the photon fluxes and irradiation times given in Table 1.

The EXAFS data in Figure 4 are averages from the appropriate samples in Table 1. The Fourier transforms of the EXAFS data in are shown in Figure 6. Back-transformation of the two major transform peaks (indicated by the filter windows in Figure 6) yields the filtered EXAFS curves which are shown superimposed on the raw EXAFS data in Figure 4. The filtered EXAFS curves for the three distinct enzyme derivatives are compared in Figure 5. Since the low R and high R Fourier transform peaks included in the filters are apparently due to Cu-N (or -O) and Cu-S interactions, respectively(5), we have performed two-shell fits on the filtered EXAFS data with one shell of nitrogens (which are indistinguishable from oxygens) and one shell of sulfurs. The results of these fits are given in Table 2.

DISCUSSION

Since there is some controversy concerning the possibility of radiation-induced oxidation state changes in cytochrome c oxidase samples used for x-ray absorption studies(8,11), we have expended considerable effort to assure that the samples described in this study were maintained at the appropriate oxidation level before, during, and after the data acquisition. Both epr and low-temperature optical reflectance spectroscopies were used to assay the samples before and after irradiation. Since radical signals are observed in all irradiated samples, the optical reflectance spectra were most informative regarding the integrity of the samples. For the resting state samples, the reflectance spectra recorded after irradiation (Figure 2) show a Soret peak at ca. 420 nm, the relatively broad α-band at 600 nm, a very distinct 655 nm band and the deep trough at ca. 700 nm. These features are all characteristic of the fully oxidized enzyme(12). The reduced samples (Figure 3a,b) show a Soret peak at ca. 440 nm, a sharp α-band at 605 nm and the clear absence of any 655 nm band, features which are characteristic of fully reduced enzyme(12). In the mixed valence-formate samples (Figure 3c-e), the Soret is distinctly split (both before and after irradiation in the case of sample MVF-A), the sharp α-band occurs at 600 nm, and the 655 nm absorption is present as is typical for this mixed valence derivative(12).

Even more convincing is the fact that the copper K-absorption edges of all three derivatives do not change during the course of data collection (Figure 1a-c). This is an important point since the edge shift upon reduction of the enzyme is substantial (Figure 1d) and any change in oxidation state would be detectable as a shift in the copper edge. In the case of the resting state, the scans collected in the first ca. 60 minutes of irradiation (from all of the resting state samples in Table 1) were averaged and the EXAFS (and edge) from this average were compared to the EXAFS (and edge) of the average of all the later scans. The edges of the "first half" and "last half" of data collection were superimposable and the EXAFS data showed insignificant changes well within the noise. In addition, EXAFS data on different samples of the same derivative are identical within the relatively low noise level. Thus, all evidence suggests that these samples were maintained in their respective oxidation states during x-irradiation.

The comparison of filtered EXAFS curves in Figure 5 indicates that the EXAFS derived from the two major transform peaks looks very similar for the mixed valence-formate derivative and the fully reduced enzyme. Both of these data sets look substantially different from the resting state EXAFS (Figures 5 and 6). As expected, these relationships carry over to the two-shell fits (Table 2). It is evident from Figure 5 that the average phase shift of the resting state EXAFS is significantly different from the average phase shift of the fully reduced and mixed valence-formate EXAFS. Since sulfur is a stronger backscattering atom than nitrogen (or oxygen), it is reasonable to expect that the phase shift observed in Figure 5 is due to a significant change in the average copper-sulfur (Cu-S) distance. The fits in Table 2 support this conclusion, the average Cu-S distance changing from 2.26 Å in the resting state to 2.32 Å in the fully reduced and mixed valence-formate states. In contrast, the average Cu-N (or -O) distance does not change significantly upon reduction. (Since it is probable that the Debye-Waller factors for Cu-S and Cu-N (or -O) interactions change appreciably upon reduction of Cu(II) to Cu(I), the differences observed in coordination numbers in Table 2 are of unknown significance.)

It is generally believed that Cu_B remains oxidized, while Cu_A is reduced in the mixed valence-formate derivative(12). With this assumption, the event that gives rise to the lengthening of the average Cu-S distance (by 0.06 Å) is the reduction of Cu_A. Upon reduction of Cu_B (conversion of the mixed valence-formate derivative to the fully reduced state), no further change is observed in the average Cu-S distance. This is strong evidence for assigning most (if not all) of the redox-sensitive sulfur ligands to Cu_A. If this assignment is correct, it would be expected that the EXAFS of the mixed valence-CO derivative should show a strong resemblance to the EXAFS of the resting state (since in the mixed valence-CO derivative, Cu_B is reduced while Cu_A is oxidized). Preliminary data on two different samples of this derivative indicate that this is indeed the case.

The EXAFS evidence leads us to postulate the following assignment of copper ligands: Cu_A has two (or three) sulfur ligands with an average Cu-S distance which changes upon reduction from 2.26 to 2.32 Å. Cu_A also has two (or one) nitrogen (or oxygen) ligands. Cu_B has mostly nitrogen (or oxygen) ligands with a coordination sphere which does not change significantly upon reduction. Work is in progress to further refine this model by collection of better data on the mixed valence-CO derivative and more confirmatory data on the other oxidation states.

We have not yet been able to conclusively identify a scatterer in the copper environment of any of the enzyme derivatives which can be assigned as an iron atom. Thus, we have as yet no confirmation of the distance between the iron of heme $\underline{a}_3$ and Cu_B suggested by Chance and coworkers(13). However, the Fourier transform peak at R=2.65 Å for the resting state data set in Figure 6a does show behavior reminiscent of metal-metal scattering (i.e., the back-transform of this peak exhibits an amplitude envelope which peaks at k=7-9 $Å^{-1}$). In addition, the presence of this transform peak seems to parallel the presence of the 655 nm band in the optical spectrum, being weaker in the mixed valence-formate and totally absent in the fully reduced spectrum. If this is the iron backscattering peak, its position indicates the Cu-Fe distance would be *ca.* 3.0 Å, much different from the distance quoted by Chance and coworkers(13). Work is continuing to resolve this question.

ACKNOWLEDGMENTS

Dr. S. P. Cramer of Exxon was instrumental in the development of the fluorescence detection equipment. Drs. Helmut Beinert and Robert Shaw prepared the samples at Madison. The work was carried out while the author was a National Institutes of Health Postdoctoral Fellow (5-F32-HL06047) at Stanford University with Dr. K. O. Hodgson. Experiments at SSRL were supported by National Science Foundation Grant DMR-07692-A02 in cooperation with the Stanford Linear Accelerator Center and the Department of Energy.

REFERENCES AND NOTES

1. For a recent review, see: Malmström, B. G.: 1979, *Biochim. Biophys. Acta* 549, pp. 281-303.
2. Tweedle, M. F.; Wilson, L. J.; García-Iñiguez, L.; Babcock, G. T.; Palmer, G.: *J. Biol. Chem.* 253, pp. 8065-71.
3. For recent reviews of x-ray absorption spectroscopy, see: Cramer, S. P.; Hodgson, K. O.: 1979, *Prog. Inorg. Chem.* 25, pp. 1-39. Shulman, R. G.; Eisenberger, P.; Kincaid, B. M.: 1978, *Ann. Rev. Biophys. Bioeng.* 7, pp. 559-78. Teo, B.-K.: 1980, *Accts. Chem. Res.* 13, pp. 412-9. Chan, S. I.; Hu, V. W.; Gamble, R. C.: 1978, *J. Mol. Struct.* 45, pp. 239-66.
4. See Procedure I of: Hartzell, C. R.; Beinert, H.; van Gelder, B. F.; King, T. E.: 1978, *Methods Enzymol.* 53, pp. 54-66.
5. Scott, R. A.; Cramer, S. P.; Shaw, R. W.; Beinert, H.; Gray, H. B.: 1981, *Proc. Natl. Acad. Sci. USA* 78, pp. 664-7.
6. For a description of the facility, see: Bienenstock, A.: 1980, *Nucl. Instrum. Methods* 172, pp. 13-20.
7. Cramer, S. P.; Scott. R. A.: 1981, *Rev. Sci. Instrum.* 52, pp. 395-9.
8. Chance B.; Angiolillo, P.; Yang, E. K.; Powers, L.: 1980, *FEBS Lett.* 112, pp. 178-82.
9. Hu, V. W.; Chan, S. I.; Brown, G. S.: 1977, *Proc. Natl. Acad. Sci. USA* 74, pp. 3821-5. Hu, V. W.; Chan, S. I.; Brown, G. S.: 1977, *FEBS Lett.* 84, pp. 287-90.
10. Powers, L.; Blumberg, W. E.; Chance, B.; Barlow, C. H.; Leigh, J. S.; Smith, J.; Yonetani, T.; Vik, S.; Peisach, J.: 1979, *Biochim. Biophys. Acta* 546, pp. 520-38.
11. Brudvig, G. W.; Bocian, D. F.; Gamble, R. C.; Chan, S. I.: 1980, *Biochim. Biophys. Acta* 624, pp. 78-89.
12. Beinert, H.; Shaw, R. W.; Hansen, R. E.; Hartzell, C. R.: 1980, *Biochim. Biophys. Acta* 591, pp. 458-70.
13. Powers, L.; Chance, B.; Ching, Y.; Angiolillo, P.: 1981, *Biophys. J.* 34, pp. 465-98.

LIST OF PARTICIPANTS

Philip Aisen, Albert Einstein College of Medicine, Yeshiva University, 1300 Morris Park Avenue, Bronx, New York U.S.A. 10461

Graham Bisset, Department of Chemistry, University of British Columbia, Vancouver, B.C., Canada V6T 1W5

Jon B. Bjarnason, Science Institute, University of Iceland, Dunhagi 3, 107 Reykjavik, Iceland.

Alison Butler, D-006 Chemistry Department, University of California, LaJolla, California, U.S.A. 92093

B. Rowe Byers, Medical Center, University of Mississippi, 2500 North State Street, Jackson, Mississippi, U.S.A. 39216

Carl J. Carrano, Department of Chemistry, University of Vermont, Burlington, Vermont, U.S.A. 05405

Anthony Cass, Inorganic Chemistry Laboratory, University of Oxford, South Parks Road, Oxford, England OX1 3QR

Chi K. Chang, Department of Chemistry, Michigan State University, East Lansing, Michigan, U.S.A. 48824

Lise Christiansen, Department of Chemistry, Odense University Campusvej 66, DK 5230 Odense M, Denmark

Scott Collins, Department of Chemistry, University of Calgary, Calgary, Alberta, Canada T2N 1N4

Robert R. Crichton, Université Catholique de Louvain, Unité de de Biochimie, Place Louis Pasteur 1, 1348 Louvain-la-Neuve, Belgium

Ron Corbett, Department of Chemistry, University of Calgary, Calgary, Alberta, Canada T2N 1N4

Cathy Coulman, Harvard Medical School, c/o 7 Story Street, Cambridge, Massachusetts, U.S.A. 02138

David Dolphin, Department of Chemistry, University of British Columbia, Vancouver, B.C., Canada V6T 1W5

H. Brian Dunford, Department of Chemistry, University of Alberta, Edmonton, Alberta, Canada T6G 2G2

Volker Eck, Fritz-Haber-Institut, der Max-Planck-Gesselschaft, Faradayweg 4-6, 1000 Berlin 33, Germany

Nils Ellfolk, Department of Biochemistry, University of Helsinki, Unioninkatu 35, SF-00170 Helsinki 17, Finland

Gavriella Gabor, Israel Institute for Biological Research, Department of Chemistry, Ness-Ziona, P.O.B. 19, Israel 70450

Philip George, Department of Biology, University of Pennsylvania, Philadelphia, Pennsylvania, U.S.A. 19104

Carlos F. Geraldes, Departamento de Quimica, Universidade de Coimbra, Coimbra, Portugal

Hans Grande, Laboratory of Biochemistry, De Dreijen 11 Agricultural University, 6703 BC Wageningen, The Netherlands

Martin C. Grootveld, Department of Chemistry, Birkbeck College, Malet Street, London WC1E 7HX England
Anders Hammershøi, Department of Chemistry, Stanford University, Stanford, California, U.S.A. 94305
Patricia Harrington, Department of Chemistry, New Mexico State University, Box 3C, Las Cruces, New Mexico, U.S.A. 88003
Wesley Harris, Laboratory for Energy-Related Health Research, University of California, Davis, California, U.S.A. 95616
Richard Haser, CRMC^2-CNRS-Campus Luminy, Case 913, 13288 Marseille cedex 2 France
Karel Heremans, Department of Chemistry, Katholieke Universiteit te Leuven, Celestijnenlaan 200D, B-3030 Heverlee, Belgium
H. Allen O. Hill, Department of Inorganic Chemistry, Oxford University, South Parks Road, Oxford, England OX1 3QR
Scott Hinman, Department of Chemistry, University of Alberta, Edmonton, Alberta, Canada T6G 2G2
Brian M. Hoffman, Department of Chemistry, Northwestern University, Evanston, Illinois, U.S.A. 60201
B.H. Vincent Huynh, Department of Physics, Emory University, Atlanta, Georgia, U.S.A. 30322
Peter Jones, School of Chemistry, University of Newcastle upon Tyne, Newcastle upon Tyne, England NE1 7RU
Jeanne Jordanov, Centre d'Etudes Nucleaires de Grenoble, Ave. des Martyrs-38, 85X - 38041 Grenoble cedex France
Richard J. Kassner, Department of Chemistry, University of Illinois, Chicago, Illinois, U.S.A. 60680
Teizo Kitagawa, Department of Molecular Physiological Chemistry, Medical School, Osaka University, 4-3-57 Nakanoshima, Kitaga, Osaka 530, Japan
Gerd N. LaMar, Department of Chemistry, University of California, Davis, California, U.S.A. 95616
Anne-Marie Lambeir, Department of Chemistry, University of Alberta, Edmonton, Alberta, Canada T6G 2G2
Jean LeGall, Department of Biochemistry, University of Georgia, Athens, Georgia, U.S.A. 30601
Bernard Lemire, Biochemistry Department, University of Alberta, Edmonton, Alberta, Canada T6G 2E1
R. Daniel Libby, Roger Adams Laboratory, Department of Biochemistry, University of Illinois, Urbana, Illinois, U.S.A. 61801
Ming-Cheh Liu, Department of Biochemistry, University of Georgia, Athens, Georgia, U.S.A. 30602
Torbjørn Ljones, Department of Chemistry-NLHT, University of Trondheim, Bjørnsonsgate 12, N-7000 Trondheim, Norway
Gilda H. Loew, Life Science Division, SRI International, Menlo Park, California, U.S.A. 94025

Kathleen Mailer, Department of Chemistry, Saint Mary's University, Halifax, Nova Scotia, Canada B3H 3C3
Daniel Mansuy, CNRS, Ecole Normale Superieure, Laboratoire de Chimie, 24 Rue Lhomond, 75231 Paris cedex France
Howard S. Mason, Department of Biochemistry, University of Oregon, Health Science Center, Portland, Oregon, U.S.A. 97201
Grant Mauk, Department of Biochemistry, University of British Columbia, Vancouver, B.C., Canada V6T 1W5
Brian Morgan, Department of Chemistry, University of British Columbia, Vancouver, B.C., Canada V6T 1W5
Isao Morishima, Department of Hydrocarbon Chemistry, Faculty of Engineering, Kyoto University, Kyoto, Japan
Isobel Moura, Centro de Quimica Estructural, Complexo Interdisciplinar, Instituto Superior Tecnico, 1096 Lisboa-codex, Portugal
José G.G. Moura, Centro de Quimica Estructural, Complexo Interdisciplinar, Instituto Superior Tecnico, 1096 Lisboa-codex, Portugal
Assanah M. Mydin, Department of Crystallography, Birkbeck College, Malet Street, London WC1E 7HX England
John Oakes, Port Sunlight Laboratory, Unilever, Port Sunlight, Wirral, Merseyside, England L62 4XN
Monica M. Palcic, 3001 College Avenue, Berkeley, California, U.S.A. 94705
Marianne G. Patch, Department of Chemistry, Cook Building University of Vermont, Burlington, Vermont, U.S.A. 05405
Vincent L. Pecoraro, Department of Chemistry, University of California, Berkeley, California, U.S.A. 94720
Jim Peterson, School of Chemical Sciences, University of East Anglia, Norwich, England NR4 7TJ
Dimitris Petridis, Greek Atomic Energy Commission, Nuclear Research Center "Demokritos", Aghia Paraskevi, Attiki, Greece
Gary Pielak, Department of Chemistry, Washington State University, Pullman, Washington, U.S.A. 99164
Kenneth Raymond, Department of Chemistry, University of California, Berkeley, California, U.S.A. 94720
Christopher A. Reed, Department of Chemistry, University of Southern California, Los Angeles, California, U.S.A. 90007
Thomas J. Reid III, Uniformed Services, University of the Health Services, Bethesda, Maryland, U.S.A. 20014
James E. Roberts, Department of Chemistry, Northwestern University, Evanston, Illinois, U.S.A. 60201
A. Edward Robinson, School of Chemical Sciences, University of East Anglia, Norwich, England NR4 7TJ
John Robinson, Department of Biochemistry, University of Alberta Edmonton, Alberta, Canada T6G 2E1

Inge Romslo, Laboratory of Clinical Biochemistry, University of Bergen, 5016 Haukerland Sykehus, Bergen, Norway
Rick Rutter, 415 Roger Adams Laboratory, Department of Biochemistry, University of Illinois, Urbana, Illinois, U.S.A. 61801
Helena Santos, Centro de Quimica Estructural, Complexo Interdisciplinar, Instituto Superior Tecnico, 1096 Lisboa-codex, Portugal
W. Robert Scheidt, Department of Chemistry, University of Notre Dame, Notre Dame, Indiana, U.S.A. 46556
Richard E. Scott, Department of Chemistry, University of Calgary, Calgary, Alberta, Canada T2N 1N4
Robert A. Scott, Department of Chemistry, Stanford University, Stanford, California, U.S.A. 94305
Judith Shelling, Department of Biochemistry, University of Alberta, Edmonton, Alberta, Canada T6G 2E1
Larry Sieker, Department of Biochemistry, University of Washington, Seattle, Washington, U.S.A. 98195
A. Simopoulos, Greek Atomic Energy Commission, Nuclear Research Center "Demokritos", Aghia, Paraskevi, Attiki, Greece
Sietske Smit, Department of Chemical Pathology, Erasmus University of Rotterdam, P.O. Box 1738, 3000 DR Rotterdam The Netherlands
Ronald E. Stenkamp, Department of Biological Structure, School of Medicine, University of Washington, Seattle, Washington, U.S.A. 98195
Nobuo Tanaka, Institute for Protein Research, Osaka University, 5311 Yamada-kami, Suita, Osaka, Japan
Arild Tangerås, Department of Biochemistry, University of Bergen, Arstadveien 19, N-5000 Bergen, Norway
Hans Toftlund, Department of Chemistry, Odense University, Campusvej 66, DK 5230 Odense M, Denmark
Alfred Trautwein, Physik Department, FB12.1 University Saarlandes, D-6600 Saarbrüken 11, West Germany
Volker Ullrich, Fachrichtung 3.3, Physiologische Chemie, Universitat des Saarlandes, 665 Homberg, West Germany
Cees Veeger, De Dreijen 11, Agricultural University, 6703 BC Wageningen, The Netherlands
Gerritt A. Veldink, Rijksuniversiteit Utrecht, Organisch Chemisch Laboratorium, Croestraat 79, 3522 AD Utrecht, The Netherlands
Gerald Wagner, Department of Biochemistry, 415 Roger Adams Laboratory, University of Illinois, Urbana, Illinois U.S.A. 61801
William J. Wallace, Department of Chemistry, University of Regina, Regina, Saskatchewan, Canada S4S 0A2
Raymond Weiss, Université Louis Pasteur, Institut le Bel 4, rue Blaise Pascal, Strasbourg, France
Ron Wever, Vakgroep Biochemie, B.C.P. Jansen Instituut, Plantage Muidergracht 12, 1018 TV Amsterdam - C, The Netherlands

Ralph G. Wilkins, Department of Chemistry, New Mexico State University, Box 3C, Las Cruces, New Mexico, U.S.A. 88003

Günther Winkelmann, Institut fur Biologie I, Universitat Tubingen, Auf der Morgenstelle 1, D-7400 Tubingen, West Germany

António Xavier, Centro de Quimica Estructural, Complexo Interdisciplinar, Instituto Superior Tecnico, 1096 Lisboa-codex, Portugal

AUTHOR INDEX

SUBJECT INDEX

GPSR Compliance

The European Union's (EU) General Product Safety Regulation (GPSR) is a set of rules that requires consumer products to be safe and our obligations to ensure this.

If you have any concerns about our products, you can contact us on ProductSafety@springernature.com

In case Publisher is established outside the EU, the EU authorized representative is:

Springer Nature Customer Service Center GmbH
Europaplatz 3
69115 Heidelberg, Germany

Batch number: 10815974

Printed by Printforce, the Netherlands